Tracer and Timescale Methods for Passive and Reactive Transport in Fluid Flows

Tracer and Timescale Methods for Passive and Reactive Transport in Fluid Flows

Editors

Eric Deleersnijder
Inga Monika Koszalka
Lisa V. Lucas

MDPI • Basel • Beijing • Wuhan • Barcelona • Belgrade • Manchester • Tokyo • Cluj • Tianjin

Editors

Eric Deleersnijder
Université catholique de Louvain
Louvain-la-Neuve
Belgium

Inga Monika Koszalka
Stockholm University
Sweden

Lisa V. Lucas
U.S. Geological Survey
USA

Editorial Office
MDPI
St. Alban-Anlage 66
4052 Basel, Switzerland

This is a reprint of articles from the Special Issue published online in the open access journal *Water* (ISSN 2073-4441) (available at: https://www.mdpi.com/journal/water/special_issues/Fluid_Flows).

For citation purposes, cite each article independently as indicated on the article page online and as indicated below:

LastName, A.A.; LastName, B.B.; LastName, C.C. Article Title. *Journal Name* **Year**, *Volume Number*, Page Range.

ISBN 978-3-0365-3521-0 (Hbk)
ISBN 978-3-0365-3522-7 (PDF)

Cover image courtesy of Eric Deleersnijder

Contents

About the Editors

Eric Deleersnijder's research interests focus on the development and use of unstructured mesh models for simulating geophysical and environmental flows (www.slim-ocean.be) and the related diagnostic methods (www.climate.be/cart). The domains of application comprise most of the hydrosphere, i.e., lakes, rivers, estuaries, coastal regions, and shelf seas. Additional pieces of information may be found on his website (www.ericd.be).

Inga Monika Koszalka has a broad background and scientific interests spanning various topics in geophysical fluid dynamics. The focus of her current research is on mesoscale and regional ocean circulation, its variability, as well as ocean interactions with the atmosphere, cryosphere, and biosphere. In her research, she employs observations, idealized ocean models and regional ocean model output, Lagrangian analysis and modeling, statistical methods, and theory. Her studies often address ocean turbulence and turbulent dispersion which are not well understood yet important elements of the climate system and require holistic and novel approaches. Her study domain considers primarily the Atlantic Ocean, the Nordic Seas, Greenland shelves and fjords, as well as the Baltic Sea, where she started her oceanographic studies as an undergraduate student. Find out more at her professional page (https://www.su.se/english/profiles/inko3680-1.408678).

Lisa V. Lucas is an ecohydrodynamicist who studies how hydrodynamics interact with biology to control water quality and ecological function in surface aquatic systems. She develops, adapts, and applies integrated, process-based models ranging from simple, algebraic "pencil and paper" expressions to 3D-coupled numerical physical–biological models. She enjoys collaborative, interdisciplinary science and relishes her partnerships with field scientists, laboratory experimentalists, and other modelers. Find out more at her professional page (https://www.usgs.gov/staff-profiles/lisa-v-lucas).

Preface to "Tracer and Timescale Methods for Passive and Reactive Transport in Fluid Flows"

Timescales and tracer-based methods have been recognized to be powerful approaches to gain quantitative insight into the underlying physical and biochemical processes in oceanography, limnology, and marine science, and have widely been used to interpret field observations and numerical model results. The concept of timescales has been instrumental to understanding the complex hydrodynamics and transport processes of reactive biochemical substances in surface waters since Bolin and Rodhe introduced timescales of age, transit time, and turnover time in 1973. In 1976, Zimmerman applied the timescale concept to the Dutch Wadden Sea and introduced 'residence time' to deal with moving individual particles for a spatially varying situation. Takeoka further studied residence time in 1984 under the same framework as Bolin and Rodhe. These timescales are holistic, in that they account for all underlying advection and mixing transport processes. The classical empirical model, introduced by Vollenweider in 1976 to describe lake eutrophication based on hydraulic residence time, was groundbreaking, demonstrating the crucial role of hydrodynamics in biochemical processes. It also demonstrated that the complex interactions between hydrodynamics and biochemical processes can be explained by a simplified model as a function of hydraulic residence time. One of the clearest benefits of timescales is that they provide a "common currency" that enables us to compare different processes under a common scale, as demonstrated by Lucas et al. in 2009 for complex phytoplankton dynamics. As the capabilities of numerical models have been enhanced with increased computer power, timescales have been applied broadly in freshwater, groundwater, coastal waterbodies, and the ocean. The two approaches commonly used to reveal transport trajectory and compute timescales in numerical models are the Lagrangian and Eulerian approaches. The Lagrangian methods, based on particle tracking, are straightforward and appealing, being much superior to their Eulerian counterparts at revealing the trajectory of fluid and material movement and conveniently computing their corresponding timescales, which have been commonly used in marine science and limnology. However, the Eulerian approaches, based on tracer movement, are widely used in surface water systems to understand transport properties and timescales. Commonly used timescales such as residence time, flushing time, freshwater flushing time, replacement time, turnover time, transit time, age, etc., are often poorly defined by users and have been used rather carelessly, which could cause misleading interpretations of the model's results and the underlying hydrodynamics and transport processes of waterbodies, which calls for a thorough fundamental study to reconcile these misinterpretations with the truth.

An influential theoretical breakthrough was presented by Deleersnijder and Delhez, who introduced the Constituent-oriented Age and Residence Time Theory (CART). CART directly estimates the timescales of age and residence time from the solutions of partial differential problems. It clearly defines different timescales and provides a solid theoretical basis for using the Eulerian approach to compute the Lagrangian properties of timescales. Under the CART framework, the spatial and temporal variations of timescales in complex surface waters can be computed efficiently by numerical models, and the results are consistent with the transport of reactive substances simulated by the model. On the other hand, Lagrangian methods have been enhanced to develop an individual-based model to incorporate particle behaviors to accurately simulate particle movement and timescales to understand the interaction between hydrodynamic and biological processes. These significant advances provide the capabilities of using timescale-, tracer-, and particle-based methods to diagnose complicated hydrodynamics and transport of reactive substances, and to further understand the important role of hydrodynamics in contributing to biochemical processes.

The holistic characteristics of timescales have been widely recognized. The applications of tracer-based and Lagrangian particle methods to quantify timescales contribute to diverse topics that include the diagnosis of complex transport processes, influences of external forcings on transport processes, material retention and fate, understanding results of complex marine models, evaluation of the contributions of physical and transport processes on biochemical processes, and explaining the aquatic conditions that are favorable for biological processes. This Special Issue collects reviews, recent advances in theoretical analysis, the use of a stable isotope-based method to verify water age, and various applications of Lagrangian and Eulerian approaches to interpolate dynamic fields as well as numerical model results to understand the underlying processes and timescales across multiple geographic regions and various scientific fields. It not only summarizes our current understanding and applications of timescale-, tracer-, and particle-based approaches in the past two decades, but also provides methodological ideas leading to future development. Most importantly, it clarifies the definitions of different timescales, broadens their applications in different scientific fields, and bridges the tracer-based and particle-based approaches. This Special Issue marks a new milestone and will be an excellent reference to the scientific community.

Jian Shen
Research Professor of Marine Science
Virginia Institute of Marine Science
The College of William & Mary
Gloucester Point, Virginia, USA

water

Editorial

Tracers and Timescales: Tools for Distilling and Simplifying Complex Fluid Mechanical Problems

Lisa V. Lucas [1],* and Eric Deleersnijder [2]

[1] Integrated Modeling and Prediction Division, Water Mission Area, U.S. Geological Survey,
      345 Middlefield Road, MS #496, Menlo Park, CA 94025, USA
[2] Institute of Mechanics, Materials and Civil Engineering (IMMC) & Earth and Life Institute (ELI),
      Université Catholique de Louvain, Bte L4.05.02, Avenue Georges Lemaître 4,
      1348 Louvain-la-Neuve, Belgium; eric.deleersnijder@uclouvain.be
*   Correspondence: llucas@usgs.gov; Tel.: +1-650-329-4588

Citation: Lucas, L.V.; Deleersnijder, E. Tracers and Timescales: Tools for Distilling and Simplifying Complex Fluid Mechanical Problems. *Water* **2021**, *13*, 2796. https://doi.org/10.3390/w13192796

Received: 17 September 2021
Accepted: 23 September 2021
Published: 8 October 2021

Publisher's Note: MDPI stays neutral with regard to jurisdictional claims in published maps and institutional affiliations.

1. Introduction

The last several decades have seen significant advances in fluid–mechanical, water-quality, and ecological observation systems, as well as in related scientific computing capabilities. These leaps forward have provided an increasingly detailed view of fluid flow and reactive transport in natural and engineered systems, thus enriching our process of understanding and improving (1) science-based resource management, (2) the design of anthropogenic structures, (3) the interpretation of field- and laboratory-scale phenomena, and (4) predictive capacity. The benefits of these advancements are easily apparent if one considers (as just a few examples):

- The ability to observe turbulent transport at scales of centimeters and seconds, enabling the more accurate characterization of oxygen, nutrient, carbon, contaminant, or other scalar fluxes through the water column;
- The capacity to model whole or multiple connected aquatic systems in multiple dimensions at a spatial resolution of tens to hundreds of meters and timesteps of seconds to minutes, helping us understand how riverine, oceanic, and atmospheric drivers influence constituent transport or habitat connectivity (e.g., [1–3]);
- The capability to remotely track hundreds of physical drifters across hundreds of kilometers and over timescales of days, revealing complex Lagrangian transport patterns not discernible from moored (Eulerian) sensor measurements ([4]).

Thus, the current technological capacity in environmental fluid mechanics and hydraulic engineering is, clearly, impressive. However, so is the massive amount of generated data requiring processing and analysis in order to transform it into useful information [5]. Furthermore, this intensive data-generating capacity enriches our awareness and understanding of individual processes operating over a broad range of spatial and temporal scales, but the net effect—the "upshot"—of those processes and their interactions may be challenging to identify. Techniques for distilling, simplifying, and extracting the essence from large, complex environmental datasets have arguably never been more necessary [5–7].

It is in this realm—the extraction of meaningful information—where tracers and timescales shine. Both are tools that can paint a simplified, digestible (yet quantitative; [6]) picture of a large number of reactive transport processes that co-exist and interact in aquatic environments. These techniques thereby help us understand and interpret those processes, identifying the most crucial ones and establishing causal relationships between them [6,7]. The contributions to this Special Issue relate to one or both of these commonly related tools, providing new methodologies, applications, or reviews of tracers or timescales in fluid mechanics.

So, what precisely are "timescales" and "tracers", as discussed in this Special Issue? A timescale is a diagnostic parameter that communicates approximately how long a process

1

takes [8]. A timescale can be defined for any process, be it physical, biological, chemical, or radiological but, because all timescales carry the same units (time), they serve as a common currency [9], allowing disparate processes to be directly compared and easily integrated. Timescales have several other virtues, including the ability to encapsulate spatial, temporal, and multi-process complexity into a single number (for example, the use of a "residence time" to capture the net effect of all transport processes operating on a transported substance (e.g., [8,10]). One might consider this timescale role as that of an integration tool. Depending on how they are estimated, though, timescales may also be devised to deconstruct a complex problem, separating key processes and assessing their relative overall contributions to observed or modeled outcomes. Importantly, timescales may also help in building simplified reactive transport models, which may be regarded as a form of model reduction (i.e., replacing a model containing many variables or parameters with a model containing far fewer variables/parameters that are deemed to be satisfactory indicators of the state of the system; [11]).

A tracer, on the other hand, is a distinguishable dye or chemical compound that is added to a fluid system [12] and is used to learn something about the physical, biological, or chemical functioning of that system. Tracers may be natural or anthropogenic, and can be introduced either intentionally, accidentally, or without any human intervention. They are used in field, laboratory, and numerical studies of aquatic habitats and, as such, may be real or virtual. As discussed in the tracer-related papers within this Issue, there are some scientific questions for which conservative (i.e., inert, non-reactive) tracers present the ideal tool, whereas in other situations, reactive (or both types of) tracers are needed.

What do tracers and diagnostic timescales have to do with each other, and why does this Special Issue incorporate papers from both realms? First, tracers and timescales are both useful diagnostic tools that can help us understand, visualize, and quantify the net effect of one or multiple interacting processes. Both tools can therefore aid us in answering practical questions, such as:

1. To where will a particle travel if it is released at Point A and Time t in a fluid body?
2. How much time will be required for a particle to travel from Point A to Point B?
3. Will the travel time between Points A and B be sufficient to allow for any particle reactions to run to completion?
4. How long will a particle or collection of particles remain within a defined fluid environment?

Second, tracers are often implemented to mark a water "mass" or "type", such as that originating in a specific region or released into the domain from a point or boundary source. As such, tracers frequently form a basis for the estimation of characteristic timescales, whether experimentally (e.g., [13]) or numerically [14]. For example, a number of the papers in this Special Issue describe studies in which transport timescales such as age, residence time, and exposure time are computed from the solution of partial differential problems ("PDPs", which include partial differential equations plus initial and boundary conditions). These PDPs describe the evolution of numerical tracers and tracer-like quantities, including tagged water types or masses, which are treated as passive tracers (e.g., [2,15–17]).

This Special Issue includes papers relevant to a variety of fluid–mechanical domains: idealized model environments, seas, gulfs, estuaries, human-made impoundments, lakes, deltas, river networks, reservoirs, marinas, and other hydrogeologic environments. Below, we provide an integrative summary of the contributions in this Special Issue, following the imperfect delineation between papers focused on tracers and those focused on timescales, acknowledging that there is overlap for the reasons mentioned above. We also attempt to identify key remaining gaps in general understanding or capability and areas ripe for future work.

2. Tracer-Focused Contributions

Two papers in this volume provide useful reviews of different classes of tracers implemented in physical experiments. Cao et al. [12] delineate tracer types at the highest level based on the degree of interaction with the aquatic system to which they are added, with: (1) "conservative" tracers displaying virtually no interaction with materials within the system, thus behaving as inert substances under the conditions characteristic of that system and flowing passively with the fluid, and (2) "reactive" tracers defined as "compounds that undergo a chemical reaction or physicochemical interaction processes in a predictable way." Those authors explain that conservative tracers are typically used to assess physical transport processes or hydromechanical properties such as porosity, whereas reactive tracers—which are used in tandem with conservative tracers—can provide information on physicochemical properties such as sorption capacity, redox condition, or microbial activity [12]. While conservative tracer compounds have been the subject of other reviews, Cao et al. [12] note that there had not previously been a systematic review of reactive tracers—a need which they fill herein, with an emphasis on subsurface and hyporheic processes. Those authors then discuss in detail the properties, behaviors, and potential applications of three major subgroups of reactive tracers: (A) partitioning tracers (whose breakthrough curves display a retardation—or time shift—relative to conservative tracers), (B) kinetic tracers (which experience degradation but no retardation), and (C) hybrid tracers (which experience both retardation and degradation). Cao et al. [12] emphasize that the selection of an optimal tracer compound for a given purpose is a complex "art" and describe a general approach for designing and creating tracers tailored for specific applications.

Bailly du Bois et al. [1] review a different class of tracers: dissolved anthropogenic radionuclides (or "artificial radiotracers"), which may be inadvertently or intentionally discharged into ocean waters and emanate from atmospheric nuclear tests, nuclear fuel reprocessing plants, or nuclear power plants (e.g., Chernobyl, Fukushima Daiichi). Regardless of how such tracers are introduced into the aquatic environment, these authors demonstrate the valuable role that dissolved radionuclides can play as oceanographic tools. Bailly du Bois et al. [1] present an extensive, updated dataset of in situ radionuclide (e.g., ^{3}H, ^{137}Cs, ^{134}Cs, ^{125}Sb) measurements collected between 1982 and 2016 aboard 80 oceanographic campaigns across the English Channel, Bay of Biscay, Celtic Sea, Irish Sea, and the North Sea. With a focus on the measured tracers behaving most conservatively in seawater (^{3}H and ^{125}Sb), the authors describe useful applications of radiotracer data for illuminating physical transport processes and pathways and for validating hydrodynamic models. Additionally, they describe how measured tracers could be used to deduce percentages of water masses originating in different regions, construct radionuclide inventories, ascertain transport timescales, or assess seawater–sediment or seawater–organism exchange. The many applications demonstrated in this paper could be relevant to other parts of the globe receiving inputs of radionuclides, such as the Gulf of Mexico and the Mediterranean, Arctic, China, and Arabian Seas, all of which are within reach of nuclear power plants [1].

Other tracer-focused papers in this issue assess or advance methodologies for measuring or modeling tracers or tracer-like quantities. For example, Staneva et al. [18] investigate processes influencing particle transport by comparing different ocean-modeling approaches to observed paths of physical drifters in the North Sea. Specifically, those authors compare particle tracks computed with a stand-alone ocean circulation model to those computed with the same ocean circulation model coupled with a wave model, examining the fidelity of each model set-up to observed drifter trajectories, as well as high-frequency radar-based observations of current velocity. It is shown that wave-induced drift significantly influences particle transport in the upper layers of the ocean. This work demonstrates that the coupling of circulation models to wave models may greatly improve simulations of the transport of marine litter, oil, larvae, or other biological materials.

So et al. [19] tackle the ongoing challenge of improving measurements of turbulent tracer transport. They note that (1) "most fluid motions in nature and engineering are turbulent", and (2) estimates of tracer transport under turbulent conditions in shallow waters can be contaminated by wave action. As So et al. [19] explain, observations of turbulence properties may be confounded by velocity fluctuations that include orbital velocities generated by waves, potentially leading to tracer transport overestimates by one or more orders of magnitude. Therein lies their motivation to improve available methods for eliminating the wave-induced contamination of ADCP (Acoustic Doppler Current Profiler) measurements. The authors propose a new method—the Harmonic Analysis or "HA" method—and demonstrate that it performs better than a previously introduced approach under conditions with energetic surface gravity waves, potentially leading to more accurate estimates of tracer flux in shallow coastal waters.

The intersection of tracer measurements and modeling is explored by Tomkovic et al. [3], who employ both approaches while pursuing a deeper understanding of source water provenance in a branching tidal river system. Their field-based approach focuses on the utilization of stable isotope compositions of oxygen and hydrogen in water samples to build a mixing model. Their numerical approach involves the implementation of a two-dimensional (vertically averaged) model of the study domain and its adjacent environments, allowing for the computational depiction of tracer transport through the domain. Results from the two approaches are compared and are shown to agree well with each other, with each approach providing validation of the other and providing the authors with an opportunity to evaluate the strengths and shortcomings of each. Although Tomkovic et al. [3] implement this dual-track study with the aim of supporting ecosystem restoration efforts in the Sacramento–San Joaquin Delta (California, USA), their approaches and findings can be applied in other systems as well.

3. Timescale-Focused Contributions

The review by Lucas and Deleersnijder [8] describes diagnostic timescale uses and methods of estimation, primarily in the coastal zone. Citing numerous examples from the literature, they demonstrate how timescales can improve the understanding of aquatic systems by helping us distill large datasets, compare systems across space or time, compare relative speeds of disparate processes, and build simple and tractable models. The authors reiterate what numerous others have before them: the one point of consistency in diagnostic timescale use across the aquatic sciences is the inconsistency in how (or whether) terms are defined and calculated. Different definitions and calculation methods can lead to substantially different values and represent different phenomena, so it is most constructive for all employing diagnostic timescales to choose, define, calculate, and present them with care and clarity.

Several papers in this volume provide examples of computationally derived transport timescales (e.g., water age, residence time, exposure time, and flushing time) for the purposes of supporting engineering design, guiding resource management, or strengthening the understanding of physical processes in aquatic systems. This selection of papers provides a window into a variety of approaches for implementing state-of-the-art numerical models for computing transport timescales.

For example, in their investigation of water renewal in La Rochelle Marina (France), Huguet et al. [17] implement a hydrodynamic and transport model with conservative numerical tracer and particle tracking capabilities to compute spatially variable residence, exposure, and *e*-folding flushing times, as well as return flow parameters. Timescales for several scenarios are compared in a sensitivity analysis assessing the effects of wind and spring–neap tidal phase. A particularly novel aspect of this study is the incorporation of floating structures (docks and moorings) into the modeling and assessment of the sensitivity of water renewal in the marina to the presence of those structures.

Additionally, using a 3D hydrodynamic and transport model, Liu et al. [16] employ a straightforward method utilizing pairs of conservative and decayable numerical tracers (each pair associated with a group of tributaries) to assess spatially variable water age in Taihu Lake (China). This environment is widely known for its large size, economic and ecological importance, and for its challenges with eutrophication and harmful cyanobacteria blooms. Liu et al. [16] conduct their study of water age to quantitatively investigate nutrient loads from different source regions to Taihu Lake and to understand the influence of wind and river discharge on transport times. It is seen that age is a useful water-quality related diagnostic parameter, which may be of use for decision making at the basin scale.

Some articles in this Issue contribute to the development and use of the Constituent-oriented Age and Residence time Theory (CART, www.climate.be/cart, accessed on 25 September 2021). This conceptual toolbox provides a consistent set of partial differential equations, along with relevant initial and boundary conditions, aimed at estimating various diagnostic timescales at every time and position, chiefly the age and the residence or exposure time. These timescales may be derived for every constituent of fresh or salty water (or any other liquid mixture) or aggregates of them (i.e., groups of constituents), including the water itself.

Deleersnijder et al. [7] investigate the boundary conditions for age calculations. In CART, the mean age is computed as the ratio of the "age concentration" to the concentration of a numerical tracer. These variables are not independent, for the former is the first-order moment of the age distribution function, whilst the latter is the zeroth order moment. Therefore, the boundary conditions cannot be prescribed independently; rather, they must be consistent with each other. This paper shows how to do this and also illustrates the impact of inconsistent boundary conditions. Based on these considerations, a strategy to design meaningful age diagnoses is outlined.

Using consistent boundary conditions, Pham Van et al. [15] apply the water renewal assessment strategy of de Brye et al. [20] to the delta of the Mahakam River (Indonesia). Water renewal timescales (i.e., age, residence, and exposure times) are quite short (i.e., a few days to a couple of weeks) and crucially depend on the river discharge in spite of the large influence of tides on the hydrodynamics. The return coefficient (a measure of the propensity of water particles to re-enter the domain after leaving it for the first time) is of the order of 0.3 far away from the boundaries, suggesting that re-entering due to tides is important almost everywhere. Overall, this study illustrates how forward- and backward-looking diagnostic timescales (residence/exposure time and age, respectively) may be combined to gain insight into the water renewal processes of a river delta whose hydrodynamics are driven mostly by tides and river discharge.

In the northern San Francisco Estuary (USA), Gross et al. [14] compare water age estimated from field data (isotopic water composition) and numerical estimates derived from a thoroughly validated hydrodynamic and tracer transport model. The discrepancies, which are relatively small, are investigated in detail, suggesting avenues of improvement for both approaches. Interestingly, a new concept is introduced—namely, the mean property experienced by a tracer—which may be seen as a generalization of CART's philosophy. This concept may be applied to non-positive-definite variables, which would have been impossible using the diagnoses based on partial differential equations developed up to now. This novelty deserves to be investigated further and, above all, put to use.

For rectangular, flat-bottomed reservoirs commonly used in urban hydraulics or river engineering, Dewals et al. [21] compute the depth-integrated, steady-state, position-dependent water age distribution function (i.e., the histogram of the water age in every water parcel). No other article in the present issue evaluates such a distribution function. It is shown that (A) simple indicators such as the ratio of domain volume to volumetric flow rate or (B) a more sophisticated diagnosis, namely the position-dependent mean water age (i.e., the mean time elapsed since entering the domain), provide insufficient information regarding the pathways of water particles in the domain. This is because particles follow both slow and fast routes from the inlet to the outlet, causing, in many instances, the age histogram to exhibit several maxima. This rather complex behavior could not have been anticipated from a simple inspection of the flow field, underscoring the need for suitably designed diagnostic timescales.

Cheng et al. [22] compute the three-dimensional, time-dependent residence time in a typical tributary bay of the reservoir of Three Gorges Dam (China). They adopt the adjoint approach of Delhez et al. [23] and perform a detailed sensitivity analysis aimed at identifying the processes that have the largest impact on the exchanges of the domain of interest with its environment. It is shown that water-level regulation of the dam and density currents crucially influence the residence time, whereas surface wind force and river discharge are much less significant. A strengthened understanding of these issues is important because of the poor water quality in the bay under study, as in many such bays. This investigation is one of those that suggests that diagnostic timescales can provide useful pieces of information for water quality management.

Hong et al. [2] divide the upstream boundary of the Pearl River Estuary (China) into several parts. Then, the three-dimensional, time-dependent age of passive tracers emanating from those sub-regions is evaluated, and a thorough sensitivity analysis is carried out. A strong correlation between river discharge and transport time is found. The ultimate objective of this study is to contribute to an understanding of the dynamics of pollution caused by terrestrial substances. The approach of Hong et al. (2020) is based on the hypothesis that the age of passive tracers is a diagnostic tool capable of providing information relevant to the fate of numerous terrestrial substances entering the domain of interest. Similar hypotheses have guided other investigations to implement conservative tracers to gain insights into the dynamics of non-conservative substances. Validation of this hypothesis (for the Pearl River Estuary and other deltas or estuaries as well) is yet to be achieved in full and is likely to require a vast amount of dedicated research. Hong et al. (2020) thus provide novel and valuable results and, by doing so, highlight important areas for future work.

4. Conclusions

The articles of this Special Issue deal with topics related to a wide variety of domains of interest that are investigated with a wide range of approaches. Clearly, tracer and timescale methods are alive and well, with the need for such tools only expanding over time and the related methods and technologies continually advancing.

A range of timescale quantification approaches exists, with many such methods relying on tracers or tracer-like substances (e.g., particles, drifters) to characterize and track fluid movement. Similarly, a range of diagnostic timescales themselves exist, with some integrating across space and/or time (e.g., the volume/flow advective flushing time) and others defined and calculated as varying fields across one to four dimensions, such as with the CART methods [23,24].

Another classification approach distinguishes timescales by the degree of "holism" or "process richness" incorporated into them [8]. At the low end of the holism spectrum, "atomistic" timescales depict the time associated with a single process (e.g., advection, diffusion, or a reaction). Atomistic timescales can be relatively easy to estimate, often relying on simple algebraic relationships (e.g., distance/velocity), and have been used for many decades (e.g., [25–28]). These, which we might think of as "Grandma and Grandpa's" timescales, can help assess the relative importance of individual terms of the governing equations and also form the basis for simple, process-based mathematical models [8,29]. However, more often than not, they cannot paint a satisfactory picture of the overall impact of transport and reaction processes taking place in a fluid environment. More holistic approaches are often needed.

"Holistic" timescales each represent the time associated with a collection of processes (e.g., multiple transport and/or reaction processes), producing a parameter conveying the net effect of those processes and their interplay with each other [8]. Over the past several decades, the ability to compute holistic timescales with numerical models, or quantify them in the field using drifters or tracers, has exploded—at least with respect to transport timescales. Additionally, the last ~20 years have brought increasingly sophisticated methods for doing so. The articles in this Special Issue collectively describe a smorgasbord of these new and powerful approaches, presenting new applications, innovative directions for validation, and novel extensions of what is now the established state of the art (these are not your grandparents' timescales.) Similarly, observational approaches for characterizing and quantifying the net effect of multiple interacting transport processes and hydraulic connectivity have experienced significant advancements and are well represented in this Issue.

Notwithstanding the exciting capabilities and continual methodological improvements demonstrated, some key questions remain:

1. Is it always necessary to use the most advanced methods? For example, is it always needed to solve PDPs to obtain tracer fields and/or timescales at every location within a fluid domain? Indeed, the most sophisticated methods are not always the best option. Such detailed information is not always necessary and, sometimes, simpler timescales are precisely what is needed, for example, to compare relative speeds of processes.

2. Are the most sophisticated methods clearly "better"? Possibly, if detailed spatial and/or temporal variations in timescales or scalar fields are needed. However, although powerful numerical approaches hold enormous potential, they also present some important limitations and critical open questions, such as:

   - Significant data requirements for driving, calibrating, and validating a complex numerical model (not to mention research-and-development resources).
   - The challenge of directly validating the new (e.g., CART- and PDP-based) methods. Diagnostic timescales cannot be directly measured in situ like salinity or temperature. Rather, in the field or laboratory, they must be deduced from other measured quantities, which will carry increased errors.
   - Remaining challenges in incorporating particle dynamics or biogeochemical reactions and transformations into models, thus limiting models from achieving 100% holism for transported reactive constituents.

How to select the most suitable holistic diagnostic approach, be it based on tracers alone, timescales closely related to tracers, or more simplistic atomistic timescales, is clearly still an open question. Additionally, in our view, the need for the full spectrum of diagnostic approaches—from the very simple to the mathematically complex and cutting edge—remains.

Author Contributions: Conceptualization, L.V.L. and E.D.; writing—original draft preparation, L.V.L. and E.D.; writing—review and editing, L.V.L. and E.D. All authors have read and agreed to the published version of the manuscript.

Funding: L.V.L. was supported by the Water Availability and Use Science and National Water Quality Programs of the U.S. Geological Survey.

Data Availability Statement: Not applicable.

Acknowledgments: We thank Ed Gross and Lily Tomkovic for their valuable comments on this Editorial, as well as Jessie Lacy and Mark Stacey for their helpful input. In addition, we would like to acknowledge the authors who contributed a fine selection of papers to this Special Issue, as well as the important contributions of our co-guest-editor Inga Koszalka, the reviewers, and the MDPI staff, notably Rachel Lu. Without them, this Special Issue could not have been assembled. E.D. is an honorary Research Associate with the Belgian Fund for Scientific Research (F.R.S.-FNRS).

Conflicts of Interest: The authors declare no conflict of interest.

References

1. Bailly du Bois, P.; Dumas, F.; Voiseux, C.; Morillon, M.; Oms, P.-E.; Solier, L. Dissolved Radiotracers and Numerical Modeling in North European Continental Shelf Dispersion Studies (1982–2016): Databases, Methods and Applications. *Water* **2020**, *12*, 1667. [CrossRef]
2. Hong, B.; Wang, G.; Xu, H.; Wang, D. Study on the transport of terrestrial dissolved substances in the Pearl River Estuary using passive tracers. *Water* **2020**, *12*, 1235. [CrossRef]
3. Tomkovic, L.A.; Gross, E.S.; Nakamoto, B.; Fogel, M.L.; Jeffres, C. Source Water Apportionment of a River Network: Comparing Field Isotopes to Hydrodynamically Modeled Tracers. *Water* **2020**, *12*, 1128. [CrossRef]
4. Pawlowicz, R.; Hannah, C.; Rosenberger, A. Lagrangian observations of estuarine residence times, dispersion, and trapping in the Salish Sea. *Estuar. Coast. Shelf Sci.* **2019**, *225*, 106246. [CrossRef]

5. Lucas, L.V.; Deleersnijder, E. Diagnostic Timescales: Old Concepts, New Methods, and the Ageless Power of Simplification. In *CERF's Up!* Coastal and Estuarine Research Federation: Seattle, WA, USA, 2021; Volume 47, pp. 14–15.

6. Deleersnijder, E.; Delhez, E.J.M. Timescale- and tracer-based methods for understanding the results of complex marine models. *Estuar. Coast. Shelf Sci.* **2007**, *74*, V–Vii. [CrossRef]

7. Deleersnijder, E.; Draoui, I.; Lambrechts, J.; Legat, V.; Mouchet, A. Consistent boundary conditions for age calculations. *Water* **2020**, *12*, 1274. [CrossRef]

8. Lucas, L.V.; Deleersnijder, E. Timescale Methods for Simplifying, Understanding and Modeling Biophysical and Water Quality Processes in Coastal Aquatic Ecosystems: A Review. *Water* **2020**, *12*, 2717. [CrossRef]

9. Lucas, L.V.; Thompson, J.K.; Brown, L.R. Why are diverse relationships observed between phytoplankton biomass and transport time? *Limnol. Oceanogr.* **2009**, *54*, 381–390. [CrossRef]

10. Lucas, L.V.; Thompson, J.K. Changing restoration rules: Exotic bivalves interact with residence time and depth to control phytoplankton productivity. *Ecosphere* **2012**, *3*, 117. [CrossRef]

11. Deleersnijder, E. *The Unreasonable Effectiveness of Dimension Reduction in Complex Geophysical Flow Modelling*; Université catholique de Louvain: Louvain-la-Neuve, Belgium, 2009; pp. 1–25. Available online: http://hdl.handle.net/2078.1/154174 (accessed on 23 September 2021).

12. Cao, V.; Schaffer, M.; Taherdangkoo, R.; Licha, T. Solute Reactive Tracers for Hydrogeological Applications: A Short Review and Future Prospects. *Water* **2020**, *12*, 653. [CrossRef]

13. Downing, B.D.; Bergamaschi, B.A.; Kendall, C.; Kraus, T.E.C.; Dennis, K.J.; Carter, J.A.; Von Dessonneck, T.S. Using Continuous Underway Isotope Measurements To Map Water Residence Time in Hydrodynamically Complex Tidal Environments. *Environ. Sci. Technol.* **2016**, *50*, 13387–13396. [CrossRef]

14. Gross, E.; Andrews, S.; Bergamaschi, B.; Downing, B.; Holleman, R.; Burdick, S.; Durand, J. The Use of Stable Isotope-Based Water Age to Evaluate a Hydrodynamic Model. *Water* **2019**, *11*, 2207. [CrossRef]

15. Pham Van, C.; de Brye, B.; de Brauwere, A.; Hoitink, A.J.F.T.; Soares-Frazao, S.; Deleersnijder, E. Numerical Simulation of Water Renewal Timescales in the Mahakam Delta, Indonesia. *Water* **2020**, *12*, 1017. [CrossRef]

16. Liu, S.; Ye, Q.; Wu, S.; Stive, M.J.F. Wind Effects on the Water Age in a Large Shallow Lake. *Water* **2020**, *12*, 1246. [CrossRef]

17. Huguet, J.-R.; Brenon, I.; Coulombier, T. Characterisation of the water renewal in a macro-tidal marina using several transport timescales. *Water* **2019**, *11*, 2050. [CrossRef]

18. Staneva, J.; Ricker, M.; Carrasco Alvarez, R.; Breivik, Ø.; Schrum, C. Effects of Wave-Induced Processes in a Coupled Wave–Ocean Model on Particle Transport Simulations. *Water* **2021**, *13*, 415. [CrossRef]

19. So, S.; Valle-Levinson, A.; Laurel-Castillo, J.A.; Ahn, J.; Al-Khaldi, M. Removing Wave Bias from Velocity Measurements for Tracer Transport: The Harmonic Analysis Approach. *Water* **2020**, *12*, 1138. [CrossRef]

20. de Brye, B.; de Brauwere, A.; Gourgue, O.; Delhez, E.J.M.; Deleersnijder, E. Water renewal timescales in the Scheldt Estuary. *J. Mar. Syst.* **2012**, *94*, 74–86. [CrossRef]

21. Dewals, B.; Archambeau, P.; Bruwier, M.; Erpicum, S.; Pirotton, M.; Adam, T.; Delhez, E.; Deleersnijder, E. Age of Water Particles as a Diagnosis of Steady-State Flows in Shallow Rectangular Reservoirs. *Water* **2020**, *12*, 2819. [CrossRef]

22. Cheng, Y.; Mu, Z.; Wang, H.; Zhao, F.; Li, Y.; Lin, L. Water Residence Time in a Typical Tributary Bay of the Three Gorges Reservoir. *Water* **2019**, *11*, 1585. [CrossRef]

23. Delhez, E.J.M.; Heemink, A.W.; Deleersnijder, E. Residence time in a semi-enclosed domain from the solution of an adjoint problem. *Estuar. Coast. Shelf Sci.* **2004**, *61*, 691–702. [CrossRef]

24. Deleersnijder, E.; Campin, J.M.; Delhez, E.J.M. The concept of age in marine modelling I. Theory and preliminary model results. *J. Mar. Syst.* **2001**, *28*, 229–267. [CrossRef]

25. Langmuir, I. The velocity of reactions in gases moving through heated vessels and the effect of convection and diffusion. *J. Am. Chem. Soc.* **1908**, *30*, 1742–1754. [CrossRef]

26. Ketchum, B.H.; Rawn, A.M. The flushing of tidal estuaries [with Discussion]. *Sewage Ind. Wastes* **1951**, *23*, 198–209.

27. Cushman-Roisin, B.; Beckers, J.-M. *Introduction to Geophysical Fluid Dynamics: Physical and Numerical Aspects*; Cushman-Roisin, B., Beckers, J.-M., Eds.; Academic Press: Waltham, MA, USA, 2011; Volume 101, pp. 1–828.

28. Fischer, H.B.; List, E.J.; Koh, R.C.Y.; Imberger, J.; Brooks, N.H. *Mixing in Inland and Coastal Waters*; Academic Press, Inc.: San Diego, CA, USA, 1979.

29. Deleersnijder, E. *Classical vs. Holistic Timescales: The Mururoa Atoll Lagoon Case Study*; Université catholique de Louvain: Louvain-la-Neuve, Belgium, 2019; pp. 1–12. Available online: http://hdl.handle.net/2078.1/224391 (accessed on 23 September 2021).

Review

Solute Reactive Tracers for Hydrogeological Applications: A Short Review and Future Prospects

Viet Cao [1,*], Mario Schaffer [2], Reza Taherdangkoo [3] and Tobias Licha [4]

[1] Faculty of Natural Sciences, Hung Vuong University, Nguyen Tat Thanh Str., Viet Tri, 35120 Phu Tho, Vietnam

[2] Lower Saxony Water Management, Coastal Defence and Nature Conservation Agency (NLWKN), Hanover-Hildesheim branch, An der Scharlake 39, 31135 Hildesheim, Germany; Mario.Schaffer@nlwkn-hi.niedersachsen.de

[3] Geoscience Centre, Department of Applied Geology, University of Goettingen, Goldschmidtstr. 3, 37077 Göttingen, Germany; reza.taherdangkoo@geo.uni-goettingen.de

[4] Applied Geology, Institute of Geology, Mineralogy and Geophysics, Ruhr University Bochum, Universitätsstr. 150, 44801 Bochum, Germany; tobias.licha@rub.de

* Correspondence: caoviet@hvu.edu.vn

Received: 30 January 2020; Accepted: 27 February 2020; Published: 28 February 2020

Abstract: Tracer testing is a mature technology used for characterizing aquatic flow systems. To gain more insights from tracer tests a combination of conservative (non-reactive) tracers together with at least one reactive tracer is commonly applied. The reactive tracers can provide unique information about physical, chemical, and/or biological properties of aquatic systems. Although, previous review papers provide a wide coverage on conservative tracer compounds there is no systematic review on reactive tracers yet, despite their extensive development during the past decades. This review paper summarizes the recent development in compounds and compound classes that are exploitable and/or have been used as reactive tracers, including their systematization based on the underlying process types to be investigated. Reactive tracers can generally be categorized into three groups: (1) partitioning tracers, (2) kinetic tracers, and (3) reactive tracers for partitioning. The work also highlights the potential for future research directions. The recent advances from the development of new tailor-made tracers might overcome existing limitations.

Keywords: reactive tracers; tailor-made tracer design; hydrogeological tracer test; kinetics; partitioning

---

1. Introduction

Tracer tests are one of the most well established techniques for site and process characterizations in the aquatic environment (i.e., in hydrology or hydrogeology). Various additives (e.g., particles, solids, solutes, and gases) and physical quantities (e.g., temperature and pressure) can be applied as tracers for interpreting hydraulic transport properties and/or reactive processes in the aquatic environment [1–4]. Some basic hydraulic properties, such as flow velocity or porosity, can be obtained by tracer tests using conservative (non-reactive) tracer compounds. The combination of a conservative tracer with at least one reactive tracer is commonly applied in order to assess additional system parameters, such as residual saturation [5,6], microbial activity [7,8], or temperature distribution [9,10]. The unique features of reactive tracers could provide valuable information on physical, chemical, and/or biological properties of the hydrological system which surpasses the capability of conservative tracers.

The application potential for tracers within the scope of advanced reservoir management, such as geothermal power generation or carbon capture and storage, has triggered the development of new tracers and tracer techniques in the past decades [11,12]. Reactive tracers used to detect specific properties and processes in the aquatic environment must generally either have distinctive physicochemical properties (e.g., sorption) or undergo specific reactions such as hydrolysis. To identify the most suitable tracer compounds for a specific system or problem, a thorough understanding of the physicochemical properties and their chemically reactive behavior in the probed system is a prerequisite.

The main objective of this overview article is to present a systematic review of existing and proposed reactive solute tracers based on current research advances conducted in different scientific fields. The focus of this work is on chemical/artificial tracers which are intentionally introduce in the tracer tests. For each subclass of tracer, the underlying process, their key properties, and possible target parameters/applications are described. Furthermore, the potential areas for the future development and exploitation of new reactive tracers are elaborated. Hereby, the new approach of producing tailor-made reactive tracers may break down currently existing limitations on the investigation potential of commercially available compounds.

2. Definition and Theoretical Background

2.1. Definition

A tracer is defined herein as a distinguishable chemical compound which is deliberately added to an aquatic system having a temporally and spatially well-known input function (e.g., pulse injection). The respective system property or information of interest is derived based on the relation of the input function to the observed response function (breakthrough curve) within the investigated system.

Two general tracer types can be defined based on the degree of interaction with the systems. First, conservative tracers show virtually no interaction with the reservoir materials, and thus they flow passively with the carrier fluids at their velocity. Furthermore, they do not suffer any chemical or biological processes. This implies that these tracers are inert under reservoir conditions. The second type of tracers can be summarized as reactive tracers. The interpretation of reactive tracers relies on their known properties, physicochemical or chemical behavior during the transport. Reactive tracers are compounds that undergo a chemical reaction or physicochemical interaction processes in a predictable way under specific boundary conditions existing in the investigated system. Consequently, using the particular features of reactive tracers could provide unique information on physicochemical properties and/or water chemistry of the hydrological system far beyond the capability of conservative tracers.

Traditionally, tracer tests were conducted using conservative tracers. These tracers can provide general physical and hydraulic parameters of the system (e.g., porosity, dispersivity, or arrival time). In order to derive these parameters with great accuracy, the compounds are desired to behave ideally. The properties of an ideal tracer are well established [2]; they (1) behave conservatively (e.g., are transported with water velocity, not degradable), (2) have a low background concentration in the system, (3) are detectable in very low concentrations, and (4) have low or no toxicological environmental impact. Nevertheless, all solute tracers are influenced to some degree by physical, chemical, and/or biological processes. This means that completely ideal tracers do not exist in reality. Therefore, some knowledge of the investigated system is required beforehand to verify the practicality of the tracer behavior and thus to avoid test failure.

2.2. Conservative Tracer Transport versus Reactive Tracer Transport

The transport behavior of a tracer compound in the aquatic environment is affected by several physical and chemical processes. These processes result in spatial and/or temporal concentration changes of the introduced tracer during its transport, which are reflected in the system response function (e.g., breakthrough curve $c(t)$). Tracer transport is commonly described based on the principle of mass conservation by means of the advection-dispersion-reaction model in the three-dimensional form as follows:

$$R\frac{\partial c}{\partial t} = -\overline{V}\cdot\nabla c + \nabla(D_H\nabla c) + S, \tag{1}$$

where R is the retardation factor, c is the tracer concentration, t is the time, $\overline{V}$ is the average pore water velocity, D_H is the hydrodynamic dispersion tensor (including mechanical dispersion and molecular diffusion), and S is the source/sink term accounting for the tracer transformation (degradation/generation).

As described above, tracer transport in water can be classified as conservative or reactive according to its interaction within the system to be studied. A conservative tracer does not interact or alter during the transport, and thus the concentration is not changed by processes other than dilution, dispersion, and partial redirection. As such, conservative tracers are expected to mimic the transport of water without retardation and transformation. They underlie only the purely hydrodynamic transport processes: advection, diffusion, and dispersion (as terms 2 and 3 in Equation (1). It should be noted that various types of mixing always exist which should be interpreted with caution as the mixing or other dilution processes may influence the results of tracer experiments [13–16]. Therefore, conservative tracers are generally used to investigate hydraulic properties (e.g., tracking connectivities, flow pathways), analyzing travel times and flow velocities, determining recharge and discharge, and estimating hydromechanical properties (e.g., dispersivity, porosity). Common examples of conservative tracers under ambient temperatures are major anions such as bromide [17,18], stable isotopes such as ^{2}H and ^{18}O [4,19], dye tracers such as uranine [20–22], and rhodamine WT [23–27].

Apart from hydrodynamic transport processes, reactive tracers additionally underlie physical, chemical, and/or biological processes during their transport (terms 1 and 4 in Equation (1)). The implementation of reactive tracers with identical and well understood interactions or reactions could implicitly provide unique information on physicochemical aquifer properties (e.g., sorption capacity), water chemistry (e.g., redox condition, pH, ion concentrations), and other influencing parameters (e.g., temperatures, microbial activity) [28–30].

In order to benefit from the selective and process specific nature of reactive tracers, it is a prerequisite to combine them with at least one conservative reference tracer by performing a multitracer experiment to account for the purely hydrodynamic transport processes that affect both tracer types in the same way. Consequently, the reactive processes can be identified and quantified. The intended information from the tracers is gained by comparing the concentration versus time curves (breakthrough curves) of the reactive tracers with the conservative tracers (reference). This can be illustrated by the schematic breakthrough curves for a simulated tracer test having a pulse input function (Figure 1). The time shift and/or the reduction of the peak area (tracer mass) of the breakthrough curves indicate retardation and/or degradation, respectively. Measured breakthrough curves can be inversely interpreted using analytical or numerical models to estimate the values of controlling parameters, such as the distribution coefficient for the sorption process, the decay rate for the sorption process, or the decay rate for the biodegradation process.

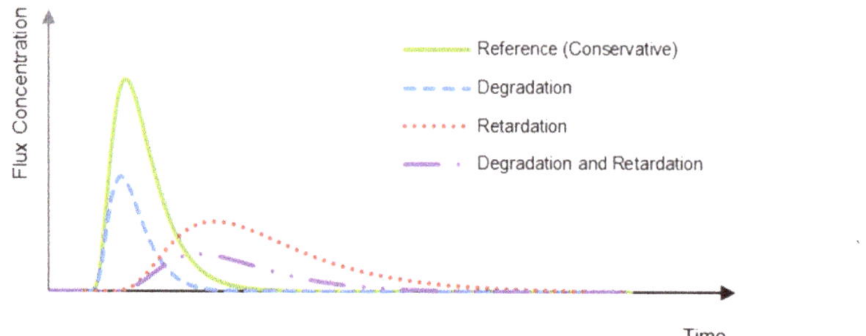

Figure 1. Schematic breakthrough curves for conservative tracer and reactive tracers after a pulse injection.

3. Types of Reactive Tracers

A generalized classification of currently existing reactive tracers and proposed reactive tracer concepts, including their required properties, possible applications, and processes is provided. Depending on their physical, chemical, and/or biological behavior, three major subgroups are distinguished (Table 1):

- Partitioning tracers: These types are based on the partitioning equilibrium between two immiscible phases or at their interfaces (fluid–solid, fluid–fluid) leading to a retardation relative to the conservative tracer remaining in (one) fluid phase.
- Kinetic tracers: These types are non-equilibrium tracers in which only the reaction kinetics are used for the parameter determination. As a result of the tracer reaction, the tracer signals are decreasing (parent compound) or increasing (daughter compound) with time (degradation). These tracers usually do not show retardation (no partitioning).
- Reactive tracers for partitioning: These tracers are a hybrid form of the preceding tracers, containing features of both: chemical reaction (degradation) of the parent compound and subsequent partitioning (retardation) of the daughter products.

Table 1. Classification of reactive tracers.

Reactive Tracer Type	Equilibrium tracers (partitioning tracers)				Reactive tracer for partitioning (volume or interface sensitive after degradation)	Kinetic tracers (decaying tracers)		
	Fluid–Solid		Fluid–Fluid			One phase		Two phases
	Sensitive for uncharged surfaces	Sensitive for charged surfaces	Volume sensitive	Interface sensitive		Degradation sensitive	Thermo-sensitive	Interface-sensitive
Determining properties of tracer molecules	Uncharged but still soluble (moderately polar) organic compounds	Organic and inorganic ions with opposite charge compared to the surface	Compounds with quantifiable partitioning between two phases (soluble in both phases)	Amphiphilic compounds (surface active agents-surfactants)	Hydrophilic compounds, susceptible to decay leading to daughter compounds with different partitioning properties (coefficients)	Degradable compounds under applied conditions	Hydrolysable and hydrophilic compounds with known kinetic parameters and decay mechanisms	Hydrophobic and hydrolysable compounds
Possible target parameters/Application	Organic carbon content, surface area to volume ratio A/V (of uncharged surfaces)	Surface charge (surface charge density, exchange capacity), Surface area to volume ratio A/V (of charged surface)	Residual saturation	Residual distribution, contact area	Residual saturation or residual distribution, contact area	Attenuation capacity, other reaction relevant boundary conditions (e.g., redox conditions, pH)	Temperature and temperature distribution (cooling fractions)	Interfacial area (development with time)
Underlying (reactive) process	Sorption due to hydrophobic interactions	Sorption due to electrostatic interactions (e.g., ion exchange, hydrogen bonding)	Phase partitioning	Interfacial adsorption (partitioning between bulk phase and interface)	In-situ decay with subsequent partitioning	Chemical and biological reactions	Hydrolysis reaction, substitution	Hydrolysis reaction leading to inter-phase mass transfer

3.1. Equilibrium Tracers

3.1.1. Fluid–Solid (Sorbing Tracers)

Sensitive for Uncharged Surfaces

A tracer compound sensitive for uncharged surfaces undergoes hydrophobic sorption onto uncharged sites of the sorbent (e.g., soil, aquifer material), particularly organic matter. Hydrophobic sorption is the result from a weak solute-solvent interaction coming from a decrease in entropy of the solution and can be explained by general interactions between sorbate and sorbent, e.g., van-der-Waals forces (dipole and/or induced-dipole interactions) [31]. The organic carbon content (f_{OC}) of the aquifer material generally correlates with the sorptivity and thus the retardation of a neutral (uncharged) organic compound [32–34]. Therefore, it is conceivable that substances, which are sensitive to uncharged surfaces, have the potential to determine the f_{OC} of a system from their observed retardation factor (R_{unc}) assuming a linear sorption isotherm:

$$R_{unc} = 1 + \frac{\rho}{n_e} K_{unc},\qquad(2)$$

where ρ is bulk density, n_e is effective porosity, and K_{unc} is the sorption coefficient. K_{unc} depends primarily on the hydrophobicity of the tracer molecules, typically characterized by the n-octanol-water partition coefficient ($\log K_{OW}$) and the f_{OC} of the geological materials. From $\log K_{OW}$ of the tracer compound, K_{unc} for a particular system can be estimated. According to the literature [35–37] $\log K_{OW}$ can empirically be related to the organic carbon normalized sorption coefficient (K_{OC}) in the form:

$$\log K_{OC} = a \log K_{OW} + b,\qquad(3)$$

$$K_{OC} = \frac{K_{unc}}{f_{OC}},\qquad(4)$$

where a and b are empirical parameters.

Thus, from known $\log K_{OW}$ and determined R_{unc}, the average f_{OC} between the injection and observation points can be estimated. By selecting non-ionic compounds with moderate $\log K_{OW}$ values between 1 and 3 (1H-benzotriazole, carbamazepine, diazepam, and isoproturon) from formerly published column experiments by Schaffer et al. [38,39] using correlation factors for non-hydrophobic compounds after Sabljic et al. (1995), the observed f_{OC} values of the columns agree very well with the independently measured ones from the bulk using total organic carbon measurements. Despite the relatively large uncertainty regarding the chosen $\log K_{OW}$ values, all deviations of the absolute values between the measured and calculated f_{OC} are within one order of magnitude (less than factor 5).

To the extent of our knowledge, this tracer type has not yet been explicitly proposed, and therefore their potential could be further investigated. Some promising examples include 8:2 fluorotelomer alcohol [40], short-chained alkyl phenols [41], or pharmaceutical compounds [42–44].

Sensitive for Charged and Hydrophilic Surfaces

A tracer compound sensitive for charged surfaces undergoes ionic sorption between a charged moiety of a tracer molecule and an oppositely charged surface of the sorbent (e.g., soil, aquifer material). In this case, there is a strong electrostatic interaction (e.g., ion exchange, hydrogen bonding, or surface complexation) between tracer sorbate and sorbent.

Retardation of a solute due to ion sorption on natural solids (R_c) can be related either to a sorbent mass (Equation (2)) or to its surface sensitivity to the surface area (A) to volume (V) ratio if the sorption coefficient (K_c) is known [45]:

$$R_c = 1 + \frac{A}{V} K_c,\qquad(5)$$

These tracers are required to be water soluble, ionized (electrically charged), and can be organic or inorganic substances. The selection of tracers for this application is based on the surface charge of the sorbents. Further, the pH condition strongly influences the charge states of organic compounds (e.g., bases, acids, and ampholytes) and the sorbent's surface [33]; thus, pH and the point of zero charge of the surface should be considered before selecting a tracer compound.

Many laboratory tests have been conducted to demonstrate the feasibility of charged surface tracers to interrogate the surface area, e.g., using safranin [46], lithium [47–49], and monoamines [50]. A couple of field tests have also demonstrated the potential use of charged surface tracers for investigating the surface area, e.g., using safranin [51] and caesium [52,53]. Furthermore, this tracer type has the potential to estimate the ion exchange capacity of sediments [54].

3.1.2. Fluid–Fluid

The fluid-fluid tracers summarize liquid–liquid tracers and liquid–gas tracers due to the similarity in the underlying processes and applications.

Volume Sensitive Tracers

A volume sensitive tracer is a compound that partitions between two immiscible fluid phases (liquid–liquid or liquid–gas). A different solubility in the two fluid phases leads to the specific phase distribution and results in a retardation of the tracer. Volume sensitive tracers are very useful in estimating the volume of the immobile phase (residual saturation). For example, one common application of this type of tracer is to characterize the source zone of non-aqueous phase liquids (NAPLs) for contaminated sites. Another popular use is to evaluate the effectiveness of treatment techniques before and after the remediation of NAPLs, thereby obtaining independent estimates on the performance of the cleanup. This tracer can also be used to identify residual gas or supercritical fluid phases, such as in carbon capture and storage applications. When sorption onto solids is negligible, the retardation factor (R_{vs}) is a function of the average residual saturation (S_r) within the tracer flow field [55,56]:

$$R_{vs} = 1 + \frac{S_r}{(1 - S_r)} K_{vs},\tag{6}$$

where K_{vs} is the partition coefficient between two fluid phases.

A large number of laboratory experiments and field-scale tests have been conducted to detect NAPL contaminations since the 1990s. The most commonly applied volume sensitive tracers are alcohols of varying chain length, such as 1-hexanol [57–60], 1-pentanol and 1-heptanol [61–63], 2-ethyl-1-butanol [5,61,64], 6-methyl-2-heptanol [65,66], 2,2-dimethyl-3-pentanol [56,65,66], 2,4-dimethyl-3-pentanol [57,63,64,67–70], substituted benzyl alcohols [6,71] and fluorotelomer alcohols [72]. Additionally, sulfur hexafluoride (SF$_6$) [73–77], perfluorocarbons [61,78], radon-222 [79–81], and fluorescent dyes (e.g., rhodamine WT, sulforhodamine B, and eosin) [82] have also been suggested for use as volume sensitive tracers. Recently, the noble gases krypton and xenon were applied successfully in the determination of the residual CO$_2$ saturation [83–87].

Interface Sensitive Tracers

An interface sensitive tracer is a compound that undergoes the accumulation (adsorption) at the interface between two immiscible fluids, typically liquid–liquid or liquid–gas, leading to the retardation of the tracer. The magnitude of adsorption at the interface is controlled by the physicochemical properties of tracer compounds and by the interfacial area, particularly the size of the specific fluid–fluid interfacial area (a_{nw}) and the interfacial adsorption coefficient (K_{if}). The retardation factor (R_{if}) defined through porous media follows [88,89]:

$$R_{if} = 1 + \frac{a_{if}}{\theta_w} K_{if},\tag{7}$$

$$K_{if} = \frac{G_{eq}}{C_{eq}}, \tag{8}$$

where a_{if} is the specific interfacial area, θ_w is the volumetric water content, and K_{if} is the interfacial adsorption coefficient (ratio between the interfacial tracer concentration in the sorbed phase at the interface (G_{eq}) and the fluid (C_{eq}) at equilibrium).

The desired compounds for this tracer class are amphiphilic molecules (containing both hydrophobic and hydrophilic groups). Information on fluid-fluid interfacial areas, along with residual saturation (assessed by volume sensitive tracers) assists the understanding of the fate and transport of contamination in the systems.

One of the most popular interface sensitive tracers that have been successfully tested in laboratory and field scales is the anionic surfactant sodium dodecylbenzene sulfonate [67,69,88,90–100]. Further potential arises for other ionic and non-ionic surfactants (e.g., marlinat [101], 1-tetradecanol [102,103], sodium dihexylsulfosuccinate [104]) and for cosurfactants (e.g., *n*-octanol and *n*-nonanol [105]).

3.2. Kinetic Tracers

3.2.1. One Phase

Degradation Sensitive Tracers

Degradation sensitive tracers are compounds that undergo biotic and/or abiotic transformations. Depending upon the nature of the tracer specific (reaction controlling boundary conditions), chemical and/or biological characteristics of the flow system can be investigated. Information on the decay mechanism and the equivalent kinetic parameters is a prerequisite for their successful application. The decay mechanism is usually desired to follow a (pseudo) first order reaction to limit the number of required kinetic parameters and to avoid ambiguity. In addition, other influencing factors on kinetics should be considered before application (e.g., pH, light, and temperature). The reaction rate constant (k_{DS}) can be estimated by measuring the extent of tracer loss of the mother compound or the associated increase of a transformation product along the flow path.

This type of tracer has been studied and tested in field-scale experiments over the past 20 years. Their main purpose is to determine microbial metabolic activity (natural attenuation processes) and/or to assess redox conditions. Numerous redox-sensitive tracers have been applied for laboratory and field scale investigations, such as inorganic electron acceptors (e.g., O_2, NO_3^-, SO_4^{2-}, CO_3^{2-}) [106–116], organic electron donors (e.g., low-molecular weight alcohols and sugars [117] and benzoate [118–120]), or the organic electron acceptor resazurin [8,121–128].

Thermo-Sensitive Tracers

Thermo-sensitive tracers are compounds undergoing chemical reactions that are well-defined and temperature driven, such as hydrolysis [129–131] or thermal decay [132,133]. Prior knowledge on their reaction mechanisms is required for each specific thermo-sensitive tracer. To avoid ambiguity, reactions following (pseudo) first order reaction are desired, and the reaction speed (expressed by the reaction rate constant (k_{TS})) is preferred to be solely controlled by temperature. For these reactions, the dependence of temperature (T) on k_{TS} is the essential factor for estimating the thermo-sensitivity expressed by Arrhenius law:

$$k_{TS} = Ae^{-\frac{E_a}{RT}}, \tag{9}$$

where A is the pre-exponential factor, E_a is the activation energy, and R is the ideal gas constant.

By knowing the corresponding kinetic parameters, the equivalent temperature (T_{eq}) and the cooling fraction (χ) can be obtained [134]. T_{eq} references the thermal state of a probed reservoir relative to an equivalent system having isothermal conditions, whereas χ has the potential to further estimate a spatial temperature distribution of the investigated system.

A typical application of these tracers is to investigate the temperature distribution of a georeservoir. The first field experiments using ester compounds (ethyl acetate and isopentyl acetate), however, were unable to determine a reservoir temperature [135–137]. The failure of the studies was attributed to the poor determination of pH dependence and the lower boiling point of the tracer compounds compared to the reservoir temperature leading to vaporization. New attempts demonstrated the successful application in the laboratory [9] and in the field [138]. Other studies using classical tracers like fluorescein [139] or Amino G [132,133] were able to identify the reservoir temperatures. Currently, extensive research has been conducted to study structure-related kinetics of defined thermo-sensitive reactions with promising results [9,10,130,131,134].

3.2.2. Two Phases

Kinetic Interface Sensitive (KIS)

KIS tracers are intended to be dissolved or mixed with a non-aqueous carrier fluid (e.g., supercritical CO_2 [11]) and injected into the reservoir. The underlying process is an interface-sensitive hydrolysis reaction at the interface between the aqueous and the non-aqueous phase. Here, the tracer saturates the interface of the evolving plume due to interfacial adsorption and reacts irreversibly with water (hydrolysis with first-order kinetics). Due to the constant (adsorbed) concentration of the reactant at the interface, the reaction kinetics is simplified to (pseudo) zero order kinetics. The formed reaction products are monitored in the water phase.

In order to have minimal partitioning into the polar water phase, the potential tracers have to be non-polar in conjunction with high log K_{OW} values. Furthermore, the KIS tracer reaction kinetics has to be adapted to the characteristics of the reservoir (T, pH) and the interfacial area dynamics in order to resolve the plume development. In contrast to the parent compound, at least one of the reaction products has to be highly water soluble resulting in low or even negative log K_{OW} values. Thus, back-partitioning into the non-aqueous phase can be avoided.

This class of reactive tracers was originally intended to characterize the fluid–fluid interfacial area (e.g., between supercritical CO_2 and formation brine during CO_2 storage experiments [140]). Currently, only limited laboratory experiments with the supercritical CO_2 analogue fluid *n*-octane are available [11].

3.3. Reactive Tracers for Partitioning

A reactive tracer for partitioning is a compound comprising the features of both partitioning tracers and kinetic tracers. This type of tracer undergoes in-situ decay of the parent tracer compounds with subsequent partitioning of the daughter compounds. The concentration of both parent and daughter compounds are determined. The separation of the arrival times of the two tracers indicates the residual saturation similar to volume sensitive tracers (see Section 3.1.2). The tracer compounds are hydrophilic and must be susceptible to decay leading to daughter compounds with different partitioning coefficients. Kinetic parameters should be evaluated in order to acquire suitable compounds for specific conditions of tracer tests (e.g., types and time scales). In contrast to kinetic tracers, the kinetic parameters are not used in the evaluation of the breakthrough curves for these tracers.

The most common fields for the application of these types of tracers are oilfields and carbon capture and storage. Esters like ethyl acetate have been proposed to determine the residual oil saturation according to Cooke [141]. By 1990 they have been successfully applied to oilfields [142,143] and are continued to be implemented today [144,145]. Myers et al. (2012) demonstrate the feasibility of using reactive ester tracers (i.e. triacetin, propylene glycol diacetate and tripropionin) to quantify the amount of residually trapped CO_2 through an integrated program of laboratory experiments and computer simulations. Later, the research was also demonstrated successfully in field experiments [146].

4. Exploitation Potential and Further Challenges of Developing Reactive Tracers

4.1. The Necessity for New Tracers–Tracer Design Approach

The use of tracers for hydrogeological applications has a long history. The first reported tracer application was around 10 A.D. to track the connection between the spring source of the Jordan River and a nearby pond [147]. Since then, the development of technology and the advances of tracer testing with a wide selection of tracer compounds have brought effective tools for investigating different properties of the aquatic environment. In general, tracer tests could be applied to any kind of natural and engineered systems. It is especially advantageous for not directly accessible systems compared to other techniques. Nevertheless, there are still many systems in which the potential of using reactive tracers is not yet fully exploited and more attention should be paid to these, including:

- The hyporheic zone, a transition zone between surface water and subsurface water, has been recognized as a hotspot for biogeochemical reactions, making the exchange of water, nutrients, and organic matter important parameters. This zone is a mixing zone which has a complex hydrological situation and heterogeneity containing dissolved gasses, oxidized and reduced species, temperature patterns, flow rates, etc. Due to the large number of variables, the quantification of processes in the hyporheic zone is still a challenge [148,149].

- Hydraulic fracturing (fracking) in shale/tight gas reservoirs has gained growing interests in the oil and gas industry during the last decade [150]. However, fracking may pose environmental risks [151,152]. During the stimulation process, fracking fluid is injected into the reservoir to create additional flow paths for the transport of hydrocarbons. Hydraulically induced fractures may connect pre-existing natural fractures and faults leading to the creation of multiple permeable pathways which may cause groundwater contamination [153]. Therefore, there is a high demand for the application of tracers to predict the risk or to track the contamination (i.e. fracking fluid) [154].

- Other fields may include karst aquifer characterization (due to the strong system heterogeneity and variability), geothermal fluids and acid-mine-drainage (due to complex water chemistry and temperature).

The design of new innovative reactive tracers requires new strategies. Molecular design has been successfully established as a methodology for producing tailor-made molecules with desired properties or effects in several scientific disciplines, especially in life sciences, such as pharmacology, biochemistry, medicine [155], and material sciences [156]. The target-oriented combination of well-studied structural elements and molecular features (e.g., functional groups, substructures, homologues, etc.) allows the creation of novel compounds with desired structures and properties. Almost an unlimited number of compounds is imaginable and can be synthesized individually for a magnitude of applications. However, molecular target design of tracer substances for studying the aquatic environment has yet to be widely considered.

4.2. Strategy for Designing Novel Reactive Tracers

Creating tracer molecules, which react in a predictable way under given physicochemical conditions, is a relatively new and very innovative concept. By knowing exactly how certain reservoir conditions drive the tracer reaction, new insights into the controlling variables may be gained. In the following, the exemplary molecular target design of thermo-sensitive and interface-sensitive tracers is described. The prerequisite for the design (selection and modification) of molecules that are able to act as thermo-sensitive and interface-sensitive tracers in reservoir studies, respectively, is a thorough understanding of their reactive behavior. In particular, it is vital to understand the role and influence of each structural element in the molecule on its reaction kinetics and its physicochemical tracer properties (e.g., detection, acidity, solubility, sorption, etc.). In Figure 2, the main steps for a successful theoretical and practical molecular target tracer design are shown schematically.

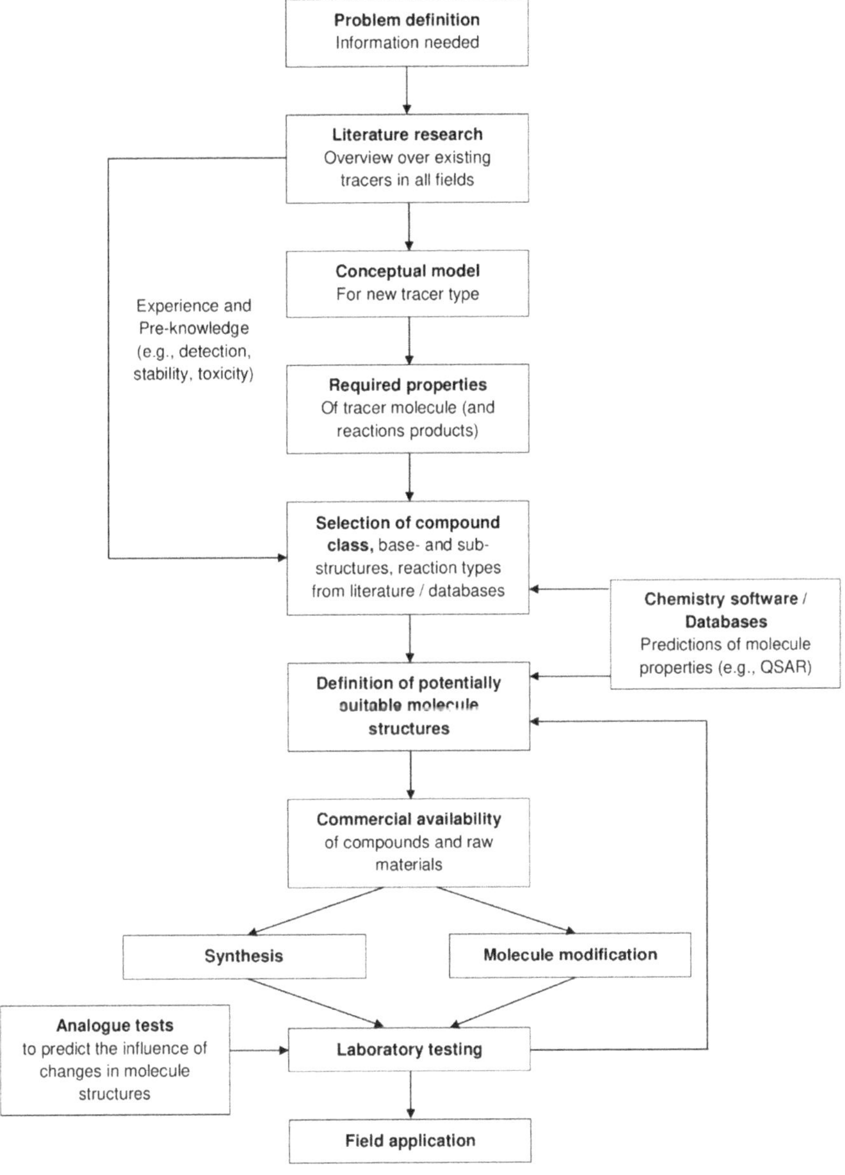

Figure 2. Schematic overview for the design of reservoir tracers.

Based on available literature and experiences from laboratory and field tests, a promising base molecule for both tracer types is believed to be the class of naphthalenesulfonates, into which thermo- and interface-sensitive groups can be incorporated (Figure 3). Several physicochemical attributes make them convenient for the selection as the backbone structure. Naphthalenesulfonates are strong acids with corresponding low logarithmic acidity constants (pK_a) of <1. Therefore, this compound class forms anions even at very low pH values and is highly water-soluble (>1000 g L^{-1}). The resulting pH-dependent log K_{OW} of −2.87 at pH > 5 (SciFinder, ACD (Advanced Chemistry Development)/Labs) is also very low, which implies a non-sorptive behavior and, thus, a high mobility in aquatic systems. Additionally, naphthalenesulfonates are stable under oxygen-free conditions and temperatures up to 250 °C [129,157]. The molecule's good fluorescence with a direct detection limit in the low µg L^{-1} range is another important feature of naphthalenesulfonates. Hence, their detection in field tests by online determination simplifies the experimental effort needed. Furthermore, (high-pressure liquid) ion pair chromatography combined with solid phase extraction and fluorescence detection (SPE-IPC-FLD) lowers the detection limit by around one order of magnitude (<1 µg L^{-1}) even in highly saline matrices, such as brines from deep reservoirs [158,159]. The chromatographic separation even allows the simultaneous analysis of several compounds and, therefore, the use of different isomers, derivatives, and homologues. Finally, naphthalenesulfonates are non-toxic [160], their use in groundwater studies is administratively non-restricted, and they are established conservative tracers for the characterization of geothermal reservoirs [158,161].

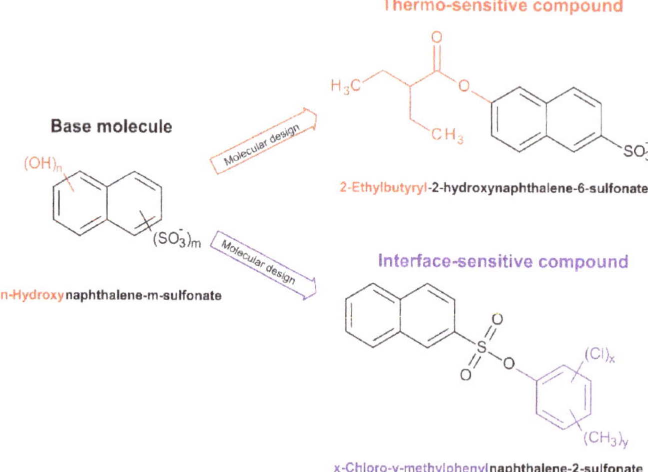

Figure 3. Design of two different types of potential reservoir tracers based on naphthalenesulfonate as common structural element.

5. Summary and Conclusions

The selection of optimal reactive tracer compounds is main challenge that needs to be considered before conducting a tracer test. For instance, when designing a thermo-sensitive tracer test, a tracer that decays too slowly under a system's temperature lengthens test duration needlessly and thus makes observing the differences in mean residence times difficult; too fast decay makes it challenging for the test implementation. Moreover, new reactive tracer compounds have been extensively developed in the past decades due to the demand in new advanced technologies. Therefore, a complete understanding of the physicochemical properties of reactive tracers and their occurring processes is essential. Depending on the biophysicochemical behavior, three types of reactive tracers can be distinguished, namely: equilibrium tracers, kinetic tracers and reactive tracer for partitioning. Equilibrium tracers are based on the partitioning equilibrium between two immiscible phases or at their interfaces. Kinetic tracers are non-equilibrium tracers in which only the reaction kinetics are used for the parameter determination. Reactive tracers for partitioning are a hybrid form of equilibrium tracers and kinetic tracers.

The complexities of natural systems, along with the large number of requirements for the tracers, make the selection and use of reactive tracers not a simple task, but an art. Based on the knowledge of tracer properties, tailor-made tracer compounds are being developed with the required properties or effects in hydrogeology. The target-oriented combination of well-studied structural elements and molecular features (e.g., functional groups, substructures, homologues) allows for the creation of novel compounds with desired structures and properties. Nearly an unlimited number of compounds can be synthesized individually for specific applications. This innovative concept can expand the potential application of tracers in different fields (e.g., quantification of processes in the hyporheic zone, prediction of environmental risks of hydraulic fracturing). Molecular design assists the preselected properties (e.g., fluorescence) of both reactants and products. This allows a mass balance, and thereby opens the opportunity of a tracer test design without an additional conservative tracer.

Author Contributions: Conceptualization, V.C., M.S., and T.L.; methodology, V.C., M.S., and T.L.; formal analysis, V.C.; writing-original draft preparation, V.C. and M.S.; writing-review and editing, V.C., M.S., R.T. and T.L.; visualization, V.C., R.T; funding acquisition, R.T. All authors have read and agreed to the published version of the manuscript.

Funding: This research was funded by the Ministry of Education and Training of Vietnam (MOET, DA-911), the European Community's 7th Framework Programme FP7/2007–2013, within the MUSTANG project (grant agreement no. 227286), the German Ministry for Environment (BMU) within the project REAKTHERM (grant no. 0325417) and the German Research Foundation (DFG) within the GEOCAT project (project no. LI 1314/3-1 and 1314/3-2.

Conflicts of Interest: The authors declare no conflict of interest.

References

1. Chrysikopoulos, C.V. Artificial tracers for geothermal reservoir studies. *Environ. Geol.* **1993**, *22*, 60–70. [CrossRef]
2. Flury, M.; Wai, N.N. Dyes as tracers for vadose zone hydrology. *Rev. Geophys.* **2003**, *41*, 2.1–2.37. [CrossRef]
3. Serres-Piole, C.; Preud'homme, H.; Moradi-Tehrani, N.; Allanic, C.; Jullia, H.; Lobinski, R. Water tracers in oilfield applications: Guidelines. *J. Pet. Sci. Eng.* **2012**, *98–99*, 22–39. [CrossRef]
4. Abbott, B.W.; Baranov, V.; Mendoza-Lera, C.; Nikolakopoulou, M.; Harjung, A.; Kolbe, T.; Balasubramanian, M.N.; Vaessen, T.N.; Ciocca, F.; Campeau, A.; et al. Using multi-tracer inference to move beyond single-catchment ecohydrology. *Earth-Sci. Rev.* **2016**, *160*, 19–42. [CrossRef]
5. Rhee, S.; Kang, J.; Park, J. Partitioning tracer method for quantifying the residual saturation of refined petroleum products in saturated soil. *Environ. Earth Sci.* **2011**, *64*, 2059–2066. [CrossRef]
6. Silva, M.; Stray, H.; Bjornstad, T. Studies on new chemical tracers for determination of residual oil saturation in the inter-well region. In Proceedings of the SPE Oklahoma City Oil Gas Symposium, Oklahoma City, OK, USA, 27–31 March 2017; pp. 1–14.
7. Hillebrand, O.; Noedler, K.; Sauter, M.; Licha, T. Multitracer experiment to evaluate the attenuation of selected organic micropollutants in a karst aquifer. *Sci. Total Environ.* **2015**, *506–507*, 338–343. [CrossRef]
8. Knapp, J.L.A.; González-Pinzón, R.; Drummond, J.D.; Larsen, L.G.; Cirpka, O.A.; Harvey, J.W. Tracer-based characterization of hyporheic exchange and benthic biolayers in streams. *Water Resour. Res.* **2017**, *53*, 1575–1594. [CrossRef]
9. Maier, F.; Schaffer, M.; Licha, T. Temperature determination using thermo-sensitive tracers: Experimental validation in an isothermal column heat exchanger. *Geothermics* **2015**, *53*, 533–539. [CrossRef]
10. Schaffer, M.; Idzik, K.R.; Wilke, M.; Licha, T. Amides as thermo-sensitive tracers for investigating the thermal state of geothermal reservoirs. *Geothermics* **2016**, *64*, 180–186. [CrossRef]
11. Schaffer, M.; Maier, F.; Licha, T.; Sauter, M. A new generation of tracers for the characterization of interfacial areas during supercritical carbon dioxide injections into deep saline aquifers: Kinetic interface-sensitive tracers (KIS tracer). *Int. J. Greenh. Gas Control* **2013**, *14*, 200–208. [CrossRef]
12. Myers, M.; Stalker, L.; Ross, A.; Dyt, C.; Ho, K.-B. Method for the determination of residual carbon dioxide saturation using reactive ester tracers. *Appl. Geochem.* **2012**, *27*, 2148–2156. [CrossRef]
13. Cornaton, F.J.; Park, Y.J.; Deleersnijder, E. On the biases affecting water ages inferred from isotopic data. *J. Hydrol.* **2011**, *410*, 217–225. [CrossRef]
14. Delhez, É.J.M.; Deleersnijder, É.; Mouchet, A.; Beckers, J.M. A note on the age of radioactive tracers. *J. Mar. Syst.* **2003**, *38*, 277–286. [CrossRef]

15. Deleersnijder, E.; Delhez, E.J.M.; Crucifix, M.; Beckers, J.M. On the symmetry of the age field of a passive tracer released into a one-dimensional fluid flow by a point-source. *Bull. la Soc. R. des Sci. Liege* **2001**, *70*, 5–21.

16. White, L.; Legat, V.; Deleersnijder, E. Tracer conservation for three-dimensional, finite-element, free-surface, ocean modeling on moving prismatic meshes. *Mon. Weather Rev.* **2008**, *136*, 420–442. [CrossRef]

17. Reimus, P.W.; Pohll, G.; Mihevc, T.; Chapman, J.; Haga, M.; Lyles, B.; Kosinski, S.; Niswonger, R.; Sanders, P. Testing and parameterizing a conceptual model for solute transport in a fractured granite using multiple tracers in a forced-gradient test. *Water Resour. Res.* **2003**, *39*. [CrossRef]

18. Yakirevich, A.; Shelton, D.; Hill, R.; Kiefer, L.; Stocker, M.; Blaustein, R.; Kuznetsov, M.; McCarty, G.; Pachepsky, Y. Transport of conservative and "smart" tracers in a first-order creek: Role of transient storage type. *Water* **2017**, *9*, 485. [CrossRef]

19. Moeck, C.; Radny, D.; Popp, A.; Brennwald, M.; Stoll, S.; Auckenthaler, A.; Berg, M.; Schirmer, M. Characterization of a managed aquifer recharge system using multiple tracers. *Sci. Total Environ.* **2017**, *609*, 701–714. [CrossRef]

20. Hillebrand, O.; Nödler, K.; Licha, T.; Sauter, M.; Geyer, T. Caffeine as an indicator for the quantification of untreated wastewater in karst systems. *Water Res.* **2012**, *46*, 395–402. [CrossRef]

21. Field, M.S.; Wilhelm, R.G.; Quinlan, J.F.; Aley, T.J. An assessment of the potential adverse properties of fluorescent tracer dyes used for groundwater tracing. *Environ. Monit. Assess.* **1995**, *38*, 75–96. [CrossRef]

22. Wolkersdorfer, C.; Göbel, J.; Hasche-Berger, A. Assessing subsurface flow hydraulics of a coal mine water bioremediation system using a multi-tracer approach. *Int. J. Coal Geol.* **2016**, *164*, 58–68. [CrossRef]

23. Zhu, Z.; Motta, D.; Jackson, P.R.; Garcia, M.H. Numerical modeling of simultaneous tracer release and piscicide treatment for invasive species control in the Chicago Sanitary and Ship Canal, Chicago, Illinois. *Environ. Fluid Mech.* **2017**, *17*, 211–229. [CrossRef]

24. Battaglia, D.; Birindelli, F.; Rinaldi, M.; Vettraino, E.; Bezzi, A. Fluorescent tracer tests for detection of dam leakages: The case of the Bumbuna dam—Sierra Leone. *Eng. Geol.* **2016**, *205*, 30–39. [CrossRef]

25. Guo, C.; Cui, Y.; Dong, B.; Liu, F. Tracer study of the hydraulic performance of constructed wetlands planted with three different aquatic plant species. *Ecol. Eng.* **2017**, *102*, 433–442. [CrossRef]

26. Olson, J.C.; Marcarelli, A.M.; Timm, A.L.; Eggert, S.L.; Kolka, R.K. Evaluating the effects of culvert designs on ecosystem processes in Northern Wisconsin Streams. *River Res. Appl.* **2017**, *33*, 777–787. [CrossRef]

27. Shih, S.S.; Zeng, Y.Q.; Lee, H.Y.; Otte, M.L.; Fang, W.T. Tracer experiments and hydraulic performance improvements in a treatment pond. *Water* **2017**, *9*, 137. [CrossRef]

28. Ptak, T.; Piepenbrink, M.; Martac, E. Tracer tests for the investigation of heterogeneous porous media and stochastic modelling of flow and transport—A review of some recent developments. *J. Hydrol.* **2004**, *294*, 122–163. [CrossRef]

29. Divine, C.E.; McDonnell, J.J. The future of applied tracers in hydrogeology. *Hydrogeol. J.* **2005**, *13*, 255–258. [CrossRef]

30. Luhmann, A.J.; Covington, M.D.; Alexander, S.C.; Chai, S.Y.; Schwartz, B.F.; Groten, J.T.; Alexander, E.C., Jr. Comparing conservative and nonconservative tracers in karst and using them to estimate flow path geometry. *J. Hydrol.* **2012**, *448–449*, 201–211. [CrossRef]

31. Hassett, J.J.; Means, J.C.; Banwart, W.L.; Wood, S.G. *Sorption Properties of Sediments and Energy-Related Pollutants*; Environmental Research Laboratory, Office of Research and Development, US Environmental Protection Agency: Athens, GA, USA, 1980.

32. Schwarzenbach, R.P.; Giger, W.; Hoehn, E.; Schneider, J.K. Behavior of organic compounds during infiltration of river water to groundwater. Field studies. *Environ. Sci. Technol.* **1983**, *17*, 472–479. [CrossRef]

33. Schaffer, M.; Licha, T. A framework for assessing the retardation of organic molecules in groundwater: Implications of the species distribution for the sorption-influenced transport. *Sci. Total Environ.* **2015**, *524–525*, 187–194. [CrossRef]

34. Maeng, S.K.; Sharma, S.K.; Lekkerkerker-Teunissen, K.; Amy, G.L. Occurrence and fate of bulk organic matter and pharmaceutically active compounds in managed aquifer recharge: A review. *Water Res.* **2011**, *45*, 3015–3033. [CrossRef] [PubMed]

35. Karickhoff, S.; Brown, D.; Scott, T. Sorption of hydrophobic pollutants on natural sediments. *Water Res.* **1979**, *13*, 241–248. [CrossRef]

36. Sabljic, A.; Gusten, H.; Verhaar, H.; Hermens, J. Corrigendum: QSAR modelling of soil sorption. Improvements and systematics of log k(OC) vs. log k(OW) correlations. *Chemosphere* **1995**, *33*, 2577.

37. Schwarzenbach, R.P.; Gschwend, P.M.; Imboden, D.M. *Environmental Organic Chemistry*; John Wiley & Sons, Inc.: Hoboken, NJ, USA, 2002; ISBN 9780471649649.

38. Schaffer, M.; Boxberger, N.; Börnick, H.; Licha, T.; Worch, E. Sorption influenced transport of ionizable pharmaceuticals onto a natural sandy aquifer sediment at different pH. *Chemosphere* **2012**, *87*, 513–520. [CrossRef] [PubMed]

39. Schaffer, M.; Kröger, K.F.; Nödler, K.; Ayora, C.; Carrera, J.; Hernández, M.; Licha, T. Influence of a compost layer on the attenuation of 28 selected organic micropollutants under realistic soil aquifer treatment conditions: Insights from a large scale column experiment. *Water Res.* **2015**, *74*, 110–121. [CrossRef] [PubMed]

40. Liu, J.; Lee, L.S. Solubility and sorption by soils of 8:2 fluorotelomer alcohol in water and cosolvent systems. *Environ. Sci. Technol.* **2005**, *39*, 7535–7540. [CrossRef]

41. Fischer, S.; Licha, T.; Markelova, E. Transportverhalten kurzkettiger Alkylphenole (SCAP) im Grundwasser und in der Umwelt. *Grundwasser* **2014**, *19*, 119–126. [CrossRef]

42. Burke, V.; Treumann, S.; Duennbier, U.; Greskowiak, J.; Massmann, G. Sorption behavior of 20 wastewater originated micropollutants in groundwater—Column experiments with pharmaceutical residues and industrial agents. *J. Contam. Hydrol.* **2013**, *154*, 29–41. [CrossRef]

43. Hebig, K.H.; Groza, L.G.; Sabourin, M.J.; Scheytt, T.J.; Ptacek, C.J. Transport behavior of the pharmaceutical compounds carbamazepine, sulfamethoxazole, gemfibrozil, ibuprofen, and naproxen, and the lifestyle drug caffeine, in saturated laboratory columns. *Sci. Total Environ.* **2017**, *590–591*, 708–719. [CrossRef]

44. Nham, H.T.T.; Greskowiak, J.; Nödler, K.; Rahman, M.A.; Spachos, T.; Rusteberg, B.; Massmann, G.; Sauter, M.; Licha, T. Modeling the transport behavior of 16 emerging organic contaminants during soil aquifer treatment. *Sci. Total Environ.* **2015**, *514*, 450–458. [CrossRef] [PubMed]

45. Freeze, R.A.; Cherry, J.A *Groundwater*; Prentice-Hall, Inc.: Englewood Cliffs, NJ, USA, 1979.

46. Leecaster, K.; Ayling, B.; Moffitt, G.; Rose, P.E. Use of safranin T as a reactive tracer for geothermal reservoir characterization. In Proceedings of the 37th Workshop on Geothermal Reservoir Engineering, Stanford, CA, USA, 30 January–1 February 2012.

47. Dean, C.; Reimus, P.W.; Newell, D.; Diagnostics, C.; Observations, E.S.; Alamos, L. Evaluation of a Cation Exchanging Tracer to Interrogate Fracture Surface Area in Egs Systems. In Proceedings of the 37th Workshop on Geothermal Reservoir Engineering, Stanford, CA, USA, 30 January–1 February 2012.

48. Reimus, P.W.; Williams, M.; Vermeul, V.; Rose, P.E.; Leecaster, K.; Ayling, B.; Sanjuan, R.; Ames, M.; Dean, C.; Benoit, D. Use of Tracers to Interrogate Fracture Surface Area in Single-well Tracer Tests in Egs Systems. In Proceedings of the 37th Workshop on Geothermal Reservoir Engineering, Stanford, CA, USA, 30 January–1 February 2012.

49. Dean, C.; Reimus, P.W.; Oates, J.; Rose, P.E.; Newell, D.; Petty, S. Laboratory experiments to characterize cation-exchanging tracer behavior for fracture surface area estimation at Newberry Crater, OR. *Geothermics* **2015**, *53*, 213–224. [CrossRef]

50. Schaffer, M.; Warner, W.; Kutzner, S.; Börnick, H.; Worch, E.; Licha, T. Organic molecules as sorbing tracers for the assessment of surface areas in consolidated aquifer systems. *J. Hydrol.* **2017**, *546*, 370–379. [CrossRef]

51. Rose, P.E.; Leecaster, K.; Clausen, S.; Sanjuan, R.; Ames, M.; Reimus, P.W.; Williams, M.; Vermeul, V.; Benoit, D. A tracer test at the Soda Lake, Nevada geothermal field using a sorbing tracer. In Proceedings of the 37th Workshop on Geothermal Reservoir Engineering, Stanford, CA, USA, 30 January–1 February 2012; pp. 30–33.

52. Neretnieks, I. A stochastic multi-channel model for solute transport-analysis of tracer tests in fractured rock. *J. Contam. Hydrol.* **2002**, *55*, 175–211. [CrossRef]

53. Hawkins, A.; Fox, D.; Zhao, R.; Tester, J.W.; Cathles, L.; Koch, D.; Becker, M. Predicting Thermal Breakthrough from Tracer Tests: Simulations and Observations in a Low-Temperature Field Laboratory. In Proceedings of the Fortieth Workshop on Geothermal Reservoir Engineering, Stanford, CA, USA, 26–28 January 2015; pp. 1–15.

54. Wilson, R.D. Reactive tracers to characterize pollutant distribution and behavior in aquifers. In *Handbook of Hydrocarbon and Lipid Microbiology*; Springer: Berlin, Germany, 2010; pp. 2465–2471, ISBN 978-3-540-77584-3.

55. Jin, M.; Delshad, M.; Dwarakanath, V.; Mckinney, D.C.; Pope, G.A.; Sepehmoori, K.; Tilburg, C.E.; Jackson, R.E.; Cherry, N. Partitioning tracer test for detection, estimation, and remediation performance assessment of subsurface nonaqueous phase liquids. *Water Resour. Res.* **1995**, *31*, 1201–1211. [CrossRef]

56. Annable, M.D.; Rao, P.S.C.; Hatfield, K.; Graham, W.D.; Wood, A.L.; Enfield, C.G. Partitioning Tracers for Measuring Residual NAPL: Field-Scale Test Results. *J. Environ. Eng.* **1998**, *124*, 498–503. [CrossRef]

57. Jawitz, J.W.; Sillan, R.K.; Annable, M.D.; Rao, P.S.C.; Warner, K. In-situ alcohol flushing of a DNAPL source zone at a dry cleaner site. *Environ. Sci. Technol.* **2000**, *34*, 3722–3729. [CrossRef]

58. Johnston, C.D.; Davis, G.B.; Bastow, T.P.; Annable, M.D.; Trefry, M.G.; Furness, A.; Geste, Y.; Woodbury, R.J.; Rao, P.S.C.; Rhodes, S. The use of mass depletion-mass flux reduction relationships during pumping to determine source zone mass of a reactive brominated-solvent DNAPL. *J. Contam. Hydrol.* **2013**, *144*, 122–137. [CrossRef]

59. Cápiro, N.L.; Granbery, E.K.; Lebrón, C.A.; Major, D.W.; McMaster, M.L.; Pound, M.J.; Löffler, F.E.; Pennell, K.D. Liquid-liquid mass transfer of partitioning electron donors in chlorinated solvent source zones. *Environ. Sci. Technol.* **2011**, *45*, 1547–1554. [CrossRef]

60. Imhoff, P.T.; Pirestani, K.; Jafarpour, Y.; Spivey, K.M. Tracer interaction effects during partitioning tracer tests for NAPL detection. *Environ. Sci. Technol.* **2003**, *37*, 1441–1447. [CrossRef]

61. Jin, M.; Jackson, R.E.; Pope, G.A.; Taffinder, S. Development of Partitioning Tracer Tests for Characterization of Nonaqueous-Phase Liquid-Contaminated Aquifers. In Proceedings of the SPE Annual Technical Conference and Exhibition; Society of Petroleum Engineers, San Antonio, TX, USA, 5–8 October 1997.

62. Young, C.M.; Jackson, R.E.; Jin, M.; Londergan, J.T.; Mariner, P.E.; Pope, G.A.; Anderson, F.J.; Houk, T. Characterization of a TCE DNAPL Zone in Alluvium by Partitioning Tracers. *Groundw. Monit. Remediat.* **1999**, *19*, 84–94. [CrossRef]

63. Brooks, M.C.; Annable, M.D.; Rao, P.S.C.; Hatfield, K.; Jawitz, J.W.; Wise, W.R.; Wood, A.L.; Enfield, C.G. Controlled release, blind tests of DNAPL characterization using partitioning tracers. *J. Contam. Hydrol.* **2002**, *59*, 187–210. [CrossRef]

64. Divine, C.E.; McCray, J.E.; Wolf Martin, L.M.; Blanford, W.J.; Blitzer, D.J.; Brusseau, M.L.; Boving, T.B. Partitioning tracer tests as a remediation metric: Case study at naval amphibious base little creek, Virginia Beach, Virginia. *Remediat. J.* **2004**, *14*, 7–31. [CrossRef]

65. Cain, R.B.; Johnson, G.R.; McCray, J.E.; Blanford, W.J.; Brusseau, M.L. Partitioning tracer tests for evaluating remediation performance. *Ground Water* **2000**, *38*, 752–761. [CrossRef]

66. Jawitz, J.W.; Annable, M.D.; Clark, C.J.; Puranik, S. Inline Gas Chromatographic Tracer Analysis: An Alternative To Conventional Sampling and Laboratory Analysis for Partitioning Tracer Tests. *Instrum. Sci. Technol.* **2002**, *30*, 415–426. [CrossRef]

67. Noordman, W.H.; De Boer, G.J.; Wietzes, P.; Volkering, F.; Janssen, D.B. Assessment of the Use of Partitioning and Interfacial Tracers To Determine the Content and Mass Removal Rates of Nonaqueous Phase Liquids. *Environ. Sci. Technol.* **2000**, *34*, 4301–4306. [CrossRef]

68. Annable, M.D.; Hatfield, K.; Cho, J.; Klammler, H.; Parker, B.L.; Cherry, J.A.; Rao, P.S.C. Field-scale evaluation of the passive flux meter for simultaneous measurement of groundwater and contaminant fluxes. *Environ. Sci. Technol.* **2005**, *39*, 7194–7201. [CrossRef]

69. Hartog, N.; Cho, J.; Parker, B.L.; Annable, M.D. Characterization of a heterogeneous DNAPL source zone in the Borden aquifer using partitioning and interfacial tracers: Residual morphologies and background sorption. *J. Contam. Hydrol.* **2010**, *115*, 79–89. [CrossRef]

70. Wang, F.; Annable, M.D.; Schaefer, C.E.; Ault, T.D.; Cho, J.; Jawitz, J.W. Enhanced aqueous dissolution of a DNAPL source to characterize the source strength function. *J. Contam. Hydrol.* **2014**, *169*, 75–89. [CrossRef]

71. Jessheim, B.; Viig, S.; Dugstad, O.; Stray, H. Tracers. WO2014096459A1, 23 December 2013.

72. Dean, R.M.; Walker, D.L.; Dwarakanath, V.; Malik, T.; Spilker, K. Use of partitioning tracers to estimate oil saturation distribution in heterogeneous reservoirs. In Proceedings of the SPE Improved Oil Recover Conference, Tulsa, OK, USA, 11–13 April 2016.

73. Wilson, R.D.; Mackay, D.M. Direct Detection of Residual Nonaqueous Phase Liquid in the Saturated Zone Using SF6 as a Partitioning Tracer. *Environ. Sci. Technol.* **1995**, *29*, 1255–1258. [CrossRef]

74. Werner, D.; Höhener, P. Diffusive Partitioning Tracer Test for Nonaqueous Phase Liquid (NAPL) Detection in the Vadose Zone. *Environ. Sci. Technol.* **2002**, *36*, 1592–1599. [CrossRef] [PubMed]

75. Vulava, V.M.; Perry, E.B.; Romanek, C.S.; Seaman, J.C. Dissolved gases as partitioning tracers for determination of hydrogeological parameters. *Environ. Sci. Technol.* **2002**, *36*, 254–262. [CrossRef] [PubMed]

76. Davis, B.M.; Istok, J.D.; Semprini, L. Push-pull partitioning tracer tests using radon-222 to quantify non-aqueous phase liquid contamination. *J. Contam. Hydrol.* **2002**, *58*, 129–146. [CrossRef]

77. Reid, M.C.; Jaffé, P.R. A push-pull test to measure root uptake of volatile chemicals from wetland soils. *Environ. Sci. Technol.* **2013**, *47*, 3190–3198. [CrossRef] [PubMed]

78. Deeds, N.; Pope, G.A.; Mckinney, D.C. Vadose Zone Characterization at a Contaminated Field Site Using Partitioning Interwell Tracer Technology. *Environ. Sci. Technol.* **1999**, *33*, 2745–2751. [CrossRef]

79. Hunkerler, D.; Hoehn, E.; Höhener, P.; Zeyer, J. 222Rn as a Partitioning Tracer To Detect Diesel Fuel Contamination in Aquifers: Laboratory Study and Field Observations. *Environ. Sci. Technol.* **1997**, *31*, 3180–3187. [CrossRef]

80. Schubert, M.; Paschke, A.; Lau, S.; Geyer, W.; Knöller, K. Radon as a naturally occurring tracer for the assessment of residual NAPL contamination of aquifers. *Environ. Pollut.* **2007**, *145*, 920–927. [CrossRef]

81. Ponsin, V.; Chablais, A.; Dumont, J.; Radakovitch, O.; Höhener, P. ^{222}Rn as Natural Tracer for LNAPL Recovery in a Crude Oil-Contaminated Aquifer. *Groundw. Monit. Remediat.* **2015**, *35*, 30–38. [CrossRef]

82. Ghanem, A.; Soerens, T.S.; Adel, M.M.; Thoma, G.J. Investigation of Fluorescent Dyes as Partitioning Tracers for Subsurface Nonaqueous Phase Liquid (NAPL) Characterization. *J. Environ. Eng.* **2003**, *129*, 740–744. [CrossRef]

83. LaForce, T.; Ennis-King, J.; Boreham, C.; Paterson, L. Residual CO_2 saturation estimate using noble gas tracers in a single-well field test: The CO_2CRC Otway project. *Int. J. Greenh. Gas Control* **2014**, *26*, 9–21. [CrossRef]

84. Stalker, L.; Boreham, C.; Underschultz, J.; Freifeld, B.; Perkins, E.; Schacht, U.; Sharma, S. Application of tracers to measure, monitor and verify breakthrough of sequestered CO2 at the CO2CRC Otway Project, Victoria, Australia. *Chem. Geol.* **2015**, *399*, 2–19. [CrossRef]

85. Rasmusson, K.; Rasmusson, M.; Fagerlund, F.; Bensabat, J.; Tsang, Y.; Niemi, A. Analysis of alternative push-pull-test-designs for determining in situ residual trapping of carbon dioxide. *Int. J. Greenh. Gas Control* **2014**, *27*, 155–168. [CrossRef]

86. Zhang, Y.; Freifeld, B.; Finsterle, S.; Leahy, M.; Ennis-King, J.; Paterson, L.; Dance, T. Single-well experimental design for studying residual trapping of supercritical carbon dioxide. *Int. J. Greenh. Gas Control* **2011**, *5*, 88–98. [CrossRef]

87. Roberts, J.J.; Gilfillan, S.M.V.; Stalker, L.; Naylor, M. Geochemical tracers for monitoring offshore CO_2 stores. *Int. J. Greenh. Gas Control* **2017**, *65*, 218–234. [CrossRef]

88. Saripalli, K.P.; Kim, H.; Annable, M.D. Measurement of Specific Fluid—Fluid Interfacial Areas of Immiscible Fluids in Porous Media. *Environ. Sci. Technol.* **1997**, *31*, 932–936. [CrossRef]

89. Saripalli, K.P.; Rao, P.S.C.; Annable, M.D. Determination of specific NAPL–water interfacial areas of residual NAPLs in porous media using the interfacial tracers technique. *J. Contam. Hydrol.* **1998**, *30*, 375–391. [CrossRef]

90. Zhong, H.; El Ouni, A.; Lin, D.; Wang, B.; Brusseau, M.L. The two-phase flow IPTT method for measurement of nonwetting-wetting liquid interfacial areas at higher nonwetting saturations in natural porous media. *Water Resour. Res.* **2016**, *52*, 5506–5515. [CrossRef]

91. Araujo, J.B.; Mainhagu, J.; Brusseau, M.L. Measuring air-water interfacial area for soils using the mass balance surfactant-tracer method. *Chemosphere* **2015**, *134*, 199–202. [CrossRef]

92. Schaefer, C.E.; DiCarlo, D.A.; Blunt, M.J. Determination of Water-Oil Interfacial Area during 3-Phase Gravity Drainage in Porous Media. *J. Colloid Interface Sci.* **2000**, *221*, 308–312. [CrossRef]

93. Annable, M.D.; Jawitz, J.W.; Rao, P.S.C.; Dai, D.P.; Kim, H.; Wood, A.L. Field Evaluation of Interfacial and Partitioning Tracers for Characterization of Effective NAPL-Water Contact Areas. *Ground Water* **1998**, *36*, 495–502. [CrossRef]

94. Kim, H.; Rao, P.S.C.; Annable, M.D. Determination of effective air-water interfacial area in partially saturated porous media using surfactant adsorption. *Water Resour. Res.* **1997**, *33*, 2705–2711. [CrossRef]

95. Kim, H.; Rao, P.S.C.; Annable, M.D. Gaseous Tracer Technique for Estimating Air-Water Interfacial Areas and Interface Mobility. *Soil Sci. Soc. Am. J.* **1999**, *63*, 1554–1560. [CrossRef]

96. Jain, V.; Bryant, S.; Sharma, M. Influence of Wettability and Saturation on Liquid—Liquid Interfacial Area in Porous Media. *Environ. Sci. Technol.* **2003**, *37*, 584–591. [CrossRef] [PubMed]

97. Narter, M.; Brusseau, M.L. Comparison of Interfacial Partitioning Tracer Test and High-Resolution Microtomography Measurements of Fluid-Fluid Interfacial Areas for an Ideal Porous Medium. *Water Resour. Res.* **2010**, *46*. [CrossRef] [PubMed]

98. Brusseau, M.L.; Narter, M.; Janousek, H. Interfacial partitioning tracer test measurements of organic-liquid/water interfacial areas: Application to soils and the influence of surface roughness. *Environ. Sci. Technol.* **2010**, *44*, 7596–7600. [CrossRef] [PubMed]

99. Anwar, F.A.H.M.; Bettahar, M.; Matsubayashi, U. A method for determining air–water interfacial area in variably saturated porous media. *J. Contam. Hydrol.* **2000**, *43*, 129–146. [CrossRef]

100. Brusseau, M.L.; Peng, S.; Schnaar, G.; Murao, A. Measuring Air—Water Interfacial Areas with X-ray Microtomography and Interfacial Partitioning Tracer Tests. *Environ. Sci. Technol.* **2007**, *41*, 1956–1961. [CrossRef]

101. Setarge, B.; Danze, J.; Klein, R.; Grathwohl, P. Partitioning and Interfacial Tracers to Characterize Non-Aqueous Phase Liquids (NAPLs) in Natural Aquifer Material. *Phys. Chem. Earth Part B Hydrol. Oceans Atmos.* **1999**, *24*, 501–510. [CrossRef]

102. Karkare, M.V.; Fort, T. Determination of the air-water interfacial area in wet "unsaturated" porous media. *Langmuir* **1996**, *12*, 2041–2044. [CrossRef]

103. Silverstein, D.L.; Fort, T. Studies in air-water interfacial area for wet unsaturated particulate porous media systems. *Langmuir* **1997**, *7463*, 4758–4761. [CrossRef]

104. Dobson, R.; Schroth, M.H.; Oostrom, M.; Zeyer, J. Determination of NAPL-water interfacial areas in well-characterized porous media. *Environ. Sci. Technol.* **2006**, *40*, 815–822. [CrossRef] [PubMed]

105. Kim, H.; Annable, M.D.; Rao, P.S.C. Influence of Air—Water Interfacial Adsorption and Gas-Phase Partitioning on the Transport of Organic Chemicals in Unsaturated Porous Media. *Environ. Sci. Technol.* **1998**, *32*, 1253–1259. [CrossRef]

106. Istok, J.D.; Humphrey, M.D.; Schroth, M.H.; Hyman, M.R.; O'Reilly, K.T. Single-well, "push-pull" test for in situ determination of microbial activities. *Ground Water* **1997**, *35*, 619–631. [CrossRef]

107. Schroth, M.H.; Istok, J.D.; Conner, G.T.; Hyman, M.R.; Haggerty, R.; O'Reilly, K.T. Spatial variability in in situ aerobic respiration and denitrification rates in a petroleum-contaminated aquifer. *Ground Water* **1998**, *36*, 924–937. [CrossRef]

108. Goldhammer, T.; Einsiedl, F.; Blodau, C. In situ determination of sulfate turnover in peatlands: A down-scaled push-pull tracer technique. *J. Plant Nutr. Soil Sci.* **2008**, *171*, 740–750. [CrossRef]

109. Michalsen, M.M.; Weiss, R.; King, A.; Gent, D.; Medina, V.F.; Istok, J.D. Push-pull tests for estimating rdx and tnt degradation rates in groundwater. *Groundw. Monit. Remediat.* **2013**, *33*, 61–68. [CrossRef]

110. Henson, W.R.; Huang, L.; Graham, W.D.; Ogram, A. Nitrate reduction mechanisms and rates in an unconfined eogenetic karst aquifer in two sites with different redox potential. *J. Geophys. Res. Biogeosci.* **2017**, *122*, 1062–1077. [CrossRef]

111. Cho, Y.; Han, K.; Kim, N.; Park, S.; Kim, Y. Estimating in situ biodegradation rates of petroleum hydrocarbons and microbial population dynamics by performing single-well push-pull tests in a fractured bedrock aquifer. *Water. Air. Soil Pollut.* **2013**, *224*, 1364. [CrossRef]

112. Schroth, M.H.; Kleikemper, J.; Bolliger, C.; Bernasconi, S.M.; Zeyer, J. In situ assessment of microbial sulfate reduction in a petroleum-contaminated aquifer using push-pull tests and stable sulfur isotope analyses. *J. Contam. Hydrol.* **2001**, *51*, 179–195. [CrossRef]

113. Burbery, L.F.; Flintoft, M.J.; Close, M.E. Application of the re-circulating tracer well test method to determine nitrate reaction rates in shallow unconfined aquifers. *J. Contam. Hydrol.* **2013**, *145*, 1–9. [CrossRef]

114. Kleikemper, J.; Schroth, M.H.; Sigler, W.V.; Schmucki, M.; Bernasconi, S.M.; Zeyer, J. Activity and diversity of sulfate-reducing bacteria in a petroleum hydrocarbon-contaminated aquifer. *Appl. Environ. Microbiol.* **2002**, *68*, 1516–1523. [CrossRef] [PubMed]

115. Urmann, K.; Gonzalez-Gil, G.; Schroth, M.H.; Hofer, M.; Zeyer, J. New Field Method: Gas Push−Pull Test for the In-Situ Quantification of Microbial Activities in the Vadose Zone. *Environ. Sci. Technol.* **2005**, *39*, 304–310. [CrossRef]

116. Addy, K.; Kellogg, D.Q.; Gold, A.J.; Groffman, P.M.; Ferendo, G.; Sawyer, C. In situ push-pull method to determine ground water denitrification in riparian zones. *J. Environ. Qual.* **2002**, *31*, 1017–1024. [CrossRef] [PubMed]

117. Rao, P.S.C.; Annable, M.D.; Kim, H. NAPL source zone characterization and remediation technology performance assessment: Recent developments and applications of tracer techniques. *J. Contam. Hydrol.* **2000**, *45*, 63–78. [CrossRef]

118. Sandrin, S.K.; Brusseau, M.L.; Piatt, J.J.; Bodour, A.A.; Blanford, W.J.; Nelson, N.T. Spatial variability of in situ microbial activity: Biotracer tests. *Ground Water* **2004**, *42*, 374–383. [CrossRef]

119. Brusseau, M.L.; Nelson, N.T.; Zhang, Z.; Blue, J.E.; Rohrer, J.; Allen, T. Source-zone characterization of a chlorinated-solvent contaminated Superfund site in Tucson, AZ. *J. Contam. Hydrol.* **2007**, *90*, 21–40. [CrossRef] [PubMed]

120. Alter, S.R.; Brusseau, M.L.; Piatt, J.J.; Ray-Maitra, A.; Wang, J.M.; Cain, R.B. Use of tracer tests to evaluate the impact of enhanced-solubilization flushing on in-situ biodegradation. *J. Contam. Hydrol.* **2003**, *64*, 191–202. [CrossRef]

121. González-Pinzón, R.; Haggerty, R.; Myrold, D.D. Measuring aerobic respiration in stream ecosystems using the resazurin-resorufin system. *J. Geophys. Res.* **2012**, *117*, G00N06. [CrossRef]

122. Stanaway, D.; Haggerty, R.; Benner, S.; Flores, A.; Feris, K. Persistent metal contamination limits lotic ecosystem heterotrophic metabolism after more than 100 years of exposure: A novel application of the Resazurin Resorufin Smart Tracer. *Environ. Sci. Technol.* **2012**, *46*, 9862–9871. [CrossRef]

123. Lemke, D.; González-Pinzón, R.; Liao, Z.; Wöhling, T.; Osenbrück, K.; Haggerty, R.; Cirpka, O.A. Sorption and transformation of the reactive tracers resazurin and resorufin in natural river sediments. *Hydrol. Earth Syst. Sci.* **2014**, *18*, 3151–3163. [CrossRef]

124. Haggerty, R.; Ribot, M.; Singer, G.A.; Marti, E.; Argerich, A.; Agell, G.; Battin, T.J. Ecosystem respiration increases with biofilm growth and bed forms: Flume measurements with resazurin. *J. Geophys. Res. Biogeosci.* **2014**, *119*, 2220–2230. [CrossRef]

125. González-Pinzón, R.; Ward, A.S.; Hatch, C.E.; Wlostowski, A.N.; Singha, K.; Gooseff, M.N.; Haggerty, R.; Harvey, J.W.; Cirpka, O.A.; Brock, J.T. A field comparison of multiple techniques to quantify groundwater—Surface-water interactions. *Freshw. Sci.* **2015**, *34*, 139–160. [CrossRef]

126. González-Pinzón, R.; Peipoch, M.; Haggerty, R.; Marti, E.; Fleckenstein, J.H. Nighttime and daytime respiration in a headwater stream. *Ecohydrology* **2016**, *9*, 93–100. [CrossRef]

127. Haggerty, R.; Argerich, A.; Marti, E. Development of a "smart" tracer for the assessment of microbiological activity and sediment-water interaction in natural waters: The resazurin-resorufin system. *Water Resour. Res.* **2008**, *44*. [CrossRef]

128. Haggerty, R.; Marti, E.; Argerich, A.; von Schiller, D.; Grimm, N.B. Resazurin as a "smart" tracer for quantifying metabolically active transient storage in stream ecosystems. *J. Geophys. Res.* **2009**, *114*, G03014. [CrossRef]

129. Nottebohm, M.; Licha, T.; Ghergut, I.; Nödler, K.; Sauter, M. Development of Thermosensitive Tracers for Push-Pull Experiments in Geothermal Reservoir Characterization. In Proceedings of the World Geothermal Congress, Bali, Indonesia, 25–29 April 2010; pp. 25–29.

130. Nottebohm, M.; Licha, T.; Sauter, M. Tracer design for tracking thermal fronts in geothermal reservoirs. *Geothermics* **2012**, *43*, 37–44. [CrossRef]

131. Cao, V.; Schaffer, M.; Licha, T. The feasibility of using carbamates to track the thermal state in geothermal reservoirs. *Geothermics* **2018**, *72*, 301–306. [CrossRef]

132. Rose, P.E.; Clausen, S. The Use of Amino G as a Thermally Reactive Tracer for Geothermal Applications. In Proceedings of the 39th Workshop on Geothermal Reservoir Engineering, Stanford, CA, USA, 24–26 February 2014; pp. 1–5.

133. Rose, P.; Clausen, S. The use of amino-substituted naphthalene sulfonates as tracers in geothermal reservoirs. In Proceedings of the 42nd Workshop on Geothermal Reservoir Engineering, Stanford, CA, USA, 3–15 February 2017; pp. 1–7.

134. Maier, F.; Schaffer, M.; Licha, T. Determination of temperatures and cooled fractions by means of hydrolyzable thermo-sensitive tracers. *Geothermics* **2015**, *58*, 87–93. [CrossRef]

135. Batchelor, A. Reservoir behaviour in a stimulated hot dry rock system. In Proceedings of the Eleventh Workshop on Geothermal Reservoir Engineering, Stanford, CA, USA, 21–23 January 1986; pp. 35–41.

136. Kwakwa, K.A. Tracer measurements during long-term circulation of the Rosemanowes HDR geothermal system. In Proceedings of the Thirteenth Workshop on Geothermal Reservoir Engineering, Standford, CA, USA, 19–21 January 1988; pp. 245–252.

137. Tester, J.W.; Robinson, B.A.; Ferguson, J.H. Inert and Reacting Tracers for Reservoir Sizing in Fractured, Hot Dry Rock Systems. In Proceedings of the Eleventh Workshop on Geothermal Reservoir Engineering, Standford, CA, USA, 21–23 January 1986; pp. 149–159.

138. Hawkins, A.J.; Fox, D.B.; Becker, M.W.; Tester, J.W. Measurement and simulation of heat exchange in fractured bedrock using inert and thermally degrading tracers. *Water Resour. Res.* **2017**, *53*, 1210–1230. [CrossRef]

139. Adams, M.C.; Davis, J. Kinetics of fluorescein decay and its application as a geothermal tracer. *Geothermics* **1991**, *20*, 53–66. [CrossRef]

140. Tatomir, A.B.; Schaffer, M.; Kissinger, A.; Hommel, J.; Nuske, P.; Licha, T.; Helmig, R.; Sauter, M. Novel approach for modeling kinetic interface-sensitive (KIS) tracers with respect to time-dependent interfacial area change for the optimization of supercritical carbon dioxide injection into deep saline aquifers. *Int. J. Greenh. Gas Control* **2015**, *33*, 145–153. [CrossRef]

141. Cooke, C.E.J. Method of Determining Fluid Saturations in Reservoirs. U.S. Patent No. 3,590,923, 06 July 1971.

142. Tang, J.; Harker, B. Mass Balance Method to Determine Residual Oil Saturation from Single Well Tracer Test Data. *J. Can. Pet. Technol.* **1990**, *29*. [CrossRef]

143. Tang, J.; Zhang, P. Determination of Residual Oil Saturation in A Carbonate Reservoir. In Proceedings of the SPE Asia Pacific Improved Oil Recovery Conference, Kuala Lumpur, Malaysia, 6–9 October 2001.

144. Pathak, P.; Fitz, D.; Babcock, K.; Wachtman, R.J. Residual Oil Saturation Determination for EOR Projects in Means Field, a Mature West Texas Carbonate Field. *SPE Reserv. Eval. Eng.* **2012**, *15*, 541–553. [CrossRef]

145. Khaledialidusti, R.; Kleppe, J.; Skrettingland, K. Numerical interpretation of Single Well Chemical Tracer (SWCT) tests to determine residual oil saturation in Snorre Reservoir. In Proceedings of the SPE Asia Pacific Improved Oil Recovery Conference, Kuala Lumpur, Malaysia, 6–9 October 2001; pp. 1–4.

146. Myers, M.; Stalker, L.; La Force, T.; Pejcic, B.; Dyt, C.; Ho, K.B.; Ennis-King, J. Field measurement of residual carbon dioxide saturation using reactive ester tracers. *Chem. Geol.* **2015**, *399*, 20–29. [CrossRef]

147. Käss, W. *Tracing Technique in Geohydrology*; Balkema: Rotterdam, The Netherlands, 1998.

148. Sophocleous, M. Interactions between groundwater and surface water: The state of the science. *Hydrogeol. J.* **2002**, *10*, 52–67. [CrossRef]

149. Palmer, M.A. Experimentation in the hyporheic Zon: Challenges and prospectus. *J. N. Am. Benthol. Soc.* **1993**, *12*, 84–93. [CrossRef]

150. Kissinger, A.; Helmig, R.; Ebigbo, A.; Class, H.; Lange, T.; Sauter, M.; Heitfeld, M.; Klünker, J.; Jahnke, W. Hydraulic fracturing in unconventional gas reservoirs: Risks in the geological system, part 2. *Environ. Earth Sci.* **2013**, *70*, 3855–3873. [CrossRef]

151. Taherdangkoo, R.; Tatomir, A.; Taylor, R.; Sauter, M. Numerical investigations of upward migration of fracking fluid along a fault zone during and after stimulation. *Energy Procedia* **2017**, *125*, 126–135. [CrossRef]

152. Tatomir, A.; Mcdermott, C.; Bensabat, J.; Class, H.; Edlmann, K.; Taherdangkoo, R.; Sauter, M. Conceptual model development using a generic Features, Events, and Processes (FEP) database for assessing the potential impact of hydraulic fracturing on groundwater aquifers. *Adv. Geosci.* **2018**, *45*, 185–192. [CrossRef]

153. Taherdangkoo, R.; Tatomir, A.; Anighoro, T.; Sauter, M. Modeling fate and transport of hydraulic fracturing fluid in the presence of abandoned wells. *J. Contam. Hydrol.* **2019**, *221*, 58–68. [CrossRef]

154. Kurose, S. Requiring the use of tracers in hydraulic fracturing fluid to trace alleged contamination. *Sustain. Dev. Law Policy* **2014**, *14*, 43–54.

155. Kuntz, I.D.; Meng, E.C.; Shoichet, B.K. Structure-Based Molecular Design. *Acc. Chem. Res.* **1994**, *27*, 117–123. [CrossRef]

156. Kang, E.T.; Zhang, Y. Surface Modification of Fluoropolymers via Molecular Design. *Adv. Mater.* **2000**, *12*, 1481–1494. [CrossRef]

157. Rose, P.E.; Benoit, W.R.; Kilbourn, P.M. The application of the polyaromatic sulfonates as tracers in geothermal reservoirs. *Geothermics* **2001**, *30*, 617–640. [CrossRef]

158. Rose, P.E.; Johnson, S.D.; Kilbourn, P.M. Tracer testing at dixie valley, nevada, using 2-naphthalene sulfonate and 2,7-naphthalene disulfonate. In Proceedings of the 26th Workshop on Geothermal Reservoir Engineering, Stanford, CA, USA, 29–31 January 2001; pp. 1–6.

159. Nottebohm, M.; Licha, T. Detection of Naphthalene Sulfonates from Highly Saline Brines with High-Performance Liquid Chromatography in Conjunction with Fluorescence Detection and Solid-Phase Extraction. *J. Chromatogr. Sci.* **2012**, *50*, 477–481. [CrossRef]

160. Greim, H.; Ahlers, J.; Bias, R.; Broecker, B.; Hollander, H.; Gelbke, H.-P.; Klimisch, H.-J.; Mangelsdorf, I.; Paetz, A.; Schön, N.; et al. Toxicity and ecotoxicity of sulfonic acids: Structure-activity relationship. *Chemosphere* **1994**, *28*, 2203–2236.

161. Sanjuan, B.; Pinault, J.-L.; Rose, P.E.; Gérard, A.; Brach, M.; Braibant, G.; Crouzet, C.; Foucher, J.-C.; Gautier, A.; Touzelet, S. Tracer testing of the geothermal heat exchanger at Soultz-sous-Forêts (France) between 2000 and 2005. *Geothermics* **2006**, *35*, 622–653. [CrossRef]

Review

Dissolved Radiotracers and Numerical Modeling in North European Continental Shelf Dispersion Studies (1982–2016): Databases, Methods and Applications

Pascal Bailly du Bois [1,*], **Franck Dumas** [2], **Claire Voiseux** [1], **Mehdi Morillon** [1], **Pierre-Emmanuel Oms** [3] **and Luc Solier** [1]

[1] Laboratoire de Radioécologie de Cherbourg, IRSN-LRC, rue Max Pol Fouchet B.P. 10, 50130 Cherbourg en Cotentin, France; claire.voiseux@irsn.fr (C.V.); mehdi.morillon@gmail.com (M.M.); luc.solier@irsn.fr (L.S.)

[2] Service Hydrographique et Océanographique de la Marine-SHOM, 29200 Brest, France; franck.dumas@shom.fr

[3] Centre d'Etudes et de Valorisation des Algues CEVA, 83 Presqu'île de Pen Lan, 22610 Pleubian, France; omspe@hotmail.fr

* Correspondence: pascal.bailly-du-bois@irsn.fr; Tel.: +33-2-33-01-41-05

Received: 2 October 2019; Accepted: 5 June 2020; Published: 10 June 2020

Abstract: Significant amounts of anthropogenic radionuclides were introduced in ocean waters following nuclear atmospheric tests and development of the nuclear industry. Dispersion of artificial dissolved radionuclides has been extensively measured for decades over the North-European continental shelf. In this area, the radionuclide measurement and release fluxes databases provided here between 1982 and 2016 represent an exceptional opportunity to validate dispersion hydrodynamic models. This work gives accessibility to these data in a comprehensive database. The MARS hydrodynamic model has been applied at different scales to reproduce the measured dispersion in realistic conditions. Specific methods have been developed to obtain qualitative and quantitative results and perform model/measurement comparisons. Model validation concerns short to large scales with dedicated surveys following the dispersion: it was performed within a two- and three-dimensional framework and from minutes and hours following a release up to several years. Results are presented concerning the dispersion of radionuclides in marine systems deduced from standalone measurements, or according to model comparisons. It allows characterizing dispersion over the continental shelf, pathways, transit times, budgets and source terms. This review presents the main approaches developed and types of information derived from studies of artificial radiotracers using observations, hydrodynamic models or a combination of the two, based primarily on the new featured datasets.

Keywords: radionuclide; tracer; data collection; antimony 125 (^{125}Sb); tritium (^{3}H); dispersion; modeling; English Channel; North Sea; Biscay Bay

---

1. Introduction

Significant amounts of anthropogenic radionuclides have been introduced in ocean waters since 1945. The main origins were the fallout from the atmospheric nuclear tests that occurred before 1980, controlled releases from the nuclear industry and the accidental releases resulting from the Chernobyl (1986) and Fukushima Dai-ichi (2011) nuclear power plants. Among these sources, in Europe, the releases from nuclear fuel reprocessing plants from Sellafield and La Hague were the most important.

Oceanographic sampling campaigns made it possible to measure the dispersion of the artificial radionuclides remaining dissolved in seawater over the past 30 years. These works have demonstrated the interest of dissolved radionuclides as tools for oceanography [1–3]. Over the European continental shelf, extensive in-situ measurements have been performed between 1950 and 2000 by English, German, Belgian and French institutes [4,5] that allowed to draw up the general circulation pathways and water masses transit times in the Irish Sea, the North Sea, the English Channel (Figure 1) and the Arctic ocean [6–11]. The Cherbourg Radioecology Laboratory (IRSN-LRC) has contributed to these studies at the scale of the English Channel and the North Sea since 1988 [12–21]. These works have been ongoing since 1994 in the North-East Atlantic waters, the Celtic Sea, Irish Sea and the Bay of Biscay. During the 1990s, it appears that dissolved radionuclide measurements represent exceptional tools to test and validate marine hydrodynamic models applied to represent short to long term dispersion processes.

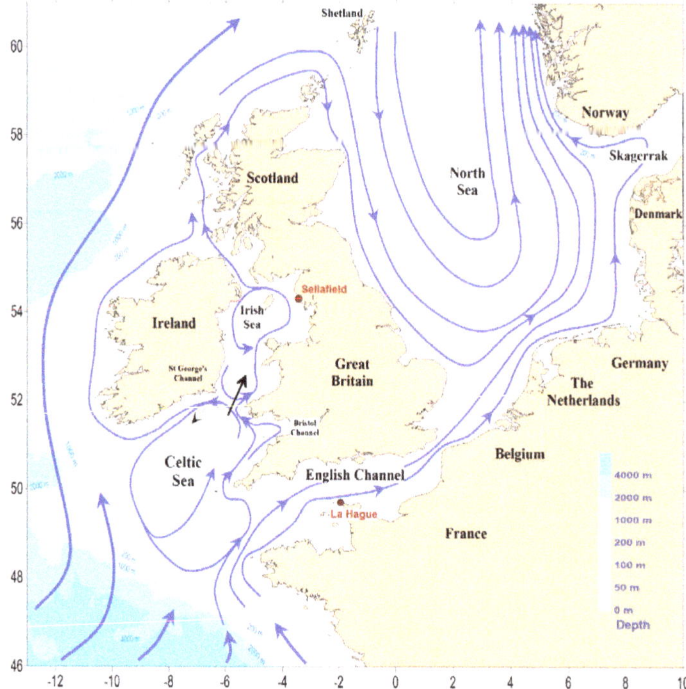

Figure 1. General circulation of water masses in the north-west of Europe, adapted from [11].

This work presents a review and an update of results from the main dissolved radionuclides that have been measured extensively by the IRSN-LRC, i.e., ^{3}H, ^{125}Sb, ^{137}Cs, ^{134}Cs, ^{106}Ru and ^{60}Co. Other radionuclides could be used as oceanographic tracers, such as ^{90}Sr, ^{99}Tc, ^{129}I, $^{238,\,239+240}$Pu. Radionuclides with a half-life lower than one year has been sparsely detected as ^{110m}Ag, ^{54}Mn, ^{58}Co and ^{131}I.

Realistic simulations of the dispersion of soluble substances in the marine environment are essential for management of the marine ecosystem. Such simulations were applied to study the fate of chronic or accidental releases into the sea; they may also be used to feed ecological models that encompass exchanges between the different compartments of the environment: seawater, living organisms and sediments. Such tools are particularly relevant in seas that are subject to strong anthropogenic pressures, such as the macro-tidal seas of north-western Europe.

Various methods have been tested for calculating the behavior of water masses, their advection and their dispersion. Models commonly simulate currents at different resolutions (from a tenth of a meter to kilometers), spatial (1 km–1000 km) and temporal coverage (from hour to decades).

While these models produce an accurate representation of tidal levels and associated currents, greater requirements are needed to simulate advection of soluble substances over periods longer than the tidal cycle. The models' ability to reproduce dispersion under realistic conditions of release, wind and tide over several days, weeks or years is a sensitive criterion for assessing their reliability.

Validation of hydrodynamic models applied for realistic simulation of mid- to long term dispersion in seawater requires field data of comparable parameters and coverage. The ideal tracer must have a conservative behavior in the water mass; that is to say, neither fixed by the environmental compartments (sediment, living species) nor modified during its stay in seawater and when subsequently diluted. It must be measurable several hundred or thousand kilometers from its input point (meaning even at very low levels of concentration). The discharge conditions and flow must be well known and it must have few properly controlled origins. The radioactive decay is easily accounted by hydrodynamic models.

Some artificial (issued from nuclear industry) radionuclides released by nuclear plants fully meet these specifications if their half-life is long enough compared to the transit-times in the studied area (from weeks to years). Among them, ^{125}Sb and ^{3}H as HTO have proven to be conserved in seawater over years at the scale of European waters. These radionuclides can be measured at very low concentrations, up to the Atlantic seawater background concentrations (concentrations 40,000 times lower than natural radioactivity for gamma-emitters).

Collaboration between Ifremer-DYNECO-PHYSED oceanographic physicians and IRSN-LRC marine radio ecologists contributes to improve the marine hydrodynamic models used [22–29]. Further studies have associated systematically in-situ measurements with model simulations in order to improve the efficiency of measurements to check model's precision and obtain more reliable and versatile models, applicable to all coastal seas of the European continental shelf. Model/measurements comparisons have been performed at the scale of the English Channel and the North Sea with targeted two-dimensional (2D) residual and instantaneous models [11,22–24]. This was done in the Bay of Biscay [26] and the Pacific [27] with three-dimensional (3D) models.

This study also included high resolution 2D and 3D model/measurements comparisons at short scale close to a release outfall [25].

The purpose of this work is to present a review and an update of field data and methods applicable to validate and improve dispersion models in European macro-tidal seas at all scales. The homogeneous database thus gathered can be used by modelers to test the reliability of their models against appropriate data. It complements the existing databases as the IAEA Marine Information System (MARIS [30]), the World Ocean Database 2013 [31,32], IFREMER–SISMER [33] or BODC [34].

This database has an historical value as releases from nuclear plants have significantly decrease for most gamma emitters (two orders of magnitudes for ^{125}Sb, ^{106}Ru, ^{137}Cs and ^{60}Co presented here); it represents a huge amount of work (80 oceanographic campaigns) and the dispersion plumes measured before 2000 will be difficult to obtain in the future. Moreover, in a context of a decreasing number of marine radio ecologists, it is valuable to share these data with a larger scientific community. We are convinced that they have not given all the knowledge they could provide.

This work is mainly focused on the dispersion of the Orano recycling plant located at La Hague in the mid English Channel. It represents the main source-term of dissolved radionuclides in this sea. Consequently, its dispersion concerns mostly the English Channel and Southern North Sea.

Nevertheless, other radiotracers source terms have been accounted as Sellafield releases in the Irish Sea or nuclear power plant releases reaching the Bay of Biscay through rivers. It could represent a metrological challenge, but tracers exist to extent the work elsewhere in the world.

This work addresses the following successive aspects: the background section describes successively the radionuclides measured, sampling and measurement methods; the measurement and release database achieved; a short description of the models used and the different model/measurement comparisons performed.

The application section focuses on the main features that could be retained from the point of view of radiotracers, hydrodynamic models and methods, from the short scale in the vicinity of an outfall, to the large scale in the English Channel, North Sea, the Biscay Bay and in the Pacific (Fukushima accident). The last section presents perspectives of applications in other areas or oceanographic domains.

2. Background

2.1. Dissolved Radionuclides as Oceanographic Tracers

To study the dispersion process over short to long periods or carry out and interpret repeated measurements for varying conditions of release or forcing, it is necessary to use tracers which fulfill the following characteristics:

1. It must originate from one or a small number of clearly identified release points;
2. The release conditions must be precisely known (time, fluxes);
3. Labeling in seawater must be significant, in particular in relation to the pre-existing background level. Labeling concerns seawaters where a significant concentration of radionuclide could be measured.
4. The tracer must be soluble and not fix onto living organisms or sediments over time (i.e., the stable element is conservative in seawater);
5. It must be possible to measure the tracer after dilution in the sea over hours, weeks, months or years after its release.

Natural tracers, such as copper, iron, nutrients, cannot generally be used because of the multiplicity of their source terms and the complexity of the phenomena governing their production and fate in the marine environment.

Even if the release conditions are known for artificial tracers, there are often many release points for each one. Such tracers are often involved in geochemical and biogeochemical processes, and therefore their conservative behavior in the marine environment are not guaranteed as shown in Section 3.2.4 for ^{60}Co and ^{106}Ru.

Radioactive tracers generally meet criteria 1 and 2. As regards criteria 3, 4 and 5, some radionuclides exhibit long term conservative behaviors in seawater, such as ^{125}Sb, ^{99}Tc and ^{3}H (tritium) [35], and to a lesser extent ^{90}Sr, ^{137}Cs and ^{134}Cs. Between 1970 and 1995 extensive measurements of these radionuclides were carried in the seas of north-western Europe [5,6,11,12,14–21,24,36]. Reductions in fluxes released during the period from 1980 to 2000 have led to significant decreases in concentrations in the marine system. Out of the radionuclides mentioned above, only tritium released from nuclear fuel re-processing plants has not undergone a reduction since 1980 (Figure 2); nevertheless, it fully satisfies the five criteria, having a radioactive half-life of 12.4 year.

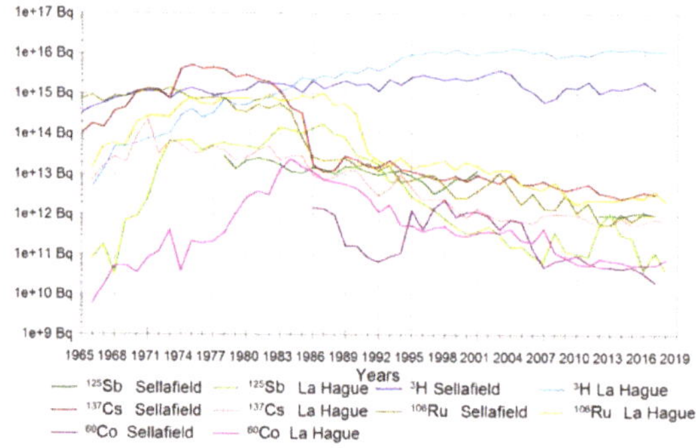

Figure 2. Annual liquid releases from Sellafield and La Hague nuclear reprocessing plants for ^{125}Sb, ^{3}H, ^{137}Cs, ^{106}Ru and ^{60}Co.

An additional interest of artificial radionuclides as environmental tools is that there is no risk of samples contamination during in-board sampling process, which is generally not the case for other chemicals that require clean conditions to prevent pollutions.

The radionuclides measured in this study that have the most conservative behavior in seawater are mainly antimony 125 (^{125}Sb) and tritium (^{3}H) in the form of tritiated water HTO (Table 1). The radioactive emission determines the technic that could be applied for each radionuclide.

Table 1. Characteristics of the radionuclides used in this work.

Radionuclide	Tritium ^{3}H	Cesium 137 ^{137}Cs	Cesium 134 ^{134}Cs	Antimony 125 ^{125}Sb	Ruthenium 106 ^{106}Ru	Cobalt 60 ^{60}Co
Radioactive decay (year)	12.3	30.2	2.1	2.8	1	5.3
Principal radioactive emission	β	γ	γ	γ	γ	γ
Conservative behavior *	100%	83%–86%	83%–86%	98%	19%–26%	8%–14%

*: Percentage of radionuclide quantities remaining dissolved in seawater at the scale of the English Channel (from [20]).

2.1.1. Gamma Emitters

Gamma emitter's (^{137}Cs, ^{134}Cs, ^{106}Ru and ^{60}Co in Table 1) concern radionuclides which the main radioactive emission is a high energy γ photon. Thus, they can be detected all together by their specific energy emission without radiochemistry for the isolation of each element. Specific methods adapted to the dilution conditions observed in the marine environment and the on-board treatment of samples was developed in order to measure artificial activities of the order of 0.3 Bq·m^{-3} [37,38]. This is in contrast with the natural radioactivity of seawater, which is of the order of 12,000 Bq·m^{-3} (mainly ^{40}K).

This work focuses mainly on ^{125}Sb and ^{3}H which are conservative in seawater; nevertheless, the database provides measurements of the radionuclides, which are measured together (^{106}Ru, ^{137}Cs, ^{134}Cs and ^{60}Co). They are useful to investigate other process as exchanges between seawater and sediments. Methods to measure ^{125}Sb, ^{106}Ru, ^{137}Cs, ^{134}Cs and ^{60}Co have been described in length in [37,38]. Analyses were carried out at the French Navy-Groupe d'études atomiques in Cherbourg (GEA), and at the IRSN-LRC laboratory in Cherbourg.

2.1.2. Tritium

^{3}H is present in all nuclear plants' liquid releases. The two main sources of tritium are the nuclear fuel reprocessing plants at Sellafield into the Irish Sea that have been active since 1952, and from La Hague into the English Channel since 1966. Liquid releases from each nuclear power plant are two orders of magnitude lower than the fluxes issued from reprocessing plants. Nevertheless, these releases must also be accounted for all along the European coasts and rivers, particularly away from the reprocessing plants influences.

Due to the small amount of seawater required for direct measurement by liquid scintillation (8 mL) a large amount of samples can be processed that can be used to have a high frequency 2D picture or even a 3D picture of the tritium dispersion at a short scale close to the La Hague outfall (Sections 2.4.1 and 2.4.2 [26,39]). The power plants labeling could be used in areas out of theses influences, with corresponding levels lower than 1 Bq·L^{-1}, in the Bay of Biscay [26] or in the Mediterranean Sea.

In the releases performed by nuclear reprocessing plants and French nuclear power plants, tritium is the form of the tritiated water molecule HTO; thus, it has a strict conservative behavior in seawater. This is not always the case: for example, the releases performed by the Amersham plant in Cardiff in the Bristol Channel concern organic compounds tagged with tritium. Organically bounded tritium (OBT) has a specific behavior in marine environment with strong assimilation by living species [40]. Due to their locations (mainly the Bristol Channel and Rhone River), these kinds of releases do not influence the results presented here.

The most used method to measure tritium is liquid scintillation, which allows up to 10,000 samples per year for one analyzer. With a detection limit around 1 Bq·L^{-1}, this method could only be relevant within the plume of reprocessing plant releases (eastern English Channel, Irish Sea, rivers). Extra suited methods exists for tackling lower levels of concentration, which requires sampling one liter of seawater. Low level ^{3}H measurements were performed with two methods:

(i)   Electrolytic enrichment of water samples [41]. The detection limit reached 0.01 Bq·L^{-1}.

(ii) ^{3}He regrowth and measurement by mass spectrometry [42]. The detection limit could reach 0.001 Bq·L^{-1}.

2.2. Database

2.2.1. Radionuclide Measurements in Seawater

The measurement database provided includes all oceanographic campaigns performed by IRSN-LRC (CEA-LRM before 2002) between 1982 and 2016 (Table A1). Data are provided for ^{3}H, ^{125}Sb, ^{106}Ru, ^{137}Cs, ^{134}Cs and ^{60}Co. The database concerns 80 oceanographic campaigns and totals 39,642 sampling locations at sea. Data concerning the coastal monitoring station the closest to the La Hague outfall were added (744 measurements between 1984 and 2018).

Table A1 in Appendix A lists the different campaigns, the number of measurements and maximum concentrations measured. Figure 3 shows the locations of all samples obtained. The measurement database is available in [43] as a supporting material for this work. Part of these data was already given available in [5] (2010 ^{137}Cs measurements), in [26,44] (14,494 tritium measurements), and sparsely in previous publications. The new database encompasses 47,387 measurements; more than 60% of these data were previously unpublished.

Figure 3. Location of radionuclide measurements samples obtained between 1982 and 2016.

As said in the introduction, existing databases include measurements of dissolved radioactivity in the European seas. The most important are the IAEA Marine Information System (MARIS [30]), the World Ocean Database 2013 [31,32], the IFREMER–SISMER [33] or the BODC [34]. The data provided here complete them in different ways.

^{137}Cs and ^{134}Cs are commonly measured in oceanographic studies, but ^{125}Sb, ^{106}Ru and ^{60}Co are sparsely detected. ^{3}H was used as a tracer in open oceans but to a lesser extent in coastal waters.

Due to their conservative behavior in seawater, ^{125}Sb and ^{3}H are choice tracers to track the dispersion of industrial releases.

The data collection was designed to appraise the dispersion of French nuclear plants releases from short to large spatial and temporal scales through repeated oceanographic campaigns since 1982. The La Hague controlled radioactive releases in seawater are the more important in Europe; the corresponding radiotracer measurement database allows a complete case study of its dispersion in marine systems.

To our knowledge, no other in-situ measurement database allows to follow individual radionuclides releases in the marine environment at short scale.

The provided database is homogeneous with the release data; it could be included further in other existing databases.

2.2.2. Radioactive Releases

Releases in seawater from La Hague reprocessing plant was transmitted by Orano company for each individual release since 1982 up to 2018. Quantity released, date-time of the beginning and end of each release are given. Prior to 1982, only annual releases were available. Part of these data was already given available in [26,45] (7840 tritium releases).

The new database encompasses 22,183 individual releases for tritium, ^{137}Cs, ^{134}Cs, ^{125}Sb, ^{106}Ru and ^{60}Co (110,915 values). In total, 93% of these data are unpublished as part of a comprehensive accessible database until now.

The Sellafield reprocessing plant annual release data have been published by [46] and in MAFF, RIFE and Sellafield Ltd. reports [47,48]. ^{125}Sb data have been available only since 1978 with an information gap between 2001 and 2012.

Annual releases from British and French nuclear power plants are issued from annual reports from MAFF, RIFE and EDF reports [47,48].

The releases database is available in [49] as a Supplementary Materials for this work. Figure 2 presents the annual fluxes released by Sellafield and La Hague reprocessing plants for ^{125}Sb, ^{137}Cs, ^{106}Ru and ^{60}Co. It shows that the main release period for ^{125}Sb, ^{3}H, ^{137}Cs, ^{106}Ru and ^{60}Co occurs before 1992, with a maximum for ^{137}Cs between 1969 and 1983 at Sellafield (1–5 PBq·y^{-1}), and for ^{106}Ru between 1969 and 1991 at La Hague (close to 1 PBq·y^{-1}). Tritium releases (in blue) show a more homogeneous temporal evolution, with fluxes in the same magnitude as Sellafield since 1969 (1–3 PBq·y^{-1}) and since 1993 from La Hague (around 10 PBq·y^{-1}). It depends mainly from the quantities of nuclear fuel processed by the plants.

Existing datasets concern mainly annual releases from selected nuclear plants. This work provides a compilation of all the release data in one accessible database for French and British nuclear plants in European seas. The knowledge of each individual release is essential to perform precise comparisons between measurements and hydrodynamic model simulations and evaluate the dispersion in the vicinity of an outfall. The detail of the La Hague controlled radioactive releases represent an unparalleled dataset in this perspective.

2.3. Hydrodynamic Models

The different models applied are based on the Model for Applications at Regional Scale (MARS) developed by the French Ifremer institute since 1987 [50]. This model was built under various assumptions presented below: using the non-stationary Saint-Venant equations (i.e., 2D); the primitive 3D equations; the "Lagrangian barycentric" method in 2D to filter out the tidal signal.

2.3.1. 2D Modeling

Numerous modeling studies [20,22–25] have demonstrated that models using two-dimensional horizontal approximation (i.e., shallow-water equations) are able to simulate a satisfactory dissolved-substance transport in the non-stratified area. These equations were solved using the finite-difference MARS2D model [50].

In the largest domain the baroclinic effects can be neglected (so that 2D models can relevantly be used) as long as they do not influence that much the targeted area. Naturally, they play a major role over the shelf of the Bay of Biscay or next to the mouth of large estuaries (Loire, Gironde, Rhine etc.). In the eastern English Channel and southern North Sea, away from rivers, the ocean is rarely thermally stratified nor stratified in terms of salinity and temperature; this is due to the strong tidally induced mixing.

2.3.2. 3D Modeling

The three-dimensional MARS3D model is described in detail in [50]. The model uses a 3D finite difference scheme, applying the Boussinesq approximation and hydrostaticity to resolve primitive equations. The model involves a nesting strategy (example Figure 4), starting from a broad region covering the entire North-West European continental shelf (with a 5.6 km grid resolution) down to a detailed domain covering a few tens of km (with a 5–100 m resolution).

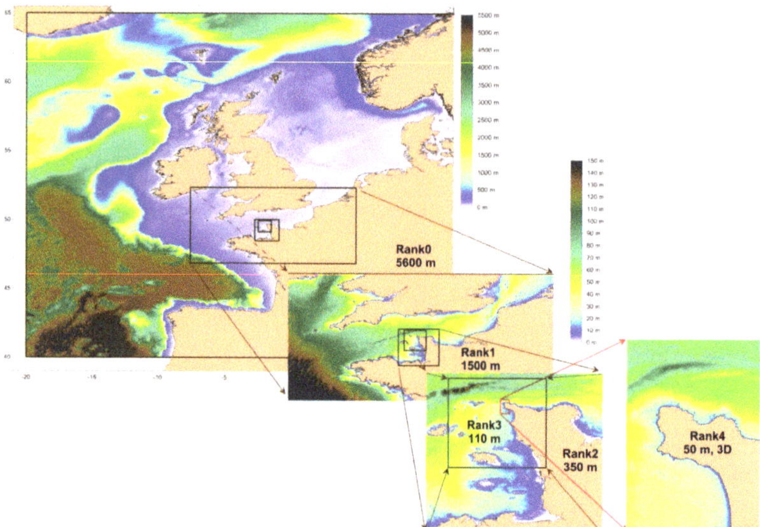

Figure 4. Example of model nesting for the La Hague Cape area, Ranks 0–2 are two-dimensional (2D), Rank3 is 2D or three-dimensional (3D), Rank4 is 3D, adapted from [26].

The bathymetry at the grid nodes of the different models is estimated with the method described in [51] from various data sources [25,52].

2.3.3. 2D "Lagrangian Barycentric" Method

A tidal residual model was designed to reproduce transport, dilution and decay phenomena over long time scales (ranging from a week to several years) and extensive spatial scales (from 30 to 1000 km).

It is based on the hypothesis that the water column is homogeneous and that barotropic phenomena will prevail over dynamic baroclinic phenomena. This model has been extensively described and applied in [18,20,25,52–54]. The interest of applying "Lagrangian barycentric" currents is to investigate the residual circulation occurring at long time scales (more than a tidal cycle of 12.42 h), and to simulate dispersion very quickly at the scale of the English Channel and the North Sea (computation time about 1000 times lower than with a similar 2D model).

2.4. Sampling Close to An Outfall

2.4.1. Model Assisted Sampling

In the vicinity of a punctual source-term (represented by the location of the end of the release pipe), the sampling strategy must be adapted to catch the dispersion of a rapidly moving narrow plume exhibiting short scale features.

During the first hours following release, it is necessary to have good knowledge of the time schedule of the release and where the plume will be located close to the outfall. The dynamics of the currents and duration of releases close to the La Hague Cape impose precise positioning in time (less than 15 min) and space (100 m), to ensure that the sampling sections of the ship's track encompass the plume area (Figure 5 [25]). The releases schedules were transmitted to the vessel by the nuclear plant unit (Service de Protection Radiologique-SPR) of the ORANO company. The MARS2D model (see Section 2.3.1, rank 3 in Figure 4) is used on-board to get some forecasts of the dispersion ahead of the release and allow precise positioning of the vessel during the hours following the start of the release.

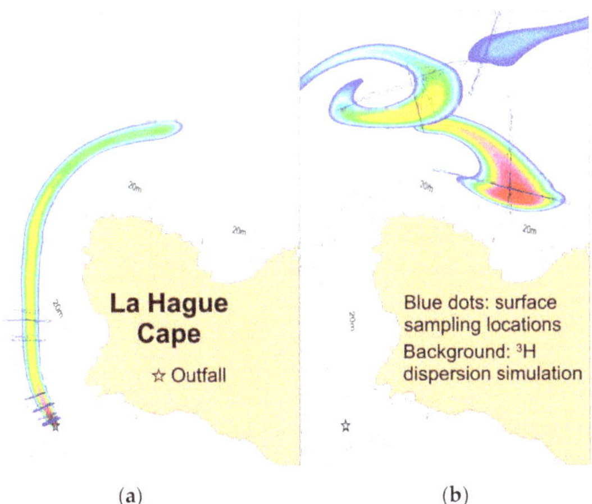

Figure 5. Sampling locations during the DISVER 2011 campaign (13,400 in-depth samples between 5 and 11 April 2011) (**a**) 0–2 h after a release (**b**) 3–8 h after a release.

The procedure made possible to sample at high frequency (30 s) an average of 300 surface stations to follow the horizontal dispersion of a given release (see for example the figure in Section 3.1.3). Varying hydrodynamic and meteorological conditions were investigated.

2.4.2. High Frequency In-Depth Sampling

Sampling at depth during oceanographic campaigns requires usually bottles deployed from the surface down to the sea floor on a manual or monitored sampling system, such as a CTD rosette. This requires the ship stops at stations during sampling event (i.e., move the sampling system downward, close bottles and move upward at around 1 m·s^{-1} speed). Overall, it takes about 15–30 min for each station in shallow waters (20–50 m). In the working area, for security reasons the ship cannot be stopped close to the coast or in areas with strong currents. Furthermore, the plume identification requires much higher sampling rates. In the context of the DISVER experiment (Section 3.1.2 [55]) for studying vertical dispersion close to an outfall, we aim to validate three-dimensional hydrodynamic models by using in-situ in-depth radiotracer measurements; this objective requires high-frequency sampling (one every 30 s) at 10 depth levels simultaneously (thus giving 1800 samples per hour).

The area off the Cap de La Hague represents one of the most difficult areas to investigate for such studies, since it combines strong currents (up to 5 m·s^{-1}), proximity of the coast (less than one km) and a complex topography with many rocky shoals and deeps varying from 25 to 90 m over a few kilometers. For operational and security reasons, the ship must remain under way normally at speeds from 1 to 5 m·s^{-1} (10 knots) while sampling. A devoted system was developed to perform safe sampling in such rough environment [39].

It comprises three main components:

- A sampling line designed to sample 10 depths simultaneously down to 65 m;
- A deep towed depressor (known as Dynalest), which maintains the line at depth close to the seabed;
- An automatic high frequency sampler with volume, flux and depth control to get 1800 samples per hour.

During sampling, the ship was operating at speeds ranging in 0.5–5 m·s^{-1} with frequent U-turns. Around 13,000 samples could be collected during a four days survey. Vertical slice pictures of the dispersion plume were thus obtained each 5–10 min (100–200 measurements each slice).

The combined system as configured is adapted for in-depth sampling when at least one of these conditions is required: high frequency sampling, proximity of reefs or coasts, strong currents, several depths simultaneous sampling.

2.5. Methods for Model/Measurement Comparisons

2.5.1. Comparison with Individual Measurements

Comparisons between measured and simulated radionuclides concentrations are performed at the same x, y, z and t (longitude, latitude, depth and time). Different statistics may be computed:
Correlation coefficient R:

$$R = \frac{\sum_{n=1}^{N}\left(\text{Meas.}_n - \overline{\text{Meas.}}\right)\left(\text{Sim.}_n - \overline{\text{Sim.}}\right)}{\sqrt{\sum_{n=1}^{N}\left(\text{Meas.}_n - \overline{\text{Meas.}}\right)^2}\ \sqrt{\sum_{n=1}^{N}\left(\text{Sim.}_n - \overline{\text{Sim.}}\right)^2}} \tag{1}$$

95th percentile of the ratio (PR) between measured and computed concentrations:

$$PR = \text{Max}\left[\frac{\text{Meas.}_n}{\text{Sim.}_n}, \frac{\text{Sim.}_n}{\text{Meas.}_n}\right] \tag{2}$$

Absolute percentage error APE:

$$APE = 100\%\frac{\left|(\text{Meas.}_n - \text{Sim.}_n)\right|}{\text{Meas.}_n} \tag{3}$$

Mean absolute percentage error MAPE:

$$MAPE = \frac{100\%}{N}\sum_{n=1}^{N}\frac{\left|(\text{Meas.}_n - \text{Sim.}_n)\right|}{\text{Meas.}_n} \tag{4}$$

where:

Meas.$_n$: Measured tritium concentration of the n sample (Bq·m^{-3});

$\overline{\text{Meas.}}$: Mean measured tritium concentration ($Bq \cdot m^{-3}$);

Sim._n: Simulated tritium concentration of the n sample ($Bq \cdot m^{-3}$);

$\overline{\text{Sim.}}$: Mean simulated tritium concentration ($Bq \cdot m^{-3}$);

N : Number of samples.

A histogram of the absolute percentage errors could be drawn in order to assess the risk of the model to misestimate the real concentration. Examples of uses of these criteria are present in [24,26,56].

In the vicinity of an outfall, in case of highly variable plume distribution a small change of location (50 m) may result in large concentration variations. Other parameters must then be applied to check for the model reliability [25].

Additional comparison criteria suited for near outfall field.

- Maximum concentration in the plume;
- Mean concentration in the plume;
- Width of plume intersected;
- Distance between maximal measured and simulated plume positions;
- Plume width discrepancy;
- Average and maximum concentration discrepancies;
- Dilution rate discrepancy.

To compare several campaigns with variable source terms, measured concentrations could be normalized by the cumulated released flux corresponding to the plume targeted. They become dilution coefficients (DC) or dilution factors with:

$$DC = \frac{Bq \cdot L^{-1}{}_M}{Bq \cdot L^{-1}{}_R} \tag{5}$$

$Bq \cdot L^{-1}{}_M$: measured concentrations

$Bq \cdot L^{-1}{}_R$: released flux

This dilution coefficient was applied for model/measurement comparisons at a short scale in the vicinity of an outfall, when individual release plumes could be distinguished [25]. At larger scales, individual releases are mixed together in seawater. A more representative approach is to compare the measured concentration ($Bq \cdot m^{-3}$) with the mean released flux in $Bq \cdot s^{-1}$. Such dilution coefficients (concentrations corresponding to a given released flux) were applied at the scale of the whole English Channel and the whole North Sea (see Section 3.2.3, [20,24]).

For 3D model/measurement comparisons, the methods described previously could be applied by accounting for the water mass stratification of the radionuclide concentrations. In case of studies in the near outfall field, particular attention should be given to obtain the same reference of x, y, z and t between measurements and simulations: an error of minutes or hundred meters could result in a concentration variation larger than an order of magnitude.

2.5.2. Radionuclides Inventories

When campaigns had a sufficient extent and sampling rate, localized radioactivity measurements were interpolated over the whole studied area at the nodes of a regular grid. A well-suited interpolation method is "kriging" [57]. This allows the visualization and comparison of the distribution plumes measured and simulated (see Section 3.2.1).

If the radionuclide distribution in seawater is not stratified (no variation from surface to sea bottom), the total inventory of radionuclides present could be simply calculated by the multiplication of the surface concentration C by the volume of seawater of each mesh (C × dx × dy × depth). In case of stratification, in-depth measurements must be performed in order to associate the right concentration to each level of the water column [26].

Quantities of measured radionuclides can be straightforwardly compared to the simulated ones at different scales [18,20,24,26,27] or compared to the known releases in order to exhibit transit times or conservative behavior [18]. Inverse calculation has also been applied to estimate the source term (e.g., in the Fukushima case) [58].

2.6. Scales for Model/Measurement Comparisons

Different time and space scales are presented in this work; the distinction of the different scales is based on the dispersion characteristics in a macro-tidal context. Super tidal scale (minutes to hours, 100 m–10 km) concerns dispersion close to an outfall when the labeled plume is not vertically averaged. The tidal scale (hours to days, 1–30 km) concerns the dispersion in the vicinity of an outfall when the labeled plumes associated to each specific release could be distinguished by surface measurements. Large scale (from week to years, 30–1000 km) concerns the dispersion when individual plumes are mixed together. This is discussed further in Section 4.3.

3. Applications

The obtained database [43] contains data from 1982 up to 2016, with all measurements acquired by the IRSN-LRC laboratory during oceanographic campaigns. It concerns mainly [125]Sb, ^{3}H, [106]Ru, ^{137}Cs, [134]Cs and [60]Co.

Such data would have been meaningless without the corresponding known releases that explain the observed labeling (Section 2.2.2 [49]).

3.1. La Hague Cape, Short Scale/High Resolution Studies

3.1.1. The La Hague Cape Main Characteristics for Dissolved Radionuclide Dispersion

The North Cotentin includes several nuclear facilities, such as the nuclear power plant at Flamanville Cape, the building of nuclear submarines at Cherbourg and the ANDRA low-level waste deposit at La Hague. The most important in term of liquid releases at sea is the Orano reprocessing plant, of which the outfall is located next to the La Hague Cape (Figure 5). The La Hague Cape forms a physical boundary between the Normandy–Brittany Gulf in the south-west, and the mid-English Channel towards the east. Because of the coastal morphology, the tidal wave coming from the Atlantic is partly blocked in the west-facing bay formed by the Normandy-Brittany Gulf. This embayment is characterized by very large tidal ranges (reaching more than 14 m near the Mont St Michel during spring tides). The Cape de la Hague works as a bottleneck for the water masses involved during the emptying and filling of this bay twice a day. This explains why the tidal currents close to the cape are among the strongest in Europe (they can reach 5 m·s^{-1} during spring tide), with highly variable tidal range around the cape (from 11 m in the south of the Cape down to 6 m in the north [59]). This area is also characterized by diverse topography exhibiting pronounced gradients (depths from 20 to 100 m), many islands, numerous bays and shallow coves.

A tidal residual currents [52] divergence zone close to the release outfall divides waters flowing into the Normandy–Brittany Gulf from the waters forming part of the general flow from west to east up the Channel and towards the Straits of Dover. As a result, small differences in the release conditions can lead to opposite directions of spreading in the medium-term [25,60]. As a consequence, simulation of the dispersion in the area is challenging for the numerical models' dispersion capability.

3.1.2. 3D Dispersion: Super Tidal Time Scale

The La Hague outfall is located on the sea floor, two kilometers off the coast (Figure 5). The vertical dispersion of the plume has been investigated during the DISVER project. In total, 19,000 in-depth samples were taken during three campaigns of four–five days [55]. Figure 5 shows an example of the sampling strategy with plume transects at different distances from the outfall. The transects are slices across the axis of the plume propagation from the surface down to 25 m depth; they are spaced at time intervals of 5 to 10 min. The database encompasses 137 vertical transects up to 65 m depth with 100–200 individual measurements per transect. This provides a rather good picture of the vertical structure of the plume along each of these transects. Figure 6 shows an example of model/measurement comparison during two series of transects at 850–1300 m and 3200–4200 m downstream the outfall. These figures show an accurate representation of the plume in space and time. Figure 6a exhibits unexpected highly complex and variable structures resulting from intense turbulent mixing with eddies of about 100 m that extend to the width of the plume rapidly. The 3D model is not able to reproduce the turbulent mixing at this scale. Further than 3000 m downstream (Figure 6b), the plume reaches the sea surface with a more homogeneous shape that is better reproduced by the model.

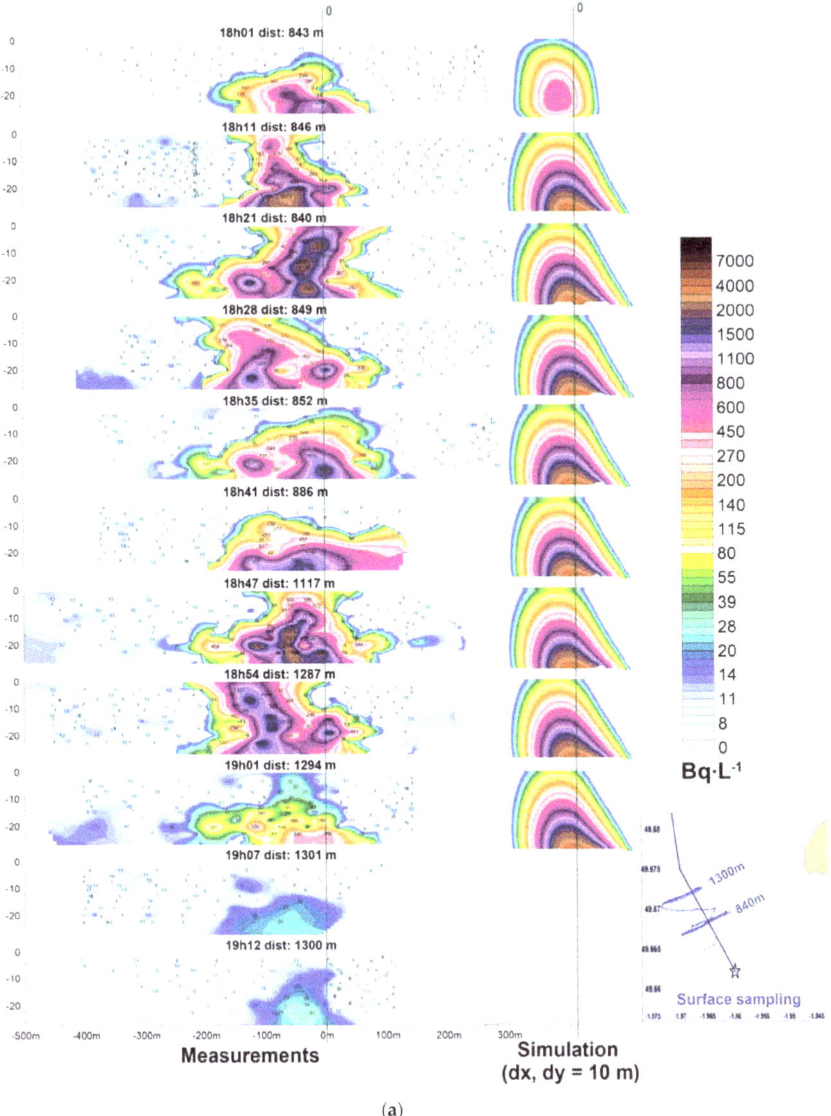

(a)

Figure 6. *Cont.*

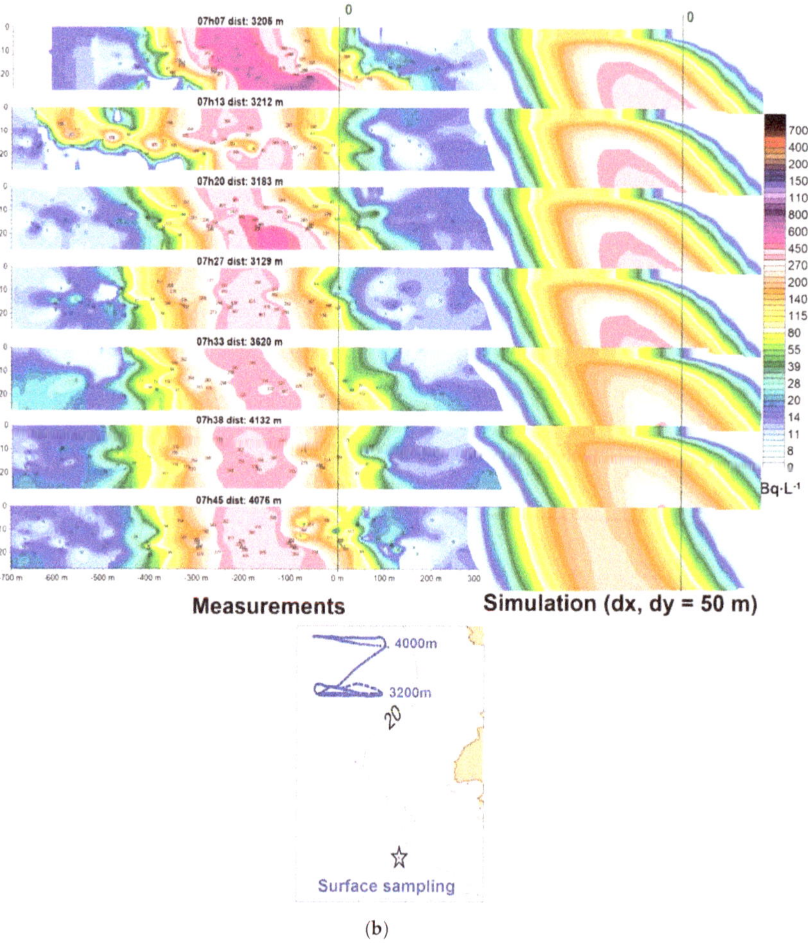

(**b**)

Figure 6. Examples of measurement/model comparison of vertical transects of the plume between 850–1300 m distance from the outfall ((**a**), 7 October 2010), and 3200–4200 m from the outfall ((**b**), 6 April 2011), adapted from [61]. Time and spatial scales (m) are the same for measurements and simulations. Insets: maps of surface sampling locations, adapted from [55].

The results obtained from all transects have been integrated by the normalization of the concentration measured with the corresponding release. Results are presented in Figure 7 with the measured and simulated extension and dilution of the plume with the distance during a constant release (steady state situation). It shows that if the model does not reproduce the instantaneous turbulence, on average, the shape and dilution coefficient are given the correct order of magnitude. Details of the dilution variations and model/measurement comparisons are given in [55].

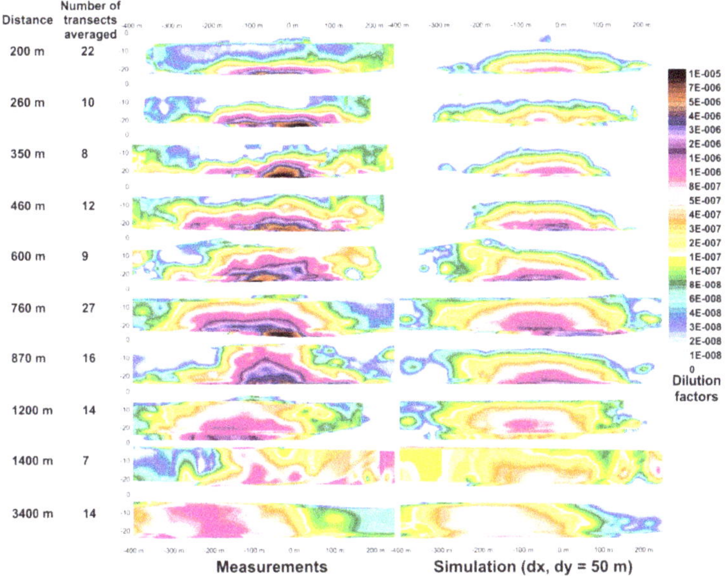

Figure 7. Variations with distance of the outfall of measured and simulated average dilution factors, adapted from [55].

As the plume shape is more complex to the north of the La Hague Cape (Figure 5b), a similar model/measurement comparison is difficult to perform. Figure 8 shows the vertical distribution of concentrations 7 km downstream and five–six hours after a release. In this area, the tritium plume crossed the La Hague trough, of which the depths range from 80 to 100 m. These observations result from current shear between surface and bottom waters: the labeling is first transported in surface waters where the currents are larger.

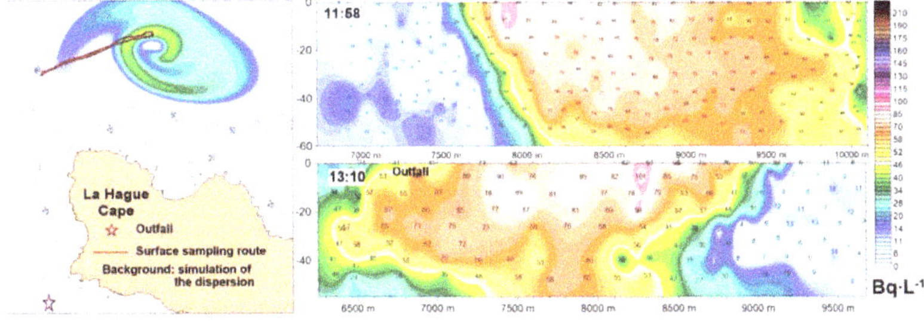

Figure 8. Vertical concentrations of the plume five and six hours after a release and at 7 km distance. Left: distribution of surface sampling locations during the survey the 6 April 2011 (red dots). Background: simulation of the expected plume at 12:15. Right: interpolation of in-depth measurements obtained during the survey at 11:58 and 13:10 the 6 April 11 (back and forth). Numbers: concentrations measured at each in-depth location.

3.1.3. 2D Dispersion: Tidal Time Scale

As shown in Figure 7, the vertical homogenization of the plume is reached at 3 km downstream the outfall, which corresponds to one hour after the release. Except at particular locations where strong bathymetry changes occurs (Figure 8), in the eastern English Channel, surface measurements of radionuclide concentrations are representative of the whole water column labeling. Investigations have been performed to test the model capability to simulate the dispersion of the La Hague plant within 1 to 48 h after the beginning of the release (DISPRO project). Beyond 48 h, it is no longer possible to distinguish two daily consecutive releases from another, because dilution and stirring mix them together. Within 48 h following a release, about 200 transects were done with sampling every 30 s (i.e., transects crossing a dispersion plume that could be traced back to a known release, and into which the width of the plume could be assessed). These transects represent roughly 3000 individual measurements, which encompass all tidal conditions (from neap to spring tides). An example of the tracking of one release together with a model/measurement comparison is given in Figure 9. Comparisons have been performed by accounting for the measured and simulated concentrations at a sampling location, and other criteria described in Section 2.5.1. Additional comparison criteria. Details of the results obtained are in [26]; Table 2 presents the main results of model/measurement comparisons for short scales model/measurements comparisons. For example, at short scales next to an outfall, the deviation between measured and simulated concentrations (expressed as mean dilution coefficient) is lower than 71% for a 95% confidence interval (Table 2).

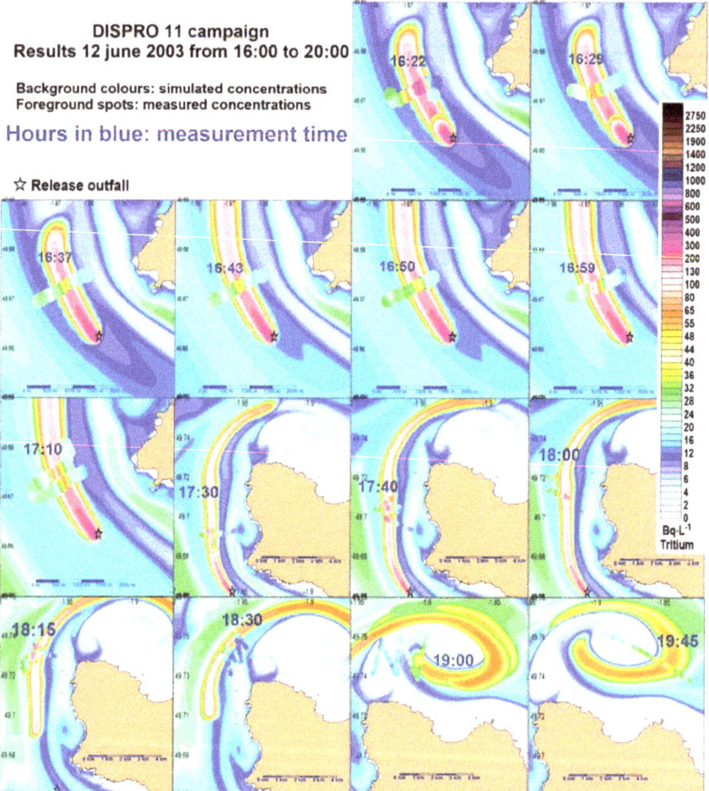

Figure 9. Example of a comparison between the measured and simulated dispersion during 4 h after release, from 16:22 to 19:45 adapted from [25].

Table 2. Characteristics of the model used to simulate soluble radioactive release dispersion in seawaters at a short scale next to an outfall around the Hague Cape.

Dispersion Characteristics 1 h to 48 h after Release	
Geographical boundaries of the model	49°17′ N–49°55′ N; 2°26′ W–1°31′ W
Hydrodynamic model	Two-dimensional: Vertically-averaged velocities and concentrations, 110 m mesh size, 20 s time step
Average discrepancy between the mean dilution coefficients measured and simulated in the plumes	9% (−66% → 70%)
Average discrepancy per transect between simulated and measured maximum dilution coefficients	3% (−72% → 73%)
Average measured/simulated plume-width discrepancy	−6% (−73% → 65%)
Average discrepancy between measured and simulated plume-position, as a function of distance from the outlet point	−1% (−22% → 22%)

Figures in brackets indicate the 95% confidence interval.

3.2. Large Scale Model/Measurement Comparisons: Multi Tidal Time Scales

3.2.1. Individual Measurements

When a sufficient amount of measurements was obtained during oceanic campaigns, we could draw up radionuclide concentration maps at the scale of each survey. Campaigns performed repeatedly covered the English Channel and the North Sea between 1988 and 1996, the north-west Atlantic and Irish Sea between 1994 and 1996, the North Atlantic between 1997 and 2004 and the Bay of Biscay between 2009 and 2016. Hydrodynamic models accounting for all source terms and oceanic forcing (tide, meteorological forcing and large-scale circulation and hydrology) allow comparing of the measured and simulated concentrations at the date of each campaign. Figure 10 gives an example of such a comparison in 1994 for ^{125}Sb and 2016 for ^{3}H.

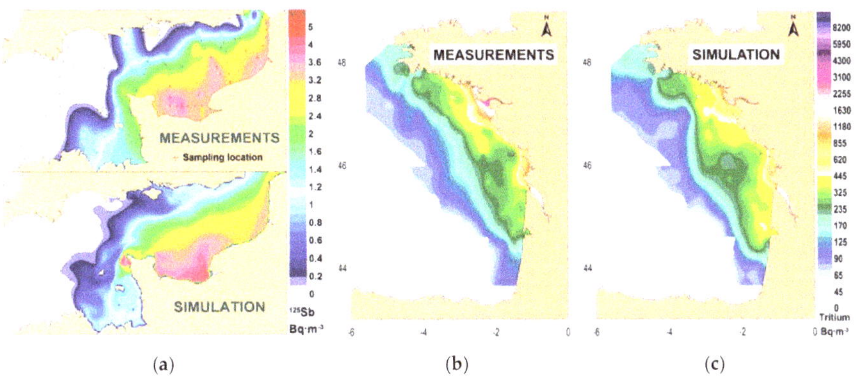

(a) (b) (c)

Figure 10. (a) Measured and simulated ^{125}Sb in the English Channel, September 1994, adapted from [24]; measured (b) and simulated (c) surface ^{3}H in the Bay of Biscay, spring 2016, adapted from [26].

Computation of the measured and simulated concentrations at each sampling location and time gives statistics of the model representativeness. For example, in the English Channel and the North Sea, application of a residual Lagrangian model gives a mean difference between the 1400 individual values calculated and measured between 1988 and 1994 of 54% [24]. The deviation at a short scale next to an outfall is lower than 70% (Table 2). At the scale of the Biscay Bay with a 3D model the deviation is 21% [62].

3.2.2. Water Masses Labeling

In case of multi-tracer studies, it is possible to associate a specific labeling of the different water masses investigated and to draw up a "picture" of the different plumes associated to the different water masses. This means that a set of initial average concentrations is affected at each water mass entering the studied area. Then, at each sampling location in this area, the resolution of an equations system can determine the contribution of each water mass that fits the measured concentrations of tracers. This implies that the number of unknown (water mass to account) does not exceed the number of independent tracers measured.

This method was particularly fruitful in 1988 in the North Sea, where it was possible to distinguish and map the distribution of the four main water masses entering the North Sea by using the simultaneous measurements of three radionuclides (^{125}Sb, ^{137}Cs, ^{134}Cs) and salinity (Figure 11) [17]. As compared with another campaign performed two years before, the contribution of the direct fallout after the Chernobyl accident and the rate of renewing of North Sea waters were also calculated.

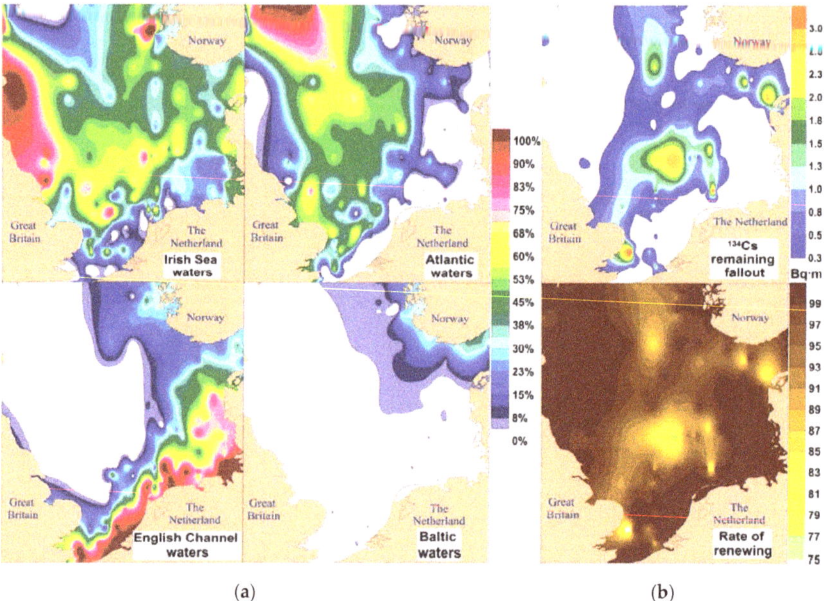

(a)   (b)

Figure 11. Percentage origins of water masses entering the North Sea in 1988 deduced from dissolved radionuclides measurements (**a**); ^{134}Cs remaining fallout from the Chernobyl accident and rate of renewing of water masses in two years (**b**), adapted from [17].

A similar approach was applied in the Bay of Biscay in 2016, by accounting for the different ^{3}H labeling of waters coming from the Loire and Gironde rivers. Therefore, it is possible to compare directly the extent of the plumes deduced from measurements and simulated separately as shown in Figure 12 [40].

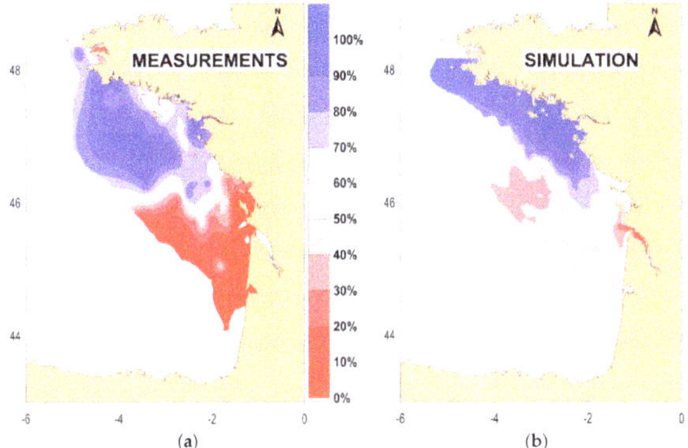

(a)   (b)

Figure 12. (**a**) Measured and (**b**) simulated freshwater contribution of the Loire River relatively to the Gironde River, adapted from [26] in spring 2016.

3.2.3. Integration of Normalized Contributions

In a similar way as presented at the end of Section 2.5.1, it is possible to compare the plume distributions obtained from different campaigns by normalizing the measured concentrations with the corresponding fluxes of releases. This method allows the average dispersion characteristics and dilution coefficient of the considered source term to be mapped. It has been applied on a short scale in 3D (Figure 7), in the English Channel [20] and the North Sea [24], as presented in Figures 13 and 14. It shows that, on average, the Lagrangian residual models are able to properly catch the dispersion process at that large scale. This result was obtained after adapting the wind stress drag, which is one of the main hydrodynamic model calibration parameters.

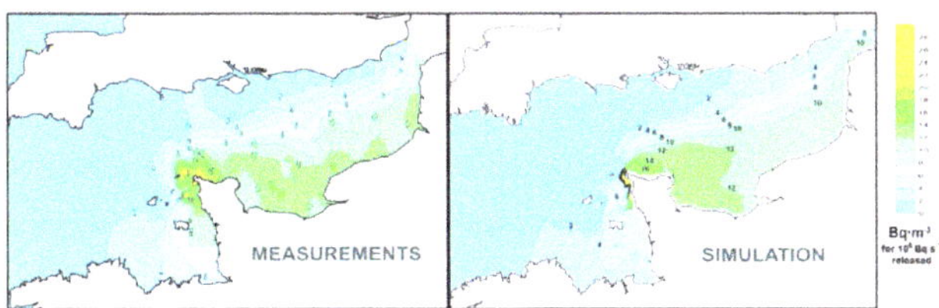

Figure 13. Normalized measured and Lagrangian residual simulated distribution of ^{125}Sb between 1983 and 1994 for a constant release of 10^6 Bq·s^{-1}, adapted from [20].

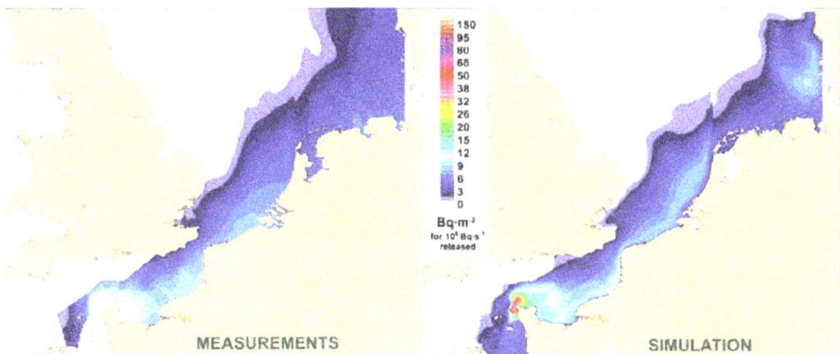

Figure 14. Normalized measured and simulated distribution of ^{125}Sb between 1988 and 1994 for a constant release of 10^6 Bq·s^{-1}, adapted from [24].

3.2.4. Inventories

As presented in Section 2.5.2, inventories of radionuclides quantities in different areas and water masses allow for the assimilation of numerous measurements in integrated quantities to account for the labeled depth and distribution of samples. The results could be compared with the known releases supposed to contribute to the labeling, and to model results in the same areas and water masses. Comparison with releases without simulation have been performed in the North Sea by slicing the southern North Sea into boxes where measured radionuclide quantities were compared with the La Hague releases [18] (Figure 15). From this correspondence it was possible to highlight the average transit time (one year from La Hague to the Skagerrak) and fluxes through the Dover Strait (97,000–195,000 m^3·s^{-1}), as well as estimations of losses for non-conservative radionuclides [18].

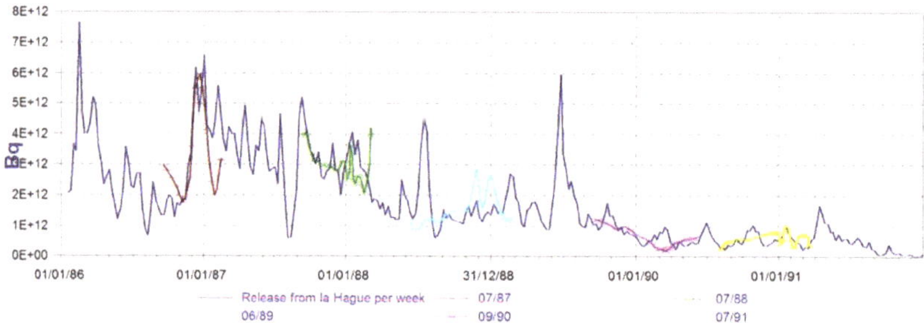

Figure 15. Quantities of ^{125}Sb measured in North Sea boxes during oceanographic campaigns compared to the corresponding releases from the La Hague plant, adapted from [18].

In the English Channel and the southern North Sea, the equilibrium between the measured and simulated quantities of radionuclides was an efficient tool to tune the wind friction that provide the best reliability of the Lagrangian residual model [24].

The loss of non-conservative radionuclides from seawater to sediments and living species (around 15% for ^{134}Cs, 75% for ^{106}Ru and 85% for ^{60}Co Table 3, Figure 16) was assessed from the measured quantity comparisons with the corresponding releases at the scale of the English Channel. It is then possible to check the environmental models accounting for the fluxes between the different environmental compartments. Such balanced budget exhibited an unknown source term at this scale, which represented twice the expected quantities measured for ^{137}Cs at the Channel scale (Table 3).

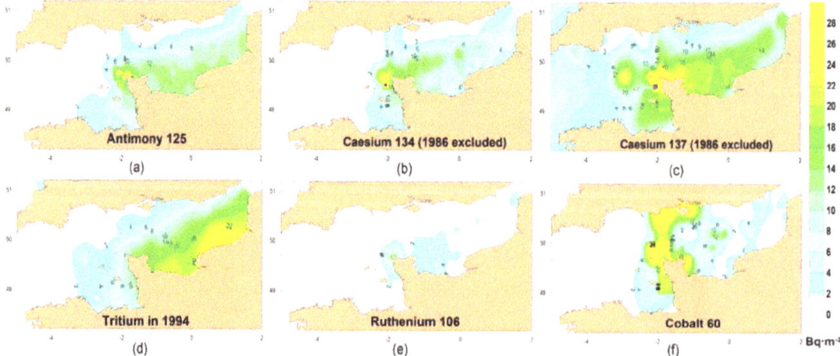

Figure 16. Average impact of the La Hague reprocessing plant in seawater corresponding to a constant release of 1 MBq·s^{-1}; (**a**) ^{125}Sb, (**b**) ^{134}Cs, (**c**) ^{137}Cs, (**d**) ^{3}H, (**e**) ^{106}Ru, (**f**) ^{60}Co, adapted from [20].

Table 3. Comparison of measured quantities of radioactivity in the English Channel with correspondent releases from La Hague from 1983 to 1994, adapted from [20]. Antimony-125 is used as reference for the calculation; the background concentrations in Atlantic surface waters (^{137}Cs, ^{3}H) and Chernobyl fallout (^{137}Cs, ^{134}Cs) has been deduced.

Volume: Equivalent Release Duration:	Whole English Channel 4702 Km^3 32 Weeks–7.3 Months						Eastern English Channel 1576 Km^3 25 Weeks–5.7 Months					
	^{137}Cs	^{134}Cs	^{106}Ru	^{125}Sb	^{60}Co	^{3}H	^{137}Cs	^{134}Cs	^{106}Ru	^{125}Sb	^{60}Co	^{3}H
Number of campaigns	3	2	3	3	3	1	5	3	5	5	2	1
Fraction of the La Hague release	233%	86%	26%	98 %	14%	103%	139%	83%	19%	98%	8%	121%

We hypothesized [18] that the ^{137}Cs excess in the English Channel resulted mainly from the influence of Sellafield releases. It was afterwards confirmed [11]; this evidences a pathway from the Irish Sea through the St George Channel in the Celtic Sea, and then a seasonal input in the English Channel as shown in Figure 1. The flux that reaches the English Channel represents around 1% of Sellafield ^{137}Cs. Figure 17 shows the English Channel areas that are the most impacted by the Sellafield releases. This corresponds to locations where sediments act as a delayed secondary source term for the water column, as demonstrated in the Irish Sea [63]. It concerns mostly the Hurd deep in the English Channel center and the French coastal areas.

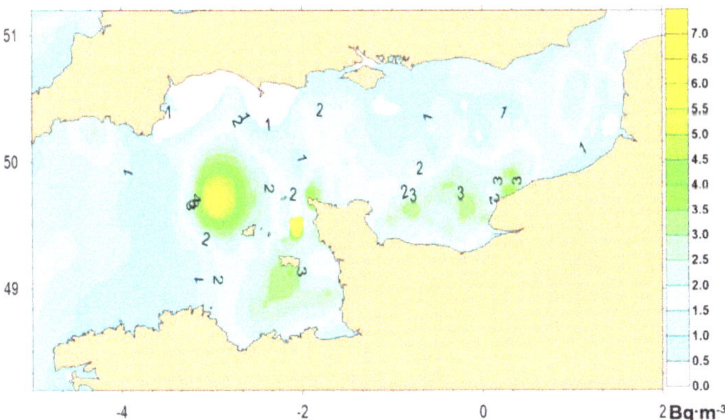

Figure 17. Calculated from Figure 16 (^{137}Cs–^{134}Cs): ^{137}Cs not coming from La Hague (average during 1986–1994).

In the Celtic Sea, comparisons between inventories and releases from La Hague and Sellafield plants give an estimate of the fluxes and residence times of water masses in the Celtic Sea and North Est Atlantic approaches, with labeling from Sellafield releases along the western Ireland coasts [11] Figure 1.

At a global scale, a comparison of the measured ^{3}H inventories and known inputs have been applied at the world oceans scale (Figure 18 [64]). It provides fluxes, estimations of residence time and values for the past and future concentration of ^{3}H in Atlantic waters entering the European waters (78 Bq·m^{-3} in 2016).

This value was applied to compare the measured and simulated radionuclide inventories in the Bay of Biscay. At this scale, the concentration simulated by the MARS3D model was compared to the measured ones by accounting for the variation of the ^{3}H concentrations with depth [26]. It provides an estimation of the residence time (around one year for the continental shelf of the Bay of Biscay), and pathways of waters coming, respectively, from the Loire and Gironde rivers [26].

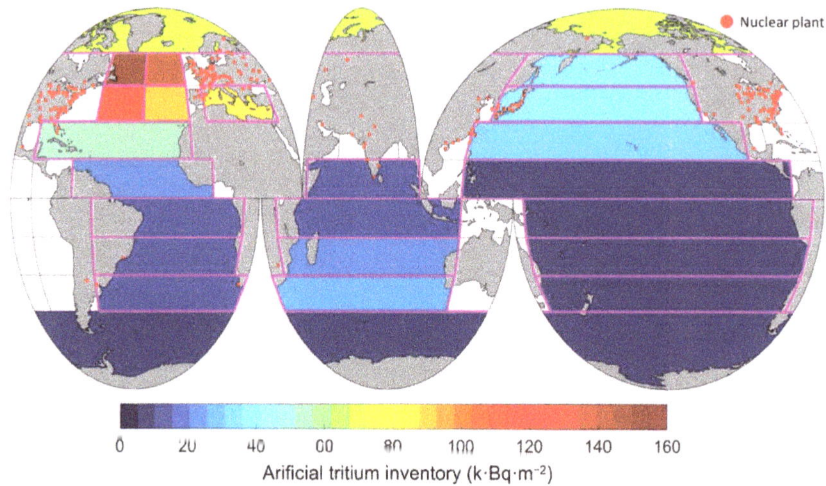

Figure 18. Inventory of anthropogenic tritium by unit of surface area (kBq·m^{-2}), adapted from [65].

Estimation of radionuclides inventories was also applied during the accidental context following the nuclear accident at Fukushima Dai-ichi power plant. Many seawater measurements of ^{137}Cs were provided continuously by Japanese authorities (Figure 19).

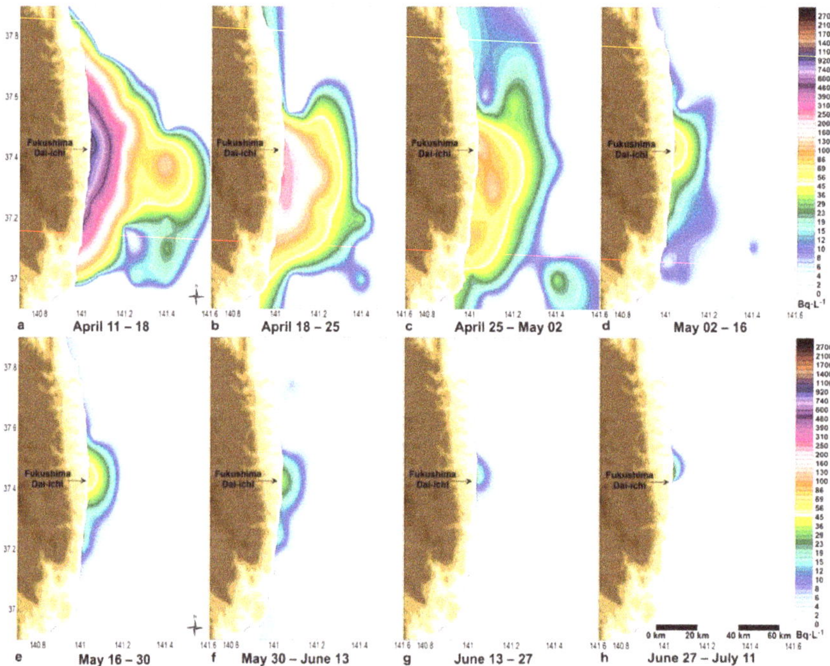

Figure 19. Concentrations of ^{137}Cs in seawater between 11 April and 11 July 2011, adapted from [59].

They enable the calculation of successive inventories of ^{137}Cs quantities in the surrounding Pacific waters and to estimate the marine source term coming at sea ((Figure 20) [58]). This first estimation of 27 PBq (12 PBq–41 PBq) appeared to be very high at the time of publication, but later estimations were in the same order of magnitude [65–67]. The rapid decrease of ^{137}Cs quantities during the weeks following the accident reveals a very rapid environmental half-life that has been reproduced by MARS3D simulation (Figure 20 [27]).

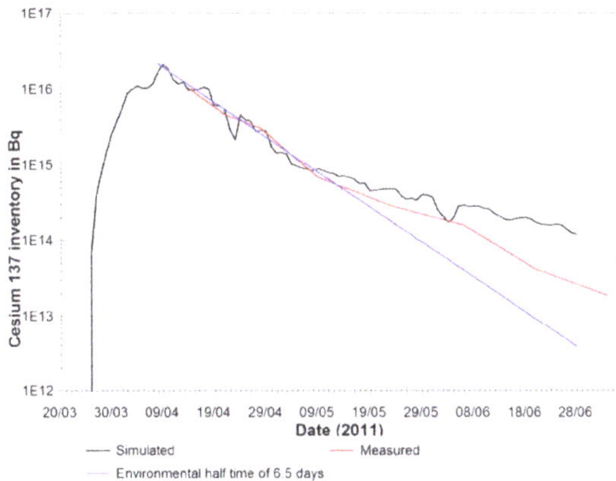

Figure 20. Time evolution of measured and simulated ^{137}Cs inventories present in the Fukushima area, adapted from [27].

4. Discussion

This work gives accessibility to a comprehensive database, including in-situ radionuclide measurements performed during oceanographic campaigns and fluxes of dissolved radioactive release between 1982 and 2018 over the North-Western European continental shelf.

The results presented show applications of in-situ radionuclide measurements coupled with hydrodynamic models. They demonstrate how they improve the knowledge of dispersion in seawater. Applied alone, in situ data provide general pathways, transit time and dilution rate at different scales. Coupled with numerical hydrodynamic models, they improve and confirm their reliability with strong constraints. Physical measurements (as current, sea level or drifter measurements) generally represent short term parameters. Dissolved radionuclides complement physical data with information associated from short to long term transport and dispersion in seawater that could not be directly measured in currents.

Use of radiotracers for model validation was initiated by Salomon and Guéguéniat in 1988 with residual Lagrangian models [68]. This pioneering work was applied at different scales with space and time adaptations to obtain results comparable to measurements and simulations.

Artificial dissolved radionuclide tracers are particularly interesting because they fulfill a detection limit close to the background level, and contain few well known source terms and low risk of pollution during sampling and treatment. These advantages allow the detection of labeled plumes from short (hours, 100 m) to large scale (years, 1000 km). It counterbalances the efforts given to radionuclide measurements such as sampling large volumes for gamma emitters and measurement constraints.

4.1. Radiotracers

The use of artificial radionuclides such as oceanographic tracers depends from the controlled liquid releases performed by nuclear industry. These releases represent opportunities of existing labeling that could be used to follow water masses. They are issued mainly from European reprocessing plants. The fluxes of gamma emitters have decreased by two orders of magnitudes during the 1990s after application of a more efficient industrial purification process to diminish the environmental risk. It results in that the concentrations measured in the vicinity of the La Hague plant in 2018 are close to the detection limit and lower than the ones measured at 1500 km distance along the Norwegian coasts in 1988 (less than 2 Bq·m^{-3}). Other radionuclides could still be applied as long term tracers, such as tritium [11,26,64], ^{99}Tc [69] ^{129}I [70,71] or ^{236}U [72–74].

^{3}H is present in all nuclear plant releases and, in the form of tritiated water HTO, it could not be retained at the source by chemical process. At the nearest monitoring station to the main outfall in the marine system at La Hague Cape, concentrations in seawater are on average 12 Bq·m^{-3} (2017–2019, O. Connan pers. comm.). This concentration is three orders of magnitude lower from the OMS maximum recommended value of 10,000 Bq·L^{-1} for drinking water. The ^{3}H labeling from the nuclear plant will remain significant as long as the nuclear plants operate, and thus, they remain a relevant long term oceanographic tracer if we have the capability to measure low levels (from the oceanic background of 0.07 Bq·L^{-1} [64]). The mapping of the dispersion of the Loire and Gironde waters at the scale of the Bay of Biscay demonstrates this approach [26]. Limitations for radiotracers measurement concern mainly the available means (low level counters, ^{3}He degassing and mass spectrometry) and delay between sampling and measurement (from days to years). As an example, the ^{3}He ingrowth method for ^{3}H measurement exists in few laboratories around the world.

4.2. Hydrodynamic Models

Oceanographic hydrodynamic models have been continuously improved following the computing capabilities enhancement. It is possible to represent the whole continental shelf in 3D with a resolution of around one kilometer. This calculation efficiency must not mask the requested improvements concerning the knowledge and representation of physical process. The representativeness and precision of forcing the parameters determines the model reliability.

It concerns open boundaries limit conditions (stratification, currents, tides), and local data such as bathymetry precision, meteorological effects, bottom nature and associated friction. A better understanding of the surface and bottom drag coefficients remains an ongoing concern in research [75], as they require significant improvements.

A demonstration that a given hydrodynamic model is reliable does not prove that other similar models are at the same level. Long term advection and dispersion are particularly discriminant, as it represents only a few percent of the instantaneous currents. The radiotracer databases will remain useful to test the next hydrodynamic model generation.

Nevertheless, the level of realism reached between measurements and simulations give way to hydrodynamic model application at the scale of the European shelf or elsewhere in the world as shown in the north-western Pacific [27].

For short scale studies in 3D, MARS3D is not able to represent the real turbulence process (Figure 6). These phenomena are outside the initial scope of this model (i.e., regional scale); it must be improved by applying other numerical schemes or discard assumptions as the hydrostatic one. The computing capabilities remain a limit to simulate together short scales (meters) and longer ones (tens of kilometers); however, methods exist to overcome these limits (AGRIF) [76].

4.3. Scales for Model/Measurement Comparisons

This work shows that different approaches must be applied to measure the dispersion from short to medium or large scales. An operational definition to distinguish short scale from longer ones in tidal seas is the capability, or lack thereof, to distinguish each individual release plume. This depends from the diffusion process and the frequency of the main forcing parameter and releases. This results in that in the center of the English Channel, short scale studies concern fewer than four tidal cycles (two days).

The short scale requires models simulating rapid plume displacements (30 km in 3 h) with a high resolution and short time step (lower than 100 m and 20 s). A difference of 10 min or 200 m could change the punctual concentration close to the plume of one order of magnitude. If we consider that the model simulates the tidal propagation over thousands of kilometers before calculating the local tide, this implies strong constraints on the numerical scheme and forcing parameters (open boundaries conditions and drag coefficients). The dispersion of individual releases could be measured, but the rapid changes during dispersion do not allow drawing 3D or 2D maps that describe the plume dispersion several times per hour. 2D and one-dimensional (1D) comparisons were performed for 3D and 2D models comparisons, respectively.

The more the model/measurement comparison is done far from a source term, the more the plumes are smoothed and the lower the deviations between the punctual measurement and equivalent simulated values. Larger scales could use, when tidal effects are smoothed, instantaneous or Lagrangian residual models with a mesh size larger than one kilometer and time steps higher than 200 s. Exhaustive measurements could be performed that allow calculating inventories with possibility to compare measured, simulated and released radionuclides quantities.

Long term-large scale models obtain good model/measurement scores without under- or overestimation of radionuclides concentrations if they account well for the residual effects of the different influences (tidal, meteorological, frictions). These influences can be adjusted such as the wind drag coefficient obtained in [24]. The determination of detailed physical phenomena hidden behind these calibration parameters is an open research field.

4.4. Sampling Tools

The measurement of radionuclide dispersion led to the development of original systems to sample seawater in highly dynamic environments. High frequency samplers allow the simultaneous collection of 10 samples each 30 s while recording the time, location, volume, flow and depth [39]. The different parts of the developed tools may be used separately or together and adapted for other purposes.

Volumes collected could be extended or sampling depth increased that allow continuous in-depth sampling of any dissolved substances with a vessel sailing. 3D sampling is not limited to seawater and could concern larvae or particulate matter with continuous sampling and measurement of vertical suspended particulate matter (SPM) profiles.

This enhances the efficiency of oceanographic campaigns. As an example, the towed depressor Dynalest was applied as a support for instruments during sailing: an acoustic current doppler profiler (ADCP) was deployed to measure currents from 3 m depth up to the bottom without being perturbed by surface waves with a ship sailing normally in the Alderney race (0.5–4 m.s^{-1}).

Variability of plume dispersion made short scales studies impossible without accurate simulation tools to forecast its location. Sampling assisted by model simulation has demonstrated its efficiency in the vicinity of an outfall in a tidal environment (Figures 5 and 9, [25,39]).

4.5. Radionuclides Inventories

By accounting for all available measurements associated with their representative area and depth, it is possible to map the radionuclide distributions and calculate radionuclides quantities existing in specific areas. Such inventories give integrated values with a better precision than individual concentrations [17,20].

Marine dissolved radionuclides give a rare example in environmental studies where balance budgets could be calculated between measured, simulated and releases inventories with precision better than 10% [20,25,26]. Comparison between radionuclides inventories quantities existing in seawater with the known sources results in determinations of environmental fluxes and transit times [18,26]. It highlights unknown sources and water masses pathways [11,20,58]. Inventories could also be compared with the simulation results and they represent robust check values to improve hydrodynamic models [20,24,27].

The balance obtained between released and simulated quantities of conservative radionuclides gives access to the loss of less conservative radionuclides. It is possible to quantify the fluxes between the different environmental compartments (i.e., seawater, living species and sediments) [18,20]. Radionuclide inventories are then powerful tools to check integrated environmental models accounting for stocks and fluxes of radionuclides at the scale of a continental sea as the English Channel.

4.6. Normalization

Normalization by a parameter representative of an environmental or anthropogenic forcing could be applied to long term time series or variable spatial distributions. This method made possible, by accounting for adequate time or space scale, to compare highly variable temporal or spatial changes of the environment as a result of variations depending on the season, meteorological effects, tides and releases. Normalization mitigates stochastics effects and highlight average characteristics of the environment difficult to quantify. It gives useful results with the measurements alone, but can also be compared with simulations. Others examples are: 3D short scale dispersion characteristics (Figure 7); average distributions and inventories in the English Channel (Figure 16 and Table 3) and North Sea (Figures 14 and 15); shape of the English Channel pathway in the southern North Sea [18]. Similar methods could be applied in areas where deterministic models fail to represent the water masses circulation as in the Kuroshio gyres in the Northern Pacific [77].

4.7. Perspectives

Tracers and methods applied here are potentially applicable elsewhere if the radionuclide labeling is sufficient to distinguish different water masses. This is obvious in the case in the Irish Sea, where the other main artificial source is the Sellafield reprocessing plant. The work performed in the Bay of Biscay demonstrates that nuclear power plant releases of tritium are sufficient to label water masses at a large scale if low level measurement methods are available [26]. This was also possible after precise evaluation of the North Atlantic surface water background concentration [64].

The possible application of ^{3}H as water masses tracer potentially concern all seas reached by nuclear power plant releases. Maps of power plant locations with the corresponding tritium releases (Figure 18 [64,78]) suggest the Mediterranean Sea, North-West Atlantic, Arctic Sea, Western and Eastern Pacific, Black Sea, Baltic Sea, China Sea, Japan Sea, Gulf of Mexico and Arabian Sea.

Radionuclide dispersion models were applied to simulate realistic behavior of controlled industrial releases from 1984 up to 2016 in the English Channel (Figure 21). Such results could be compared with monitoring measurements and give a way to understand and interpret them. It was applied by IRSN to optimize the location of monitoring stations. Another application is the forecasting of accidental situations as it was performed after shipwrecks in the English Channel (Ievoli Sun in 2000, Ece in 2006) or after the Fukushima accident [27]. An operational tool is in course of development at IRSN to forecast the consequences of accidental situations (STERNE tool [79]).

Knowledge of dissolved transport (currents, advection and dispersion) is the basis of other oceanographic studies. It made possible pollutant transfer studies between seawater and living species [56,80–82] or particulate matter and physical transport of sediments [83–85]. Radio-ecological models accounting together for biological and hydro sedimentary process are accessible. The first tests have also shown the possibility to follow the sea/atmosphere exchanges by using ^{3}H in water vapor [85].

Radionuclides behavior could be studied at the earth scale [65], as a tool for testing climate changes models (renewing of ocean water masses with ^{3}H, water cycle, carbon cycle with ^{14}C). Radionuclides were widely applied during the 1980s, but are not outdated environmental tracers.

Our wishes are that the provided database will contribute to these works and will be complemented.

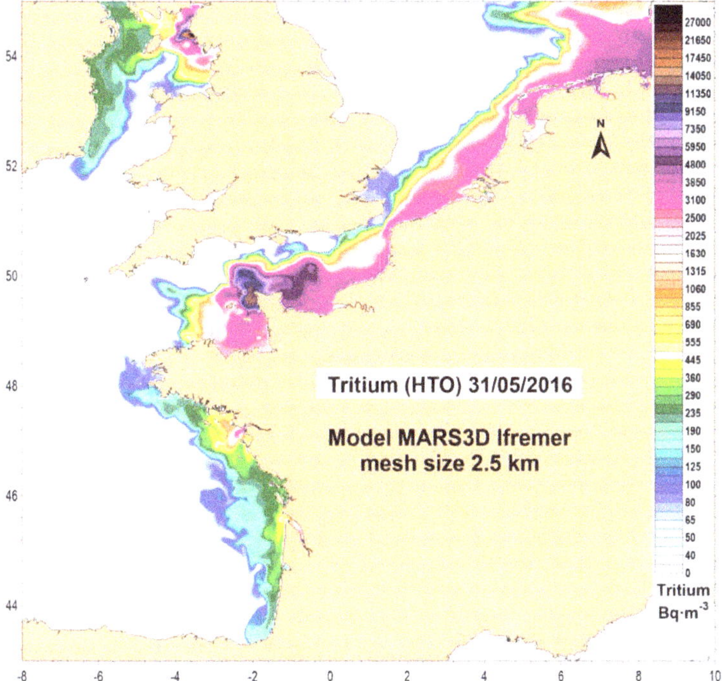

Figure 21. MARS3D simulation of the dispersion of tritium industrial releases in seawater over the European continental shelf.

Supplementary Materials: The following are available online at: Bailly du Bois, P. IRSN measurements of dissolved radioactivity in seawater, 1982–2016, 39642 stations. Available online: https://doi.pangaea.de/10.1594/PANGAEA.906541 (accessed on 02 October 2019). Bailly du Bois, P. IRSN measurements of dissolved radioactivity in seawater, 1982–2016, including liquid releases database issued from British and French nuclear plants. Available online: https://doi.pangaea.de/10.1594/PANGAEA.906749 (accessed on 2 October 2019).

Author Contributions: Conceptualization and Methodology, Formal analysis, P.B.d.B.; Data curation and databases, P.B.d.B., C.V. and P.-E.O.; Funding acquisition, P.B.d.B.; Investigation, P.B.d.B., F.D. and P.-E.O.; Project administration, P.B.d.B.; Software, M.M. and P.B.d.B.; Supervision, P.B.d.B.; Validation, P.B.d.B. and P.-E.O. Visualization, P.B.d.B. Writing—original draft preparation, P.B.d.B. Writing—review and editing, P.B.d.B. and F.D. Oceanographic campaigns, P.B.d.B., C.V., M.M., F.D. and P.-E.O. Radionuclide measurements, L.S., P.B.d.B. and C.V. Model development, F.D. and P.B.d.B. Technical development, P.B.d.B. All authors have read and agreed to the published version of the manuscript.

Funding: Most of the oceanographic campaign uses vessels and crew provided by the French oceanographic fleet operated by CNRS/INSU (Centre National de la Recherche Scientifique, France), Ifremer-Genavir (Institut Français de Recherche pour l'Exploitation de la Mer, France) and GEA-Navy (Groupe d'études atomiques de la Marine nationale). British "Cirolana" and German "Validivia" oceanographic vessels participate also to the sampling collection. This research was partly funded by Orano Company for oceanographic data acquired between 1990 and 2011.

Acknowledgments: It is not possible to mention all people that have contributed to the collection of samples during oceanographic campaigns. In particular, we want to thank: the crews from the CNRS/INSU and Genavir vessels, Pierre Guéguéniat, Dominique Boust, Rémi Gandon, René Léon, Marianne Lamotte, Bruno Fiévet, Olivier Connan, Pascal Morin, Bernard Le Cann, Louis Marié, Robert Lafite, Kins Leonard, Peter Kershaw and Jürgen Hermann. The Orano La Hague SPR team must also be thanked for providing detailed liquid radioactive releases during and after oceanographic campaigns. Lastly, we thank the editors and anonymous reviewers for their fruitful comments during the review process.

Conflicts of Interest: The authors certify that they have no conflicts of interest to declare.

Appendix A

Table A1. List of oceanographic campaigns accounted in the database. Positions are referenced in WGS84 in decimal degrees (N, E: +; S, W: −). Radionuclide concentrations are in Bq·m^{-3}. Abbreviations: LH: La Hague Cape; EC: English Channel; CI: Channel Islands; CS: Celtic Sea; IS: Irish Sea; NS: North Sea; MS: Mediterranean Sea; BB: Biscay Bay; NA: North Atlantic.

Campaign Name	Year	Area abr.	Stat. nb.	Lon. Min.	Lon. Max.	Lat. min.	Lat. max.	H3		Cs137		Cs134		Sb125		Ru106		Co60	
								Nb.	Max. meas.	Nb.	Max. meas.	Nb.	Max. meas.	Nb.	Max. meas.	Nb.	Max. meas.	Nb.	Max. meas.
Cirolana0582	1982	EC, CS, NS	42	−7.0	2.0	49.5	58.7			39	642								
Cirolana0583	1983	EC, CS, IS	54	−7.5	−1.9	49.3	54.7			44	13,283	22	68	14	459	36	7511	1	21.8
Clione0983	1983	EC, CS	36	−8.5	1.3	48.6	51.6			35	37.0								
GEA0983	1983	EC, CI	12	−2.5	−2.2	49.3	49.6			12	19.2	6	5.0	9	23.8	9	69		
Thalassa1283	1983	EC, BB	56	−4.8	0.3	47.4	50.0			55	59.2	31	10.0	45	149	44	207	8	44.4
Pluteus0583	1983	EC, CI, NS	119	−2.7	2.9	48.9	51.3			115	69.2	21	15.5	105	1668	79	3256		
Pluteus0686	1986	EC, CI	185	−4.8	2.0	48.3	51.3			184	34.5	141	1.7	163	211	134	686	29	13.0
Ferry1986	1986	EC, CS	31	−8.3	−4.0	48.8	51.8			31	15.4	7	5.3						
Pluteus1286	1986	EC	47	−1.8	−1.4	49.7	50.3			47	18.6	27	5.2	44	89.7	38	180	27	8.8
GEA1986	1986	EC, CS	97	−6.0	1.9	48.7	51.2			97	53.3	80	71.8	45	46.3	34	315	13	8.9
Nies0187	1987	NS	29	1.5	9.2	51.0	55.0			29	84.0	23	3.0						
Aurélia0387	1987	NS	31	3.0	4.7	51.7	53.0			31	15.5	31	5.7	30	57.1	29	213		
Sépia1987	1987	EC, NS	14	1.3	2.3	50.3	51.2			14	11.2	4	1.8	14	48.5	14	121	1	0.4
Manche0687	1987	EC	50	−1.9	0.3	49.6	50.8			50	10.0	10	3.0	50	84.1	40	495	7	10.4
Luctor0687	1987	NS	63	2.0	3.9	51.1	51.9			63	31.2	19	4.0	61	62.2	19	120		
Pluteus0887	1987	EC, CI, NS	165	−4.0	7.9	48.7	54.2			164	90.5	84	39.6	152	70.1	139	180	14	2.5
Aurélia1187	1987	NS	6	3.5	4.6	52.1	52.7			6	12.8	4	1.6	4	57.4	6	104		
GEA 1987	1987	EC	16	−2.0	2.5	49.5	51.2			16	11.8	11	2.6	16	44.8	13	143	5	2.2
Aurélia0488	1988	NS	9	3.5	4.6	52.1	52.9			9	12.6			6	58.0				
Pluteus0488	1988	EC	44	−1.7	1.6	49.5	51.1			44	11.1	1	1.1	43	43.6	28	116		
Pluteus0688	1988	EC	105	−5.1	1.6	49.4	51.1			104	12.2			81	112	63	222		
Pluteus0788	1988	EC, CI	103	−4.0	−1.1	48.6	49.8			103	12.5	4	1.4	94	464	42	324	6	2.7

Table A1. *Cont.*

Campaign Name	Year	Area abr.	Stat. nb.	Lon. Min.	Lon. Max.	Lat. min.	Lat. max.	H3		Cs137		Cs134		Sb125		Ru106		Co60	
								Nb.	Max. meas.	Nb.	Max. meas.	Nb.	Max. meas.	Nb.	Max. meas.	Nb.	Max. meas.	Nb.	Max. meas.
GEA0788	1988	EC	105	−5.0	1.6	49.4	51.1			104	12.2	21	1.9	81	180	64	222	5	2.6
Tramanor1	1988	EC, NS	256	−2.0	9.0	49.7	60.2			242	70.7	192	13.3	186	48.9	107	189	2	2.0
Gedynor	1989	EC, NS	127	−4.0	8.5	49.2	58.9			119	38.8	37	4.3	89	57.1	33	151	1	1.6
RadeBrest0689	1989	CI	23	−4.6	−4.3	48.3	48.4			23	5.9								
RadeBrest0789	1989	CI	23	−4.6	−4.3	48.3	48.4			23	6.3								
GEA0989	1989	EC, CI	88	−2.6	−0.9	49.4	50.2			87	31.5	57	7.8	88	107.4	80	355	23	15.9
Pluteus0989	1989	EC, CI	124	−2.8	−1.0	49.2	50.3			124	37.0	71	4.8	123	203.0	100	985	53	13.9
RadeBrest0390	1990	CI	46	−4.6	−4.2	48.3	48.4			46	4.1								
Tramanor2	1990	EC, CI, NS	205	−6.0	8.5	48.3	60.1			171	23.7	59	2.6	137	14.5	116	90	39	15.2
FluxManche0990	1990	EC	32	1.2	1.6	50.7	51.1			24	14.8	9	57.0	26	23.1	17	56		
Cirolana1	1991	EC, NS	55	−1.7	4.5	49.8	55.2			44	21.9	11	1.1	35	13.8	33	67	24	2.7
Tramanor3	1991	EC, NS	162	−1.8	9.1	49.4	63.3			144	28.0	65	1.7	116	18.9	73	37	41	1.4
Cirolana2	1992	EC, NS	42	−1.1	5.6	49.7	55.2			38	18.5	3	0.5	29	11.4	18	8	11	3.3
GolfeNormand Breton0492	1992	EC, CI	39	−3.0	−1.7	48.7	49.8			37	7.5	24	0.5	37	21.0	7	14	15	2.4
Valdivia1993	1993	EC, CS, IS, NS	63	−6.0	11.0	49.8	58.5			62	259	60	2.1	60	13.8	58	10	56	3.9
Arrho1994	1994	MS	9	4.7	4.9	43.3	43.4			9	5.1	5	1.1	5	3.6	7	68	5	0.4
Gedymac1994	1994	EC, CI, CS, IS	222	−8.4	1.5	48.0	54.5	95	8870	211	222	55	1.4	95	30.8	33	9	43	2.5
Dymanche1994	1994	EC	81	−1.9	1.5	49.3	51.1			81	27.7	45	2.4	47	5.8	6	13	18	1.3
RadialeCh-Wh0295	1995	EC	8	−1.7	−1.0	49.7	50.7	4	5860	8	5.6			1	2.4	2	5	5	0.9
RadialeCh-Wh0395	1995	EC	7	−1.7	−1.0	49.7	50.7			6	4.4			1	1.5			2	0.5
RadialeCh-Wh0595	1995	EC	8	−1.7	−1.0	49.7	50.7			8	4.7	1	0.3	3	1.3	1	4	3	0.5
Omex 0695	1995	CI, NA	7	−13.7	−6.6	47.4	56.6			7	1.7								
Omex 0895	1995	CI, NA	8	−14.1	−7.4	47.7	50.6			8	3.3								
RadialeCh-Wh0795	1995	EC, CI	31	−2.7	−1.0	49.3	50.6			31	9.0	25	0.7	28	1.6	21	2	25	0.9
RadialeCh-Wh0995	1995	EC	8	−1.9	−1.1	49.7	50.6			8	7.2	4	0.3	5	3.6	4	10	8	0.6
RadialeCh-Wh1195	1995	EC, CI	37	−2.6	−1.6	49.0	49.9			37	5.8	21	0.2	35	2.0	33	34	28	0.7
Ferry0396	1996	EC, CS	23	−8.3	−4.0	48.7	51.8	4	719	23	2.7							3	0.7

Table A1. *Cont.*

Campaign Name	Year	Area abr.	Stat. nb.	Lon. Min.	Lon. Max.	Lat. min.	Lat. max.	H3		Cs137		Cs134		Sb125		Ru106		Co60	
								Nb.	Max. meas.	Nb.	Max. meas.	Nb.	Max. meas.	Nb.	Max. meas.	Nb.	Max. meas.	Nb.	Max. meas.
FondDeBaie0496	1996	EC	15	−2.3	−1.6	48.6	48.8	5	3710	14	3.4	1	0.1	13	0.7	3	1	9	0.6
Ferry0696	1996	EC	8	−1.9	−1.7	49.7	50.7			8	5.1	2	0.5	5	2.8	4	3	6	0.5
FondDeBaie0796	1996	EC	10	−2.4	−2.0	48.6	48.7	10	3200	10	3.3			7	0.7			4	0.4
Irma1996	1996	EC	83	−6.0	−1.3	48.1	50.5	60	21,000	71	6.1	1	0.3	9	2.1	10	11	19	6.1
Arcane1997	1997	CS, NA	52	−12.8	−3.3	39.1	45.6	22	268	43	2.5								
Ferry1997	1997	EC	19	−4.2	−1.6	48.8	50.7	11	484	19	4.2							1	0.9
FluxSed1998	1998	EC, CI	66	−6.3	0.1	48.6	50.2			66	5.9			5	1.9	9	13	8	1.0
Atmara1998	1998	EC, CS, NA	196	−14.0	−3.6	45.5	55.7	130	1930	191	16.1								
Cirolana2000	2000	CS, BB	35	−11.5	1.5	47.3	52.3	35	1323										
Ovide2002	2002	NA	60	−42.6	−6.2	35.8	59.8	45	377	15	2.7								
Dispro08	2002	LH, EC, CI	1029	−2.4	0.9	49.4	50.0	845	45,103										
Dispro09	2002	LH, EC, CI	1058	−2.4	−1.6	49.5	49.8	1046	2,355,208										
Dispro10	2002	LH, EC, CI	1899	−2.2	−1.6	49.6	49.8	1890	3,604,897										
Dispro11	2003	LH, EC, CI	2398	−2.1	−1.7	49.5	49.8	2397	15,32,137										
Dispro12	2003	LH, EC, CI	640	−2.9	−1.7	48.7	49.9	587	12,111										
Ovide2004	2004	NA	31	−42.9	−9.8	40.3	59.9	17	236	14	2.4								
Dispro2004	2004	LH, EC, CI	3539	−2.9	−1.2	48.6	50.0	3537	392,028										
Dispro2005	2005	LH, EC, CI	4143	−2.4	1.5	49.3	50.7	4103	268,816										
Disver2008	2008	LH, EC, CI	796	−2.0	−2.0	49.7	49.7	784	190,660										
Aspex2009	2009	BB	36	−6.0	−1.5	44.0	47.8	34	336										
Ovide2010	2010	NA	1	−19.1	−19.1	45.8	45.8	1	76										
Aspex2010	2010	BB	62	−6.4	−1.6	44.0	48.3	62	266										
Disver2010	2010	LH	5556	−2.0	−2.0	49.7	49.7	5314	7,029,776										
Disver2011	2011	LH	12498	−2.0	−1.9	49.7	49.8	12498	8,739,130										
Aspex2011	2011	BB	74	−6.0	−1.5	44.0	47.8	62	211										
Trimadu2013	2013	NA	18	−13.2	55.5	−32.4	34.9	17	80	12	1.1								
Traces2014	2014	EC, CI	1205	−3.0	−1.6	48.6	49.8	1170	125,159										
Traces2015	2015	EC, CI	547	−3.0	−1.7	48.8	49.8	546	38,320										

Table A1. *Cont.*

Campaign Name	Year	Area abr.	Stat. nb.	Lon. Min.	Lon. Max.	Lat. min.	Lat. max.	H3		Cs137		Cs134		Sb125		Ru106		Co60	
								Nb.	Max. meas.	Nb.	Max. meas.	Nb.	Max. meas.	Nb.	Max. meas.	Nb.	Max. meas.	Nb.	Max. meas.
Dynsedim2016	2016	BB	31	−5.3	−2.6	45.6	47.9	30	640										
Pelgas2016	2016	BB	130	−5.8	−1.3	43.7	47.9	125	583										
Plume2016	2016	BB	254	−5.2	−1.1	44.7	48.6	177	19,072	12	1.3								
Goury (time series)	1984 – 2018	LH, EC	744	−1.9	−1.9	49.7	49.7	266	41,999	615	79.9	260	62.3	498	223.1	484	1745	398	50.7
Nb. total (campaigns):			39,642					35,663		3492		1295		2242		1606		568	
Nb. Total (all):	80		40,386					35,929		4107		1555		2740		2090		966	

References

1. Guary, J.C.; Guéguéniat, P.; Pentreath, J. *Radionucléides: A Tool for Oceanography*; Elsevier Applied Science: Amsterdam, The Netherlands, 1988; p. 461.
2. Kershaw, J.P.; Woodhead, D.S. *Radionuclides in the Study of Marine Processes*; Elsevier Applied Science: Amsterdam, The Netherlands, 1991; p. 393.
3. Germain, P.; Guary, J.C.; Guéguéniat, P.; Métivier, H.E. Radionuclides in the Oceans. In Proceedings of the RADOC 96–97, Part 1, Inventories, Behaviour and Processes, Cherbourg-Octeville, France, 7–11 October 1996; Radioprotection Numéro Spécial, Colloques Volume 32, C2. Les Editions de Physique: Les Ulis, France, 1997; p. 422.
4. Herrmann, J.; Kershaw, P.; Bailly du Bois, P.; Guéguéniat, P. The distribution of artificial radionuclides in the English Channel, southern North Sea, Skagerrak and Kattegat, 1990–1993. *J. Mar. Syst.* **1995**, *6*, 427–456. [CrossRef]
5. Povinec, P.P.; Bailly du Bois, P.; Kershaw, P.; Nies, H.; Scotto, P. Temporal and spatial trends in the distribution of 137Cs in surface waters of Northern European Seas—A record of 40 years of investigations. *Deep. Sea Res. Part II Top. Stud. Oceanogr.* **2003**, *50*, 2785–2801. [CrossRef]
6. Kershaw, P.; McCubbin, D.; Leonard, K. Continuing contamination of north Atlantic and Arctic waters by sellafield radionuclides. *Sci. Total Environ.* **1999**, *237*, 1119–1132.
7. Kautsky, H. Investigations on the distribution of137Cs,134Cs and90Sr and the water mass transport times in the Northern North Atlantic and the North Sea. *Ocean Dyn.* **1987**, *40*, 49–69.
8. Dahlgaard, H. Transfer of European coastal pollution to the arctic: Radioactive tracers. *Mar. Pollut. Bull.* **1995**, *31*, 33–37.
9. Mitchell, P.I.; Holm, E.; Dahlgaard, H.; Boust, D.; Leonard, K.S.; Papucci, C.; Salbu, B.; Christensen, G.; Strand, P.; Sánchez-Cabeza, J.A.; et al. *Radioecological Assessment of the Consequences of Contamination of Arctic Waters: Modelling the Key Processes Controlling Radionuclide Behaviour under Extreme Conditions (ARMARA)*; EC Nuclear Fission Safety Programme 1995-99 Contract No. F14P-CT95-0035 Final Report; Department of Experimental Physics University College Dublin: Dublin, Ireland, 1999; p. 151.
10. Nies, H.; Albrecht, H.; Herrmann, J. *Radionuclides in Water and Suspended Particulate Matter from the North Sea, Proceedings of the Radionuclides in the Study of Marine Processes, Norwich, UK, 10–13 September 1991*; Kershaw, J.P., Woodhead, D.S., Eds.; Elsevier Applied Science: Amsterdam, The Netherlands, 1991; pp. 24–36.
11. Bailly du Bois, P.; Germain, P.; Rozet, M.; Solier, L. *Water Masses Circulation and Residence Time in the Celtic Sea and English Channel Approaches, Characterisation Based on Radionuclides Labelling from Industrial Releases, Proceedings of the International Conference on Radioactivity in Environment, Monaco, Principality of Monaco, 1–5 September 2002*; Borretzen, P., Jolle, T., Strand, P., Eds.; Norwegian Radiation Protection Authority: Østerås, Norway, 2002; pp. 395–399.
12. Guéguéniat, P.; Salomon, J.; Wartel, M.; Cabioch, L.; Fraizier, A. Transfer pathways and transit time of dissolved matter in the eastern English channel indicated by Space-Time radiotracers measurement and hydrodynamic modelling. *Estuarine Coast. Shelf Sci.* **1993**, *36*, 477–494. [CrossRef]
13. Guéguéniat, P.; Bailly du Bois, P.; Gandon, R.; Salomon, J.; Baron, Y.; Leon, R. Spatial and temporal distribution (1987–1991) of 125Sb used to trace pathways and transit times of waters entering the north sea from the English channel. *Estuarine Coast. Shelf Sci.* **1994**, *39*, 59–74. [CrossRef]
14. Guéguéniat, P.; Herrmann, J.; Kershaw, P.; Bailly du Bois, P.; Baron, Y. Artificial radioactivity in the English Channel and the North Sea. In *Radionuclides in the Oceans, Inputs and Inventories, RADOC 96–97*; Guéguéniat, P., Germain, P., Métivier, H., Eds.; Les Editions de Physique: Les Ulis, France, 1997; pp. 121–154.
15. Guéguéniat, P.; Kershaw, P.; Hermann, J.; Bailly du Bois, P. New estimation of La Hague contribution to the artificial radioactivity of Norwegian waters (1992–1995) and Barents Sea (1992–1997). *Sci. Total Environ.* **1997**, *202*, 249–266. [CrossRef]
16. Bailly du Bois, P.; Guéguéniat, P.; Gandon, R.; Leon, R.; Baron, Y. Percentage contribution of inputs from the Atlantic, Irish sea, English channel and baltic into the north sea during 1988: A Tracer-Based evaluation using artificial radionuclides. *Neth. J. Sea Res.* **1993**, *31*, 1–17. [CrossRef]
17. Bailly du Bois, P.; Salomon, J.; Gandon, R.; Guéguéniat, P. A quantitative estimate of English channel water fluxes into the north sea from 1987 to 1992 based on radiotracer distribution. *J. Mar. Syst.* **1995**, *6*, 457–481. [CrossRef]

18. Bailly du Bois, P.; Rozet, M.; Thoral, K.; Salomon, J.C. *Improving Knowledge of Water-Mass Circulation in the English Channel Using Radioactive Tracers, Proceedings of the Radioprotection-Colloques, April 1997, Numéro spécial "Radionuclides in the Oceans", RADOC 96–97, Proceedings Part 1 "Inventories, Behaviour and Processes", Cherbourg-Octeville, France, 7–11 October 1996*; Germain, P., Guary, J.C., Guéguéniat, P., Métivier, H., Eds.; Les Editions de Physique: Les Ulis, France, 1997; Volume 32, pp. 63–69.

19. Bailly du Bois, P.; Guéguéniat, P. Quantitative assessment of dissolved radiotracers in the English channel: Sources, average impact of la Hague reprocessing plant and conservative behaviour (1983, 1986, 1988, 1994). *Cont. Shelf Res.* **1999**, *19*, 1977–2002. [CrossRef]

20. Guéguéniat, P.; Bailly du Bois, P.; Salomon, J.; Masson, M.; Cabioch, L. Fluxmanche radiotracers measurements: A contribution to the dynamics of the English channel and north sea. *J. Mar. Syst.* **1995**, *6*, 483–494. [CrossRef]

21. Salomon, J.C.; Guegueniat, P.; Breton, M. *Mathematical Model of 125Sb Transport and Dispersion in the Channel, Proceedings of the Radionuclides in the Study of Marine Processes, Norwich, UK, 10–13 September 1991*; Kershaw, J.P., Woodhead, D.S., Eds.; Elsevier Applied Science: London, UK, 1991; pp. 74–83.

22. Salomon, J.C.; Breton, M.; Fraizier, A.; Bailly du Bois, P.; Guéguéniat, P. A Semi-Analytic Mathematical Model for Dissolved Radionuclides Dispersion in the Channel Isles Region. *Radioprotect. Colloq.* **1997**, *32*, 375–380.

23. Bailly du Bois, P.; Dumas, F. Fast hydrodynamic model for medium-and Long-Term dispersion in seawater in the English channel and southern north sea, qualitative and quantitative validation by radionuclide tracers. *Ocean Model.* **2005**, *9*, 169–210. [CrossRef]

24. Bailly du Bois, P.; Dumas, F.; Solier, L.; Voiseux, C. In-Situ database toolbox for Short-Term dispersion model validation in Macro-Tidal seas, application for 2D-Model. *Cont. Shelf Res.* **2012**, *36*, 63–82. [CrossRef]

25. Oms, P.-E. Transferts Multi-Échelles des Apports Continentaux dans le Golfe de Gascogne. Ph.D. Thesis, Université de Bretagne Occidentale, Brest, France, June 2019.

26. Bailly du Bois, P.; Garreau, P.; Laguionie, P.; Korsakissok, I. Comparison between modelling and measurement of marine dispersion, environmental Half-Time and 137Cs inventories after the Fukushima Daiichi accident. *Ocean Dyn.* **2014**, *64*, 361–383. [CrossRef]

27. Garreau, P.; Bailly du Bois, P. Transportation of radionuclides in celtic sea a possible mechanisms. *Radioprotect. Colloq.* **1997**, *32*, 381–385

28. Perianez, R. Modelling the tidal dispersion of 137Cs and 239240Pu in the English channel. *J. Environ. Radioact.* **2000**, *49*, 259–277.

29. IAEA. MARIS. Available online: https://maris.iaea.org/Home.aspx (accessed on 23 September 2019).

30. Boyer, T.P.; Antonov, J.I.; Baranova, O.K.; Garcia, H.; Grodsky, A.; Johnson, D.; Locarnini, R.A.; Mishonov, A.; O'Brien, T.; Paver, C.; et al. *World Ocean Database 2013*; NOAA: Silver Spring, MD, USA, 2013; p. 15.

31. NOAA. World Ocean Database. Available online: https://www.nodc.noaa.gov/OC5/WOD/pr_wod.html (accessed on 23 September 2019).

32. Ifremer. SISMER. Available online: https://data.ifremer.fr/SISMER (accessed on 23 September 2019).

33. British Oceanographic Data Centre (BODC). Available online: https://www.bodc.ac.uk/ (accessed on 23 September 2019).

34. Gandon, R.; Bailly du Bois, P.; Baron, E.Y. Caractère conservatif de l'antimoine 125 dans les eaux marines soumises à l'influence des rejets de l'usine de retraitement des combustibles irradiés de La Hague. *Radioprotection* **1998**, *33*, 457–482. [CrossRef]

35. Kautsky, H. *Artificial Radioactivity in the North Sea and the Northern North ATLANTIC during the Years 1977 to 1986*; INIS-MF—11049; International Atomic Energy Agency (IAEA): Hamburg, Germany, 1988.

36. Gandon, R.; Guéguéniat, P. Preconcentration 125Sb of onto MnO_2 from Seawater Samples for Gamma-Ray Spectrometric Analysis. *Radiochim. Acta* **1992**, *57*, 159–164. [CrossRef]

37. Gandon, R.; Boust, D.; Bedue, O. Ruthenium complexes originating from the purex process: Coprecipitation with copper ferrocyanides via ruthenocyanide formation. *Radiochim. Acta* **1993**, *61*, 41–45.

38. Bailly du Bois, P.; Pouderoux, B.; Dumas, F. System for High-Frequency simultaneous water sampling at several depths during sailing. *Ocean Eng.* **2014**, *91*, 281–289. [CrossRef]

39. McCubbin, D.; Leonard, K.S.; Bailey, T.A.; Williams, J.; Tossell, P. Incorporation of organic tritium (3H) by marine organisms and sediment in the severn estuary/Bristol channel (UK). *Mar. Pollut. Bull.* **2001**, *42*, 2852–2863.

40. Ostlund, H.G.; Dorsey, H.G. *Rapid Electrolytic Enrichment and Hydrogen Gas Proportional Counting of Tritium*; Slovenske Pedagogicke Nakladatelstvo: Bratislava, Czechoslovakia, 1977.

41. Clarke, W.; Jenkins, W.; Top, Z. Determination of tritium by mass spectrometric measurement of 3He. *Int. J. Appl. Radiat. Isot.* **1976**, *27*, 515–522. [CrossRef]

42. Bailly du Bois, P. *IRSN Measurements of Dissolved Radioactivity in Seawater, 1982–2016, 39642 Stations*; PANGAEA Data Publisher: Bremerhaven, Germany, 2011; Available online: https://doi.pangaea.de/10.1594/PANGAEA.906541 (accessed on 2 October 2019).

43. Bailly du Bois, P.; Dumas, F.; Solier, L.; Voiseux, C. *Tritium Sampled and Measured in Surface Water Along Cruise Tracks*; PANGAEA Data Publisher: Bremerhaven, Germany, 2011; Available online: https://doi.pangaea.de/10.1594/PANGAEA.762261 (accessed on 2 October 2019).

44. Bailly du Bois, P.; Dumas, F.; Solier, L.; Voiseux, C. *Controlled Tritium Liquid Releases from Areva-NC Reprocessing Plant*; PANGAEA Data Publisher: Bremerhaven, Germany, 2011; Available online: https://doi.pangaea.de/10.1594/PANGAEA.762428 (accessed on 2 October 2019).

45. Gray, J.; Jones, S.R.; Smith, A.D. Discharges to the environment from the Sellafield site, 1951–1992. *J. Radiol. Prot.* **1995**, *15*, 99–131. [CrossRef]

46. RIFE. *RIFE-23 Radioactivity in Food and the Environment, 2017*; Environment Agency; Food Standards Agency; Food Standards Scotland; Natural Resources Wales; Northern Ireland Environment Agency; Scottish Environment Protection Agency Preston: Lancashire, UK, 2018; p. 260.

47. Sellafield, L. *Monitoring Our Environment, Discharges and Environmental Monitoring Annual Report 2017*; Sellafield Ltd.: Sellafield, UK, 2018; p. 260.

48. Bailly du Bois, P. *IRSN Measurements of Dissolved Radioactivity in Seawater, 1982–2016, Including Liquid Releases Database Issued from British and French Nuclear Plants*; PANGAEA Data Publisher: Bremerhaven, Germany, 2011; Available online: https://doi.pangaea.de/10.1594/PANGAEA.906749 (accessed on 2 October 2019).

49. Lazure, P.; Dumas, F. An External–Internal mode coupling for a 3D hydrodynamical model for applications at regional scale (MARS). *Adv. Water Resour.* **2008**, *31*, 233–250. [CrossRef]

50. Bailly du Bois, P. Automatic calculation of bathymetry for coastal hydrodynamic models. *Comput. Geosci.* **2011**, *37*, 1303–1310. [CrossRef]

51. Salomon, J.C.; Breton, M. An atlas of Long-Term currents in the Channel. *Oceanol. Acta* **1993**, *16*, 439–448.

52. Salomon, J.; Breton, M.; Guéguéniat, P. A 2D long term Advection—Dispersion model for the Channel and southern North Sea Part B: Transit time and transfer function from Cap de La Hague. *J. Mar. Syst.* **1995**, *6*, 515–527. [CrossRef]

53. Orbi, A.; Salomon, J.C. Dynamique de marée dans le Golfe Normand-Breton. *Oceanol. Acta* **1988**, *11*, 55–64.

54. Bailly du Bois, P.; Morillon, M.; Solier, L. *Mesure et Modélisation de la Dispersion Verticale dans le raz Blanchard (Projet DISVER)*; IRSN/Prp-Env/SERIS2014-21; IRSN: Cadarache, France, 2014; p. 56.

55. Fiévet, B.; Bailly du Bois, P.; Laguionie, P.; Morillon, M.; Arnaud, M.; Cunin, P. A dual pathways transfer model to account for changes in the radioactive caesium level in demersal and pelagic fish after the Fukushima Daï-ichi nuclear power plant accident. *PLoS ONE* **2017**, *12*, e0172442. [CrossRef]

56. Matheron, G. Les Concepts de base et L'Evolution de la geostatistique miniere. *Adv. Geostat. Min. Ind.* **1976**, 3–10. [CrossRef]

57. Bailly du Bois, P.; Laguionie, P.; Boust, D.; Korsakissok, I.; Didier, D.; Fiévet, B. Estimation of marine Source-Term following Fukushima Dai-Ichi accident. *J. Environ. Radioact.* **2012**, *114*, 2–9. [CrossRef]

58. Bailly du Bois, P.; Dumas, F.; Morillon, M.; Furgerot, L.; Voiseux, C.; Poizot, E.; Méar, Y.; Bennis, A.C. The Alderney Race: General hydrodynamic and particular features. Philosophical Transactions of the Royal Society of London A, accepted March 2020. (in press)

59. Bailly du Bois, P.; Dumas, F. Dissolved radionuclide Measurements Used for Qualitative and Quantitative Calibration of Hydrodynamic Models in the English Channel and the North Sea; Validation of "TRANSMER" Model. In Proceedings of the 34th International Liege Colloquium on Ocean Hydrodynamics, Tracer Methods in Geophysical Fluid Dynamics, Liege, Belgium, 6–10 May 2002; p. 7.

60. Bailly du Bois, P. *Dispersion des Radionucléides dans les mers du Nord-Ouest de l'Europe: Observations et Modélisation, Collection HDR ed.*; IRSN: BP 17-92262; IRSN: Fontenay-aux-Roses, France, 2014; p. 329.

61. Oms, P.E.; Bailly du Bois, P.; Dumas, F.; Lazure, P.; Morillon, M.; Voiseux, C.; Le Corre, C.; Solier, L.; Maire, D.; Caillaud, M. Tritium as an original continental runoffs tracer in the bay of biscay: Measurements and modelling. In Proceedings of the ISOBAY, Anglet, France, 7 June 2018.

62. Hunt, J.; Leonard, K.; Hughes, L. Artificial radionuclides in the Irish Sea from sellafield: Remobilisation revisited. *J. Radiol. Prot.* **2013**, *33*, 261–279. [CrossRef]

63. Oms, P.-E.; Bailly du Bois, P.; Dumas, F.; Lazure, P.; Morillon, M.; Voiseux, C.; Le Corre, C.; Cossonnet, C.; Solier, L.; Morin, P. Inventory and distribution of tritium in the oceans in 2016. *Sci. Total Environ.* **2019**, *656*, 1289–1303. [CrossRef]

64. Charette, M.A.; Breier, C.F.; Henderson, P.B.; Pike, S.M.; Rypina, I.I.; Jayne, S.R.; Buesseler, K.O. Radium-Based estimates of cesium isotope transport and total direct ocean discharges from the fukushima nuclear power plant accident. *Biogeosciences* **2013**, *10*, 2159–2167. [CrossRef]

65. Buesseler, K. Fukushima and Ocean Radioactivity. *Oceanography* **2014**, *27*, 92–105.

66. Aoyama, M.; Hamajima, Y.; Inomata, Y.; Kumamoto, Y.; Oka, E.; Tsubono, T.; Tsumune, D. Radiocaesium derived from the TEPCO Fukushima accident in the North Pacific Ocean: Surface transport processes until 2017. *J. Environ. Radioact.* **2018**, *189*, 93–102. [CrossRef]

67. Salomon, J.C.; Guéguéniat, P.; Orbi, A.; Baron, Y. *A Lagrangian Model for Long Term Tidally Induced Transport and Mixing, Verification by Artificial Radionuclide Concentrations, Proceedings of the Radionucléides: A Tool for Oceanography, Cherbourg, France, 1–5 June 1987*; Guary, J.C., Guéguéniat, P., Pentreath, R.J., Eds.; Elsevier Applied Science: London, UK, 1988; pp. 384–394.

68. Simonsen, M.; Saetra, Ø.; Isachsen, P.E.; Lind, O.C.; Skjerdal, H.K.; Salbu, B.; Heldal, H.E.; Gwynn, J.P. The impact of tidal and mesoscale eddy advection on the long term dispersion of 99 Tc from Sellafield. *J. Environ. Radioact.* **2017**, *177*, 100–112. [CrossRef]

69. Alfimov, V.; Aldahan, A.; Possnert, G. Water masses and 129I distribution in the Nordic Seas. *Nucl. Instru. Methods Phys. Res. Sect. Beam Interact. Mater. Atoms* **2013**, *294*, 542–546. [CrossRef]

70. Daraoui, A.; Tosch, L.; Michel, R.; Goroncy, I.; Nies, H.; Synal, H.-A.; Alfimov, V.; Walther, C.; Gorny, M.; Herrmann, J. Iodine-129, Iodine-127 and Cesium-137 in seawater from the North Sea and the Baltic Sea. *J. Environ. Radioact.* **2016**, *162*, 289–299. [CrossRef]

71. Christl, M.; Casacuberta, N.; Lachner, J.; Herrmann, J.; Synal, H.-A. Anthropogenic 236U in the North Sea–A Closer Look into a Source Region. *Environ. Sci. Technol.* **2017**, *51*, 12146–12153. [CrossRef]

72. Castrillejo, M.; Casacuberta, N.; Christl, M.; Vockenhuber, C.; Synal, H.-A.; García-Ibáñez, M.I.; Lherminier, P.; Sarthou, G.; Garcia-Orellana, J.; El Zrelli, R. Tracing water masses with 129I and 236U in the subpolar North Atlantic along the GEOTRACES GA01 section. *Biogeosc. Discuss.* **2018**, *2018*, 1–28. [CrossRef]

73. Christl, M.; Casacuberta, N.; Vockenhuber, C.; Elsässer, C.; Bailly du Bois, P.; Herrmann, J.; Synal, H.-A. Reconstruction of the 236U input function for the northeast Atlantic Ocean—Implications for 129I/236U and 236U/238U-based tracer ages. *J. Geophys. Res. Ocean.* **2015**. [CrossRef]

74. Bennis, A.C.; Furgerot, L.; Bailly du Bois, P.; Dumas, F.; Odaka, T.; Lathuilière, C.; Filipot, J.F. Numerical modelling of Three-Dimensional Wave-Current interactions in extreme hydrodynamic conditions: Application to Alderney Race. *Appl. Ocean. Res.* **2019**, *95*, 102021.

75. Debreu, L.; Vouland, C.; Blayo, E. AGRIF: Adaptive grid refinement in Fortran. *Comput. Geosci.* **2008**, *34*, 8–13. [CrossRef]

76. Science-Council-of-Japan. *A Review of the Model Comparison of Transportation and Deposition of Radioactive Materials Released to the Environment as a Result of the Tokyo Electric Power Company's Fukushima Daiichi Nuclear Power Plant Accident*; Science Council of Japan: Tokyo, Japan, 2014; p. 111.

77. Usman, S.M.; Jocher, G.; Dye, S.; McDonough, W.F.; Learned, J. AGM2015: Antineutrino Global Map 2015. *Sci. Rep.* **2015**, *5*, 13945. [CrossRef]

78. Duffa, C.; Bailly du Bois, P.; Caillaud, M.; Charmasson, S.; Couvez, C.; Didier, D.; Dumas, F.; Fiévet, B.; Morillon, M.; Renaud, P.; et al. Development of emergency response tools for accidental radiological contamination of French coastal areas. *J. Environ. Radioact.* **2016**, *151*, 487–494. [CrossRef]

79. Fiévet, B.; Plet, D. Estimating biological Half-Lifes of radionuclides in marine compartments from environmental time-series measurements. *J. Environ. Radioact.* **2003**, *65*, 91–107. [CrossRef]

80. Fiévet, B.; Voiseux, C.; Rozet, M.; Masson, M.; Bailly du Bois, P. Transfer of radiocarbon liquid releases from the AREVA La Hague spent fuel reprocessing plant in the English Channel. *J. Environ. Radioact.* **2006**, *90*, 173–196. [CrossRef]

81. Fiévet, B.; Pommier, J.; Voiseux, C.; Bailly du Bois, P.; Laguionie, P.; Cossonnet, C.; Solier, L. Transfer of tritium released into the marine environment by french nuclear facilities bordering the English channel. *Environ. Sci. Technol.* **2013**, *47*, 6696–6703. [CrossRef]

82. Blanpain, O.; Bailly du Bois, P.; Cugier, P.; Lafite, R.; Lunven, M.; Dupont, J.; Le Gall, E.; Legrand, J.; Pichavant, P. Dynamic sediment profile imaging (DySPI): A new field method for the study of dynamic processes at the sediment-water interface. *Limnol. Oceanogr. Methods* **2009**, *7*, 8–20. [CrossRef]

83. Blanpain, O. Dynamique Sédimentaire Multiclasse: De L'étude des Processus À La Modélisation en Manche. Ph.D. Thesis, Rouen University, Mont-Saint-Aignan, France, October 2009.

84. Rivier, A.; Le Hir, P.; Bailly du Bois, P.; Laguionie, P.; Morillon, M. Numerical modelling of heterogeneous sediment transport: New insights for particulate radionuclide transport and deposition. In Proceedings of the Coastal Dynamics, Helsingør, Denmark, 12–16 June 2017; pp. 1767–1778.

85. Bacon, G. *Etude du Transfert Entre L'eau et L'atmosphère D'un Rejet Marin de Tritium (HTO) en Zone Côtière. Etude Préliminaire*; Rapport 3ème Année Ecole D'ingénieur; CESI: Caen, France, 2011.

Article

Effects of Wave-Induced Processes in a Coupled Wave–Ocean Model on Particle Transport Simulations

Joanna Staneva [1,*], Marcel Ricker [1,2], Ruben Carrasco Alvarez [1,3], Øyvind Breivik [4,5] and Corinna Schrum [1]

[1]   Institute of Coastal Systems, Helmholtz-Zentrum Geesthacht, Max-Planck-Straße 1, 21502 Geesthacht, Germany; Marcel.Ricker@hzg.de (M.R.); corinna.schrum@hzg.de (C.S.)
[2]   Institute for Chemistry and Biology of the Marine Environment, University of Oldenburg, Carl-von-Ossietzky-Straße 9–11, 26111 Oldenburg, Germany
[3]   Institute of Coastal Ocean Dynamics, Helmholtz-Zentrum Geesthacht, Max-Planck-Straße 1, 21502 Geesthacht, Germany; Ruben.Carrasco@hzg.de
[4]   Norwegian Meteorological Institute, Henrik Mohns Plass 1, 0313 Oslo, Norway; oyvindb@met.no
[5]   Geophysical Institute, University of Bergen, Allegt. 70, 5007 Bergen, Norway
*   Correspondence: joanna.staneva@hzg.de; Tel.: +49-4152-87-1804

Abstract: This study investigates the effects of wind–wave processes in a coupled wave–ocean circulation model on Lagrangian transport simulations. Drifters deployed in the southern North Sea from May to June 2015 are used. The Eulerian currents are obtained by simulation from the coupled circulation model (NEMO) and the wave model (WAM), as well as a stand-alone NEMO circulation model. The wave–current interaction processes are the momentum and energy sea state dependent fluxes, wave-induced mixing and Stokes–Coriolis forcing. The Lagrangian transport model sensitivity to these wave-induced processes in NEMO is quantified using a particle drift model. Wind waves act as a reservoir for energy and momentum. In the coupled wave–ocean circulation model, the momentum that is transferred into the ocean model is considered as a fraction of the total flux that goes directly to the currents plus the momentum lost from wave dissipation. Additional sensitivity studies are performed to assess the potential contribution of windage on the Lagrangian model performance. Wave-induced drift is found to significantly affect the particle transport in the upper ocean. The skill of particle transport simulations depends on wave–ocean circulation interaction processes. The model simulations were assessed using drifter and high-frequency (HF) radar observations. The analysis of the model reveals that Eulerian currents produced by introducing wave-induced parameterization into the ocean model are essential for improving particle transport simulations. The results show that coupled wave–circulation models may improve transport simulations of marine litter, oil spills, larval drift or transport of biological materials.

Keywords: Lagrangian transport modelling; coupled wave–ocean models; ocean drifters; wave-induced processes; model skills

Citation: Staneva, J.; Ricker, M.; Carrasco Alvarez, R.; Breivik, Ø.; Schrum, C. Effects of Wave-Induced Processes in a Coupled Wave–Ocean Model on Particle Transport Simulations. *Water* **2021**, *13*, 415. https://doi.org/10.3390/w13040415

Academic Editor: Inga Monika Koszalka

Received: 21 July 2020
Accepted: 25 January 2021
Published: 5 February 2021

Publisher's Note: MDPI stays neutral with regard to jurisdictional claims in published maps and institutional affiliations.

1. Introduction

A rapid increase in marine litter in the ocean has recently been recognized as a serious environmental problem. The role of the physical factors contributing to it (e.g., atmosphere–ocean–wave interaction processes) has been not yet fully understood. Lagrangian analyses represent the natural approach to studying oceanic transport based on model simulations and observational data [1,2]. The transport and accumulation of floating marine debris and the assessment of different scenarios of marine plastic distribution, along with a comprehensive synopsis of the tools currently available for tracking virtual particles. A detailed review of the Lagrangian ocean analysis, associated problems, sources of errors and validation issues was presented in [2,3]. Still, an improved understanding of the physical processes influencing the transport of particles is required [2]. The Lagrangian simulation can be assessed by performing a time-evolving analysis of the separation distance between

67

the real track and the simulated ones. A skill score, based on the separation distance normalized by the length of the trajectory, has recently been proposed [4]. The separation distance between model simulations and observed trajectories has been estimated, showing that one day after initialization the distance was about 15–25 km, and five days later this increased to about 60–180 km. It was found that the root-mean-square error (RMSE) 13 days after the start of the integration was about 5 km [5]. A separation distance from model simulations versus observed trajectories in the first days after initialization of about 15 km was found [6].

Sea-state dependent processes affect the ocean circulation and thus also the results from Lagrangian transport models. A considerably enhanced momentum transfer from the atmosphere to the wave field is found [7] during growing sea state (young sea). Recently, a wind stress formulation depending on wind stress and wind–wave momentum released to the ocean was proposed [8]. Swell waves can even cause momentum transfer from the ocean to the atmosphere [9]. In growing sea states, waves extract momentum from the atmosphere, so that the ocean receives less momentum from the atmosphere than if waves were not considered [10]. Using stand-alone ocean or atmosphere models, the surface waves that represent the air–sea interface are not taken into account. This can cause biases in the upper ocean due to insufficient or, in some cases, too strong mixing [11], or because the momentum transfer is shifted in time and space compared to how the fluxes would behave in the presence of waves. Several parameterizations were recently proposed for momentum flux that is sea state-dependent, e.g., [12,13]. Recently, the role of the Stokes drift and wave-induced transport of floating marine litter was studied in [14], showing that accounting for the wave-induced Eulerian-mean flow significantly alters predictions of transport of floating marine litter by waves. The skill scores between the model and observations were improved by adding the Stokes drift in [15], postulating that for Lagrangian simulations, the Stokes drift's contribution can be at the same order or even higher than the accuracy of the Eulerian circulation. However, the accuracy of Lagrangian simulations also depends on the hydrodynamic and Lagrangian model, demonstrating the need to tune Lagrangian models for the specific setup [16].

In a large and complex system such as the North Sea (Figure 1), minor perturbations may displace particles to very different drift regimes, causing strongly divergent particle trajectories [17]. Therefore, the particle distribution analysis is not intended to fully reproduce or explain the observations (the latter are extremely limited, particularly during extreme weather conditions). Instead, we aim to indicate the differences that can be due to wave-induced forcing processes. Although tide and wind-driven circulation in the North Sea seem to have been widely studied in the past, further research from a Lagrangian perspective with respect to coupling with waves and biogeochemical processes may be particularly relevant.

The importance of wave forcing for ocean circulation and sea-level predictions has been demonstrated [18–23], showing that the predictive skill of ocean circulation and sea level could be significantly enhanced by considering wave-induced processes. In extreme storm surge conditions over the North Sea, due to the strong non-linearity of wave–ocean–tidal interactions, wave–ocean coupling is considered to be significant for correct model predictions [18,24]. The impact of Stokes-related drift effects (Stokes–Coriolis forcing and Stokes drift advection on tracers and mass) were studied on the North Sea and Baltic Sea regions [25]. For tracer distribution and upwelling, the direct sea-state dependent momentum and energy fluxes are of higher importance than the Stokes drift processes implemented in the NEMO circulation model [23].

In this study, we investigate the role of wave-induced interaction processes in a fully coupled ocean–wind wave model system on Lagrangian transport modelling. Data from drifters deployed in the southern North Sea were used to assess the particle model simulations [26]. Lagrangian simulations were studied [27] based on the same drifter observations as in the present study and corresponding Lagrangian model simulations by using two circulation models. While only a stand-alone ocean model was used in the previous study [28], here, the same Ocean General Circulation Model (OGCM) is coupled to a wave model. The observational drifters [26] used to assess our simulations have previously been described in detail [27,29].

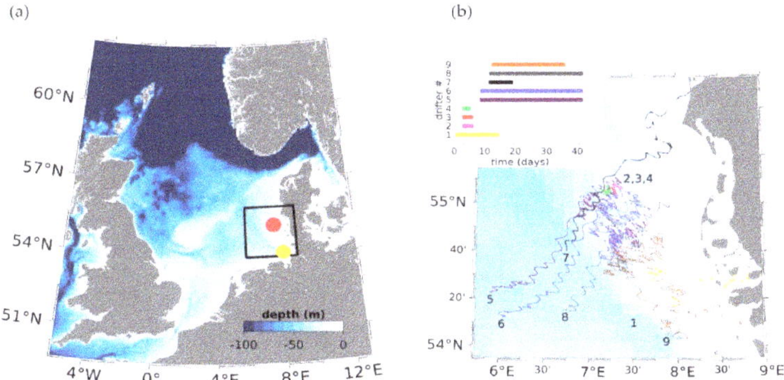

Figure 1. (**a**) The North Sea (depth in m) topography as used for the model simulations and buoy locations (red circle: Fino3, yellow circle: Elbe). (**b**) Magnification of the German Bight showing the HZG drifter trajectories released at the location of their respective numbers. Deployment periods of the HZG drifters are shown on the top left corner.

This paper is organized as follows. In Section 2, we describe the circulation, wave and Lagrangian transport models and the experimental setup. The evaluation of the model simulations is described in Section 3. Further, we assess the model results with direct comparisons of the simulations with the drifter data (Section 4). The drifter trajectories were modelled to investigate the importance of the wave effects, e.g., for search-and-rescue applications. We performed sensitivity experiments to investigate the impact of wave-induced forcing in the ocean model. The Lagrangian model used only the Eulerian current as provided by the hydrodynamical model simulations, by neglecting or considering the contributions from the Stokes drift or the wind drift corrections. The wave effects on the general circulation in the North Sea are studied. The paper ends with a discussion of the findings in Section 5 and the conclusions in Section 6.

2. Methods

2.1. Models and Set-Up

The Geesthacht coupled coastal model system (GCOAST) [30,31] was built upon a flexible and comprehensive coupled model system, integrating the most important key components of regional and coastal models. GCOAST encompasses (i) atmosphere-ocean–wave interactions, (ii) dynamics and fluxes in the land–sea transition, and (iii) coupling of the marine hydrosphere and biosphere. In our study, we used the GCOAST circulation, wave and drift model components to investigate the role of coupling in particle transport simulations in the North Sea. Those particles can be considered, for example, as simple representations of either oil fractions, fish larvae or search-and-rescue objects [32,33]. The wave–current interaction processes are momentum and energy sea state dependent fluxes, wave-induced mixing and Stokes–Coriolis forcing.

2.1.1. The Circulation Model NEMO

NEMO (Nucleus for European Modelling of the Ocean, [34]) is a framework of ocean-related computing engines, from which we use the OPA (Océan Parallélisé) package (for the ocean dynamics and thermodynamics) and the LIM3 (Louvain-la-Neuve Sea Ice Model) sea-ice dynamics and thermodynamics package [34]. In OPA, six primitive equations (momentum balance, the hydrostatic equilibrium, the incompressibility equation, the heat and salt conservation equations and an equation of state) are solved, where the

Arakawa C grid is used in the horizontal. In the vertical, terrain-following coordinates, z coordinates or hybrid z-s coordinates can be chosen. Previously, NEMO was applied to the Baltic Sea and the North Sea area in uncoupled mode [35], coupled to atmospheric models [36] and forced with a wave model [18,19,25]. For the northwestern European Shelf, NEMO is used as a forecasting model in the COPERNICUS Marine Services [23,37,38]. The horizontal model resolution is about 2 nm with 51 σ-levels in the vertical providing instantaneous hourly surface (0.6 m uppermost level thickness) velocity fields. The study domain is 48.0–62.5° N and −4.7–13.2° E, which includes the North Sea, Skagerrak and Kattegat (Figure 1). The hourly atmospheric forcing is taken from subsequent short-range forecasts from the regional atmospheric model COSMO-EU, operated by the German Weather Service (DWD). Atmospheric pressure and tidal potential are included in the model forcing. River run-off is provided in the form of a daily climatology based on river discharge datasets. Lateral open boundary and initial condition fields (temperature, salinity, velocities and sea level) are derived from the MetOffice Forecasting Ocean Assimilation Model (FOAM) AMM7 (7 km horizontal resolution [23]), currently used by the Copernicus Marine Environment and Monitoring Service (CMEMS) as an operational service.

2.1.2. The Wave Model WAM

The wave model WAM [39,40] is a third-generation wave model, which solves the action balance equation without any a priori restriction on the evolution of spectrum. It is based on the spectral description of the wave conditions in frequency and directional space at each of the active model sea grid points of a certain model area. The version used in this study is the WAM Cycle 4.7, which is described in [41–43]. The source function integration scheme is made by [44], and the updated source terms of [45] are incorporated (Appendix A). The new version considers the wave-induced processes needed for coupling and are described in the next section. The spectrum in WAM is discretized with 24 directions and 25 frequencies. The wave model boundary information used at the open boundaries is taken from the regional wave model EWAM for Europe, which is run twice daily in operational routine at DWD.

2.1.3. Wave-Induced Processes

Ocean waves influence the circulation through a number of processes: turbulence due to breaking and non-breaking waves, momentum transfer from breaking waves to currents in deep and shallow water, wave interaction with planetary and local vorticity, Langmuir turbulence. The NEMO ocean model has been modified to take into account the following wave effects [11,18]: (1) the Stokes–Coriolis forcing; (2) sea-state dependent momentum flux, set as a scalar dependence of the flux from the atmosphere to waves and ocean or as a vector; and (3) a sea-state dependent energy flux. A description of the wave-induced forcing and the processes of wave interaction with the ocean circulation is given in Appendix A.1.

2.1.4. The Lagrangian Model

OpenDrift [46] is a freely available open-source off-line Lagrangian particle trajectory model that contains several modules for the advection of, e.g., oil spills [47], larvae and passive tracers [48]. For this study, the passive tracer module is used, which advects tracers only due to currents and winds. The sea and wind drift input is described in the next section. To investigate the influence of the wind, three different windage (referred to as leeway) coefficients when the bulk effect of waves and wind on the object is considered [32] (L_w) are used (0.1%, 0.5% and 1.0%). The real wind drag of the drifters is not known, so these values are used as estimates. The windage coefficients are multiplied by the wind velocity and added to the current velocity. The advection scheme is a 2nd-order Runge-Kutta scheme, and no additional diffusion is added because we only want to investigate the wave effects without any additional random "disturbances" [49]. Each particle thus follows a trajectory influenced by the sea surface currents and the equations read:

$$\Delta x(t) = \left(u\left(t + \frac{\Delta t}{2} \right) + L_w u_{wind}\left(t + \frac{\Delta t}{2} \right) \right) \Delta t \qquad (1)$$

$$\Delta y(t) = \left(v\left(t + \frac{\Delta t}{2} \right) + L_w v_{wind}\left(t + \frac{\Delta t}{2} \right) \right) \Delta t \qquad (2)$$

here $\Delta x(t)$, $\Delta y(t)$ are the particle displacements and $u(t + \Delta t/2)$, $v(t + \Delta t/2)$ are the horizontal velocity components at the particle's position at $t + \Delta t/2$. If a particle reaches land, it is not further advected and considered beached. These particles are not further taken into account. If a particle leaves the domain through an open boundary, e.g., from the North Sea into the Atlantic, this particle is treated the same way. The model time step Δt and the particle seeding strategies are dependent on the specific experiment (provided in Section 3).

2.2. Observational Data

2.2.1. Drifter Data

The HE 445 cruise was performed between May and July 2015 (see Figure 1b for the trajectories and deployment period of the drifters) [26]. The R/V Heincke deployed nine Albatros drifters corresponding to two models (Figure S1): MD03i (drifters 1–6) and ODi (drifters 7–9). The drifters provided their current positions by a Global Positioning System (GPS), which were transmitted to the R/V Heincke via Iridium (a bi-directional satellite communication network). The MD03i is a cylinder-shaped drifter, with a diameter of 0.1 m and length of 0.32 m, but only approximately 0.08 m remains above the water surface. The MD03 drag ratio is 33.2, and according to the parametrization [26] the MD03 slippage is around 1.1 to 1.6 cm/s, for 10 m/s wind speed and velocity difference across the drogue (ΔU) equal to 0.1 cm/s. The ODi is a spherical drifter with a diameter of 0.2 m, but only approximately 0.1 m is above the water surface. A sail (0.5 m in length and diameter, Figure S1) was attached to every drifter to enhance the drag below the water surface, and it was 0.5 m below the surface. Due to the small drifter surface above the water compared to the sail surface below the water, the drifter is designed to follow the ambient current in the upper meter of the water column. Due to the meteorological conditions, only some of the drifters were recovered at the end of the campaign. The recovered drifters corresponded to the short data-set, while the non-recovered drifters from the long data-set transmitted data until the batteries drained (Figure 1b).

2.2.2. HF Radar Data

HFR surface current data were acquired by three radar stations in the German Bight [50,51]. The radar systems are based on linear antenna arrays installed near the shoreline. More details on the required processing can be found in [52]. For the considered system, the spatial radar resolution is 1.5 km in range. The radar system in the German Bight can reach out to about 120 km off the coast in favorable conditions and provides measurements with a 9 min averaging window every 20 min. The observations are available as interpolated to a 2 km Cartesian grid. Through a combination of the radial components from the different antenna stations, meridional and zonal current components can be derived; however, the original radial components were used for the subsequent data assimilation procedure. Further details of the system can be found in [51].

2.3. Model Experiments

For the control experiment (REF), the velocity is taken from the stand-alone circulation model NEMO (Section 2.1), and the wave–current interaction processes are not included. In the coupled wave-ocean experiments (CPL), the wave-induced processes described in Section 2.1.3 calculated by the wave model WAM are introduced in NEMO to simulate the Eulerian velocity, needed for the OpenDrift. We performed the following experiments, in which the individual or combined effects of the wave-induced processes are included: (i) sea state dependent momentum flux (CPL-TAUOC); (ii) Stokes–Coriolis forcing (CPL-STCOR) and (iii) both the sea state dependent momentum flux and the Stokes–Coriolis (CPL-TAUST).

In order to study the role of the windage, two additional sets of experiments have been performed. For these, in addition to the Eulerian velocity from REF and CPL experiments, the windage is included in OpenDrift. We will name these experiments WD-REF and WD-CPL, respectively. The different windage contributions that we consider are 0.1, 0.5 and 1% (WD-REF and WD-CPL with _0.1, _0.5, and _1.0, respectively). All experiments are summarized in Table 1.

Table 1. List of experiments.

Experiment	REF	CPL	WD-Ref	WD-CPL
NEMO-only	Yes	No	Yes	
NEMO-WAM	No	Yes	No	
Windage	No	No	0.1%/0.5%/1%	0.1%/0.5%/1%

3. Evaluation of Model Simulations

3.1. Methodology

The evaluation of the model results consisted of two parts. The first part (Section 3.2) aimed at studying the model runs statistically with direct comparisons of the simulations with the drifter data. In the follow-up part (Section 5), the drifter trajectories were modelled with OpenDrift to investigate the importance of the wave effects, e.g., for search-and-rescue applications.

The drifter velocities along the trajectories are calculated by dividing the distance between each drifter position by its time difference. The velocity is located in the middle between the two positions in time and space. Afterwards, the model velocities are interpolated trilinearly (lat, lon, time) to the drifter velocity positions and times. The root-mean-square error (*rmse*), standard deviation (*std*), bias (*bia*) and linear correlation coefficient (*cor*) are calculated to assess the model and drifter currents. Trilinear interpolation to the drifter position is performed for the wind and water depth and for each model experiment. For the analysis of meteorological conditions, periods with weak winds (25–27 June 2015) and strong southerly winds (01–03 June 2015) were chosen. In order to assign the errors of model velocities to different sources, a multi-linear regression was performed, solving the following equations:

$$u_{drifter} = u_{model}a_{1i} + u_{wind}b_{1i} + c_{1i},$$
$$v_{drifter} = v_{model}a_{2i} + v_{wind}b_{2i} + c_{2i}, \tag{3}$$

Here, u_{model} is the model velocity, u_{wind} is the wind velocity, $U = \sqrt{u^2 + v^2}$ denotes the velocity magnitude, i corresponds to the drifter number and a, b, c are the coefficients. Coefficient b represents direct windage acting on the GPS drifters. The coefficient a can indicate if the model under- or over-estimates surface velocity. Coefficient c includes deviations that cannot be explained by the model currents or the wind, like inaccurate directions; it can be considered an "offset".

The drifter starting positions are used as initial positions in OpenDrift. In the trajectory simulations, the particles are initialized at the start location and start time of the Albatros drifters and move under the influence of the forcing until the real drifters expire. Drifter simulations of 25 h drift paths that were initialized every day from 0 to 53 at 13:00 UTC were previously performed [27]. It is important to stress that inall of our experiments, after starting the drifter simulations, we did not re-initialize the modelled drifters. The time steps in the OpenDrift experiments and the output frequency are both 10 min. In the first set of experiments, only the currents of REF and CPL were used in OpenDrift. To study the impact of the wind, REF and CPL current data were used together with the windage coefficients WD-REF and WD-CPL, respectively.

As a measure of the modelled trajectories, the skill score ss [4] is chosen. First, an index s is estimated as

$$s = \sum_{i=1}^{N} d_i \Big/ \sum_{i=1}^{N} l_{oi} \qquad (4)$$

where N is the total number of time steps, d_i is the distance of real and modelled drifter at time step i and l_{oi} is the total length of the trajectory of the real drifter at time step i. This index is used to calculate ss with

$$ss = \begin{cases} 1 - \frac{s}{n}, & (s \leq n) \\ 0, & (s > n) \end{cases} \qquad (5)$$

where n is the tolerance threshold, equal to 1 [4], which means that the cumulative separation distance is not larger than the cumulative trajectory length. If s and ss are close to 1, the modelled trajectory is close to the observed trajectory. It assumes that the model performance is better with higher ss and the cumulative separation distance is less significant than the cumulative trajectory length. If the skill score is close to zero, the model simulations have no skill.

The skill of the particle transport model depends on the quality of the ocean circulation, but the skill score has limitations [53]; e.g., in situations with small currents, it yields too small cumulative distances. It can thus give a too high estimate of s and a low skill score ss. To overcome these limitations, it has been suggested [4] to have an appropriate definition of the tolerance threshold n. In the German Bight study area, due to the strong tidal currents and wind forcing, the skill score can be an applicable measure for the drift model performance.

3.2. Surface Currents

3.2.1. HF-Radar versus Drifter Observations

Figure 2 shows the drifter velocity magnitude versus the HF radar velocity as scatter plots generated for the whole period of drifter deployment. For the study period, 334 collocations of radar and drifter data were available for comparison. The velocity data of the drifters were filtered using a Savitzky–Golay filter [54] with a quadratic fit and a window length of ~2.3 days in order to remove outliers. The velocity of all nine drifters (regarding their deployment period or quality) was used in the scatter diagram to demonstrate the statistical robustness of the method. A detailed analysis of the separate trajectories of the Albatros drifters for the deployment periods was reported [27].

The HF radar currents should be interpreted as Eulerian currents [55] (i.e., without including the Stokes drift). However, it has also been argued that the HF radar velocity partially contains the Stokes drift, in which case the HF radar measurements contain a filtered component of the Stokes drift [56]. On the other side, comparisons of HF radar velocity measurements with drifter observations demonstrated that the presence of the Stokes drift in HF measurements is not settled. In Figure 2 and Figure S2 we compare HF radar velocities interpreted as Eulerian currents against the velocity estimated by the nine Albatros drifters. With a standard deviation of 15 cm/s and a bias of 2.7 cm/s, the drifter and the HF radar are in reasonably good agreement. A noticeable feature is a slight underestimation of HF velocity at higher values. There are several reasons for the disagreement between HF and drifters when the speed is over 0.6 m/s: (i) the HF radar retrieves the currents at a depth of around 1 m, but the drifter's sails are located at 0.5 m. An underestimation of the HF radar regarding the drifter measurements may be expected. (ii) Due to strong wind stress during storm events, taking into account that the current profile over depth is logarithmic, a stronger current gradient between 1 m (HF) and 0.5 m (drifters) depth is expected. (iii) During breaking events in storm conditions, the drifter and the sail can surf over the sea surface, while the HF radar measures the current speed under the surface.

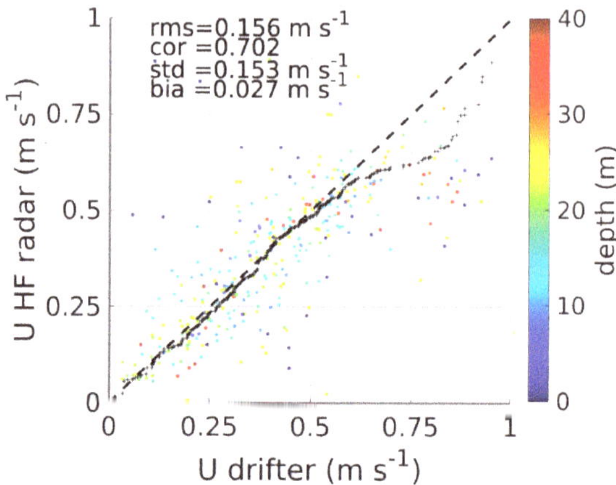

Figure 2. Velocity scatter plots: drifter versus HF radar velocity magnitude (m/s). The black dots indicate the quantile–quantile (*q-q*) plot. The black dashed line is the diagonal. The model topography is also interpolated to the drifter positions and depicted with colors.

3.2.2. Assessment of Model Velocity

The simulated model velocities of the stand-alone NEMO (REF) and wave-circulation coupled model (CPL) are compared with the matching drifter velocity in Figure 3a. REF and CPL model velocities are well represented in the scatter plots and show good agreement with the drifter data. Above 30 cm/s, the modelled velocities are lower than the drifter velocities, as seen in the zonal and meridional components. The range of the *q-q* plot is smoother and fits better in moderate and high velocities in the coupled model simulations (Figure 3). The comparisons demonstrate a better fit of the meridional velocity to the observations by the coupled wave–current model for strong winds in both directions (Figure S3). The Taylor diagram shows a good agreement between both simulations and observations with slightly improved skills of the CPL experiments. It is noteworthy that despite the coarse resolution of the model and the complex bathymetry and coastline in our study area, the comparison is satisfying, and a general improvement of the surface currents in both directions is observed due to the wind–wave–ocean coupling. This is consistent with previous findings [21] in which coupled and stand-alone model circulation against Acoustic Doppler Current Profiler (ADCP) demonstrated an intensification of velocities due to coupling with waves, leading to a reduction in prediction errors.

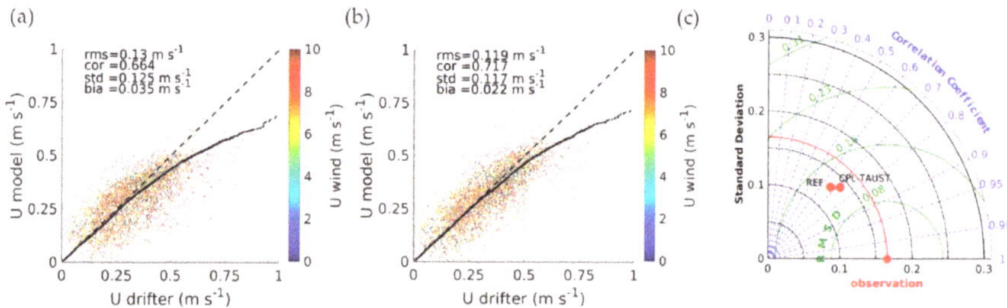

Figure 3. Scatter plots of velocity magnitude for the REF experiment (**a**) and CPL-TAUST experiment (**b**) of the HZG drifters vs. model data colored to wind velocity component and the quantile–quantile (*q-q*) plot in black. The black dashed line is the diagonal. Taylor Diagram of the velocity magnitude (**c**).

A multi-linear regression is performed, including model velocities, wind velocities and an offset to obtain insight into the error sources. Figure 4 shows these coefficients separately for each of the nine drifters. The model velocities are reduced in the zonal component by about 2.5% and in the meridional component by about 8%. While reducing the model velocities, the wind to be taken into account is around 1.0% for both velocity components. This value can also be interpreted as a guess about the wind drag of the drifters. For the same drifters, different windage and Stokes drift contributions were tested [27], in a combination of the Eulerian velocity of two hydrodynamic models. The parameterization used in [57] predicted a slippage of 1.1 to 1.6 cm/s for 10 m/s winds, and the expected windage should thus be close to 0.135%. However, as demonstrated in [58] this value depends strongly on the drifters and the chosen models and usually ranges between 0.1 and 1%. A wind drag of 0.27% was calculated from drifter surface ratio and 0.3% with Stokes drift was found to be the best combination in model simulations [49]. The offset of the zonal components is slightly negative (about -2 cm/s), while the offset of the meridional ones is slightly positive (about 2 cm/s).

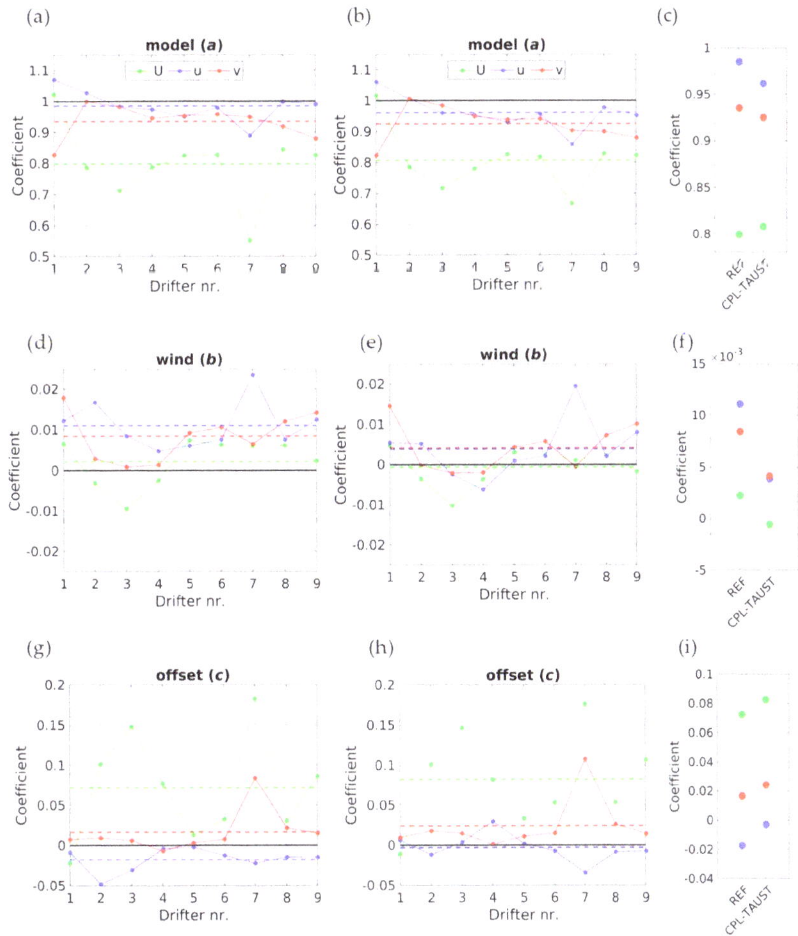

Figure 4. A multi-linear regression. The coefficients for each drifter and velocity component (solid lines) and the mean of all drifters (dashed lines). Without wave coupling (REF) (**a**,**d**,**g**) and with wave coupling (CPL-TAUST) (**b**,**e**,**h**). The averages for all drifters are shown in (**c**,**f**,**i**).

By considering the wave coupling in the circulation model (Figure 4, right panels), the windage halves and model velocities are taken less into account. Instead, the offset increases. Note that drifter 7 shows substantial differences compared to the others, which may be due to technical problems. Drifters 1 and 9 also show increased/decreased coefficients. Together with drifter 7, these are the only drifters that are beached. A possible explanation is that the drogue was damaged during the storm and then landed on the beach. Thus, model uncertainties at the boundaries or grounding of drifters could be a possible reason.

Figure 5 shows the time series for the chosen periods of drifter velocity and wind conditions for the stand-alone NEMO model and coupled model experiments considering wave-induced processes described in Section 2. The tidal cycle is simulated relatively well by all model experiments. There is a good fit between the observations and the model velocities during calm wind and wave conditions. The magnitude of the velocity in the CPL-TAUOC experiments is higher on 1 June than in REF, due to the veering of the wind, which influences the sea-state dependent momentum flux forcing. In periods of strong winds (e.g., 03/06), the velocity of the REF model is under-estimated, while the coupled NEMO-WAM currents performed better than the currents from the stand-alone NEMO mode. The wave model compares very well against in situ and satellite observations [43,59], see Figure S4.

(a) (b)

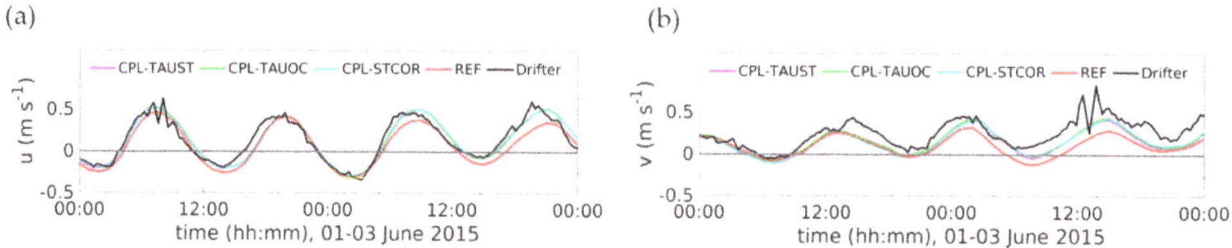

Figure 5. Velocity from drifter #5 and model runs. Zonal (**a**) and meridional component (**b**). Time period is 01–03 June 2015.

4. Model Trajectories

We performed several Lagrangian experiments to investigate the impact of wave-induced forcing on the ocean model. In the first set of experiments, OpenDrift used only the Eulerian current fields as provided by NEMO (REF and CPL experiments) without adding any wind drift correction. In the second set of experiments, an additional windage was included.

4.1. Time Series of Separation Distance, Skill Scores and Standard Deviation

As a separation metric, the normalized cumulative Lagrangian separation [4] was applied. First, we focused on the model skill in drifter trajectories to demonstrate the impact of the local wind and wave-induced velocity correction terms. Further, we quantified the sensitivity of the simulated particles to the individual or combined effect of the wave-induced process that were implemented in our coupled wave-circulation model system.

During the whole integration period, the separation distance of the drifters (Figure 6) remained very low, even though we did not perform any re-initialization of the drifter model after the initial release. Drifters 2–4 travelled for only a few days in the German Bight (Figure 1). It is noticeable that during the lifetime of these drifters, the simulated separation distance was kept within the model grid resolution. The model performance for drifters 5 and 6 (Figure S5) was high for all coupled experiments, and the separation distance was kept below 20 and 40 km, respectively, even one month after the start of integration. The separating distance of the REF (drifters 5–6) sharply increased after 20 June, showing significant deviation from the observations. This result coincides with earlier findings [27], demonstrating that on 16 and 22 June the wind direction sharply changed. Consequently, under these transitional conditions, signs of directional errors substantially differed, and the model performance was unsatisfactory. Implementing wave-induced processes into NEMO and using these Eulerian currents in OpenDrift led to a decrease in the separation distance between the CPL model experiments and the observations. In [18] was found that intensification of zonal velocity towards the coast, simulated by the coupled model, fits better with ADCP measurements. The present results show that for the CPL experiments, even during periods with moderate significant wave height, the inclusion of wave parameterizations improved the model performance. This demonstrates that separately adding a contribution of the Stokes drift to the Eulerian currents to OpenDrift in many cases might be insufficient to simulate the drifters under changing sea state conditions appropriately. The external Stokes drift can also be inconsistent with simulated ocean currents. For drifter 3 (Figure 6a), the distance error was of the same order as the resolution of our ocean model. For integration periods longer than a month, the distance error was less than twice the order of the grid resolution for the one-month simulation of drifter 5 (Figure 6c). Thus, the sea-state contribution of the momentum flux helps bring down the separation distance during most of the simulation periods.

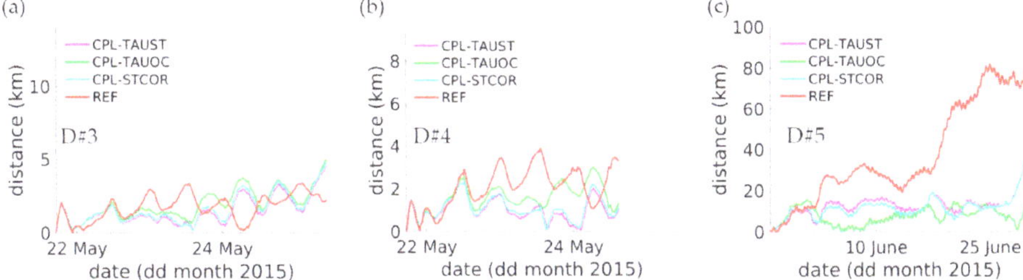

Figure 6. Time series of the separation distance (km) between the observed and model drifter #3 (**a**), #4 (**b**) and #5 (**c**) trajectories of REF (red line), CPL-TAUOC (green line), CPL-TAUST (magenta line) and CPL-STCOR (blue line) experiments.

Sensitivity experiments on different wind drift (leeway) percentages (Figures 7 and 8) demonstrated that only very few of them managed to reach higher skill scores than the CPL experiments. The results show the significance of producing the Eulerian velocity by the circulation models forced by sea state dependent momentum fluxes. The skill scores of the simulated drifters were generally high in all experiments, above 0.8 (Figure 9). The CPL experiments managed to keep the std low even after this period without considering any windage (see Figures S6–S8).

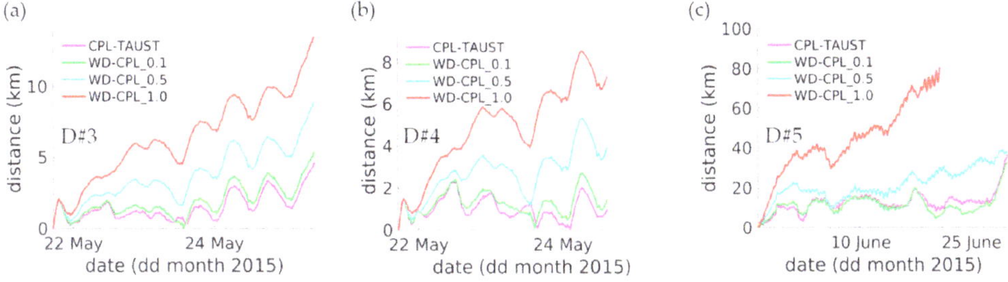

Figure 7. Time series of the separation distance (km) between the observed and model drifter #3 (**a**), #4 (**b**) and #5 (**c**) trajectories of WD-CPL experiments with wind drag contributions of 0.1, 0.5 and 1.0%.

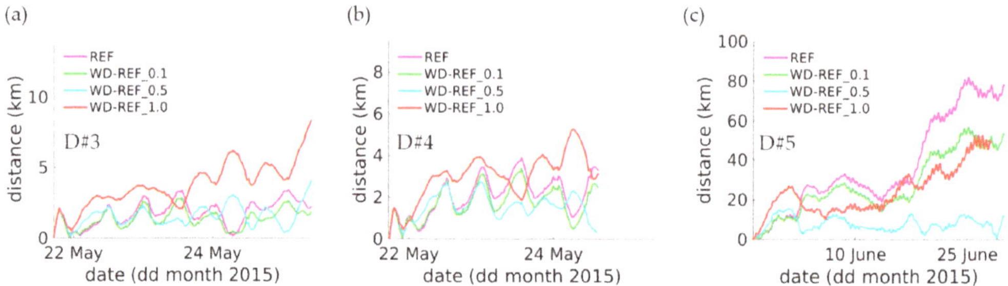

Figure 8. Time series of the separation distance (m) between the observed and model drifter #3 (**a**), #4 (**b**) and #5 (**c**) trajectories of WD-REF experiments with wind drag contributions of 0.1, 0.5 and 1.0%.

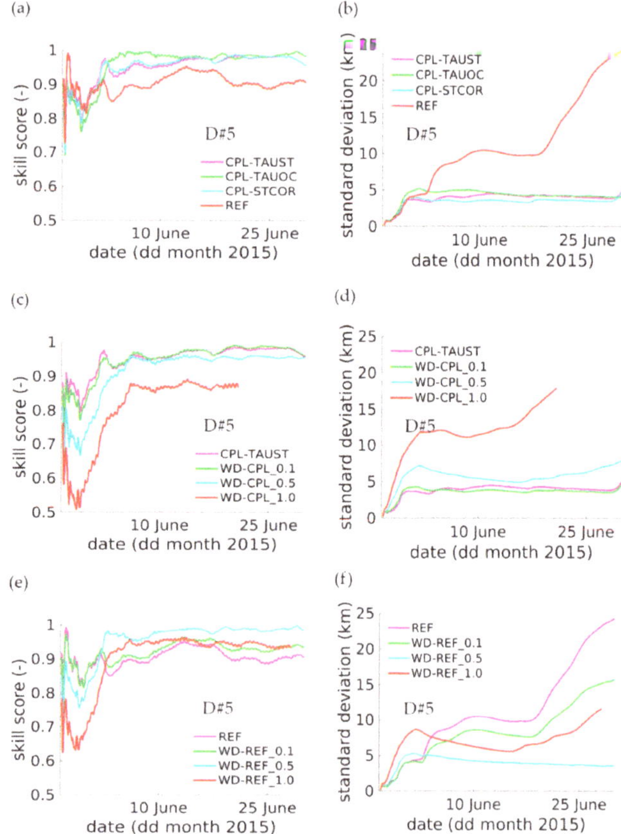

Figure 9. Time series of the skill score and the standard deviation of the distance (km) between the observed and model drifter trajectories of the different experiments (see the Table 1 and the legend): (CPL (**a,b**), WD-CPL (**c,d**) and WD-REF (**e,f**).

Applying wind drift correction of 0.5% or higher to WD-CPL experiments reduced the skill score (Figure 9c,d). In the first days of the integration period, the skill score of drifter #5 by windage correction of 0.5% and 1.0% dropped to 0.7 and 0.5, respectively. The skill score of the experiment with windage correction of 0.1% was closer to the CPL experiments but still slightly lower in the period from 26 May until 2 June (see also Figures S7 and S8).

These differences were illustrated even better by the standard deviations of the WD-CPL experiments (Figure 9d). The standard deviations of D#5 stayed almost constant through the integration, at about 0.5 and 5 km, respectively, for CPL and WD-CPL plus 0.1% (Figure 9). Adding 0.5% windage yielded a higher standard deviation for all experiments. By assuming 1.0% windage, a trend of the systematic increase was observed for all drifters. Neither of the WD-CPL experiments (in Figure 9c,d) improved the skill scores and reduced the deviation between the trajectories of the Lagrangian model and observations by additionally considering the windage contribution to the Eulerian currents obtained by the coupled wave–circulation ocean model.

Sensitivity to the windage contribution was also determined for the REF runs (Figure 9e,f). The skill score and standard deviation of WD-REF were better than those of WD-CPL for all experiments. On the other side, the skill scores/standard deviations were lower/higher than those of CPL. Only by considering 0.5% windage of WD-REF are the *ss* and *dd* similar to the CPL.

It was demonstrated in [53] that simulated trajectories obtained by considering the wind drag of current velocities or the Stokes drift's contribution showed better skills against the observations than without the corrections. They postulated that considering 1.0% windage or adding the contribution of the Stokes drift gives almost identical results. Our sensitivity analysis, however, showed that the CPL experiments performed best. From the rest of the experiments, only WD-REF with 0.5% windage provided similar skill. The model experiments showed that using Eulerian velocity estimated by considering Stokes–Coriolis and the sea state momentum provided similar results.

4.2. Particle Trajectories of the Albatros Drifters

The trajectory of drifter 5 in the REF experiment deviated in a northwesterly direction. (Figure 10a) The trajectories obtained by the CPL experiments correctly reproduced the drifter direction and were in good agreement with observations. By north and northwesterly winds on 10 June and the anticyclonic circulation [27], the CPL experiments reproduced the windage of drifter 5 well. CPL-TAUOC separation distance was on the order of the model grid resolution (Figure 6). The low *dd* values remained until 29 June. The trajectory of WD-CPL_1.0 (Figure 10b) made a higher loop towards the north and northeast, deviating from the observations. By a 0.5% contribution of windage, the drifter was unrealistically transported towards the east and was beached. The trajectory of WD-CPL with 0.1% started deviating for 2.3 days after starting the simulation. WD-REF with 0.5% leeway showed a smaller separation distance (Figure 8), and the trajectory remained closer than that of WD-CPL with 0.5%. By adding 1.0% leeway, the simulated trajectories of WD-CPL and WD-REF can be considered as wrong. Skill scores and STDs for all experiments and during the deployment period of the drifters are given in Tables S1 and S2.

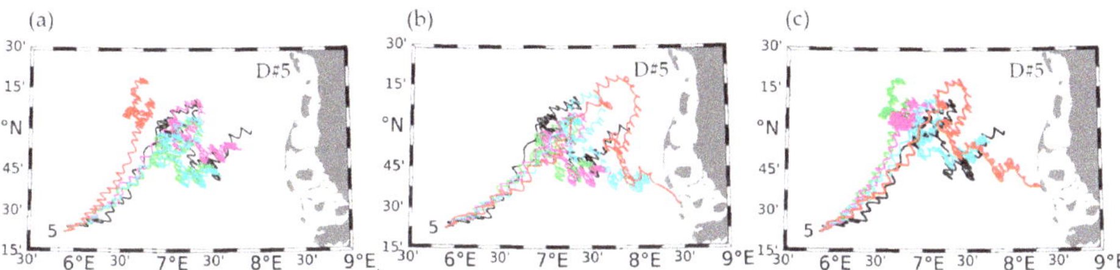

Figure 10. Drifter trajectories: observed (black line) and modelled. The colors of the different experiments are given in the legend in Figure 6 for (**a**), in Figure 7 for (**b**) and in Figure 8 for (**c**).

5. Discussion

Coupled ocean–wave models together with a Lagrangian transport model can be beneficial for drift and transport studies, ranging from search and rescue modelling of specific objects [33], backtracking [60], larval drift [61] and marine plastics [62,63] to the connectivity between different marine protected areas [16]. Accurate measures can potentially have a strong impact on biodiversity risk assessments (e.g., connected to marine litter or oil spills). In [16], several models of the North Sea were compared to study the variability and resulting uncertainty and differences between the models.

In our study, Eulerian currents from a coupled ocean-wave model were used to perform CPL and WD-CPL experiments. In this way, additional tuning of the contributing factors of the Stokes drift by the drifter model can be excluded in CPL experiments. Our results are in line with the study of [56]. The latter concluded that implementing the Stokes drift as a simple additive component of drift velocity, parameterized in terms of wind forcing, can be inconsistent (a violation of momentum and energy conservation) if Eulerian currents were simulated without taking into account the reservoir of wave momentum and energy. Our results indicate the relevance of the role of waves for redistribution of momentum, especially in periods of changing wind direction (Figures 6–8).

The Eulerian currents in most of the Lagrangian transport models, e.g., for search and rescue operations, are taken from, for example, operational model output. The wind stress is parameterized to drive the dynamics of the upper ocean directly if the wave model is not included. However, part of the wind stress is supported by the flux of momentum from wind to waves. These processes were considered in our CPL-TAUOC and CPL-TAUST experiments to simulate the Eulerian velocity needed for OpenDrift. Due to the non-linearity of wave–current interactions, the individual effect of the wave-induced coupled STCOR and TAUOC processes are not superimposed on TAUST [18]. This result shows that the Eulerian currents by the coupled run provided the best fit to the observed particle trajectories. The worsening of the Lagrangian model skill, especially at the beginning of the drifter simulation, can have an impact on applications like search and rescue, in which skilled Lagrangian transport forecasts are needed at the very beginning of the operation. In summary, the use of additional wind corrections cannot be justified in WD-CPL experiments, since neither of the additional windage correction experiments demonstrated an improvement compared with the CPL experiments.

Displacements of the more offshore drifters, 5 and 6, were observed on 3–6 June [27], and the models turned out to be largely overestimated, concluding that neither simulated currents, windage fields nor Stokes drift by their Lagrangian model were able to reproduce the spatial gradients. In the present study, CPL experiments showed good agreement for these drifters with the observations. A deviation of the trajectory, and consequently an increase in the separating distance, was observed, but stayed within the same grid resolution. On the other side, the WD experiments deviations from the simulated drifter trajectories from the observations and separating distances were high. This result proves the importance of improving the Eulerian current simulations needed for Lagrangian transport modelling by implementing the sea state momentum forcing and Stokes–Coriolis forcing in the numerical model.

We did not aim to assess all specific differences between individual drifters and discuss in detail the wind forcing and the North Sea circulation during the different periods of their deployment, as this was done by [27]. Neither did we aim to tune the model to find the optimal percentage of the contribution of either the Stokes drift or windage that can be taken into account in OpenDrift simulations (as was recently done by [64]. Additional errors of the drifter simulations can be due to the errors in the atmospheric forcing, but also the boundary or tidal forcing. Another reason may be errors in the vertical resolution and mixing and bottom friction parameterization; the latter is essential for shallow water dynamics in regions like the German Bight. By using the same model configuration as in the present work, the interactions between barotropic tides and mesoscale processes were studied by [65] showing that barotropic tides affect diapycnal mixing substantially. In our study, we aimed to investigate the potential of a different approach, namely the contribution of wave-induced processes in the NEMO model.

It is important to stress here again that the North Sea circulation is very complex. The drifters were deployed in the shallow German Bight, a coastal ocean where the currents are dominated by tidal and wind forcing and are steered by the bathymetry and coastline. Other possible error sources are circulation features such as inertial oscillations, sub-mesoscale dynamics and baroclinic effects. Further studies are needed to quantify the combined effects of the mesoscale variability and resolution of the ocean and particle transport model, especially in the coastal areas of the German Bight (the Wadden Sea) as well as for other regions. For these aims, the GCOAST framework will be taken into account and tested for its best performance in terms of trajectory simulations over regions with different oceanographic properties (e.g., Baltic Sea, the North-East Atlantic).

The advantage of the method here is that by coupling the circulation model to the wave model, no further sensitivity experiments on the percentage of the contribution of windage or Stokes drift to improve the skills of the Lagrangian model are needed.

6. Conclusions

The results of our experiments lead us to the following conclusions:

1. Comparing currents from coupled and stand-alone model simulations demonstrated that the coupled model velocities fit better with the observations, especially moderate and high values. In calm wind and wave conditions, the differences are not pronounced. By using a fully coupled model, consistent atmosphere–wave–ocean forcing is applied to simulate the Eulerian currents needed for particle transport simulation. Besides, the bias of the directions in the wind–wave–ocean currents simulations is reduced compared to that in the stand-alone model. It is noteworthy that despite the coarse resolution of the NEMO model and the coupled NEMO-WAM setup and the complex bathymetry and coastline in our study area, the comparisons are satisfactory, also for the stand-alone model simulations. A general improvement in surface currents for both directions is observed due to the wind–wave–ocean coupling.

2. The multi-linear regression analysis showed that for CPL, windage is halved (from 1.0 to 0.5%), and the model velocities are taken less into account. Instead, the offset increases. These results show that the wind drift is better accounted for in the coupled NEMO-WAM model than in the stand-alone NEMO. It was also shown that the regression can reveal technical problems of the drifter, like probable drogue loss.

3. We showed the particle analysis by calculating values such as separating distance, skill score and standard deviation between model experiments and observations. We demonstrated that the skill score, based on the cumulative Lagrangian separation distance standardized by the associated cumulative trajectory length, proved to be a useful parameter to evaluate the overall model performance, rather than using a daily validation metric. Although the skill score and standard deviation of the ODi drifter are slightly lower than those of the MD03i, both types demonstrate good predictive skill. The MD03i drifter shows better skill for CPL than ODi. The skill scores with wind showed similar behavior by both types of drifters. For drifters that reached land, the model performance was low, which might be due to the insufficient model resolution and parameterization together with higher model uncertainty at the boundaries or beaching of drifters.

4. The skill scores and separation distances improved when wave-induced processes were taken into account in the ocean-only simulations. By considering sea state momentum dependencies or Stokes–Coriolis forcing in the coupled model, the skill scores are quite similar. Adding the contribution of windage or Stokes drift to currents produced by the fully coupled run with waves might lead to over-parameterization.

5. In the CPL experiments, it turned out that no additional tuning of the wind drift factor was needed for best fit with the observations. This also indicates that no additional contribution of windage or separate consideration of external Stokes drift was needed to predict the drifter trajectories satisfactorily. For some drifters, the skill scores of the WD-REF or WD-CPL experiments were similar to the CPL by adding a direct windage

contribution to the REF. However, these percentages varied between different drifters and experiments.

The results based on the drifters used in this study showed that, in some cases, it might be favorable to implement full coupling of waves and circulation models to produce the currents needed for drifter simulations. This could lead to a more consistent approach instead of trying to tune the windage factor or percentage of external Stokes drift, separately or combined. Such tuning of the contribution coefficients is typically restricted by the need for availability and testing a large number of drifter observations (that are normally lacking), which is required to improve wind–wave–current forcing dependencies in the Lagrangian model. We note, however, that this conclusion is based on a limited period and a small area (over the German Bight). Nevertheless, the model simulations showed that the results are promising for better understanding and prediction of Lagrangian transport by using Eulerian currents simulated by coupled wave–ocean model simulations instead of a stand-alone ocean model. The results revealed that the newly introduced wave effects are essential for the drift model performance.

Supplementary Materials: The following are available online at https://www.mdpi.com/2073-4441/13/4/415/s1, Table S1: Skill score over the deployment period of drifter; Table S2: STDs average over the deployment period of stds (km); Figure S1. (a)The experiment site in the German Bight (a). The shaded colours show the number of the HF radar measurements from the three radar antennas. The trajectories of the drifter #1–9 are plotted with the colour lines. Sail illustration of the MD03i frifter (b); MD03i (drifter #1–6) (c) and ODi (drifter 7–9)-HZG drifters (d). Figure S2. Zonal (a) and meridional (b) velocity scatter plots: drifter versus HF radar data (m/s). The black dots indicate the quantile-quantile (*q-q*) plot. The black dashed line is the diagonal. The model topography is also interpolated to the drifter positions and depicted with colours. Figure S3. Scatter plots of velocity magnitude, as well as of the zonal and meridional components for the REF experiment (a), (d) and (g), (and CPL-TAUST experiment (b), (e) and (h) of the HZG drifters vs. model data coloured to wind velocity and the quantile-quantile (*q-q*) plot in black. The black dashed line is the diagonal. Taylor Diagram of the velocity magnitude, as well as of the zonal and meridional components (c), (f) and (i). Figure S4. Significant wave heights (m) at Elbe (top) and FINO-3 (bottom) buoy station in in May and June 2015. The blue dots are the in-situ observations; the red line corresponds to the WAM simulations. For the position see Figure 1a. Figure S5. Time series of the distance (km) between the observed and model drifter #1–9 (a–i), # trajectories of REF (red line), CPL-TAUOC (green line), CPL-TAUST (magenta line) and CPL-STCOR (blue line) experiments. Figure S6. Time series of the skill score and the standard deviation of the distance (km) between the observed and model drifter #3–9 (a–f) trajectories of the REF (red line), CPL-TAUOC (green line), CPL-TAUST (magenta line) and CPL-STCOR (blue line) experiments. Figure S7. Time series of the skill score and the standard deviation of the distance (km) between the observed and model drifter #3–9 (a–f) trajectories of the WD-CPL experiments with wind drag contribution of 0.1, 0.5 and 1.0%. Figure S8. Time series of the skill score and the standard deviation of the distance (km) between the observed and model drifter #3–9 (a–f) trajectories of the WD-REF experiments with wind drag contribution of 0.1, 0.5 and 1.0%.

Author Contributions: J.S. conceived the study and wrote major parts of the manuscript. M.R. conducted the model simulations and provided parts of the text, R.C.A. made the GPS drifter experiments, provided their data and contributed to their description. Ø.B. helped develop the model coupling and contributed to the manuscript. C.S. contributed to discussion. J.S. and Ø.B. contributed to the project administration. All authors have read and agreed to the published version of the manuscript.

Funding: J.S. and Ø.B. gratefully acknowledge funding from CMEMS through the Service Evolution 2 project WaveFlow. Ø.B. is grateful for funding from the Joint Rescue Coordination Centres which has helped fund the development of the OpenDrift model. The Research Council of Norway has also helped fund the development of OpenDrift through the projects CIRFA (grant no. 237906) and Retrospect (grant no. 244262). M.R. has been supported by the project "Macroplastics Pollution in the Southern North Sea–Sources, Pathways and Abatement Strategies" (grant no. ZN3176) funded by the German Federal State of Lower Saxony. C.S. and J.S acknowledge the funding by Initiative and Networking Fund of the Helmholtz Association through the project "Advanced Earth System

Modelling Capacity (ESM)" and CLICCS Center of Excellence, J.S. is grateful for funding from BMBF bilateral project "Ocean Currents_MoDA" (grant no: 03F0822A).

Data Availability Statement: The raw data of observed drifter locations are freely available from [26]. German Weather Service data are available from: https://opendata.dwd.de/weather/nwp/icon-eu/grib/. The HF-radar data are available through the Coastal Observing System for Northern and Arctic seas (COSYNA) [54,55]. OpenDrift [46] is a freely available open-source off-line Lagrangian particle trajectory model.

Acknowledgments: The authors thank the UK Met Office for providing the NEMO model setup. We thank Arno Behrens, Oliver Krüger and Sebastian Grayek for technical support with the models. We thank to Gerhard Gayer for the wave model validations. We are thankful to Sabine Hartmann or technical support with the figures. We appreciate the use of the German Weather Service (DWD) atmospheric data, provided by the German Weather Service.

Conflicts of Interest: The authors declare that there are no conflict of interest.

Appendix A

Appendix A.1. Wave–Current Interaction Processes

As described in Section 2, the NEMO ocean model has been modified to take into account the following wave effects [11,18]: (1) the Stokes–Coriolis forcing; (2) sea-state dependent momentum flux, set as a scalar dependence of the flux from the atmosphere to waves and ocean or as a vector; and (3) a sea-state dependent energy flux. Below is the description of the wave-induced forcing and the processes of wave interaction with the ocean circulation.

Stokes Drift

The surface Stokes drift u_{st} is defined by the following integral expression in the WAM model:

$$\vec{u_{st}} = \int_0^\infty \frac{2gk}{\omega \tanh{(2kD)}} \vec{k} E(f,\theta) \, df \, d\theta. \tag{A1}$$

Here $E = E(\omega,\theta)$ is the two-dimensional wave spectrum which gives the energy distribution of the ocean waves over angular frequency ω and propagation direction θ. Particle trajectories in water waves do not form entirely closed orbits because the particles spend more time forward under wave crests than backwards under wave troughs. This sets up a second-order effect, which leads to a discrepancy between the average Lagrangian flow velocity of a fluid parcel and the Eulerian flow velocity known as the Stokes drift. As is the case for the wind-induced currents, the Stokes drift also interacts with the Earth's rotation. This adds an additional veering to the ocean currents known as the Stokes–Coriolis force [66],

$$\frac{Du}{Dt} = -\frac{1}{\rho}\nabla p + (\mathbf{u}+\mathbf{v_s}) x \, f\hat{\mathbf{z}} + \frac{1}{\rho}\frac{\partial \tau}{\partial z} \tag{A2}$$

where $\mathbf{v_s}$ is the Stokes drift vector, p is the pressure, τ is the surface stress and $\hat{\mathbf{z}}$ is the upward unit vector. Because calculating the full vertical profile is costly, the Stokes drift velocity profile was first calculated with an approximation from [10]. The Stokes–Coriolis force is also included in the tracer advection equations as described by [28]. In the present approach, the Stokes drift velocity profile is also considered [67,68].

Appendix A.2. Momentum and Energy Flux from the Wave Model

Provided that current gradients are sufficiently weak, the energy and momentum fluxes can be calculated from the energy balance equation (with an approximation from [36]:

$$\frac{\partial}{\partial t}E + \frac{\partial}{\partial \vec{x}} \cdot \left(\vec{v_g}E\right) = S_{in} + S_{nl4} + S_{diss} + S_{bot} + S_{br} \tag{A3}$$

where $E = E(\omega,\theta)$ is the two-dimensional wave spectrum which gives the energy distribution of the ocean waves over angular frequency ω and propagation direction θ, v_g is the group velocity. On the right-hand side of the action balance equation are the source terms that represent physical processes which generate, redistribute or dissipate wave energy in the WAM model. These terms denote, respectively, wave growth by the wind S_{in}, non-linear transfer of wave energy through four-wave interactions S_{nl4} and wave dissipation caused by white capping S_{diss} and bottom friction S_{diss}. In the present calculations, we also took into account depth-induced wave breaking S_{br}.

Making use of the energy balance in the Equation (A4) the wave-induced stress is given by

$$\overrightarrow{\tau_{in}} = \rho_w g \int_0^{2\pi} \int_0^{\infty} \frac{\overrightarrow{k}}{\omega} S_{in}\, d\omega\, d\theta \tag{A4}$$

while the dissipation stress is given by

$$\overrightarrow{\tau_{diss}} = \rho_w g \int_0^{2\pi} \int_0^{\infty} \frac{\overrightarrow{k}}{\omega} S_{diss}\, d\omega\, d\theta. \tag{A5}$$

Similarly, the energy flux from wind to waves is given by

$$\Phi_{in} = \rho_w g \int_0^{2\pi} \int_0^{\infty} S_{in}\, d\omega\, d\theta \tag{A6}$$

and the energy flux from waves to the ocean, Φ_{diss}, is defined as

$$\Phi_{diss} = \rho_w g \int_0^{2\pi} \int_0^{\infty} S_{diss}\, d\omega\, d\theta. \tag{A7}$$

Under stationary and homogenous conditions the momentum and energy balance reduces to

$$\int_0^{2\pi} \int_0^{\infty} \frac{k}{\omega}(S_{in} + S_{diss} + S_{NL})\, d\omega\, d\theta = 0 \tag{A8}$$

and

$$\int_0^{2\pi} \int_0^{\infty} (S_{in} + S_{diss} + S_{NL})\, d\omega\, d\theta = 0. \tag{A9}$$

The momentum flux to the ocean column, denoted by τ_{oc}, is the sum of the flux transferred by turbulence across the air–sea interface which was not used to generate waves $\tau_a - \tau_{in}$ and the momentum flux transferred by the ocean waves due to wave breaking τ_{diss}. This leads to

$$\overrightarrow{\tau_{oc}} = \overrightarrow{\tau_a} - \rho_w g \int_0^{2\pi} \int_0^{\omega_c} \frac{\overrightarrow{k}}{\omega}(S_{in} + S_{diss} + S_{NL})\, d\omega\, d\theta. \tag{A10}$$

The contribution to the energy flux is

$$\Phi_{oc} = \rho_w g \int_0^{2\pi} \int_{\omega_c}^{\infty} S_{in}\, d\omega\, d\theta - \rho_w g \int_0^{2\pi} \int_0^{\omega_c} (S_{diss} + S_{NL})\, d\omega\, d\theta. \tag{A11}$$

It is important to note that while the momentum fluxes are mainly determined by the high-frequency part of the wave spectrum, the energy flux is to some extent also determined by the low-frequency waves.

The high frequency ($\omega > \omega_c$) contribution to the energy flux (first term of Equation (A11)) is

$$\Phi_{oc_{hf}} = \rho_w g \int_0^{2\pi} \int_{\omega_c}^{\infty} S_{in}\, d\omega\, d\theta. \tag{A12}$$

In NEMO, the wave-induced turbulent kinetic energy (TKE) flux introduced at the sea surface depends on the wave energy factor α [69] and is set to a constant value regardless of the sea state. Authors in [69] argued that the turbulent kinetic energy flux is relatively insensitive to the sea state and is well approximated by αu^3_{w*} (u_{w*} is the water-side friction velocity), and $\alpha = 100$ was thought to be representative of a mid-range of sea states between young wind seas and fully developed situations. As shown above, using the full spectral wave model, it is possible to estimate both the momentum energy and energy fluxes directly from the wave breaking source terms [11,18,25].

References

1. Van Sebille, E.; Griffies, S.M.; Abernathey, R.; Adams, T.P.; Berloff, P.; Biastoch, A.; Blanke, B.; Chassignet, E.P.; Cheng, Y.; Cotter, C.J.; et al. Lagrangian Ocean Analysis: Fundamentals and Practices. *Ocean Modell.* **2018**, *121*, 49–75. [CrossRef]
2. Van Sebille, E.; Aliani, S.; Law, K.L.; Maximenko, N.; Alsina, J.M.; Bagaev, A.; Bergmann, M.; Chapron, B.; Chubarenko, I.; Cózar, A.; et al. The Physical Oceanography of the Transport of Floating Marine Debris. *Environ. Res. Lett.* **2020**, *15*, 023003. [CrossRef]
3. Van Sebille, E.; van Leeuwen, P.J.; Biastoch, A.; Barron, C.N.; de Ruijter, W.P.M. Lagrangian Validation of Numerical Drifter Trajectories Using Drifting Buoys: Application to the Agulhas System. *Ocean Modell.* **2009**, *29*, 269–276. [CrossRef]
4. Liu, Y.; Weisberg, R.H. Evaluation of Trajectory Modeling in Different Dynamic Regions Using Normalized Cumulative Lagrangian Separation. *J. Geophys. Res. Oceans* **2011**, *116*. [CrossRef]
5. Sotillo, M.G.; Fanjul, E.A.; Castanedo, S.; Abascal, A.J.; Menendez, J.; Emelianov, M.; Olivella, R.; García-Ladona, E.; Ruiz-Villarreal, M.; Conde, J.; et al. Towards an Operational System for Oil-Spill Forecast over Spanish Waters: Initial Developments and Implementation Test. *Mar. Pollut. Bull.* **2008**, *56*, 686–703. [CrossRef]
6. Huntley, H.S.; Lipphardt, B.L.; Kirwan, A.D. Lagrangian Predictability Assessed in the East China Sea. *Ocean Modell.* **2011**, *36*, 163–178. [CrossRef]
7. Janssen, P.A.E.M. Wave-Induced Stress and the Drag of Air Flow over Sea Waves. *J. Phys. Oceanogr.* **1989**, *19*, 745–754. [CrossRef]
8. Janssen, P.A.E.M. Quasi-Linear Theory of Wind-Wave Generation Applied to Wave Forecasting. *J. Phys. Oceanogr.* **1991**, *21*, 1631–1642. [CrossRef]
9. Semedo, A.; Saetra, Ø.; Rutgersson, A.; Kahma, K.K.; Pettersson, H. Wave-Induced Wind in the Marine Boundary Layer. *J. Atmos. Sci.* **2009**, *66*, 2256–2271. [CrossRef]
10. Breivik, Ø.; Janssen, P.A.E.M.; Bidlot, J.-R. Approximate Stokes Drift Profiles in Deep Water. *J. Phys. Oceanogr.* **2014**, *44*, 2433–2445. [CrossRef]
11. Breivik, Ø.; Mogensen, K.; Bidlot, J.-R.; Balmaseda, M.A.; Janssen, P.A.E.M. Surface Wave Effects in the NEMO Ocean Model: Forced and Coupled Experiments. *J. Geophys. Res. Ocean.* **2015**, *120*, 2973–2992. [CrossRef]
12. Guan, C.; Xie, L. On the Linear Parameterization of Drag Coefficient over Sea Surface. *J. Phys. Oceanogr.* **2004**, *34*, 2847–2851. [CrossRef]
13. Wu, L.; Rutgersson, A.; Sahlée, E.; Larsén, X.G. Swell Impact on Wind Stress and Atmospheric Mixing in a Regional Coupled Atmosphere-Wave Model. *J. Geophys. Res. Ocean.* **2016**, *121*, 4633–4648. [CrossRef]
14. Higgins, C.; Vanneste, J.; Bremer, T.S. Unsteady Ekman-Stokes Dynamics: Implications for Surface Wave-Induced Drift of Floating Marine Litter. *Geophys. Res. Lett.* **2020**, *47*. [CrossRef]
15. Röhrs, J.; Christensen, K.H.; Hole, L.R.; Broström, G.; Drivdal, M.; Sundby, S. Observation-Based Evaluation of Surface Wave Effects on Currents and Trajectory Forecasts. *Ocean Dyn.* **2012**, *62*, 1519–1533. [CrossRef]
16. Hufnagl, M.; Payne, M.; Lacroix, G.; Bolle, L.J.; Daewel, U.; Dickey-Collas, M.; Gerkema, T.; Huret, M.; Janssen, F.; Kreus, M.; et al. Variation That Can Be Expected When Using Particle Tracking Models in Connectivity Studies. *J. Sea Res.* **2017**, *127*, 133–149. [CrossRef]
17. Röhrs, J.; Christensen, K.H.; Vikebø, F.; Sundby, S.; Saetra, Ø.; Broström, G. Wave-Induced Transport and Vertical Mixing of Pelagic Eggs and Larvae. *Limnol. Oceanogr.* **2014**, *59*, 1213–1227. [CrossRef]
18. Staneva, J.; Alari, V.; Breivik, Ø.; Bidlot, J.-R.; Mogensen, K. Effects of Wave-Induced Forcing on a Circulation Model of the North Sea. *Ocean Dyn.* **2017**, *67*, 81–101. [CrossRef]
19. Alari, V.; Staneva, J.; Breivik, Ø.; Bidlot, J.-R.; Mogensen, K.; Janssen, P. Surface Wave Effects on Water Temperature in the Baltic Sea: Simulations with the Coupled NEMO-WAM Model. *Ocean Dyn.* **2016**, *66*, 917–930. [CrossRef]
20. Brown, J.M.; Bolaños, R.; Wolf, J. The Depth-Varying Response of Coastal Circulation and Water Levels to 2D Radiation Stress When Applied in a Coupled Wave–Tide–Surge Modelling System during an Extreme Storm. *Coast. Eng.* **2013**, *82*, 102–113. [CrossRef]
21. Brown, J.M.; Wolf, J. Coupled Wave and Surge Modelling for the Eastern Irish Sea and Implications for Model Wind-Stress. *Cont. Shelf Res.* **2009**, *29*, 1329–1342. [CrossRef]
22. Staneva, J.; Wahle, K.; Koch, W.; Behrens, A.; Fenoglio-Marc, L.; Stanev, E.V. Coastal Flooding: Impact of Waves on Storm Surge during Extremes—A Case Study for the German Bight. *Nat. Hazards Earth Syst. Sci.* **2016**, *16*, 2373–2389. [CrossRef]

23. Lewis, H.W.; Castillo Sanchez, J.M.; Siddorn, J.; King, R.R.; Tonani, M.; Saulter, A.; Sykes, P.; Pequignet, A.-C.; Weedon, G.P.; Palmer, T.; et al. Can Wave Coupling Improve Operational Regional Ocean Forecasts for the North-West European Shelf? *Ocean Sci.* **2019**, *15*, 669–690. [CrossRef]

24. Cavaleri, L.; Abdalla, S.; Benetazzo, A.; Bertotti, L.; Bidlot, J.-R.; Breivik, Ø.; Carniel, S.; Jensen, R.E.; Portilla-Yandun, J.; Rogers, W.E.; et al. Wave Modelling in Coastal and Inner Seas. *Prog. Oceanogr.* **2018**, *167*, 164–233. [CrossRef]

25. Wu, L.; Breivik, Ø.; Rutgersson, A. Ocean-Wave-Atmosphere Interaction Processes in a Fully Coupled Modeling System. *J. Adv. Modeling Earth Syst.* **2019**, *11*, 3852–3874. [CrossRef]

26. Carrasco, R.; Horstmann, J. German Bight Surface Drifter Data from Heincke Cruise HE 445. 2015. Available online: https://doi.pangaea.de/10.1594/PANGAEA.874511 (accessed on 28 January 2020).

27. Callies, U.; Groll, N.; Horstmann, J.; Kapitza, H.; Klein, H.; Maßmann, S.; Schwichtenberg, F. Surface Drifters in the German Bight: Model Validation Considering Windage and Stokes Drift. *Ocean Sci.* **2017**, *13*, 799–827. [CrossRef]

28. Ricker, M.; Stanev, E.V. Circulation of the European Northwest Shelf: A Lagrangian Perspective. *Ocean Sci.* **2020**, *16*, 637–655. [CrossRef]

29. Callies, U.; Carrasco, R.; Floeter, J.; Horstmann, J.; Quante, M. Submesoscale Dispersion of Surface Drifters in a Coastal Sea near Offshore Wind Farms. *Ocean Sci.* **2019**, *15*, 865–889. [CrossRef]

30. Von Schuckmann, K.; Traon, P.-Y.L.; Smith, N.; Pascual, A.; Djavidnia, S.; Gattuso, J.-P.; Grégoire, M.; Nolan, G.; Aaboe, S.; Aguiar, E.; et al. Copernicus Marine Service Ocean State Report, Issue 3. *J. Oper. Oceanogr.* **2019**, *12*, S1–S123. [CrossRef]

31. Ho-Hagemann, H.T.M.; Hagemann, S.; Grayek, S.; Petrik, R.; Rockel, B.; Staneva, J.; Feser, F.; Schrum, C. Internal Model Variability of the Regional Coupled System Model GCOAST-AHOI. *Atmosphere* **2020**, *11*, 227. [CrossRef]

32. Breivik, Ø.; Allen, A.A. An Operational Search and Rescue Model for the Norwegian Sea and the North Sea. *J. Mar. Syst.* **2008**, *69*, 99–113. [CrossRef]

33. Breivik, Ø.; Allen, A.A.; Maisondieu, C.; Olagnon, M. Advances in Search and Rescue at Sea. *Ocean Dyn.* **2013**, *63*, 83–88. [CrossRef]

34. Madec, G. *Note du Pôle de modélisation, Institut Pierre-Simon Laplace (IPSL)*; NEMO Ocean Engine: Paris, France, 2008.

35. Hordoir, R.; Axell, L.; Höglund, A.; Dieterich, C.; Fransner, F.; Gröger, M.; Liu, Y.; Pemberton, P.; Schimanke, S.; Andersson, H.; et al. Nemo-Nordic 1.0: A NEMO-Based Ocean Model for the Baltic and North Seas—Research and Operational Applications. *Geosci. Model Dev.* **2019**, *12*, 363–386. [CrossRef]

36. Dietrich, J.C.; Zijlema, M.; Westerink, J.J.; Holthuijsen, L.H.; Dawson, C.; Luettich, R.A.; Jensen, R.E.; Smith, J.M.; Stelling, G.S.; Stone, G.W. Modeling Hurricane Waves and Storm Surge Using Integrally-Coupled, Scalable Computations. *Coast. Eng.* **2011**, *58*, 45–65. [CrossRef]

37. O'Dea, E.; Furner, R.; Wakelin, S.; Siddorn, J.; While, J.; Sykes, P.; King, R.; Holt, J.; Hewitt, H. The CO5 Configuration of the 7 km Atlantic Margin Model: Large-Scale Biases and Sensitivity to Forcing, Physics Options and Vertical Resolution. *Geosci. Model Dev.* **2017**, *10*, 2947–2969. [CrossRef]

38. O'Dea, E.J.; Arnold, A.K.; Edwards, K.P.; Furner, R.; Hyder, P.; Martin, M.J.; Siddorn, J.R.; Storkey, D.; While, J.; Holt, J.T.; et al. An Operational Ocean Forecast System Incorporating NEMO and SST Data Assimilation for the Tidally Driven European North-West Shelf. *J. Oper. Oceanogr.* **2012**, *5*, 3–17. [CrossRef]

39. Group, T.W. The WAM Model—A Third Generation Ocean Wave Prediction Model. *J. Phys. Oceanogr.* **1988**, *18*, 1775–1810. [CrossRef]

40. ECMWF. *IFS Documentation CY40R1*; IFS Documentation; ECMWF: Readink, UK, 2014.

41. Komen, G.J.; Cavaleri, L.; Donelan, M.; Hasselmann, K.; Hasselmann, S.; Janssen, P.A.E.M. *Dynamics and Modelling of Ocean Waves*; Cambridge University Press: Cambridge, UK, 1994; ISBN 978-0-521-57781-6.

42. Günther, H.; Hasselmann, S.; Janssen, P.A.E.M. *The WAM Model Cycle 4.0*; Deutsches Klimarechenzentrum: Hamburg, Germany, 1992; p. 102.

43. Staneva, J.; Behrens, A.; Wahle, K. Wave Modelling for the German Bight Coastal-Ocean Predicting System. *J. Phys. Conf. Ser.* **2015**, *633*, 012117. [CrossRef]

44. Hersbach, H.; Janssen, P.A.E.M. Improvement of the Short-Fetch Behavior in the Wave Ocean Model (WAM). *J. Atmos. Ocean. Technol.* **1999**, *16*, 884–892. [CrossRef]

45. Bidlot, J.-R.; Janssen, P.; Abdalla, S. *A Revised Formulation of Ocean Wave Dissipation and Its Model Impact*; ECMWF: Reading, UK, 2007; p. 27.

46. Dagestad, K.-F.; Röhrs, J.; Breivik, Ø.; Ådlandsvik, B. OpenDrift v1.0: A Generic Framework for Trajectory Modelling. *Geosci. Model Dev.* **2018**, *11*, 1405–1420. [CrossRef]

47. Jones, C.E.; Dagestad, K.-F.; Breivik, Ø.; Holt, B.; Röhrs, J.; Christensen, K.H.; Espeseth, M.; Brekke, C.; Skrunes, S. Measurement and Modeling of Oil Slick Transport. *J. Geophys. Res. Ocean.* **2016**, *121*, 7759–7775. [CrossRef]

48. Christensen, K.H.; Breivik, Ø.; Dagestad, K.-F.; Röhrs, J.; Ward, B. Short-Term Predictions of Oceanic Drift. *Oceanography* **2018**, *31*, 59–67. [CrossRef]

49. Meyerjürgens, J.; Badewien, T.H.; Garaba, S.P.; Wolff, J.-O.; Zielinski, O. A State-of-the-Art Compact Surface Drifter Reveals Pathways of Floating Marine Litter in the German Bight. *Front. Mar. Sci.* **2019**, *6*. [CrossRef]

50. Baschek, B.; Schroeder, F.; Brix, H.; Riethmüller, R.; Badewien, T.H.; Breitbach, G.; Brügge, B.; Colijn, F.; Doerffer, R.; Eschenbach, C.; et al. The Coastal Observing System for Northern and Arctic Seas (COSYNA). *Ocean Sci.* **2017**, *13*, 379–410. [CrossRef]

51. Stanev, E.V.; Schulz-Stellenfleth, J.; Staneva, J.; Grayek, S.; Grashorn, S.; Behrens, A.; Koch, W.; Pein, J. Ocean Forecasting for the German Bight: From Regional to Coastal Scales. *Ocean Sci.* **2016**, *12*, 1105–1136. [CrossRef]

52. Gurgel, K.W. Remarks on Signal Processing in HF Radars Using FMCW Modulation. *IRS* **2009**, 63–67. Available online: http://wera.cen.uni-hamburg.de/pub_70.pdf (accessed on 28 January 2020).

53. De Dominicis, M.; Pinardi, N.; Zodiatis, G.; Archetti, R. MEDSLIK-II, a Lagrangian Marine Surface Oil Spill Model for Short-Term Forecasting – Part 2: Numerical Simulations and Validations. *Geosci. Model Dev.* **2013**, *6*, 1871–1888. [CrossRef]

54. Savitzky, A.; Golay, M.J.E. Smoothing and Differentiation of Data by Simplified Least Squares Procedures. *Anal. Chem.* **1964**, *36*, 1627–1639. [CrossRef]

55. Röhrs, J.; Sperrevik, A.K.; Christensen, K.H.; Broström, G.; Breivik, Ø. Comparison of HF Radar Measurements with Eulerian and Lagrangian Surface Currents. *Ocean Dyn.* **2015**, *65*, 679–690. [CrossRef]

56. Ardhuin, F.; Marié, L.; Rascle, N.; Forget, P.; Roland, A. Observation and Estimation of Lagrangian, Stokes, and Eulerian Currents Induced by Wind and Waves at the Sea Surface. *J. Phys. Oceanogr.* **2009**, *39*, 2820–2838. [CrossRef]

57. Niiler, P.P.; Sybrandy, A.S.; Bi, K.; Poulain, P.M.; Bitterman, D. Measurements of the Water-Following Capability of Holey-Sock and TRISTAR Drifters. *Deep Sea Res. Part I Oceanogr. Res. Pap.* **1995**, *42*, 1951–1964. [CrossRef]

58. Stanev, E.V.; Badewien, T.H.; Freund, H.; Grayek, S.; Hahner, F.; Meyerjürgens, J.; Ricker, M.; Schöneich-Argent, R.I.; Wolff, J.-O.; Zielinski, O. Extreme Westward Surface Drift in the North Sea: Public Reports of Stranded Drifters and Lagrangian Tracking. *Cont. Shelf Res.* **2019**, *177*, 24–32. [CrossRef]

59. Wiese, A.; Staneva, J.; Schulz-Stellenfleth, J.; Behrens, A.; Fenoglio-Marc, L.; Bidlot, J.-R. Synergy of Wind Wave Model Simulations and Satellite Observations during Extreme Events. *Ocean Sci.* **2018**, *14*, 1503–1521. [CrossRef]

60. Breivik, Ø.; Bekkvik, T.C.; Wettre, C.; Ommundsen, A. BAKTRAK: Backtracking Drifting Objects Using an Iterative Algorithm with a Forward Trajectory Model. *Ocean Dyn.* **2012**, *62*, 239–252. [CrossRef]

61. Strand, K.O.; Vikebø, F.; Sundby, S.; Sperrevik, A.K.; Breivik, Ø. Subsurface Maxima in Buoyant Fish Eggs Indicate Vertical Velocity Shear and Spatially Limited Spawning Grounds. *Limnol. Oceanogr.* **2019**, *64*, 1239–1251. [CrossRef]

62. Kukulka, T.; Proskurowski, G.; Morét-Ferguson, S.; Meyer, D.W.; Law, K.L. The Effect of Wind Mixing on the Vertical Distribution of Buoyant Plastic Debris. *Geophys. Res. Lett.* **2012**, *39*. [CrossRef]

63. Van Sebille, E.; England, M.H.; Froyland, G. Origin, Dynamics and Evolution of Ocean Garbage Patches from Observed Surface Drifters. *Environ. Res. Lett.* **2012**, *7*, 044040. [CrossRef]

64. Sutherland, G.; Soontiens, N.; Davidson, F.; Smith, G.C.; Bernier, N.; Blanken, H.; Schillinger, D.; Marcotte, G.; Röhrs, J.; Dagestad, K.-F.; et al. Evaluating the Leeway Coefficient of Ocean Drifters Using Operational Marine Environmental Prediction Systems. *J. Atmos. Ocean. Technol.* **2020**, *37*, 1943–1954. [CrossRef]

65. Stanev, E.V.; Ricker, M. Interactions between Barotropic Tides and Mesoscale Processes in Deep Ocean and Shelf Regions. *Ocean Dyn.* **2020**, *70*, 713–728. [CrossRef]

66. Hasselmann, K. Wave-driven Inertial Oscillations. *Geophys. Fluid Dyn.* **1970**, *1*, 463–502. [CrossRef]

67. Breivik, Ø.; Bidlot, J.-R.; Janssen, P.A.E.M. A Stokes Drift Approximation Based on the Phillips Spectrum. *Ocean Model.* **2016**, *100*, 49–56. [CrossRef]

68. Li, Q.; Fox-Kemper, B.; Breivik, Ø.; Webb, A. Statistical Models of Global Langmuir Mixing. *Ocean Model.* **2017**, *113*, 95–114. [CrossRef]

69. Craig, P.D.; Banner, M.L. Modeling Wave-Enhanced Turbulence in the Ocean Surface Layer. *J. Phys. Oceanogr.* **1994**, *24*, 2546–2559. [CrossRef]

Article

Removing Wave Bias from Velocity Measurements for Tracer Transport: The Harmonic Analysis Approach

Sangdon So [1,*], Arnoldo Valle-Levinson [1], Jorge Armando Laurel-Castillo [2], Junyong Ahn [3] and Mohammad Al-Khaldi [4]

1   Civil and Coastal Engineering, University of Florida, Gainesville, FL 32611, USA
2   National Water Commission, Mexico City 04340, Mexico
3   Mechanical and Civil Engineering, Florida Institute of Technology, Melbourne, FL 32901, USA
4   Kuwait Institute for Scientific Research, Safat 13109, Kuwait
*   Correspondence: sangdon@ufl.edu

Received: 4 September 2019; Accepted: 13 April 2020; Published: 16 April 2020

Abstract: Estimates of turbulence properties with Acoustic Doppler Current Profiler (ADCP) measurements can be muddled by the influence of wave orbital velocities. Previous methods—Variance Fit, Vertical Adaptive Filtering (VAF), and Cospectra Fit (CF)—have tried to eliminate wave-induced contamination. However, those methods may not perform well in relatively energetic surface gravity wave or internal wave conditions. The Harmonic Analysis (HA) method proposed here uses power spectral density to identify waves and least squares fits to reconstruct the identified wave signals in current velocity measurements. Then, those reconstructed wave signals are eliminated from the original measurements. Datasets from the northeastern Gulf of Mexico and Cape Canaveral, Florida, are used to test this approach and compare it with the VAF method. Reynolds stress estimates from the HA method agree with the VAF method in the lower half of the water column because wave energy decays with depth. The HA method performs better than the VAF method near the surface during pulses of increased surface gravity wave energy.

Keywords: turbulence; ADCP measurement; wave bias; Reynolds stress

---

1. Introduction

The range of fluid and tracer dynamics in aquatic systems extends from basin scale, up to approximately 10,000 km in the ocean, to turbulence scale, approximately 1 mm, which is related to the dissipation of Turbulent Kinetic Energy (m^2/s^3 or Watt/kg) [1–3]. Turbulent Kinetic Energy has a critical effect in transporting and changing the local concentration of ocean tracers such as dissolved oxygen, carbon dioxide, nutrients, plankton, and pollutants. Processes at the air–water interface can influence the transport and transformation of tracers in the water column. For instance, wind-driven waves can affect tracer concentration and distribution through advection and diffusion. In particular, diffusive processes should be dominated by turbulence, which may be biased by waves because of their overlap in frequency and scale. Thus, the study of diffusion and turbulence in the water column, and its identification from waves' influence, should allow an understanding of vertical exchange processes that determine the fate of tracers in aquatic environments. It follows that estimating Turbulent Kinetic Energy, including its transport, production, and dissipation, is obscured by the presence of surface waves because, as mentioned above, waves and turbulence share spectral energy bands. A reliable approach for velocity measurements is needed to distinguish the signal related to waves and to turbulence.

Most fluid motions in nature and engineering are turbulent [4]. Measurements of turbulence in the coastal environment can bolster our understanding of vertical exchange of momentum, energy, heat, and any tracer (e.g., nutrients), and they can help us parameterize turbulence in boundary layers [5–8]. Microstructure measurements have been used to estimate turbulence in the marine environment with devices that either fall, rise, or move horizontally (on an AUV). However, these microstructure measurements are labor intensive and require a dedicated platform [9,10]. Although this approach can produce reliable results, it is logistically and financially unfeasible to collect continuous long-term spatial—in the water column—series of turbulence [11]. The use of anchored shipboard or bottom-mounted Acoustic Doppler Current Profilers (ADCPs) has allowed measurements of time series of turbulent parameters in the water column [9,12,13].

Estimates of tracer transport under turbulent conditions in shallow coastal regions throughout the world may be contaminated by wave action. Turbulence properties may be confused by velocity fluctuations that include orbital velocities generated by multi-scale and multi-directional waves. Unless the effects of wave orbital velocities are eliminated from turbulence assessments, descriptions of tracer transport will be overestimated by at least one order of magnitude. Therefore, in studies of tracer transport, it is essential to implement a reliable approach to eliminate, remove, or at least reduce wave bias.

2. Previous Methods for Removing Wave Bias

Direct turbulence measurements can be overwhelmed by wave motion in coastal and ocean environments [14–16]. Observational studies have recognized that velocity covariance produced by waves can be one order of magnitude larger even than the covariance generated by an instrument tilt [16,17]. Several methods have been introduced to remove wave contamination from turbulent velocity measurements. The Velocity Differencing method, introduced by Lohrmann et al. (1990), is based on the assumption that the correlation scale of wave-induced velocity is larger than that of the turbulence. Wave contamination can be reduced by contrasting measurements between two velocimeters that are separated by a distance where the turbulent fluctuations are uncorrelated [17]. This differencing method was extended to ADCPs and named the Variance Fit method [13].

The Variance Fit method [13,14] assumes that the wave-induced velocities along ADCP beams are in phase and that velocities at two different vertical positions are scaled by a constant. An adaptive filtering method was introduced by Shaw and Trowbridge (2001) [15] assuming the wave-induced fluctuations are 100% coherent in space. Rosman et al. (2008) [18] applied the adaptive filtering method to ADCP measurements. Vertical Adaptive Filtering (VAF) and Horizontal Adaptive Filtering (HAF) methods relax the assumptions of the Variance Fit method [18]. These adaptive filtering methods assume that the velocities at a position are a linear function of the velocities at another position rather than a constant [18]. Both of these methods reduce wave bias to an acceptable level in relatively "mild" wave climates [16] by differencing velocities measured by two sensors. Gerbi et al. (2008) [19] proposed an alternative approach showing that the cospectra of the Reynolds stress ($u'w'$, where u' and w' are fluctuating parts of horizontal and vertical velocity, respectively) are approximately constant at low frequencies and roll off with a $-7/3$ slope in the inertial subrange [16]. This method was later extended to ADCP observations by Kirincich et al. (2010) [16]. Scannell et al. (2017) [20] introduced a method to remove the wave bias in the Turbulent Kinetic Energy Dissipation Rate (ε), which is estimated by a modified structure function. From ADCP data at three depths, they demonstrated that the modified structure function is effective in removing wave bias, compared to the standard structure function [10]. However, this method is limited to the estimation of a structure function.

Free fall profilers recording at high frequencies (>100 Hz) have been developed to observe turbulence. However, data collection is labor-intensive and limits observational periods (Williams and Simpson, 2004). While ADCPs were not ideal to measure turbulence in coastal environments because of their coarse sampling frequency, they have been used widely to derive Reynolds stresses $\overline{\rho u'w'}$ and $\overline{\rho v'w'}$ [12,13]. The variance method for extracting Reynolds stresses from ADCP observations has been applied by different researchers [13,16,18]. In the absence of instrument misalignment and waves, the Reynolds stresses can be computed from the difference of beam velocity variance along opposing beams [12,21]:

$$-\overline{u'w'} = \frac{\overline{u'^2_2} - \overline{u'^2_1}}{4sin\theta cos\theta} \tag{1}$$

$$-\overline{v'w'} = \frac{\overline{u'^2_4} - \overline{u'^2_3}}{4sin\theta cos\theta} \tag{2}$$

where u'_i is the demeaned along-beam velocity of 4 beams which are positive toward the transducer, and θ is the angle formed by the beam and the vertical. By comparing with results from Acoustic Doppler Velocimeters (ADVs), the variance method was validated in Souza and Howarth (2005) [22] and Nidzieko et al. (2006) [23]. However, the variance method failed in the presence of surface waves, because their orbital velocities are orders of magnitude larger than turbulent velocities and bias the estimates [16].

In the presence of surface waves, the velocity components can be decomposed as:

$$u_i = \overline{u}_i + \widetilde{u}_i + u'_i \tag{3}$$

where an overbar denotes a temporal mean value, and tilde and prime are orbital velocities and turbulent fluctuations, respectively. Substituting Equation (3) into Equation (2) yields

$$\frac{\overline{\left(\widetilde{u}_4 + u'_4\right)^2} - \overline{\left(\widetilde{u}_3 + u'_3\right)^2}}{4sin\,\theta\,cos\,\theta} = \underbrace{-\overline{v'w'}}_{} \underbrace{-\overline{\widetilde{v}\widetilde{w}} + \phi_p\left(\overline{\widetilde{v}^2} - \overline{\widetilde{w}^2}\right) - \phi_R\overline{\widetilde{u}\widetilde{v}}}_{E_{tilt}} + \underbrace{\phi_p\left(\overline{v'^2} - \overline{w'^2}\right) - \phi_R\overline{u'v'}}_{E_{turb}} \tag{4}$$

$$\underbrace{\phantom{-\overline{\widetilde{v}\widetilde{w}}}}_{E_{ws}}$$

with the corresponding expression for Equation (1), where ϕ_P and ϕ_R are instrument tilts, which are commonly referred to as pitch and roll, respectively. The derivation of wave-induced bias is illustrated in detail by Rosman et al. (2008) [18] and Trowbridge (1998) [17]. On the right-hand side of Equation (4), the first term is the unbiased Reynolds stress [18], which is the quantity of interest. The second term E_{ws} is the error due to wave stresses. The third term E_{tilt} includes errors of the interaction between waves and the instrument misalignment. The fourth term, the turbulence bias E_{turb}, is insignificant if the sensor misalignment is small [17]. For tilts <3°, the proposed E_{turb} is less than 50% of the true Reynolds stress in the absence of waves [18]. The wave-induced errors, E_{ws} and E_{tilt}, are one order of magnitude larger than the Reynolds stresses even if the tilt is small. E_{ws} can be removed by rotating the principal axis where waves propagate [17]. To reduce the error to 10% of the Reynolds stress, one must be able to establish the principal axis, which is not feasible [17] in an open coastal area. Therefore, it is essential to remove wave contamination from the beam velocities before applying the variance method.

The bias produced by surface waves can be reduced to an acceptable level by differencing velocities measured by two sensors [17]. In order to be effective, the two sensors must be separated by a distance shorter than the predominant wavelength and longer than the distance at which the turbulent fluctuations are correlated [15,17]. Hereafter, this method is referred to as the Differencing method. An extension of the Differencing method to ADCP measurements [13] consists of vertically differencing beam velocities (Δu) at two locations. One of the velocities is scaled by a parameter β that accounts for wave attenuation, in accordance with linear wave theory.

$$\Delta u_3 \equiv u_{3B} - \beta u_{3A} = u'_{3B} - \beta u'_{3A} + \widetilde{u}_{3B} - \beta\widetilde{u}_{3A} \tag{5}$$

$$\Delta u_4 \equiv u_{4B} - \beta u_{4A} = u'_{4B} - \beta u'_{4A} + \widetilde{u}_{4B} - \beta\widetilde{u}_{4A} \tag{6}$$

where u_3 and u_4 denote the demeaned beam velocities, and A and B are two different vertical positions with A being farther from the transducer than B. The scaling parameter β can be chosen to minimize the difference of wave orbital velocity variance between two levels:

$$\beta \approx \sqrt{\frac{\overline{\widetilde{u}_B^2}}{\overline{\widetilde{u}_A^2}}} \tag{7}$$

where wave variances $\widetilde{u}_a^2$ and $\widetilde{u}_b^2$ are computed from linear wave theory. Wave height (H), frequency (ω), and wave number (k) are determined by fitting a model variance vertical profile, $\overline{\widetilde{u}_{beam}^2}$, to the observations of velocity variance (Whipple et al., 2006)

$$\overline{\widetilde{u}_{beam}^2} = \left(-\frac{H^2\omega^2}{16}\right)[\operatorname{csch}(kh)]^2[\cos(2\theta) - \cosh 2k(h+z)] \tag{8}$$

where z is the vertical coordinate (zero at the water surface and positive upward), and θ is the angle formed by the beam and the vertical (approximately 20°).

Assuming β is the same for opposing beams, the variance of the remainder yields:

$$\overline{\Delta u_3^2} = \overline{u'^2_{3B}} + \beta^2\overline{u'^2_{3A}} - 2\beta\overline{u'_{3B}u'_{3A}} \tag{9}$$

$$\overline{\Delta u_4^2} = \overline{u'^2_{4B}} + \beta^2\overline{u'^2_{4A}} - 2\beta\overline{u'_{4B}u'_{4A}} \tag{10}$$

If the distance between A and B is chosen to be such that the turbulent fluctuations are uncorrelated, the third terms on the right hand side of Equations (9) and (10) are zero. The application of the variance method of Equation (2) gives:

$$\frac{\overline{\Delta u_3^2} - \overline{\Delta u_4^2}}{4\sin\theta\cos\theta} \approx \frac{\overline{u'_{3B}}^2 - \overline{u'_{4B}}^2}{4\sin\theta\cos\theta} + \frac{\beta^2\left(\overline{u'_{3A}}^2 - \overline{u'_{4A}}^2\right)}{4\sin\theta\cos\theta}. \tag{11}$$

Therefore, an average value of the Reynolds stress between positions A and B is:

$$\overline{v'w'}^{(B-A)} \approx \frac{\overline{\Delta u_3^2} - \overline{\Delta u_4^2}}{4\sin\theta\cos\theta\left(1 + \beta^2\right)} \tag{12}$$

with the corresponding equation along beams 1 and 2.

Under "high wave energy", the Differencing method in Equations (5) and (6) can fail to reduce the wave bias to an acceptable level due to the differences of amplitude and phase between two locations [15]. To minimize the wave-induced differences in Equations (5) and (6), Shaw and Trowbridge (2001) [15] used linear filtering techniques. Based on the Differencing method with linear filtering, Rosman et al. (2008) [18] developed vertical and horizontal Differencing methods with Adaptive Filtering, which are henceforth referred to as VAF and HAF, respectively. Instead of using the constant parameter β, Rosman et al. (2008) [18], and Shaw and Trowbridge (2001) [15] described the wave velocity at z^A as a linear function (L) of the wave velocity at z^B by assuming that wave-induced velocities are spatially coherent.

$$\widetilde{u}_{beam}^A(t) = L\left(\widetilde{u}_{beam}^B\right) = \int_{-\infty}^{\infty} s(t^*)\widetilde{u}_{beam}^B(t - t^*)dt^* \tag{13}$$

where t is time, t^* is an integration variable, and $s(t^*)$ is a continuous function that relates $\widetilde{u}_{beam}^B$ to $\widetilde{u}_{beam}^A$. The VAF and HAF method are explained in detail by Rosman et al. (2008) [18].

The Cospectra-Fit (CF) method [16] assumes that the momentum and heat in the bottom boundary layer of the atmosphere are transported in the upper ocean with similar turbulence scales as predicted by Kaimel et al. (1972) [24]. Based on observations of the atmospheric boundary layer, Kaimel et al. (1972) [24] determined that the cospectra of Reynolds stress are approximately constant at low frequencies but roll off as a $-7/3$ power law in the inertial subrange. According to Gerbi et al. (2008) [19], the spectral shape of turbulence cospectral energy Co^*_{uw} can be expressed as a function of wave number k, where $k = 2\pi/\lambda$ and λ is a turbulence length scale, as follows:

$$Co^*_{uw}(k) = \overline{u'w'}^* \left(\frac{7}{3\pi} \sin\frac{3\pi}{7}\right) \frac{1/k_0}{1 + (k/k_0)^{7/3}} \tag{14}$$

where the cospectra are denoted by an asterisk. The model turbulence cospectrum Co^*_{uw} [16] can be described as the integral of the cospectrum $(\overline{u'w'}^*)$, the covariances of two signals u' and w', and a "roll off" wave number k_0 (a measurement of the dominant length scale of turbulent fluctuations). The CF method involves fitting the model cospectrum (Equation (14)) to the observed cospectrum at frequencies below those of surface gravity waves [16]. If the observed cospectra are integrated in the entire frequency range, the resulting covariance $\overline{u'w'}^*$ is one or two orders of magnitude larger than the expected covariance, which is likely caused by a combination of instrument misalignment and waves [19]. From ADV data obtained at 1–3 m below the sea surface, Gerbi et al. (2008) [19] successfully estimated the unbiased Reynolds stress with the CF method.

By applying the CF method to ADCP data, unbiased Reynolds stresses were also computed by Kirincich et al. (2010) [16]. The variance method for one-side cospectra of horizontal and vertical velocities can be expressed as

$$Co_{uw}(\omega) = \frac{S_{u_1 u_1}(\omega) - S_{u_2 u_2}(\omega)}{4\cos\theta\sin\theta} \tag{15}$$

$$Co_{vw}(\omega) = \frac{S_{u_3 u_3}(\omega) - S_{u_4 u_4}(\omega)}{4\cos\theta\sin\theta} \tag{16}$$

where $S_{u_i u_i}$ is the velocity spectra of *i*th beam velocities. Integration of Equations (15) and (16) in the entire frequency range, i.e., the covariances, yields the Reynolds stresses $\overline{u'w'}$ and $\overline{v'w'}$. The wave-band cut-off frequency (ω_{wc}) for fitting to the below-wave band cospectra was estimated [16] by comparing the beam velocity spectrum to the vertical velocity spectrum derived by linear wave theory. Kirincich et al. (2010) [16] showed that the CF method minimizes wave-induced errors and yields more realistic estimates of near-bottom Reynolds stress than the VAF method. The method introduced here, the Harmonic Analysis (HA) method, relaxes the assumptions (turbulence scales < wave scales) that are used in the Differencing methods and the VAF method. The method also overcomes the assumptions of constant phase and linearity. Without considering correlation and coherence, the HA method is likely to preserve the wave-induced turbulence, which may be correlated between two locations. The HA method is tested by using a 1200 kHz RDI workhorse ADCP dataset from the northeastern Gulf of Mexico, in the Florida Big Bend region and by using a 1000 kHz RDI Sentinel V dataset from a shoal off Cape Canaveral, Florida.

3. Harmonic Analysis (HA) Method

Wave bias is only partially removed with simple filtering approaches because waves often overlap in frequency range with turbulence [18]. The approach proposed here, the HA method, identifies wave harmonics from ADCP measurements by using Power Spectral Density (PSD) analysis and then fitting, via least squares [25], those harmonics identified from the PSD distribution to the ADCP measurements. The HA method is illustrated with a synthetic signal (Supplementary Material, Section 2) and with an approximately 8-day time series of sea surface height (m) and its PSD (m^2/Hz) (Figure 1A,D). Wave energy is mostly distributed at frequencies higher than 0.09 Hz (Figure 1D). The highest significant wave height (Figure 1B) is approximately 0.4 m with a dominant period of 6 s at 20:00, on 22 November

2009. The dominant periods of the waves vary from 2 to 10 s, and the average periods change from 3 to 6.4 s (Figure 1C).

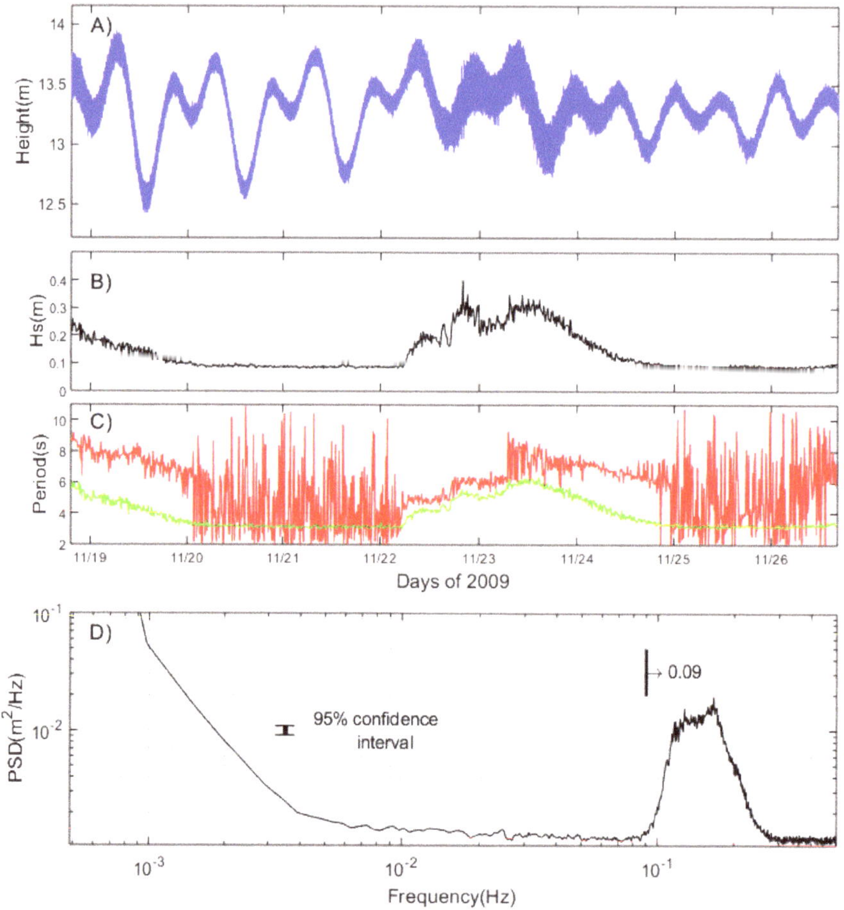

Figure 1. Data from the northeastern Gulf of Mexico, in the Florida Big Bend region. (**A**) Time series of water surface height (m), (**B**) Significant wave heights (Hs), (**C**) Dominant wave period (DPD) in red and Average wave period (APD) in green, and (**D**) Power Spectral Density (m^2/Hz) of the time series shown in (**A**).

Frequencies associated with wave energy are identified from the PSD of the four beam velocities at the bin closest to the surface (top bin, 0.5 m thick). All wave frequencies identified from each PSD are then fitted to each of the four beam velocities for the top bin. This fit provides the amplitude and phase of the wave orbital velocities that are then reconstructed and subtracted from the beam velocities recorded by the ADCP.

The HA method is applied to beam velocities because they resolve the rapidly changing orbital velocities better than the surface height. Details of the HA method are now illustrated with a segment (approximately 40 min) of the top-bin Beam1 velocity data (Figure 2A). The blue line is the top-bin Beam1 velocity measured at 1 Hz and the red lines delimit 600-s ensembles used to calculate turbulence parameters. Absolute values of pitch and roll tilts remained below 3 degrees throughout the deployment.

The HA method works under the assumption that turbulence is stationary for 10 min. Although PSDs can be estimated with 10-min ensembles, the actual PSDs are calculated with ensembles of 1024 samples (approximately 17 min or 1024 s) to avoid padding with zeros. Each PSD ensemble has overlaps of 424 s with each other, extending 212 s before and after each 10-min ensemble (e.g., Figure 2A). The black solid lines in Figure 2A indicate the window or ensemble size for the PSD calculation. The PSD for each black-line window of Figure 2A appears in Figure 2B. The shaded area in Figure 2B (frequencies > 0.09 Hz) denotes the wave-energy frequency band, which is the portion in need of wave-bias removal.

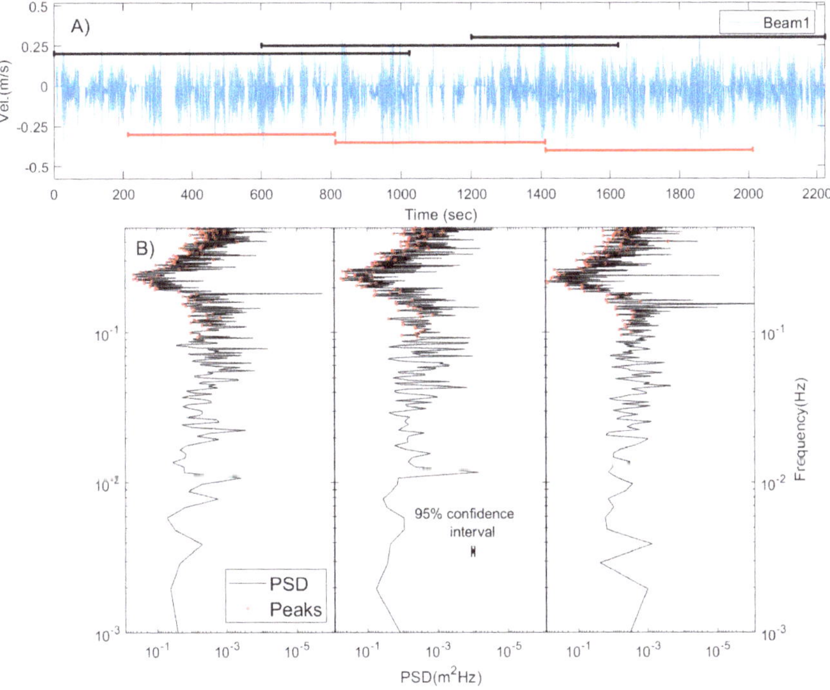

Figure 2. (**A**) Beam1 velocities at 9.04 m above the bottom (blue line) during a 37-min segment. Data windows to compute Power Spectral Density (PSDs) (m^2/s, black lines), and ensembles used to calculate Reynolds stresses ($-\overline{u'w'}$, red line). (**B**) PSDs calculated with the data windows denoted by black lines in (A); the shaded area indicates the frequency band in need of wave-bias removal. The red circles denote peaks at the frequencies identified to reconstrut wave signals.

Subsequently, approximately 8 days of top-bin beam velocities (Figure 3A for Beam1) are used to calculate PSDs for all 10-min segments, as explained in Figure 2. The first step, after PSD calculation of each ensemble, is to find spectral peaks (red circles in Figure 2B) in the frequency range between 0.09 and 0.5 Hz. Each peak is assumed to be related to wave orbital velocities and identified as having larger spectral density than two neighboring values (please also see the Supplementary Material, Section 2). The next step is to sort the spectral peaks by descending order of spectral value (Figure 3B). This is done to identify X% of the wave energy, e.g., 95%, and remove it from the beam velocity. The vertical axis in Figure 3B–D indicates the number of peaks, or harmonics, identified in each 10-min ensemble. The shaded contour plot of sorted spectral peaks (Figure 3B) shows increased spectral energy at times with more harmonics containing the highest PSD (yellow-shaded areas). Increased spectral energies should appear at times of largest Beam1 velocity fluctuations (e.g., between 11/22 and 11/24; between 11/19 and 11/20; in the middle of 11/20; and in the second half of 11/25). The frequency corresponding to each sorted spectral peak is presented in Figure 3C. The largest spectral peaks between 11/22 and 11/24 correspond to frequencies between 0.1 and 0.2 Hz (blue regions in the contour plot). Similar frequency values are observed during other periods with largest variability in Beam1 velocity.

The following step is to remove X% of wave energy (e.g., 95%) from each 10-min ensemble. For this purpose, the PSDs are normalized with the sum of all spectral peaks (between 0.09 and 0.5 Hz) of that ensemble. In other words, each normalized PSD represents the ratio of each PSD value to the sum of all spectral peaks between 0.09 and 0.5 Hz. Then, normalized PSD values are integrated throughout all peaks of each 10-min ensemble to represent the Cumulative Normalized Power Spectral Density (CNPSD, Figure 3D). Thus, a value of 1 denotes 100% of the wave energy related to each ensemble. The white contour line in Figure 3C,D shows the number of peaks (or harmonics) representing 95% of the total wave energy from each ensemble (95% of the largest peaks in the PSDs of each ensemble). Overall, around 75 peaks represent 95% of the total energy of the peaks (Figure 3C,D). This number of peaks changes over time according to wave conditions. Relatively fewer spectral peaks contain 95% of the wave energy when the beam velocities are relatively higher. The crux of this method is to fit all harmonic frequencies, up to the white line, to each 10-min ensemble and thus obtain the amplitude and phase of each oscillatory harmonic, the wave signal. The fitted wave signal is finally subtracted from each 10-min beam velocity ensemble. This final procedure is as follows.

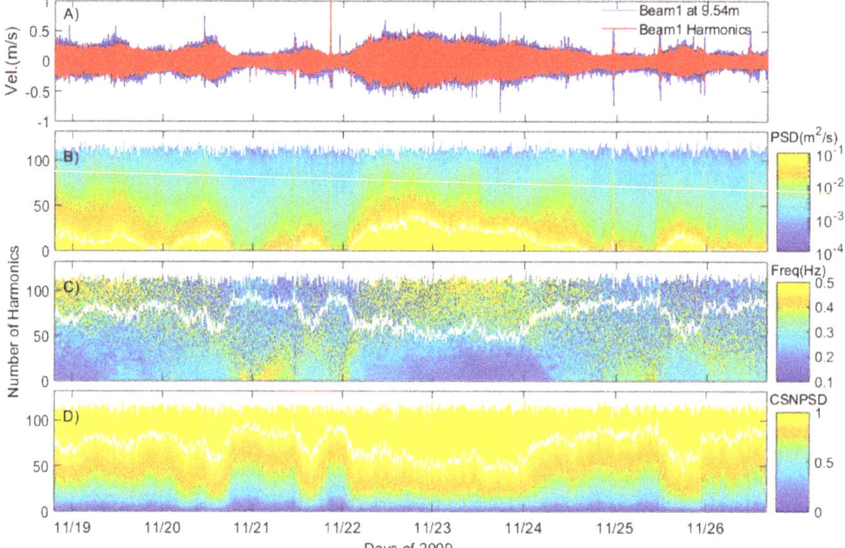

Figure 3. (**A**) Beam1 velocities (blue line) and wave orbital velocities (red line) computed by the Harmonic Analysis (HA) method; (**B**) PSDs (m²/s) of the peaks sorted by descending order; and 0.08 PSD (m²/s) in the white line (**C**) Frequencies corresponding with the sorted PSDs and (**D**) Cumulative Sum of Normalized Power Spectral Density (CNPSD). In both C and D, the white line represents the number of harmonics with 95% energy.

The wave orbital velocity may be represented by the sum of M harmonics plus a mean (u_0):

$$\widetilde{u}(t) = u_0 + \sum_{j=1}^{M} A_j \sin\left(\omega_j t + \phi_j\right) \tag{17}$$

where $\widetilde{u}(t)$ is an approximation to the wave orbital velocity of each 10-min ensemble. A_j, ω_j, and ϕ_j are the amplitude, angular frequency, and phase, respectively, associated with the spectral peaks. Values of ω_j are gleaned from the PSD peaks of the beam velocity 10-min ensemble u_{beam} (Figure 3C). Values of u_0, A_j, and ϕ_j are obtained by fitting the harmonics ω_j to u_{beam}. For a more in-depth explanation of the least squares fit to Equation (17), see the Supplementary Material, Section 1. The harmonic orbital velocities are reconstructed with u_0, A_j, and ϕ_j from Equation (17) (red line in Figure 3A). This signal is taken as the wave orbital velocities $\widetilde{u}(t)$, and then subtracted from the u_{beam}, i.e.,

$$u_{cor} = u_{beam} - \widetilde{u}(t). \tag{18}$$

The remaining signal (u_{cor}, or the difference between red and blue lines in Figure 3A) is assumed to be related predominantly to turbulent fluctuations and noise.

4. Comparison between VAF and HA Method

Power Spectral Densities derived from the VAF and HA methods are compared to the PSDs of original velocities over the entire water column for all 8 days of measurements (Figure 4). To evaluate the methods, PSDs are computed for every 10-min ensemble of the Beam1 velocities. The PSDs of original Beam1 velocities are shown in Figure 4A–D for different distances from the bottom. As shown in Figure 1B, surface gravity wave energy is predominantly distributed between 0.09 and 0.5 Hz, and it decreases with depth (Figure 4A–D). Surface gravity waves are evident between 11/19 and 11/21, evolving from low to high frequencies, and between 11/22 and 11/24, evolving from high to low frequencies. The shortest waves (>0.3 Hz) do not reach the bin closest to the bottom, while the longest waves (near 0.1 Hz) do reach those depths. This is expected from linear wave theory [26].

The PSDs corrected by VAF are shown in Figure 4E–H. A 2 m vertical separation is chosen to determine the continuous function in Equation (13), together with a 10-s window length for the weight function. Near the surface (9.04 m from the bottom, i.e., at the top bin) the VAF performs well at frequencies > 0.1 Hz. However, longer period waves are artificially amplified, especially between 11/22 and 11/23, when significant wave heights reach approximately 0.4 m (Figure 1B). The amplification of longer period waves also occurs at a distance of 7.04 m, although it is less prominent than near the surface. This artifact is caused by the least squares fit of the VAF method. Least squares Adaptive Filters amplify or generate signals, especially in longer period waves, to minimize the sum of squared residuals between two locations. Rosman et al. (2008) [18] explained that the linear transform does not sufficiently resolve the abrupt change of wave directions and phase difference between two locations during a 10-min averaging period. As a result, the VAF method does not perform well near the surface.

While the VAF method performs unreliably near the surface, the HA method removes most of the wave bias in the beam velocities (Figure 4I). The wave energy is drastically reduced between 0.09 and 0.5 Hz. Moreover, the VAF method loses vertical resolution at the surface because of its requirement to relate beam velocities at two locations. In this case, 2 m is used as the vertical separation to compute the linear function. Therefore, unbiased beam velocities are unattainable at 2 m below the surface. The wave energy is successfully removed at 1.04 m above the bottom (Figure 4L), but the VAF method generates pseudo-longer period waves in Figure 4H because VAF depends on the quality of the linear function [18] in Equation (13).

The wave signal removal by VAF and HA methods is illustrated on specific PSD plots (Figure 4M–P) corresponding to the black vertical lines of Figure 4A through L. Wave biases do not seem to be removed entirely between 11/22 and 11/24. However, the HA method reduces wave energies in the dominant peaks by approximately one order of magnitude relative to the original data. In addition, the wave energy resulting from the HA method is not amplified at low frequencies (<0.15 Hz), in contrast to the VAF method. This is particularly conspicuous in the signals at 7.04 m (Figure 4N) and 9.04 m (Figure 4M).

One way to evaluate the performance of the methods is to generate average PSD contours (Figure 5A) of all 10-min ensembles at each depth (Figure 4). The original average PSDs show wave energy predominantly at 0.17 Hz and at heights from the bottom above 5 m (Figure 5A). Deviations of the VAF and HA method from original ensemble-averaged PSDs are shown in Figure 5B,C, respectively. The deviations show that both VAF and HA methods remove the predominant wave energy at frequencies > 0.1 Hz. However, low-frequency wave energy is amplified with the VAF approach, especially at the surface and the bottom (Figure 5B). This is caused by the assumption that the wave velocity at A is a linear function of the wave velocity at B (Equation (13)). Wave velocities are unlikely to decay linearly with depth. The wave bias at frequencies > 0.1 Hz is approximately 10% less marked with the HA method than with the VAF. Moreover, spurious low frequency waves are not present in results obtained with the HA method (Figure 5C).

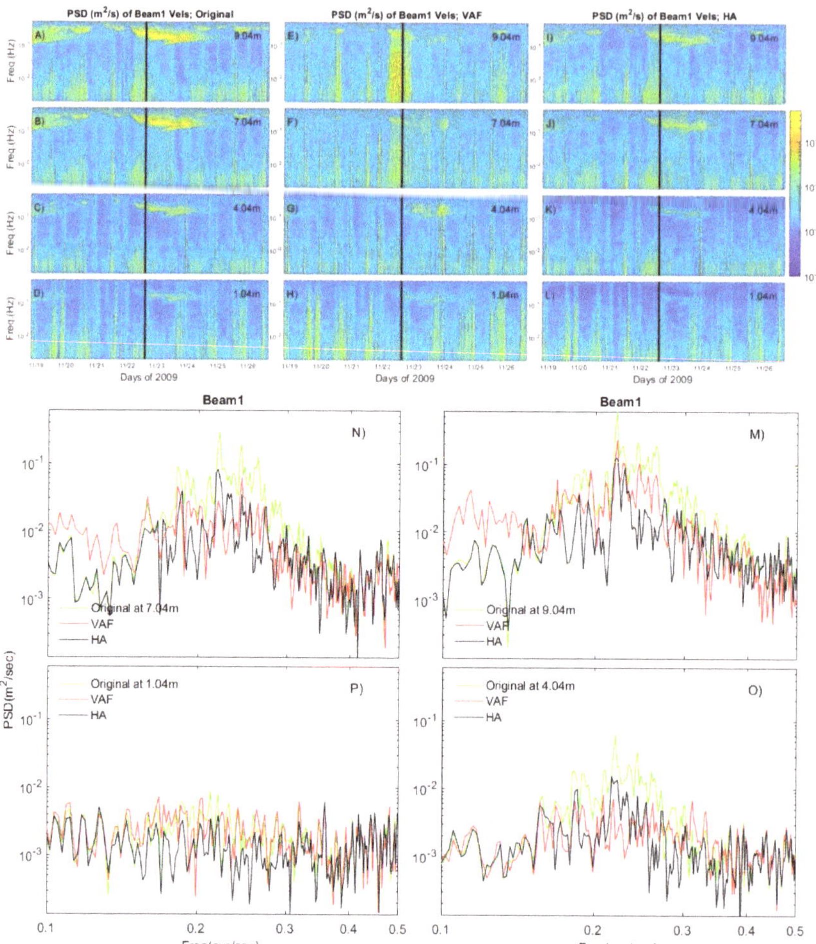

Figure 4. PSD (m^2/s) of original Beam1 velocities (m/s) at (**A**) 9.04 m, (**B**) 7.04 m, (**C**) 4.04 m, and (**D**) 1.04 m above the bottom; PSD of Beam1 velocities of the Vertical Adaptive Filtering (VAF) method at (**E**) 9.04 m, (**F**) 7.04 m, (**G**) 4.04 m, and (**H**) 1.04 m and PSD of Beam1 velocities of the HA method (**I**) 9.04 m, (**J**) 7.04 m, (**K**) 4.04 m, and (**L**) 1.04 m. Line plots: (**M–P**), show the PSD corresponding to the black vertical line in the contour plots.

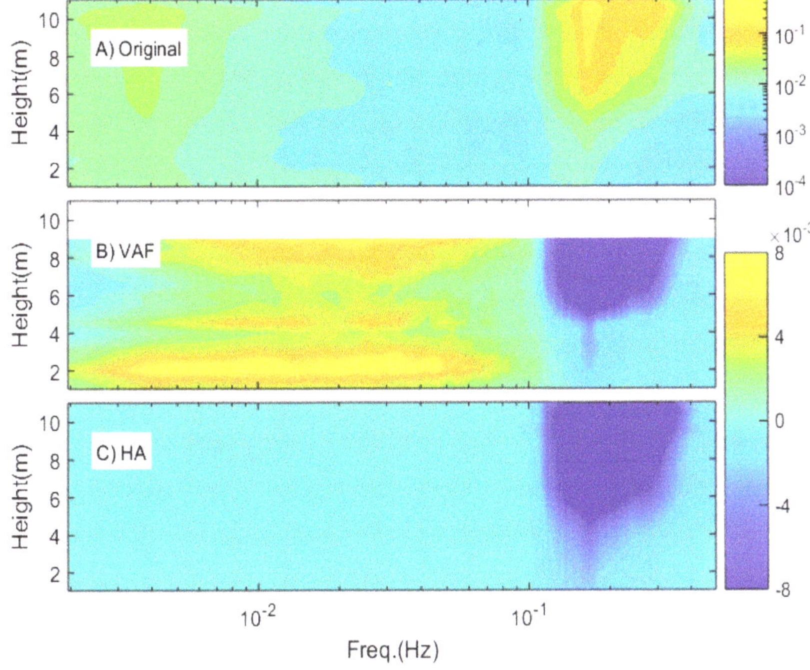

Figure 5. (**A**) Time-averaged PSD (m^2/s) as a function of height above bottom with original Beam1 velocities. (**B**) Difference between the VAF method and the original time-averaged PSD. (**C**) Difference between the HA method and original time-averaged PSD.

Another indicator of wave-bias removal is the cospectra of the beam velocities. Rosman et al. (2008) [18] indicated that the cospectra of u and w can be computed as

$$P_{-uw} = \frac{P_{u2u2} - P_{u1u1}}{4 \sin\theta \cos\theta} \tag{19}$$

where P_{u1u1} and P_{u2u2} are the Power Spectral Densities of Beam1 and Beam2, respectively. The cospectra of original beam velocities have increased energy at the surface and in the middle of the water column at frequencies > 0.1 Hz (Figure 6A). As expected, the wave energy is highest at the surface and decreases with depth. In addition, increased wave energy appears at an approximately 5 m height. The VAF method performs well at frequencies > 0.1 Hz (Figure 6B) but loses vertical resolution as it relates velocities between two vertical positions to obtain the linear function described in Section 3. The cospectra computed by the HA method provides reasonable estimates at the surface (Figure 6C). Decreased cospectra indicated that the peaks where u' and w' have the same frequencies are largely removed. The magnitude of the HA method cospectra is approximately 64% and 36% smaller than the original cospectra at the surface and bottom, respectively.

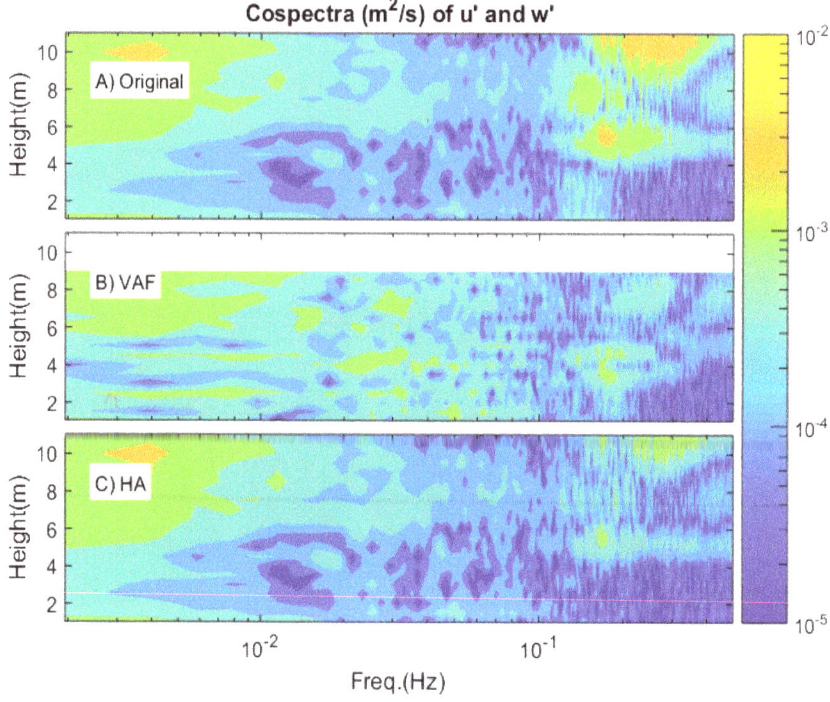

Figure 6. Time-averaged cospectra ($\mathrm{Co}_{u'w'}$, m²/s) of u' and w' (**A**) original; (**B**) corrected by VAF method; and (**C**) corrected by HA method.

5. Discussion

Turbulent Kinetic Energy was calculated with the original data and with the VAF and HA methods (Figure 7). The HA method was now applied by removing 95% and also 100% of the wave harmonics obtained from the PSD. The labels "HA-95%" and "HA-100%" in Figure 7C,D denote the Harmonic Analysis method that removes 95% and 100% of the wave energy. The sum of Turbulent Kinetic Energy (TKE) at 9.04 m was approximately 60% lower with the VAF method (Figure 7B) and approximately 67% lower with HA-95% (Figure 7C) relative to the original data. At 1.04 m, the TKE sum was approximately 24% less with the VAF method and approximately 55% less with the HA-95% method than with the original data. At 5.04 m, the TKE sum was approximately 34% less with the VAF method and approximately 63% less with the HA-95% method than that calculated with the original data (see also Table 1). The result of the TKE sum indicated that the HA method reduced more wave energy throughout the water column than the VAF method. Looking at the HA-100% approach, the TKE sum with HA-95% was approximately 15% more than the TKE sum with HA-100%. This shows a small change in the TKE budgets.

Table 1. Performance indicators of wave-removal methods in the estimates of turbulence parameters at different distances from the bottom. The HA method was more effective in reducing wave-related values of Turbulent Kinetic Energy (TKE) and Reynolds stress than the VAF method.

Method	TKE (% Reduction)			Reynolds Stress (Standard Dev $\times 10^{-3}$ m²/s²)	
	1.04 m	5.04 m	9.04 m	1.04 m	9.04 m
VAF	24	34	60	0.55	0.87
HA	55	63	67	0.46	0.58

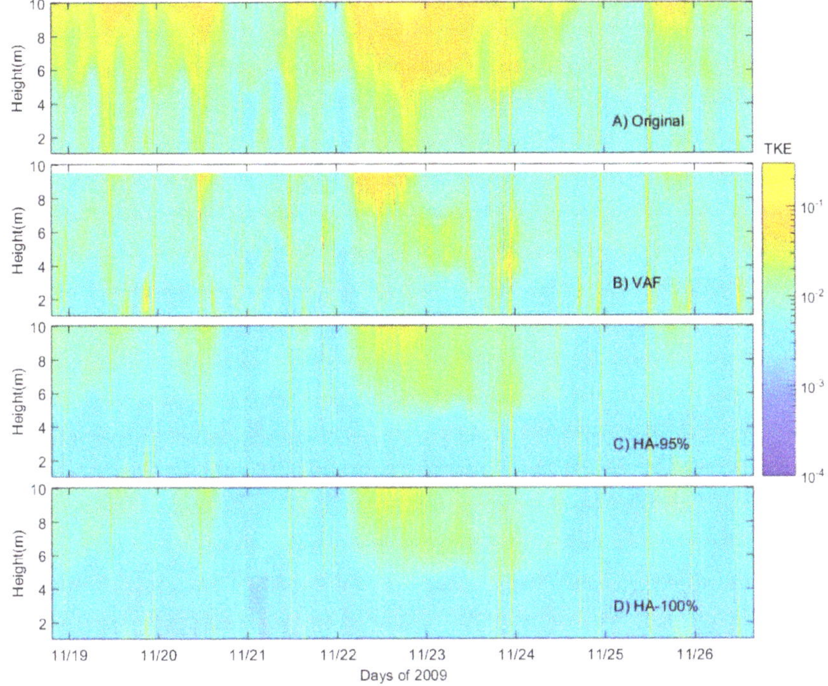

Figure 7. Turbulent Knetic Energy (TKE) estimated by (**A**) Original data, (**B**) VAF, (**C**) HA method with 95% wave energy removed, and (**D**) HA method with 100% wave energy removed.

In addition, Reynolds stresses ($-\overline{u'w'}$) were calculated with original velocities, with the VAF method and with the HA method at 3 different heights above the bottom (Figure 8). Compared to the original estimates and the VAF methods, the HA method gives the smallest standard deviation of Reynolds stresses at all depths. The standard deviation of Reynolds stresses for the HA method is approximately 0.58×10^{-3} m^2/s^2 at 9.04 m (close to the surface) and 0.46×10^{-3} m^2/s^2 at 1.04 m (closest to the bottom). Comparatively, the standard deviation of Reynolds stresses for the VAF method is approximately 0.87×10^{-3} and 0.55×10^{-3} m^2/s^2 at the same depths (Table 1). In other words, the Reynolds stresses standard deviation of the VAF method tends to be 50% greater than that of the HA method near the surface.

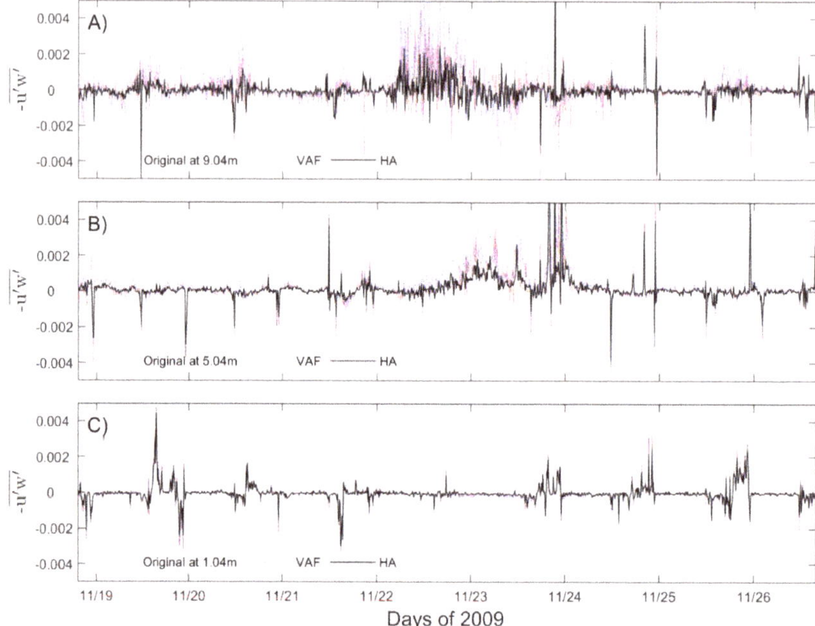

Figure 8. Reynolds stresses ($-\overline{u'w'}$) computed by original data (blue dotted line), VAF method (red dotted line), and HA method (black solid line) at (**A**) 9.04 m; at (**B**) 5.04 m; and at (**C**) 1.04 m.

Furthermore, an estimate of the variance of Reynolds stresses ($-u'w'$) with the VAF method [18] is given by:

$$var\left(\widehat{u'w'}\right) = \frac{var\left(\widehat{u_1\Delta u_1}\right) + var\left(\widehat{u_2\Delta u_2}\right)}{16\sin^2\theta\cos^2\theta} = \frac{1}{N}\frac{var(u_1\Delta u_1) + var(u_2\Delta u_2)}{16\sin^2\theta\cos^2\theta} \tag{20}$$

where $\widehat{u'w'}$ denotes the estimator of the Reynolds stress. For comparison, the variance of Reynolds stresses can be estimated for the original velocity perturbations, for the VAF method, and for the HA method following Equation (20). In particular, standard deviations of Reynolds stresses derived from the VAF and HA methods are normalized with those obtained from the original velocities to compare the reliability of the VAF method with that of the HA method (Figure 9). Contour values >1 indicate that the estimator ($u'w'$) of the VAF method is more broadly distributed than that of the original data (Figure 9A). Values > 1 from the VAF method are mostly found in the lower water column. This trend is caused by the difference between the velocity variance of the original beam velocities and the VAF method. As shown in Figure 4D,H, the Power Spectral Density for the original beam velocities is smaller than the Power Spectral Density for the VAF estimates. Based on the normalized standard deviation, the reliability in Reynolds stress estimates for the VAF method decreases toward the bottom of the water column. In contrast, contours of normalized standard deviations for the HA method are <1 everywhere (Figure 9B). This indicates that the reliability of the HA method increases in the entire water column, relative to the original estimates. The normalized standard deviations of the HA method are smallest at the top of the water column because the wave energy decreases with depth. Thus, the reliability of the Reynolds stresses is greater under relatively strong wave conditions (between 11/22 and 11/24) than under relatively weak wave conditions, as also noted in Figure 3A.

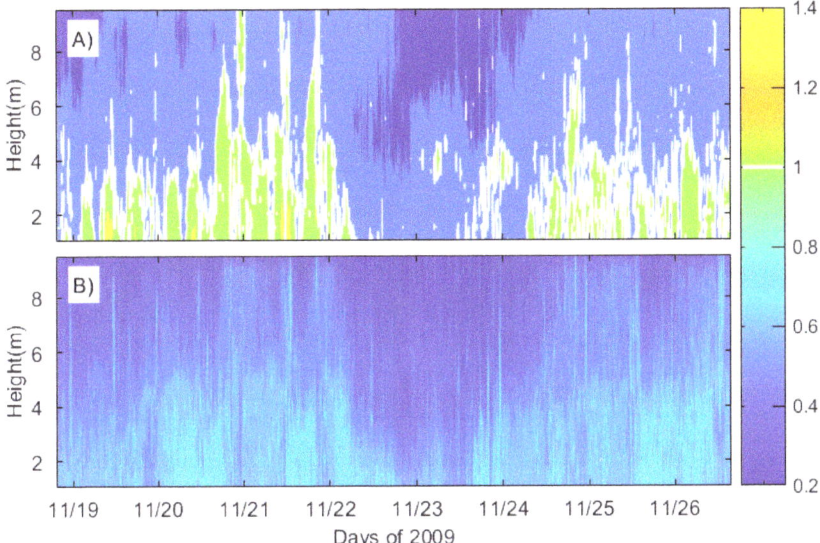

Figure 9. Standard deviation (**A**) of VAF method Reynolds stress estimates ($-u'w'$) and (**B**) of HA method Reynolds stress estimates, normalized by that of original Reynold stress estimates. The white line represents the value 1.

Rosman et al. (2008) [18] and Kirincich et al. (2010) [16] indicated that the VAF method is limited to mild wave climates, which would explain the difference with the HA method. To further evaluate the HA method in relatively more energetic wave conditions, the HA method is tested with a 4+ days, 2 Hz frequency dataset (Figure 10) over an approximately 8.5 m-depth water column near Cape Canaveral, Florida. The mean significant wave height (*Hs*) was approximately 1.0 m with a 10.2 s period. The highest significant wave height was 1.84 m with an approximately 12 s period at 16:05, August 30, 2015 (Figure 10A). The highest waves *Hs* > 1.5 m were steadily recorded from 11:30 on August 30 to 4:00 on August 31. The PSD of Beam1 velocity was estimated with a 10-min ensemble during the highest *Hs* (Figure 10B). The PSD of 95% wave-removed data (black line) decreases by approximately one order of magnitude at each peak identified, compared to the PSD of original data (green line). Moreover, the energy spectrum of the 95% wave-removed data cascades down with Kolmogorov's −5/3 slope (red line, consistent with turbulence dissipation). Contours of PSD with 95% of waves removed show energy reduction at all depths (Figure 10F–H).

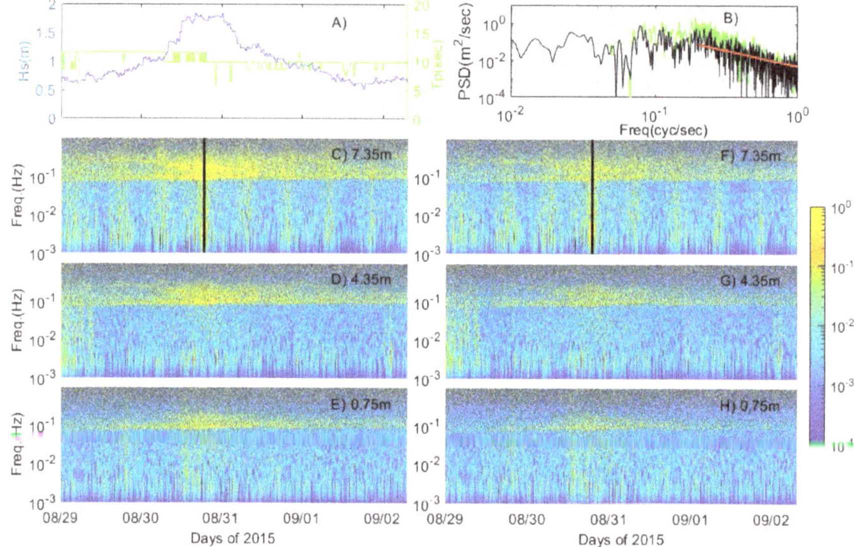

Figure 10. Data from a shoal off Cape Canaveral, Florida. (**A**) Significant wave heights (H$_s$) with a blue solid line marked on the left of the *y*-axis and dominant wave period (T$_P$) marked on the right of the *y*-axis. (**B**) The PSD corresponding to the black vertical line in (**C**) and (**F**). The green and black line denotes the PSD of original data and 95% wave-removed data, respectively. Red line is a reference line for –5/3. The PSD of original data are shown (**C**) at 7.35 m, (**D**) at 4.35 m, and (**E**) at 0.75 m height. The PSD of 95% wave-removed data are given (**F**) at 7.35 m, (**G**) at 4.35 m, and (**H**) at 0.75 m height.

Turbulent Kinetic Energy estimated with 95% and 100% wave removal (for all depths) indicates the sensitivity of the method to the percentage of wave removal. This is assessed through a rough TKE budget (Figure 11). The black dotted line has a slope of 1. Data falling on that line would indicate no wave elimination from the total TKE budget. The red and blue circles denote that the HA method reduces wave energy from the total TKE because they fall below the straight line. Turbulent Kinetic Energy with corrected data is reduced by approximately 50% for every 10-min ensemble, compared with original data (Figure 11). The difference between a 95% and a 100% wave removal is determined by the linear fit lines (red and blue). The mean slope is approximately 0.02 larger for the 95% removal (red line) than the 100% removal (blue line). This indicates that the HA method of 100% wave removal reduces 2% more wave energy than the 95% wave removal. However, the time to reconstruct harmonics with HA-100% was 10.5 times longer than the time taken by HA-95%. In this study, the difference between HA-95% and 100% was barely noticeable. The user of the HA approach may determine the most convenient implementation of wave percentage to remove. From these results, the HA method performs well under energetic wave conditions and in the surface boundary layer, where the VAF method would be inadequate.

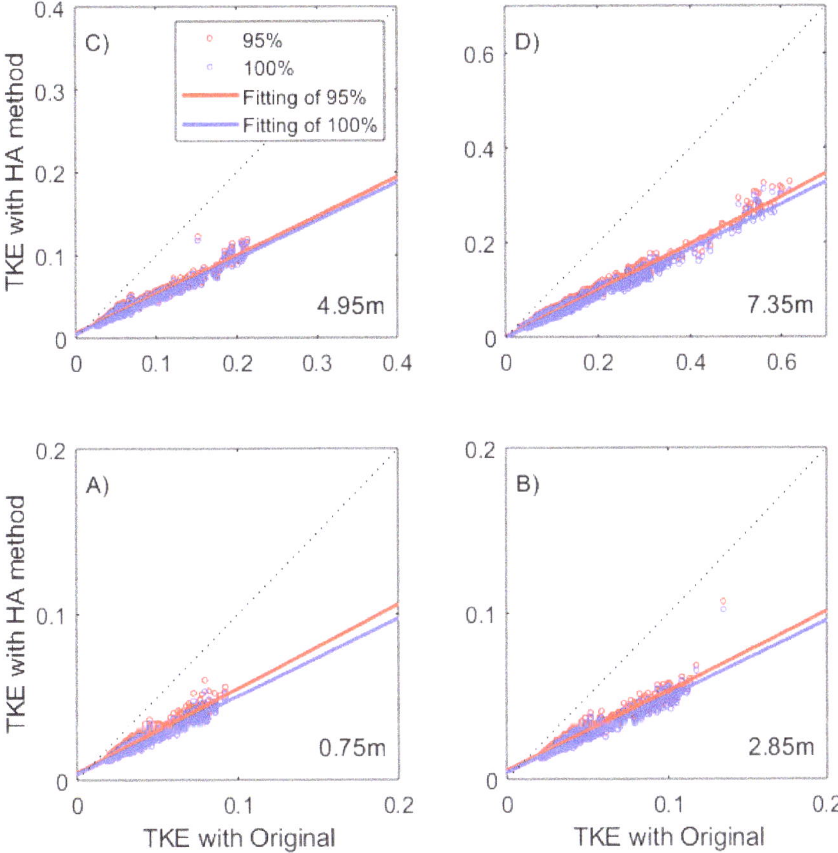

Figure 11. Comparison of TKE with original data to TKE with the HA method (**A**) at 0.75 m, (**B**) at 2.85 m, (**C**) at 4.95 m, and (**D**) at 7.35 m. Red and blue circles denote the estimation of 95% and 100% wave-removed data, respectively. Red and blue solid lines denote the corresponding linear-fitting line of the estimations. Black dotted line indicates the reference line of equilibrium between TKE with original data and TKE with the HA method.

6. Conclusions

Recent progress has been made to remove wave contamination from ADCP velocity observations for improved estimates of turbulence properties. The Variance Fit method [13] and the VAF method [18] have had some success in "mild" wave climates [16]. Moreover, the CF method [16,19] has provided accurate derivations of Reynolds stress in the presence of surface gravity waves. The CF method relies on the assumption that the momentum in the atmospheric boundary layer is transferred to the upper ocean layer at a similar scale of the velocity cospectra. This approach can allow reduction of the wave bias under energetic wave conditions. However, the cospectra may be dissimilar in the middle of the water column, especially when buoyancy may exist via a pycnocline and show internal waves. In contrast, the HA method, introduced here, relaxes the assumptions and limiting conditions (restricted to relatively weak surface waves). The HA method consists of (a) using Power Spectral Densities to identify dominant wave harmonics and (b) applying Least Squares fits of ADCP beam velocities to those harmonics to identify the wave signal. This wave signal is subsequently subtracted from the ADCP beam velocities.

The estimation of Power Spectral Densities in Figure 4 shows that the HA method reduces wave-related spectral densities by one order of magnitude. The Turbulent Kinetic Energy estimates in Figure 7 compare original data to the VAF and HA methods. It is evident that the HA method reduces more wave energy than the VAF method. The sum of TKE at 9.04 m is 20% larger with the VAF method than with the HA-95% method. This is because the HA method separated more wave energy than the VAF method. Furthermore, the HA method performs well in energetic wave conditions ($Hs > 1$ m, Figures 10 and 11). The HA method reduces approximately 50% TKE energy for every 10-min ensemble in energetic wave conditions. The u' and w' cospectra derived with the HA method near the surface and bottom are approximately 64% and 36% smaller than the original (wave-contaminated) cospectra. This implies that the wave energy of u and w is removed with the HA method. In addition, the HA method preserves the vertical resolution for the estimation of turbulence properties, while VAF loses vertical resolution as it relates velocities between two locations to obtain a required linear function.

Reynolds stress estimations and their reliability values show that the HA method performs better than the VAF method (Figures 8 and 9). The Reynolds stress standard deviations derived with the HA method are approximately 45% and 16% smaller than the original data standard deviation near the surface and bottom, respectively. Reynolds stresses calculated near the bottom show similar values among the three estimates (original, VAF, and HA) because wave energy decays with depth. When compared via regression analysis (Figure 8, regressions not shown), the Reynolds stresses of the VAF method are consistent with those of the HA method only in the lower half of the water column. However, there is a discrepancy between the two methods near the surface. This is because the VAF does not perform well under energetic wave conditions [16]. Reliability analysis based on the standard deviation of the Reynolds stresses indicates that deviations of the HA method are smaller at all depths than the standard deviations from original estimates and from the VAF method. Therefore, the HA method should dramatically reduce wave contamination under surface and internal wave conditions. This approach should allow an improvement of estimates of the diffusive component of tracer transport by removing wind-wave effects that share spectral bands with turbulence. The diffusive transport of tracers is likely to develop throughout the world in nearshore regions, around tidal inlets, and in estuaries where wave action can influence transport processes.

Supplementary Materials: The following are available online at http://www.mdpi.com/2073-4441/12/4/1138/s1, Figure S1: (A) Ten-minute synthetic signals of turbulence (red), waves (blue) and waves plus turbulence (green). The green line would be analogous to an ADCP beam-velocity ensemble. This green line is the signal to which we want to remove the wave influence. (B) Power spectral density (PSD) of the waves+turbulence line in A) is shown in green, and the PSD of waves (blue line in A) is shown in blue, featuring the 10 harmonics prescribed. Figure S2: (A) The blue line is the same wave signal as in S1A and the red line is that reconstructed from fitting the frequencies identified by the PSD in S1B (green line) to the original signal (green line in S1A). (B) PSD of the turbulence prescribed in S1A (red line) and PSD of the turbulence, or signal, that arises from subtracting the red line in S2A from the original signal (green line in S1A). Figure S3: Same as in Figure S2 but using 95% of the wave energy from the PSD in Figure S1B (green line). Figure S2 uses 100% of the wave energy from the PSD in Figure S1B. As seen, the differnce between Figures S2 and S3 is minimal, as expounded in the main text with the actual records.

Author Contributions: Conceptualization, S.S., A.V.-L. and J.A.L.-C.; Data curation, S.S.; Formal analysis, S.S., J.A.L.-C., J.A. and M.A.-K.; Funding acquisition, A.V.-L.; Investigation, S.S. and A.V.-L.; Methodology, S.S.; Project administration, A.V.-L.; Resources, A.V.-L.; Supervision, A.V.-L.; Validation, J.A.-C., J.A. and M.A.-K.; Visualization, S.S. and J.A.L.-C.; Writing—original draft, S.S.; Writing—review & editing, A.V.-L. All authors have read and agreed to the published version of the manuscript.

Acknowledgments: We appreciate the efforts by Z. Biesinger and W. Lindberg in the instrument deployments and retrievals. This research was made possible in part by a grant from The Gulf of Mexico Research Initiative. AVL acknowledges support from NSF project OCE-1332718.

Conflicts of Interest: The authors declare no conflicts of interest.

References

1. Smyth, W.D. Dissipation-range geometry and scalar spectra in sheared stratified turbulence. *J. Fluid Mech.* **1999**, *401*, 209–242. [CrossRef]
2. Lewis, D. A simple model of plankton population dynamics coupled with a LES of the surface mixed layer. *J. Theor. Boil.* **2005**, *234*, 565–591. [CrossRef]
3. Smith, K.M.; Hamlington, P.E.; Fox-Kemper, B. Effects of submesoscale turbulence on ocean tracers. *J. Geophys. Res. Oceans* **2016**, *121*, 908–933. [CrossRef]
4. Tennekes, H.; Lumley, J.L. *A First Course in Turbulence*; MIT Press: Cambridge, MA, USA, 1972; p. 300.
5. Burchard, H.; Petersen, O.; Rippeth, T.P. Comparing the performance of the Mellor-Yamada and the k-ϵ two-equation turbulence models. *J. Geophys. Res.* **1998**, *103*, 10543–10554. [CrossRef]
6. MacKinnon, J.A.; Gregg, M.C. Mixing on the late-summer New England shelf—Solibores, shear, and stratification. *J. Phys. Oceanogr.* **2003**, *33*, 1476–1492. [CrossRef]
7. Sharples, J.; Moore, C.M.; Abraham, E. Internal tide dissipation, mixing, and vertical nitrate flux at the shelf edge of NE New Zealand. *J. Geophys. Res. Space Phys.* **2001**, *106*, 14069–14081. [CrossRef]
8. Simpson, J.H.; Crawford, W.R.; Rippeth, T.P.; Campbell, A.R.; Cheok, J.V.S. The Vertical Structure of Turbulent Dissipation in Shelf Seas. *J. Phys. Oceanogr.* **1996**, *26*, 1579–1590. [CrossRef]
9. Williams, E.; Simpson, J.H. Uncertainties in Estimates of Reynolds Stress and TKE Production Rate Using the ADCP Variance Method. *J. Atmos. Ocean. Technol.* **2004**, *21*, 347–357. [CrossRef]
10. Wiles, P.J.; Rippeth, T.P.; Simpson, J.H.; Hendricks, P.J. A novel technique for measuring the rate of turbulent dissipation in the marine environment. *Geophys. Res. Lett.* **2006**, *33*. [CrossRef]
11. Lueck, R.G.; Huang, D.; Newman, D.; Box, J. Turbulence Measurement with a Moored Instrument. *J. Atmos. Ocean. Technol.* **1997**, *14*, 143–161. [CrossRef]
12. Stacey, M.T.; Monismith, S.G.; Burau, J.R. Observations of turbulence in a partially stratified estuary. *J. Phys. Oceanogr.* **1999**, *29*, 1950–1970. [CrossRef]
13. Whipple, A.C.; Luettich, R.A.; Seim, H.E. Measurements of Reynolds stress in a wind-driven lagoonal estuary. *Ocean Dyn.* **2006**, *56*, 169–185. [CrossRef]
14. Whipple, A.C.; Luettich, R.A. A comparison of acoustic turbulence profiling techniques in the presence of waves. *Ocean Dyn.* **2009**, *59*, 719–729. [CrossRef]
15. Shaw, W.J.; Trowbridge, J.H. The Direct Estimation of Near-Bottom Turbulent Fluxes in the Presence of Energetic Wave Motions. *J. Atmos. Ocean. Technol.* **2001**, *18*, 1540–1557. [CrossRef]
16. Kirincich, A.; Lentz, S.; Gerbi, G.P. Calculating Reynolds Stresses from ADCP Measurements in the Presence of Surface Gravity Waves Using the Cospectra-Fit Method. *J. Atmos. Ocean. Technol.* **2010**, *27*, 889–907. [CrossRef]
17. Trowbridge, J. On a Technique for Measurement of Turbulent Shear Stress in the Presence of Surface Waves. *J. Atmos. Ocean. Technol.* **1998**, *15*, 290–298. [CrossRef]
18. Rosman, J.H.; Hench, J.L.; Koseff, J.R.; Monismith, S.G. Extracting Reynolds Stresses from Acoustic Doppler Current Profiler Measurements in Wave-Dominated Environments. *J. Atmos. Ocean. Technol.* **2008**, *25*, 286–306. [CrossRef]
19. Gerbi, G.P.; Trowbridge, J.H.; Edson, J.B.; Plueddemann, A.J.; Terray, E.A.; Fredericks, J.J. Measurements of Momentum and Heat Transfer across the Air–Sea Interface. *J. Phys. Oceanogr.* **2008**, *38*, 1054–1072. [CrossRef]
20. Scannell, B.; Rippeth, T.P.; Simpson, J.H.; Polton, J.; Hopkins, J.E. Correcting Surface Wave Bias in Structure Function Estimates of Turbulent Kinetic Energy Dissipation Rate. *J. Atmos. Ocean. Technol.* **2017**, *34*, 2257–2273. [CrossRef]
21. Lohrmann, A.; Hackett, B.; Røed, L.P. High Resolution Measurements of Turbulence, Velocity and Stress Using a Pulse-to-Pulse Coherent Sonar. *J. Atmos. Ocean. Technol.* **1990**, *7*, 19–37. [CrossRef]
22. Souza, A.; Howarth, M.J. Estimates of Reynolds stress in a highly energetic shelf sea. *Ocean Dyn.* **2005**, *55*, 490–498. [CrossRef]
23. Nidzieko, N.J.; Fong, D.A.; Hench, J.L. Comparison of Reynolds Stress Estimates Derived from Standard and Fast-Ping ADCPs. *J. Atmos. Ocean. Technol.* **2006**, *23*, 854–861. [CrossRef]
24. Kaimel, J.C.; Wyngaard, J.C.; Izumi, Y.; Coté, O. Spectral characteristics of surface-layer turbulence. *Q. J. R. Meteorol. Soc.* **1972**, *98*, 563–589. [CrossRef]

25. Thomson, R.E.; Emery, W.J. *Data Analysis Methods in Physical Oceanography*, 3rd ed.; Elsevier: Amsterdam, The Netherlands, 2014; p. 716.
26. Dean, R.G.; Dalrymple, R.A. Water wave mechanics for engineers and scientists. *Adv. Ser. Ocean Eng.* **1991**, *2*, 353.

Article

Source Water Apportionment of a River Network: Comparing Field Isotopes to Hydrodynamically Modeled Tracers

Lily A. Tomkovic [1,*], Edward S. Gross [1], Bobby Nakamoto [2], Marilyn L. Fogel [2] and Carson Jeffres [1]

[1]   Center for Watershed Sciences, University of California Davis, 1 Shields Ave, Davis, CA 95616, USA; edward.s.gross@gmail.com (E.S.G.); cajeffres@ucdavis.edu (C.J.)
[2]   EDGE Institute, University of California Riverside; 900 University Ave, Riverside, CA 92521, USA; bnaka004@ucr.edu (B.N.); marilyn.fogel@ucr.edu (M.L.F.)
*   Correspondence: latomkovic@ucdavis.edu

Received: 7 March 2020; Accepted: 13 April 2020; Published: 15 April 2020

Abstract: Tributary source water provenance is a primary control on water quality and ecological characteristics in branching tidal river systems. Source water provenance can be estimated both from field observations of chemical characteristics of water and from numerical modeling approaches. This paper highlights the strengths and shortcomings of two methods. One method uses stable isotope compositions of oxygen and hydrogen from water in field-collected samples to build a mixing model. The second method uses a calibrated hydrodynamic model with numerical tracers released from upstream reaches to estimate source-water fraction throughout the model domain. Both methods were applied to our study area in the eastern Sacramento–San Joaquin Delta, a freshwater tidal system which is dominated by fluvial processes during the flood season. In this paper, we show that both methods produce similar source water fraction values, implying the usefulness of both despite their shortcomings, and fortifying the use of hydrodynamic tracers to model transport in a natural system.

Keywords: hydrodynamic model; stable isotopes; source water fingerprinting; floodplain

1. Introduction

Across the globe, people have settled and established communities beside estuaries and river deltas that have proven to be ideal habitats for humans, flora, and fauna [1]. Estuaries are unique habitats in a myriad of ways—they are the location where tidal influence extends into the riverine landscape, where saline seawater and fresh river water mix, and where geomorphic effects from marine processes meet fluvial erosion and deposition. All of these factors produce an ecosystem that is specially adapted to growth and success in this confluence of conditions. Even though estuaries support a large diversity of species, they are threatened habitats [2,3]. Numerous anthropogenic effects have impacted the quality of estuarine habitat [4]. For example, pollutants and anthropogenic inputs in the form of chemicals, human and agricultural wastes, and sediment have been introduced into the watersheds along their reaches [5,6]. Projected sea level rise [7] and massive diversions from natural riverine inputs [8] have impacted salinity dynamics, affecting ecosystem health [9]. Entire landscapes have been shifted from land to sea [10,11], affecting numerous processes from the headwaters of tributaries down to their deltas, all impacting the health of the estuary.

To combat these problems, humans have sought ways to balance their needs for the natural resources provided by estuaries with the long-term survival and fitness of those habitats. Better practices have been established to control pollution into rivers, estuaries, and marine environments. Furthermore, restoration efforts have been made in order to address the degradation of physical habitat surrounding river deltas and estuaries. Combinations of these efforts have been undertaken in order to produce multi-pronged benefits within these habitats. Often, the goal of restoration and management actions are to allow for the physical and biological processes important in estuaries to resume functioning. For instance, the restoration of off-channel habitat in the tributaries leading into an estuary provides the physical parameters (such as turbidity, dissolved oxygen, temperature, and light availability) to generate primary and secondary producers (phyto- and zoo-plankton) [12]. With proper connectivity, that productivity can flow into riverine habitat and be transported downstream throughout the estuary. Following the overall impacts of estuarine restoration is complicated, however, because of the complex physical properties of deltas and estuaries, such as reconnecting or braided river networks, tidal pumping effects [13], and salinity gradients.

Understanding where water flows and where it originates is useful for understanding what impacts a restoration project has as well as what the character of water is coming into a potential restoration site. Identifying source water fraction (SWF) of a river network addresses the spatiotemporal distribution of source waters and allows the investigation of tributary water fate. The properties of tributaries, such as the stable isotope compositions of water, have been used to identify source water distribution, but these studies have typically been applied to lakes or groundwater [14–17] Other studies have used isotope signatures to identify which tributaries have dominant effects on geochemical properties of the mainstem [18], to investigate snowmelt versus rainfall contribution [19], and to identify water origin [20]. Our study aimed to use two methods to find the SWF at each location within the study area over time; one method used is similar to that of Halder et al. [21] and Marchina et al. [20], which evaluates stable hydrogen (^2H) and oxygen (^{18}O) isotopes at sample locations and compares those to the tributary source waters of interest.. Another direct application of using physical properties of tributary water to find the distribution of source waters in a river network was presented by Peter et al. [22] in a physically modeled hypothetical watershed. In the study, high-resolution mass spectrometry of organic contaminants was used to confirm the source hydrology of a hypothetical river system with known flows, and thus known mixtures. This physical model study demonstrated the potential of using physical and chemical signatures in water to identify source water, but was limited to systems with an exact knowledge of tributary discharge and mixing patterns, limiting the study to a discussion of methodology instead of an application of a real-world scenario in a river network. We address this limitation using a separate method to investigate the hydraulics and transport processes of a system in order to provide estimated flows and mixing patterns.

Hydrodynamic models are numerical tools that solve equations of motion in fluid mechanics and are widely used to replicate flood events and complex physical processes in river networks. Sridharan et al. [23] used a one-dimensional streamline-following junction model to evaluate particle paths through the Sacramento–San Joaquin Delta of California (Delta), giving insight to source-water fate at a large and broad scale. Bai et al. [24] used a three-dimensional hydrodynamic and water quality model to source-apportion different tributary contributions to nitrogen and phosphorous loads. The source-apportionment indicates the chemical impact of tributaries, but not their relative volume at any particular location. In order to test the validity of a field-based approach to identifying SWF in a river network, we developed a hydrodynamic model coupled with a transport model. This study's comparison of methods can help address limitations of stable isotope methods of finding the SWF of a particular site, as well as demonstrating a field validation of source water tracking using hydrodynamic models.

This study centered on the McCormack-Williamson Tract (MWT), an island protected by a ring levee in the Delta, and the surrounding area. Slated for restoration, the site lies downstream of several other floodplain restoration sites along the Cosumnes River and Dry Creek, making understanding the distribution of incoming source waters helpful to improve the overall efficacy of the MWT restoration. Specifically, water that has had access to floodplains would likely carry more allochtonous material, providing more carbon and nutrients to kickstart in situ production at the study site. In this paper, we characterize the spatiotemporal distribution of SWF at different sites in and around the island while it was flooded. In the absence of measurements of individual tributary flows into the study area, indirect methods must be applied to estimate source water distribution. Our study aimed to evaluate the utility of the two methods discussed, to compare hydrodynamically derived values to in situ data, and to discuss the implication of these approaches to the Delta, as well as other riverine systems.

2. Methods

2.1. Study Area

The Sacramento–San Joaquin Delta (Delta), located in California, USA is a large, reconnecting, freshwater–tidal river network that leads into the San Francisco Bay Estuary. The Delta is a major economic asset to the state of California. Its major diversions of freshwater, mainly from the Sacramento River, provide irrigation for the state's thriving agricultural industry, as well as drinking water for its cities [25]. These large diversions from the Delta, supported by an extensive system of reservoirs, have altered transport and seasonal salt intrusion trends in the estuary [26,27]. The nutrient-rich soils in the off-channel habitat of the Delta have been predominantly converted to agricultural plots [10], while at the same time over 95% of floodplain and intertidal habitat [28] have been channelized, thereby eliminating shallow-water habitat from the ecosystem. These and many other anthropogenic impacts on the Delta have motivated restoration efforts to mitigate the altered ecosystem. The restoration of shallow-water habitat can boost productivity, and understanding the transport of these enriched source waters is an important factor in maximizing restoration efforts [29], making Delta an ideal location for this study.

The specific area studied in this paper is in the North-East Delta near the confluence of the Mokelumne and Cosumnes Rivers (see Figure 1). It is a freshwater tidal system dominated by fluvial processes during the flood season by the upstream Dry Creek, Cosumnes, and Mokelumne watersheds. A major feature of the study site just west of the confluence is the McCormack-Williamson Tract (MWT), which was historically wetland habitat [30] but is now a 1400-acre ring-leveed property used for agricultural purposes. The MWT is slated for restoration to tidal and floodplain habitat. During the period of this study, an accidental breach occurred near the planned breach, allowing for flood flows through MWT along with intertidal habitat in non-flooding periods.

Just west of the MWT are the Delta Cross Channel Gates, which are a major feature of the Delta, controlling diversion of freshwater flows from the Sacramento River to the Southern Delta's major pumping plants for municipal, agricultural, and other export across California. North of the MWT are Middle, Lost, and Snodgrass Sloughs. Snodgrass Slough is the drainage for Morrison Creek and the Stone Lake National Wildlife Refuge, where the stream enters the Delta. Middle and Lost Sloughs are dead-end sloughs during non-flooding periods and convey Cosumnes overland flow during floods. The Cosumnes River is the only major river coming out of the Sierra Nevada Range without a major dam, allowing for a relatively natural hydrograph with natural overbank/floodplain flow [31]. Due to this hydrograph and a variety of restoration actions, the Cosumnes River provides floodplain rearing for a variety of native juvenile fishes and exports productivity downstream [32]. This relatively natural system dramatically differs from the other major stream source to the study area—the Mokelumne River. The Mokelumne River is a 5550 km^2 watershed and has its headwaters in the Sierra Nevada Range extending down to the San Joaquin River in the Delta. River flow in the lower 55 km is regulated by Camanche Reservoir and conveyed to the confluence of the San Joaquin River through a leveed channel.

For this study, we developed two types of modeling methods to investigate the SWF of Mokelumne River water and Cosumnes River Water (locations shown in Figure 2).

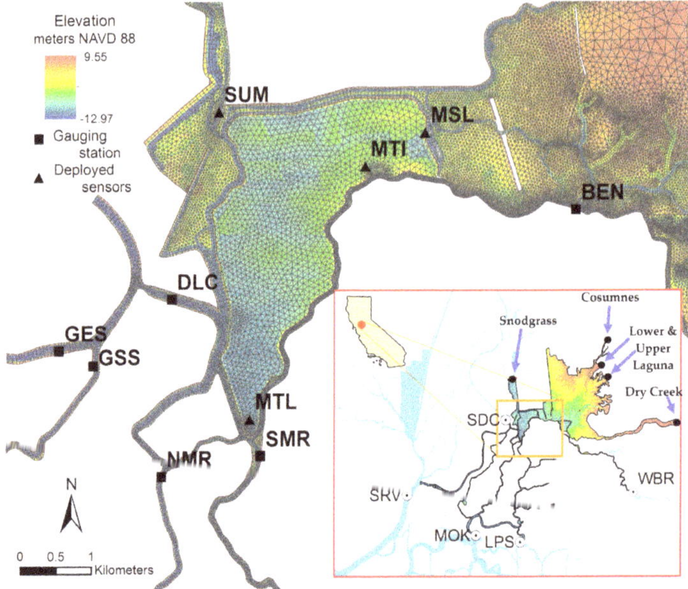

Figure 1. Study area with calibration gage stations shown as black points. Sensors deployed specifically for this study are triangles, and agency gages (either United States Geological Survey (USGS) or the California Department of Water Resources (DWR)) are squares. The model domain is shown (bottom right, location shown as red dot in California) overlying the Delta waterways in blue. Boundary gages are shown in the bottom right figure with agency gages shown as white and black icons and estimated boundaries labeled with blue arrows.

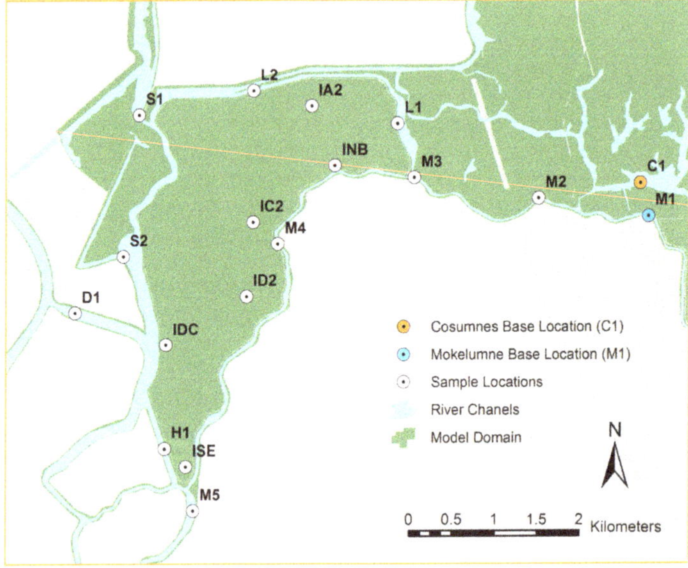

Figure 2. Study area showing isotope sample locations in white, with the Cosumnes and Mokelumne Base locations shown in orange and blue, respectively.

2.2. SWF Method I: Isotope Mixing Model

2.2.1. Mixing Model Theory

Following Gibson et al. [14], a mass balance approach was used to estimate the volumetric fraction of two tributary waters at downstream sample site (Figure 3). The mixing model asserts that the isotopic composition at the sample site is a volume-weighted average of tributary isotopic composition.

$$f_A \times \delta X_A + f_B \times \delta X_B = f_{Site} \times \delta X_{Site} = \delta X_{Site} \tag{1}$$

where f_A, f_B, and f_{Site} are the volumetric fraction of River "A", River "B", and whichever site was being analyzed, respectively. The isotopic composition, in delta notation, of each location is denoted as δX, where X is either ^2H or ^{18}O. After assuming that $f_A + f_B = f_{Site} = 1$, the SWF for River "A" or "B" at any given site is:

$$f_A = \frac{\delta X_{Site} - \delta X_B}{\delta X_A - \delta X_B} \tag{2}$$

$$f_B = \frac{\delta X_{Site} - \delta X_A}{\delta X_B - \delta X_A} \tag{3}$$

This mixing model applies to samples with $\delta_A < \delta_{Site} < \delta_B$ or $\delta_B < \delta_{Site} < \delta_A$. In the event that this condition was not met the sample was not included in the analysis. In addition, the mixing model assumes that the isotopic composition of rivers A and B are stationary, that local effects on δ^2H (e.g., isotopic fractionation via evaporation) are negligible [33], and that the sampled water is comprised entirely of contributions from the two tributaries.

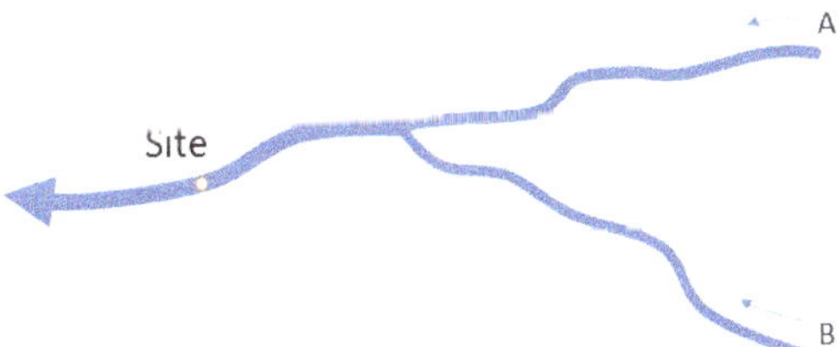

Figure 3. Schematic river system with single confluence.

2.2.2. Field Data—Collecting Isotope Samples

The field campaign at the MWT sampled the water biweekly from August 2016 to June 2019 at 18 locations (Figure 2). Not all locations were sampled throughout the entire period, due to a variety of reasons. The MWT locations (those that start with "I") were only sampled during the MWT's breach from February 2017 to June 2017. At each location, a grab sample of water was taken from the top ~0.25 m of water, filtered through glass fiber (GF/F) filters (nominal pore size: 0.7 microns) and stored refrigerated until isotopic analysis was performed.

2.2.3. Isotope Processing

We used delta notation ($\delta^{18}O$ and δ^2H) to characterize our samples. $\delta^{18}O$ and δ^2H values change with precipitation elevation, evaporation, and other hydrologic factors, often making them differ significantly between watersheds. Delta notation describes the relative composition of stable isotopes in the sample through the formula: $\delta^nX = \_\_‰ = (R_{sample}/R_{standard} - 1) \times 1000$; where n is the atomic number of the heavy isotope in question and R_{sample} and $R_{standard}$ represent the ratio of the heavy isotope (here, ^2H and ^{18}O) to the light isotope (here, ^1H and ^{16}O) in the sample and the international standard, respectively. We determined $\delta^{18}O$ and δ^2H via off-axis integrated cavity output spectroscopy (Los Gatos Research (LGR), Liquid Water Isotope Analyzer) at the University of California Merced. Working standards from LGR, calibrated against National Institute of Standards and Technology (U.S. Department of Commerce) reference materials (Vienna Standard Mean Ocean Water (VSMOW), Greenland Ice Sheet Precipitation (GISP), and Standard Light Antarctic Precipitation (SLAP)), were analyzed throughout and used to correct measured isotope ratios, so they could be expressed relative to the international standard: VSMOW. Four different standards ranging from $\delta^2 H = -9.2‰; \delta^{18}O = -2.7‰$ to $(^2H = -154.0‰; \delta^{18}O = -19.5‰$ were used as internal standards. Raw data was processed using LWIA post-analysis software (LGR).

Each time we determined the isotopic ratio of a water sample or standard, it was analyzed via repeated injections. The standard deviation of our measured values (n = 89) for standards was $0.3 \pm 0.1‰$ and $0.06 \pm 0.01‰$ for $\delta^2 H$ and $\delta^{18}O$, respectively. Averaging multiple injections per sample was effective at reducing uncertainty originating from small injection-to-injection differences, indicating high precision. Furthermore, our mean measured values differed from known standard values by only $0.03‰$ and $0.01‰$ for $\delta^2 H$ and $\delta^{18}O$ respectively; indicating high accuracy, even when the above uncertainties are considered. Propagation of uncertainty into our model is addressed in Section 4.

2.2.4. Application of Mixing Model

Addressing the C1 and M1 sites (Figure 2) as equivalent to the River "A" and River "B" hypothetical diagram (Figure 3), we modified Equations (2) and (3) to find the fraction of Cosumnes and Mokelumne water at any given site.

$$f_{Cos} = \frac{\delta X_{Site} - \delta X_{Moke}}{\delta X_{Cos} - \delta X_{Moke}} \tag{4}$$

$$f_{Moke} = \frac{\delta X_{Site} - \delta X_{Cos}}{\delta X_{Moke} - \delta X_{Cos}} \tag{5}$$

2.3. SWF Method II: Hydrodynamic Model

The second method of finding SWF in a natural system uses a two-dimensional hydrodynamic model coupled with a transport model that can transport a contaminant tracer with advection and diffusion. In this approach, the transport model releases conservative tracers in the tributaries of interest (Rivers "A" and "B" in Figure 3), which are tracked throughout the model domain in time. The summation of all tracer fractions at any given location is 1.0, allowing for a simple numerically based volumetric fraction of the tributary source waters.

2.3.1. Hydrodynamic Model Development

A two-dimensional hydrodynamic model was built and calibrated using UnTRIM [34]. The two-dimensional version of UnTRIM uses a semi-implicit algorithm for the depth-averaged shallow water equations under a hydrostatic assumption [35]. The mesh was generated using Preprocessor Janet (http://www.smileconsult.de) and contains quadrilateral and triangular elements, Figure 1) with a total of 109,129 elements. The average edge length was around 30 m with the smallest element being 3 m and the largest being over 200 m in an area insignificant to this study.

The three stage boundaries at the downstream end were set using observed gages. On the Sacramento River the boundary was set within 3 km of the United States Geological Survey (USGS) gage SRV-Sacramento River at Rio Vista-11455420 and the gage data was used directly. The other two downstream boundaries' gages (USGS gages MOK-Mokelumne R A Andrus Island NR Terminous CA-11336930 and LPS-Little Potato Slough at Terminous-11336790) were located at the model boundary directly, but did not use a known datum, so the reported time series were adjusted using calibration offsets from a larger Delta-wide model.

2.3.2. Boundary Condition Development

The seven upstream flow boundaries used either gaged data or were calculated. The USGS discharge gages SDC-Sacramento R above Delta Cross Channel-11447890 and WBR-Mokelumne R A Woodbridge CA-11325500 were used at the upstream ends of the Sacramento and Mokelumne Rivers, respectively. The discharge boundary at Snodgrass Slough was calculated using lagged discharge data from USGS MFR-Morrison C NR Sacramento CA-1136580. The boundaries at the Cosumnes River, Upper Laguna Creek, Lower Laguna Creek, and Dry Creek were all estimated using the results from a hydrologic study performed on the area by David Ford Consulting Engineers (DFCE) [36]. In order to estimate flows for the four unknown flow boundaries, a lag and amplitude relationship was taken from the DFCE study and then the flow was routed using a simple Muskingum Routing method in order to attenuate flows to the model boundary locations based on stream lengths from the study locations to the boundary locations. The estimated flow magnitudes were calibrated against combined tidally-filtered North Fork Mokelumne and South Fork Mokelumne flow gages, since these two gages are the only outlets to the system during flood flows they account for all flow derived from the estimated boundaries as well as the Mokelumne River.

2.3.3. Model Calibration

For the hydrodynamic model calibration, we deployed 10 Solinst LevelLogger pressure sensors, which were RTK surveyed and barometrically compensated using a nearby Solinst BaroLogger.

The model was run for a period of December 2016–June 2017 and was calibrated using USGS and California Department of Water Resources (DWR) gages, as well as the LevelLoggers (Figure 1). The performance of the model is demonstrated with Table 1. The model metrics used are the correlation coefficient R^2 and the Willmott Skill Index of agreement [37] (skill). The skill demonstrates a metric for comparing observed and computed time series data and ranges between 0 and 1, with 1 being perfect agreement.

$$Skill = 1 - \frac{\sum_{i=1}^{n}|P_i - O_i|^2}{\sum_{i=1}^{n}\left(\left|P_i - \overline{O}\right| - \left|O_i - \overline{O}\right|\right)^2} \tag{6}$$

where P and O are the predicted and observed values, respectively.

Table 1. Model calibration metrics for the gages shown in Figure 2.

	Gage Name	Gage ID *	R2	Skill	Date Range
	Mokelumne R @ Benson's Ferry NR Thornton	BEN [C]	0.976	0.991	1 December 2016–1 July 2017
	Delta Cross Channel BTW Sac R & Snodgrass	DLC [U]	0.963	0.989	1 December 2016–1 July 2017
	Sacramento River Below Georgiana Slough	GES [U]	0.977	0.99	1 December 2016–1 July 2017
	Georgiana Slough at Sacramento River	GSS [U]	0.976	0.989	1 December 2016–1 July 2017
Stage	Middle Slough	MSL [L]	0.959	0.986	20 January 2017–1 July 2017
	McCormack-Williamson Tract – Inlet	MTI [L]	0.932	0.974	16 February 2017–31 May 2017
	McCormack-Williamson Tract – Lower	MTL [L]	0.914	0.976	16 February 2017–31 May 2017
	North Mokelumne R at W Walnut Grove Rd	NMR [U]	0.985	0.996	1 December 2016–1/11/2017
	South Mokelumne R at W Walnut Grove Rd	SMR [U]	0.969	0.99	1 December 2016–1 July 2017
	Snodgrass Slough Upstream of Meadow	SUM [L]	0.922	0.979	1 January 2017–1 July 2017
	Sacramento River Below Georgiana Slough	GES [U]	0.993	0.997	1 December 2016–1 July 2017
Flow	Georgiana Slough at Sacramento River	GSS [U]	0.974	0.973	1 December 2016–1 July 2017
	South Mokelumne R at W Walnut Grove Rd	SMR [U]	0.914	0.965	1 December 2016–1 July 2017
	North Mokelumne R at W Walnut Grove Rd	NMR [U]	0.958	0.98	1 December 2016–11 January 2017

* Gage data sources are denoted as [C]—DWR, [U]—USGS gage, [L]—LevelLogger sensor deployed for this study.

A 72-h model spin-up period was eliminated from the calibration metrics and plots due to the initial uniform water surface elevation being very low to ensure no water was initialized within the MWT ring levee. Figure 4 demonstrates model agreement with observations over time, along with a tidally dominant period in the inset to show resolution of the tidal regime.

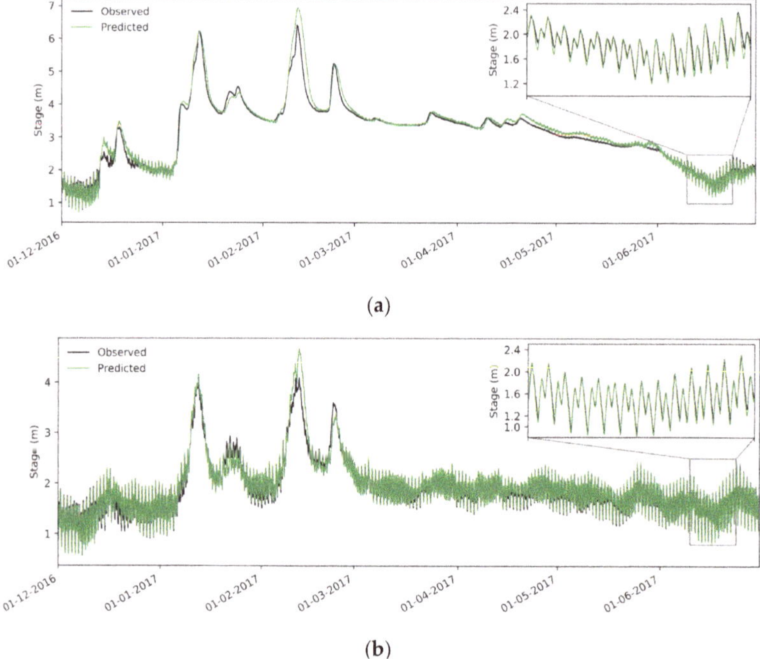

(**a**)

(**b**)

Figure 4. Calibration plot for the BEN (**a**), and SMR (**b**) gages (Table 1) showing the observed (black) and modeled (green) water surface elevation for the period simulated. The inset shows the comparison of observed versus modeled during a tidally dominated period.

2.3.4. Model Tracers

The UnTRIM modeling platform has the capacity to solve for the distribution of tracers with a transport scheme allowing for advection, diffusion, and source/sink terms. Conservative tracers were released within the model mesh and tracked across the domain with a horizontal diffusion coefficient of 0.0, considering only transport via advection using UnTRIM's built-in transport solver [38]. The tracers act similarly to an inert dye and allow the model to follow the "traced" water throughout the model domain, allowing for tracking of SWF in space and time.

The vertically integrated transport equation for the tracers used is:

$$\frac{\partial CH}{\partial t} + \frac{\partial UCH}{\partial x} + \frac{\partial VCH}{\partial y} = \frac{\partial}{\partial x}\left(HK_{xx}\frac{\partial C}{\partial x} + HK_{xy}\frac{\partial C}{\partial y}\right) + \frac{\partial}{\partial y}\left(HK_{yx}\frac{\partial C}{\partial x} + HK_{yy}\frac{\partial C}{\partial y}\right) \tag{7}$$

where C is the concentration of the conservative scalar tracer. In this study we considered the numerical diffusion introduced by the model to be on the same order as that of physical diffusion processes and set the right hand side of Equation (7) was set to zero by using horizontal diffusion coefficients (K_{xx}, *etc.*) of 0.0.

Throughout the simulation, tracers were released at the Cosumnes and Mokelumne Rivers with "Cosumnes" and "Mokelumne" tracers, corresponding to the C1 and M1 site locations, respectively. The tracers were released in all cells that spanned the rivers laterally at the locations where isotopes were collected for both rivers. An "Other" tracer was released at all flow boundaries and for all initial water. Upon passing through the Cosumnes release cells, the Mokelumne and Other tracers were set to zero. Similarly, for the Mokelumne release cells, the other and Cosumnes tracers were set to zero. After the simulation ran, the volumetric fractions of each individual tracer (Cosumnes, Mokelumne, and Other) were queried for each sample location (Figure 2) and checked to ensure that the sum of all three tracers was 1.0 over time, confirming mass balance in the transport model.

In this study, there was no field validation of the transport mechanisms. This is a limitation of the confidence in accuracy of the hydrodynamic approach, despite good hydraulic calibration.

3. Results

3.1. Isotope Mixing Model Results

At each sample location and time, a mixing model was performed for both $\delta^2 H$ and $\delta^{18}O$ ((Figure 5). With an instrument uncertainty of 0.3 ± 0.1 ‰ and 0.02 ± 0.01 ‰ for $\delta^2 H$ and $\delta^{18}O$, respectively, 95% confidence intervals were calculated from each mixing model result. The error bars were then calculated using the minimum and maximum error on each isotope value used in Equation (4). This calculation of instrument-derived error is the only measurable quantity of error that we addressed in this study due to the limitations associated with quantifying the error from breaking assumptions inherent in the mixing model. These assumptions include stationarity in the tributaries, local effects assumed to be negligible, and the sampled water was composed of only the two tributaries. At the time of sampling, only one water grab sample was taken for each site and date, meaning we were unable to compute a standard deviation in space or time of isotope composition at each site to accompany instrument error.

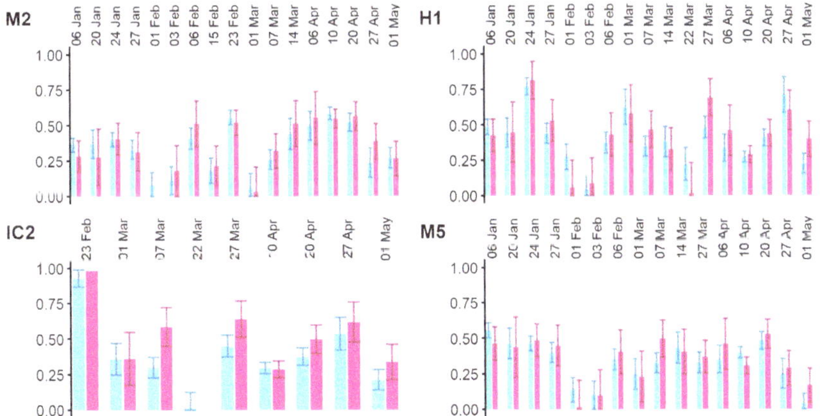

Figure 5. Isotope mixing model results showing fraction of Cosumnes on the y-axis for $\delta^2 H$ (blue) and $\delta^{18}O$ (pink), along with instrument error bars at four example locations (depicted in Figure 3) in 2017. M2 Downstream of confluence. H1: Dead Horse Cut. IC2: Island site C2. M5: South Fork Mokelumne.

As shown in the Supplementary Table S1, differences between the C1 and M1 isotope values ranged from 7.47‰ and 22.45‰ for $\delta^2 H$, and from 0.79‰ and 2.96‰ for $\delta^{18}O$. Given that the instrument error could account for anywhere from 5% to 51% of that range, the mixing model uncertainty ended up being quite significant (Figure 5).

Any isotopic values which did not lie between the C1 and M1 (Cosumnes and Mokelumne sample locations, respectively), were discarded as they would produce fractions larger than 1.0. Out of the 312 $\delta^2 H$ and $\delta^{18}O$ mixing model results, only 13 were outside of the bounds set by the C1 and M1 values for those sample dates. After eliminating invalid data points, a total of 299 mixing model results were produced across 156 unique sample dates and times.

3.2. Hydrodynamic Tracer Results

The model tracer simulation predicts the proportion of Cosumnes, Mokelumne, and Other tracers at the sample locations (Figure 6). The modeled distribution of tracers can also be displayed spatially for any moment in time during the simulation (Figure 7). The model produced time-varying SWF time series, where all tracers summed to one at all points in time, confirming mass conservation across all tracers. The time series demonstrate that over time, there is a trade-off of the three tracers, and that certain periods are dominated by a single tracer.

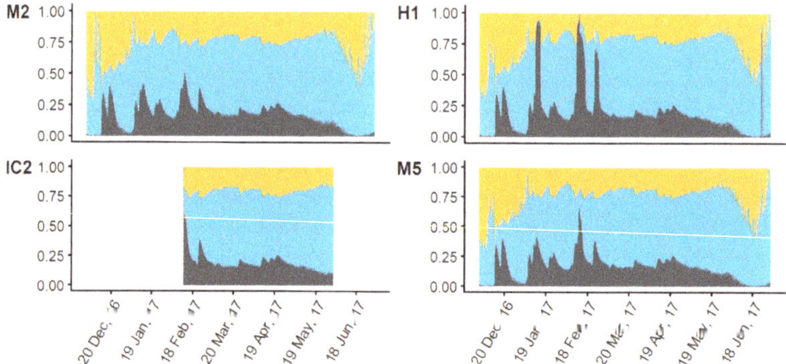

Figure 6. Source water fraction time series at four example locations (depicted in Figure 3) where all fractions add to 1.0. The yellow portion represents the Cosumnes tracer, the blue portion represents the Mokelumne tracer, and the grey represents the other tracer. M2: Downstream of confluence. H1: Dead Horse Cut. IC2: Island site C2. M5: South Fork Mokelumne.

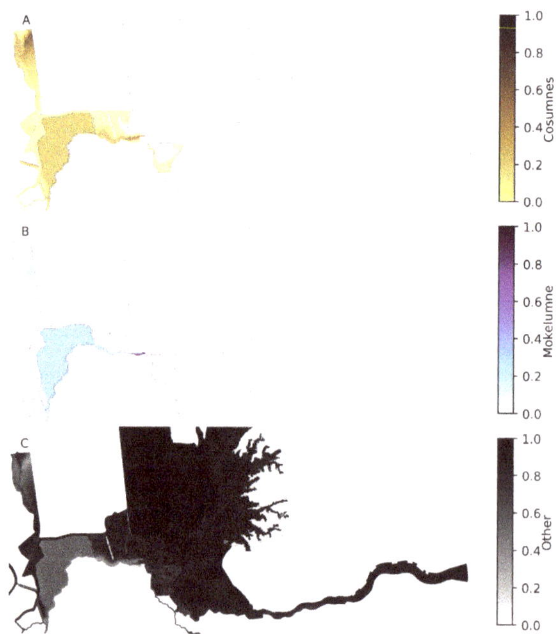

Figure 7. Mapping of the distribution of the three numerical tracers within the model domain. The Cosumnes tracer is shown in orange (**A**), the Mokelumne tracer is blue (**B**), and the other tracer is shown in black/grey (**C**). The snapshot is at 12 February, 2017 02:30, and demonstrates how the Cosumnes and Mokelumne tracers are released and transported downstream (leftwards).

3.3. Mixing Model and Hydrodynamic Model Comparison

In comparing the difference between the model tracer-based estimation of source water versus the isotope mixing model approach (Figure 8), we assume the "Other" tracer would have a similar isotopic signature to the sample taken at the Cosumnes (C1, Figure 2) sample location. This is potentially due to the fact that the Cosumnes and its nearby tributaries (Deer Creek, Upper and Lower Laguna Creek, and Dry Creek) all have watersheds at similar elevations and inland extents to the Cosumnes watershed; presumably, these waterways would receive precipitation of a similar isotopic composition. Furthermore, the water within the channel at the Cosumnes sample location is a continuation of the larger pulses of floodwaters coming through the Cosumnes River Floodplain following rain events, that would be considered "Other" by the hydrodynamic approach. Because of this assumption, we use the combined "Other" and "Cosumnes" tracer results to make a comparison with the averaged δ^2H and $\delta^{18}O$ mixing model results. Interestingly, this pattern of apparent isotopic similarity between the Cosumnes sample location and other water is consistent at the S1 and S2 sites (Figure 2), which have a significant amount of Snodgrass Slough water. That Snodgrass water was tagged as other in the model, as it is one of the flow boundaries where all water is tagged as other. Where the isotope mixing model predicts Cosumnes water, the hydrodynamic model output presents that fraction as a combination of other and Cosumnes water. This could be due to the source water for Snodgrass Slough (Morrison Creek) having similar watershed characteristics that lead to the isotope signature for the Cosumnes and its tributaries.

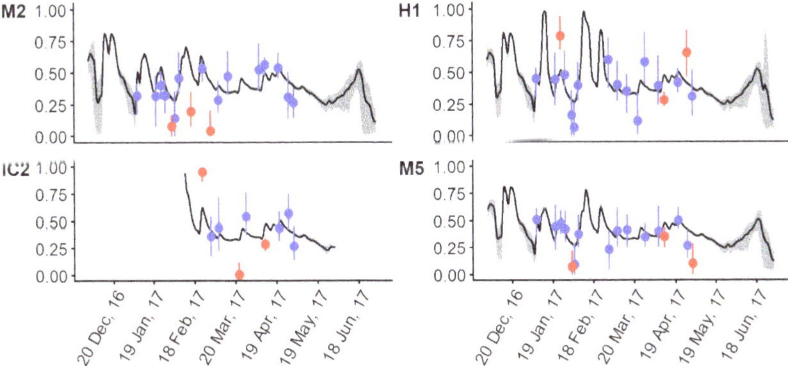

Figure 8. Time series at four locations showing the modeled Cosumnes and Other concentrations with the tidal average (black line) and tidal envelope (grey ribbon). The isotope mixing model values are shown as points with the isotope instrument 95% confidence intervals extending from the mixing model value. Blue points are those whose isotopic confidence intervals overlap with the modeled tidal envelope, while red points do not.

The hydrodynamic "Other" and "Cosumnes" tracers were combined into an instantaneous time series for each site, then the 12-h running mean, minima, and maxima were tidally averaged using a Godin filter to produce a tidal envelope to compare to the isotope mixing model results (Figure 8). At each site, there were varying degrees of overlap (Table 2), with only two sites (ISE and L1) (Figure 2) under 60%.

Table 2. Summary statistics for each site (Figure 2) showing the percentage of points overlapping as well as the count.

Site	Total Samples	Overlapping Samples	Percentage Overlapping Samples
D1	5	8	82.4%
H1	14	17	62.5%
IA2	9	11	82.4%
IC2	6	9	81.8%
ID2	1	1	66.7%
IDC	8	11	100.0%
INB	1	2	72.7%
ISE	4	10	50.0%
L1	10	11	40.0%
L2	8	9	90.9%
M2	14	17	75.0%
M3	16	17	88.9%
M4	6	8	94.1%
M5	13	16	83.3%
S1	5	6	100.0%

An overwhelming majority of the sites were considered overlapping in uncertainty, implying that both methods produced similar results within the bounds of their uncertainty (Figure 9). In total, 123 out of the 156 samples compared were considered overlapping (78.9% overall). Although the methods largely agreed, the isotopic results alone produced a good estimate with a lot of uncertainty from instrument error, while the hydrodynamic results provided improved accuracy, as well as finer spatiotemporal resolution.

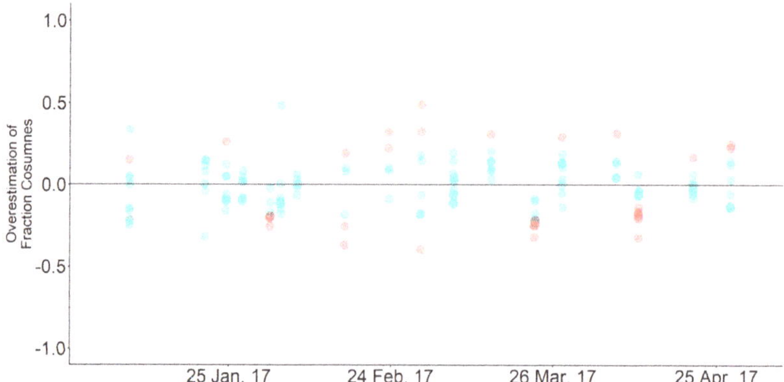

Figure 9. Time series of all locations with overestimation of the fraction Cosumnes found in the isotope method versus the hydrodynamic model method. The blue dots represent sample points whose uncertainty overlapped with the tidal envelope of the model results, and the red dots did not overlap (as in Figure 8).

4. Discussion and Conclusions

This study compared two techniques from two fields of study in order to arrive at estimates of source water fraction (SWF) within a natural system. The hydrodynamic model provided estimated flows and mixing patterns to complement a field-based isotope approach that would otherwise be unsubstantiated (as opposed to previous field-based studies [21,22]). Similarly, the isotope findings act as a field-validation for a numerically-based method of finding SWF in surface-water systems. Both methodologies can be applied to other riverine systems. In order to confidently use the isotope mixing model approach, one would want to ensure that the system is fluvially dominated, that each source tributary has a distinct isotopic signature, and that some stationarity exists in the isotope composition of those source waters [33]. Both methods provide insight to the larger issue of tracing the composition of water in a system in order to better understand the larger mixing processes and patterns in a complex river network. Inherent in the methods are a few assumptions.

Throughout the course of the 2017 sampling done for this paper, our river system was dominated by fluvial processes, unlike typical conditions in which the study area experiences a tidal signal. To compare to 2017, which was a wet year, we briefly compared the two SWF methods in 2018 and 2019 floods, which were below normal years, and found there was little agreement, likely because of the flows not being strong or persistent enough to maintain a fluvial character to the study site. During the 2017 period of this study, the two methods outlined produced similar results, despite the major assumption in the isotope mixing model ($f_A + f_B = f_{Site} = 1$) being imprecise due to widespread overland flow and the high "Other" fractions modeled at the sample sites. Presumably, an increased spatial extent for isotopic sampling could partially alleviate this limitation, although when the number of sources exceeds two, solutions to an expanded Equations (1), (8) and (9) become non-unique. There are additional limitations to the methodologies that may affect the validity of results.

The assumption that endpoint (C1 and M1) samples are representative of cross-sectional averaged isotopic composition may also be imprecise. At each of these locations one sample was collected at one location in the cross-section per sampling date. To evaluate the impacts of this limitation in single-sample field locations, we evaluated model output for all cells across the channel for sites M2, M3, and IA2 and found that the Cosumnes and Mokelumne River waters were well-mixed laterally at the time of sampling, indicating that the lack of replicate samples in the lateral did not necessarily present errors when compared to the model results. However, because this study was performed using a depth-averaged model we cannot investigate sensitivity to the water sample being taken near the top of the water column, but we can assume that the vertical mixing past the confluence is thorough [13]. Although the model accurately predicts cross-sectional flows and stage, the same issue arises as in the study on the model watershed by Peter et al. [22]: the exact flow and mixing patterns are unknown.

The calibrated hydrodynamic model coupled with a transport model provides more spatiotemporally specific output that has proven to be reliable within the bounds of the field data approach this study used. Additionally, the number of water sources traced is unlimited, unlike the isotope-based approach that is limited to two source tributaries. Another advantage of the hydrodynamic model approach over the isotope mixing model is there is no necessity for the source waters to be isotopically distinct. Lastly, the hydrodynamic model predicts additional variables such as depth and velocity that could provide habitat characteristic values (e.g., Pasternack et al. 2004 [39], Whipple 2018 [40]) to be combined with SWF results to further evaluate restoration potential.

Before exploring the significance of applications to other systems, let us explore the significance to the McCormack-Williamson Tract (MWT). There have been a few examples of multiple restoration sites along a river having cumulative benefits as one moves downstream. These have been demonstrated in gravel restoration sites along rivers in the Sacramento–San Joaquin River system [39,41,42], and the multitude of restoration sites across freshwater tributaries of the Chesapeake Bay [43] where restored sites along a river continuum act as a "string of pearls" of aquatic habitat. Ultimately, as water moves through rehabilitated habitat its ecological value or potential cumulatively improves [44,45] It follows that the restoration-laden Cosumnes River carries with it the potential to boost restoration value within the MWT restoration: the last "pearl" in the string. Understanding the spatiotemporal patterns of SWF allows us to identify regions that retain source waters from the Cosumnes, and thus water that has a higher likelihood to contain nutrients or productivity from upstream restoration sites [31,46,47]. Being downstream of completed and ongoing restoration projects is not a unique trait of the MWT within the Delta. Many restoration sites are slated or under construction in the Delta [48]. Understanding how these sites are connected and how they can be synergistic may help to better understand how a single site can have regional ecosystem benefits.

Given that the hydrodynamic method of finding SWF discussed in this paper could be applied to any river network, insights on riverine restoration potential can be gained in any system. For simple fluvial systems that satisfy the assumptions of the isotope mixing model method of finding SWF, isotope samples could suffice in place of developing hydrodynamic models. With careful regard for the limitations discussed, the implications for understanding SWF distribution can be further explored in any river network, and perhaps other connections to aquatic ecology or water quality processes can be found. For instance, Farly et al. used an isotope mixing model to investigate fish diet composition in terms of floodplain-produced or channel-produced resources [49]. This could be coupled with hydrodynamic models to investigate possible effects of drift versus autochthonous production. These methods (hydrodynamic tracers and isotope mixing models) can also be used to investigate habitat quality and its linkages to SWF, using a number of biological or water quality data.

Supplementary Materials: The following are available online at http://www.mdpi.com/2073-4441/12/4/1128/s1. Table S1: Field δ^2H and $\delta^{18}O$ isotope data for all sites.

Author Contributions: Conceptualization, B.N., C.J., and L.A.T.; Methodology, B.N., L.A.T., and E.S.G.; Supervision, C.J., E.S.G. and M.L.F.; Writing—original draft, L.T.; Writing—reviewing & editing, L.A.T., B.N., E.S.G., M.L.F., and C.J. All authors have read and agreed to the published version of the manuscript.

Funding: This research was funded by the Delta Stewardship Council, grant 1471.

Acknowledgments: Christopher "Rusty" Holleman for coding and modeling support, William "Bill" Fleenor for guidance and institutional knowledge on the MWT, Thomas Handley for bathymetric support, Fabián Bombardelli for advice and guidance in modeling and analysis, Alison Whipple for helping secure funding and for sharing hydrologic methods to develop boundary conditions, David Ford Engineering Consulting for the hydrologic work to support the boundary conditions, cbec eco engineering (especially Chris Campbell, Chris Hammersmark) for providing previous modeling work, and The Nature Conservancy for their support of the project that allowed this research to be possible.

Conflicts of Interest: The authors declare no conflict of interest.

References

1. Day, J.W.; Hall, C.A.S.; Kemp, W.M.; Yanez-Arancibia, A. *Estuarine Ecology*, 2nd ed.; Day, J.W.J., Crump, B.C., Kemp, W.M., Yanez-Arancibia, A., Eds.; John Wiley & Sons, Inc.: Hoboken, NJ, USA, 1988. [CrossRef]

2. Ruiz, G.M.; Carlton, J.T.; Grosholz, E.D.; Hines, A.H. Global invasions of marine and estuarine habitats by non-indigenous species: Mechanisms, extent, and consequences. *Am. Zool.* **1997**, *37*, 621–632. [CrossRef]

3. Worm, B.; Barbier, E.B.; Beaumont, N.; Duffy, J.E.; Folke, C.; Halpern, B.S.; Jackson, J.B.C.; Lotze, H.K.; Micheli, F.; Palumbi, S.R.; et al. Impacts of Biodiversity Loss on Ocean Ecosystem Services. *Science* **2006**, *314*, 787–790. [CrossRef] [PubMed]

4. Lotze, H.K.; Lenihan, H.S.; Bourque, B.J.; Bradbury, R.H.; Cooke, R.G.; Kay, M.C.; Kidwell, S.M.; Kirby, M.X.; Peterson, C.H.; Jackson, J.B.C. Depletion, degradation, and recovery potential of estuaries and coastal seas. *Science* **2006**, *312*, 1806–1809. [CrossRef] [PubMed]

5. Bowen, J.L.; Valiela, I. The ecological effects of urbanization of coastal watersheds: Historical increases in nitrogen loads and eutrophication of Waquoit Bay estuaries. *Can. J. Fish. Aquat. Sci.* **2001**, *58*, 1489–1500. [CrossRef]

6. Kuivila, K.; Hladik, M.L. Understanding the occurrence and transport of current-use pesticides in the San Francisco estuary watershed. *San Fr. Estuary Watershed Sci.* **2008**, *6*. [CrossRef]

7. Hong, B.; Shen, J. Responses of estuarine salinity and transport processes to potential future sea-level rise in the Chesapeake Bay. *Estuar. Coast. Shelf Sci.* **2012**, *104–105*, 33–45. [CrossRef]

8. Nilsson, C.; Reidy, C.A.; Dynesius, M.; Revenga, C. Fragmentation and flow regulation of the world's large river systems. *Science* **2005**, *308*, 405–408. [CrossRef]

9. Drinkwater, K.F. On the Role of Freshwater Outflow on Coastal Marine Ecosystems—A Workshop Summary. In *The Role of Freshwater Outflow in Coastal Marine Ecosystems*; Skreslet, S., Ed.; Springer: Berlin/Heidelberg, Germany, 1986; pp. 429–438.

10. Whipple, A.A.; Grossinger, R.M.; Rankin, D.; Stanford, B.; Askevold, R.A. *Sacramento-San Joaquin Delta Historical Ecology Investigation: Exploring Pattern and Process*; SFEI Contribution No. 672; San Francisco Estuary Institute-Aquatic Science Center: Richmond, CA, USA, 2012.

11. Dahl, T.E.; Allord, G.J. *Technical Aspects of Wetlands: History of Wetlands in the Conterminous United States*; United States Geological Survey Water Supply Paper 2425; US Geological Survey: Reston, VA, USA, 1996.

12. Ahearn, D.S.; Viers, J.H.; Mount, J.F.; Dahlgren, R.A. Priming the productivity pump: Flood pulse driven trends in suspended algal biomass distribution across a restored floodplain. *Freshw. Biol.* **2006**, *51*, 1417–1433. [CrossRef]

13. Fischer, H.B.; List, E.J.; Koh, R.C.Y.; Imberger, J.; Brooks, N.H. *Mixing in Inland and Coastal Waters*; Academic Press: San Diego, CA, USA, 1979.

14. Gibson, J.J.; Reid, R. Water balance along a chain of tundra lakes: A 20-year isotopic perspective. *J. Hydrol.* **2014**, *519*, 2148–2164. [CrossRef]

15. Sherman, L.S.; Blum, J.D.; Dvonch, J.T.; Gratz, L.E.; Landis, M.S. The use of Pb, Sr, and Hg isotopes in great lakes precipitation as a tool for pollution source attribution. *Sci. Total Environ.* **2015**, *502*, 362–374. [CrossRef]

16. Krabbenhoft, D.P.; Bowser, C.J.; Kendall, C.; Gat, J.R. Use of oxygen-18 and deuterium to assess the hydrology of groundwater-lake systems. *Environ. Chem. Lakes Reserv.* **1994**, *90*. [CrossRef]

17. Vitvar, T.; Burns, D.A.; Lawrence, G.B.; Mcdonnell, J.J.; Wolock, D.M. Estimation of baseflow residence times in watersheds from the runoff hydrograph recession: Method and application in the Neversink Watershed, Catskill Mountains, New York. *Hydrol. Process.* **2002**, *16*, 1871–1877. [CrossRef]

18. Marchina, C.; Bianchini, G.; Natali, C.; Pennisi, M.; Colombani, N.; Tassinari, R.; Knoeller, K. The Po river water from the Alps to the Adriatic Sea (Italy): New insights from geochemical and isotopic (Δ18O-ΔD) data. *Environ. Sci. Pollut. Res.* **2015**, *22*, 5184–5203. [CrossRef] [PubMed]

19. Penna, D.; Zuecco, G.; Crema, S.; Trevisani, S.; Cavalli, M.; Pianezzola, L.; Marchi, L.; Borga, M. Response time and water origin in a steep nested catchment in the Italian dolomites. *Hydrol. Process.* **2017**, *31*, 768–782. [CrossRef]

20. Marchina, C.; Lencioni, V.; Paoli, F.; Rizzo, M.; Bianchini, G. Headwaters' isotopic signature as a tracer of stream origins and climatic anomalies: Evidence from the Italian Alps in summer 2018. *Water* **2020**, *12*, 309. [CrossRef]

21. Halder, J.; Decrouy, L.; Vennemann, T.W. Mixing of Rhône River water in Lake Geneva (Switzerland–France) inferred from stable hydrogen and oxygen isotope profiles. *J. Hydrol.* **2013**, *477*, 152–164. [CrossRef]

22. Peter, K.T.; Wu, C.; Tian, Z.; Kolodziej, E.P. Application of nontarget high resolution mass spectrometry data to quantitative source apportionment. *Environ. Sci. Technol.* **2019**, *53*, 12257–12268. [CrossRef]

23. Sridharan, V.K.; Monismith, S.G.; Fong, D.A.; Hench, J.L. One-dimensional particle tracking with streamline preserving junctions for flows in channel networks. *J. Hydraul. Eng.* **2018**, *144*, 1–10. [CrossRef]

24. Bai, H.; Chen, Y.; Wang, D.; Zou, R.; Zhang, H.; Ye, R.; Ma, W.; Sun, Y. Developing an EFDC and numerical source-apportionment model for nitrogen and phosphorus contribution analysis in a lake basin. *Water* **2018**, *10*, 1315. [CrossRef]

25. Lund, J.R. *Envisioning Futures for the Sacramento-San Joaquin Delta*; Public Policy Institute of California: San Francisco, CA, USA, 2007.

26. Mount, J.; Bennett, W.; Durand, J.; Fleenor, W.; Hanak, E.; Lund, J.; Moyle, P. *Aquatic Ecosystem Stressors in the Sacramento–San Joaquin Delta*; The Public Policy Institute of California (PPIC): San Francisco, CA, USA, 2012.

27. Gross, E.S.; Hutton, P.H.; Draper, A.J. A comparison of outflow and salt intrusion in the pre-development and contemporary San Francisco Estuary. *San Fr. Estuary Watershed Sci.* **2018**, *16*. [CrossRef]

28. Lund, J.; Hanak, E.; Fleenor, W.; Bennett, W.; Howitt, R.; Mount, J.; Moyle, P. *Comparing Futures for the Sacramento-San Joaquin Delta*; University of California Press: Berkely, CA, USA, 2010.

29. Moyle, P.B.; Crain, P.K.; Whitener, K. Patterns in the use of a restored california floodplain by native and alien fishes. *San Fr. Estuary Watershed Sci.* **2007**, *5*, 1–29. [CrossRef]

30. Brown, K.J.; Pasternack, G.B. The geomorphic dynamics and environmental history of an upper deltaic floodplain tract in the Sacramento-San Joaquin Delta, California, USA. *Earth Surf. Process. Landforms* **2004**, *29*, 1235–1258. [CrossRef]

31. Whipple, A.A.; Viers, J.H.; Dahlke, H.E. Flood regime typology for floodplain ecosystem management as applied to the unregulated Cosumnes River of California, United States. *Ecohydrology* **2017**, *10*, 1–18. [CrossRef]

32. Jeffres, C.A.; Opperman, J.J.; Moyle, P.B. Ephemeral floodplain habitats provide best growth conditions for juvenile Chinook salmon in a California River. *Environ. Biol. Fishes* **2008**, *83*, 449–458. [CrossRef]

33. Gat, J.R. Oxygen and hydrogen isotopes in the hydrologic cycle. *Annu. Rev. Earth Planet. Sci.* **1996**, *24*, 225–262. [CrossRef]

34. Casulli, V.; Walters, R.A. An unstructured grid, three-dimensional model based on the shallow water equations. *Int. J. Numer. Methods Fluids* **2000**, *32*, 331–348. [CrossRef]

35. Walters, R.A.; Casulli, V. A robust, finite element model for hydrostatic surface water flows. *Commun. Numer. Methods Eng.* **1998**, *14*, 931–940. [CrossRef]

36. David Ford Consulting Engineers. *Cosumnes and Mokelumne River Watersheds—Design Storm Runoff Analysis*; David Ford Consulting Engineers: Sacramento, CA, USA, 2004.

37. Willmott, C.J. On the validation of models. *Phys. Geogr.* **1981**, *2*, 184–194. [CrossRef]

38. Casulli, V.; Zanolli, P. Semi-implicit numerical modeling of nonhydrostatic free-surface flows for environmental problems. *Math. Comput. Model.* **2002**, *36*, 1131–1149. [CrossRef]

39. Pasternack, G.B.; Wang, C.L.; Merz, J.E. Application of a 2D hydrodynamic model to design of reach-scale spawning gravel replenishment on the Mokelumne River, California. *River Res. Appl.* **2004**, *20*, 205–225. [CrossRef]

40. Whipple, A.A. Managing Flow Regimes and Landscapes Together: Hydrospatial Analysis for Evaluating Spatiotemporal Floodplain Inundation Patterns with Restoration and Climate Change Implications. Ph.D. Thesis, University of California, Davis, CA, USA, June 2018

41. Kondolf, G.M.; Angermeier, P.L.; Cummins, K.; Dunne, T.; Healey, M.; Kimmerer, W.; Moyle, P.B.; Murphy, D.; Patten, D.; Railsback, S.; et al. Projecting cumulative benefits of multiple river restoration projects: An example from the Sacramento-San Joaquin River System in California. *Environ. Manag.* **2008**, *42*, 933–945. [CrossRef] [PubMed]

42. Parker, S.S.; Remson, E.J.; Verdone, L.N. Restoring conservation nodes to enhance biodiversity and ecosystem function along the Santa Clara River. *Ecol. Restor.* **2014**, *32*, 6–8. [CrossRef]

43. Hassett, B.; Palmer, M.; Bernhardt, E.; Smith, S.; Carr, J.; Hart, D. Restoring watersheds project by project: Trends in Chesapeake Bay tributary restoration. *Front. Ecol. Environ.* **2005**, *3*, 259–267. [CrossRef]

44. Vannote, R.L.; Minshall, G.W.; Cummins, K.W.; Sedell, J.R.; Cushing, C.E. The river continuum concept. *Can. J. Fish. Aquat. Sci.* **1980**, *37*, 130–137. [CrossRef]

45. Palmer, M.A.; Bernhardt, E.S.; Allan, J.D.; Lake, P.S.; Alexander, G.; Brooks, S.; Carr, J.; Clayton, S.; Dahm, C.N.; Follstad Shah, J.; et al. Standards for ecologically successful river restoration. *J. Appl. Ecol.* **2005**, *42*, 208–217. [CrossRef]

46. Swenson, R.O.; Whitener, K.; Eaton, M. Restoring floods on floodplains: Riparian and floodplain restoration at the Cosumnes River Preserve. In *California Riparian Systems: Processes and Floodplain Management, Ecology and Restoration*; Riparian Habitat Joint Venture: Sacramento, CA, USA, 2003; pp. 224–229.

47. Opperman, J.J.; Luster, R.; McKenney, B.A.; Roberts, M.; Meadows, A.W. Ecologically functional floodplains: Connectivity, flow regime, and scale. *J. Am. Water Resour. Assoc.* **2010**, *46*, 211–226. [CrossRef]

48. California Natural Resources Agency; California Department of Food and Agriculture; California Environmental Protection Agency. *California Water Action Plan: 2016 Update*; California Natural Resources Agency: Sacramento, CA, USA; California Department of Food and Agriculture: Sacramento, CA, USA; California Environmental Protection Agency: Sacramento, CA, USA, 2016.

49. Farly, L.; Hudon, C.; Cattaneo, A.; Cabana, G. Seasonality of a floodplain subsidy to the fish community of a large temperate river. *Ecosystems* **2019**, *22*, 1823–1837. [CrossRef]

Review

Timescale Methods for Simplifying, Understanding and Modeling Biophysical and Water Quality Processes in Coastal Aquatic Ecosystems: A Review

Lisa V. Lucas [1,*] and Eric Deleersnijder [2]

[1] U.S. Geological Survey (Integrated Modeling and Prediction Division, Water Mission Area), 345 Middlefield Road, MS #496, Menlo Park, CA 94025, USA

[2] Institute of Mechanics, Materials and Civil Engineering (IMMC) & Earth and Life Institute (ELI), Université catholique de Louvain, Bte L4.05.02, Avenue Georges Lemaître 4, 1348 Louvain-la-Neuve, Belgium; eric.deleersnijder@uclouvain.be

* Correspondence: llucas@usgs.gov; Tel.: +1-650-329-4588

Received: 26 May 2020; Accepted: 13 September 2020; Published: 29 September 2020

Abstract: In this article, we describe the use of diagnostic timescales as simple tools for illuminating how aquatic ecosystems work, with a focus on coastal systems such as estuaries, lagoons, tidal rivers, reefs, deltas, gulfs, and continental shelves. Intending this as a tutorial as well as a review, we discuss relevant fundamental concepts (e.g., Lagrangian and Eulerian perspectives and methods, parcels, particles, and tracers), and describe many of the most commonly used diagnostic timescales and definitions. Citing field-based, model-based, and simple algebraic methods, we describe how physical timescales (e.g., residence time, flushing time, age, transit time) and biogeochemical timescales (e.g., for growth, decay, uptake, turnover, or consumption) are estimated and implemented (sometimes together) to illuminate coupled physical-biogeochemical systems. Multiple application examples are then provided to demonstrate how timescales have proven useful in simplifying, understanding, and modeling complex coastal aquatic systems. We discuss timescales from the perspective of "holism", the degree of process richness incorporated into them, and the value of clarity in defining timescales used and in describing how they were estimated. Our objective is to provide context, new applications and methodological ideas and, for those new to timescale methods, a starting place for implementing them in their own work.

Keywords: timescale; transport; hydrodynamic; ecological; biogeochemical; coastal; estuary; residence time; age; flushing time

---

1. Introduction

> *"Nature is pleased with simplicity. And nature is no dummy."*
>
> —Commonly attributed to Isaac Newton

A common refrain of environmental scientists is: "Environmental science isn't rocket science. It's harder than rocket science." Understanding, predicting, and managing the workings of environmental systems is a grand challenge, due in no small part to the intricate interactions between physical, biological, and geochemical processes that are, individually, complex enough for whole careers to be spent deciphering them. Moreover, those processes—and the interactions between them—operate and vary over a daunting range of temporal and spatial scales, from milliseconds to millennia, and from the microscopic to scales visible from space. Fortunately, technological advancements in field and laboratory instrumentation, remote sensing, and computing have permitted us to measure and model environmental systems with ever-increasing extent and resolution. More complex tools, thus, facilitate our understanding of the complexity. Simplicity, also, has a role to play in unraveling the complexity, by reducing it to its essential parts and giving it shape, so it can be more easily grasped. Diagnostic timescales represent one such simplifying tool.

1.1. What Are Timescales?

"Timescale" is generally defined as "the amount of time that something takes or during which something happens" [1]. In practice, a timescale often denotes an *estimate* expressing a representative or overall magnitude, as opposed to a precise value [2,3]. Similar to length, velocity, and other commonly used scales, timescales are thus often presented in order-of-magnitude terms [2].

The term "timescale" may carry many subtly different meanings, including [3]: (1) a typical period of fluctuation in system forcing or response (e.g., [4–11]); (2) a period of system adjustment or response to low-frequency forcing [11–14]; (3) the period of variability captured by measurements or models [15–17]; (4) the temporal lens through which processes are examined [9,18–20]; (5) a diagnostic parameter with units of time whose inverse characterizes the rate at which a process or collection of processes unfolds [21–26].

Herein, we primarily use "timescale" in the sense of the last definition above, i.e., to convey approximately how long a process takes or, inversely, the speed of a process. Rates of physical, biological, or chemical processes are often represented by parameters with different and mixed units (e.g., velocity (length/time), diffusivity (length^2/time), water discharge (length^3/time), growth, decay or uptake (1/time), water column production (mass/(area-time)), ingestion (mass food/(mass tissue-time))). Timescales can be defined and quantified for each of these processes by a variety of methods to be detailed in later sections. Regardless of the approach for estimating values for timescales, the following holds when using them in the fifth sense above: A smaller (or "shorter") timescale indicates a faster process, whereas a larger (or "longer") timescale suggests that the process is slower [27].

In the water realm generally, timescales are often invoked as an explanatory concept or order-of-magnitude diagnostic tool to help illuminate how natural or managed systems work. They are used to describe the flows through and/or functioning of aquifers [28]; lakes, reservoirs, and freshwater embayments [29–35]; streams, rivers, and floodplains [36,37]; hydrologic catchments [38–41]; estuaries and other coastal or tidal systems [42–45]; wetlands [46,47]; the continental shelf and open ocean [48–50]; and the atmosphere [51]. This review primarily focuses on estuaries and coastal systems, with some references to other domains as well.

Timescales may be estimated to represent the time for completion of a process [3], which in aquatic systems may include diffusive mixing over the water column depth or a fraction of it [4,25,52]; traversal of a water body or reach [53,54] or between two locations of interest [55,56]; flushing or "renewal" of an estuary by river flow, tides, wind, and/or other forcings [45,57,58]; settling of particles through a water column or layer thereof [59,60]; growth or decay by a specified factor, such as *e* [25,27]; or filtration of a water column or water body volume by benthic organisms [23,25,61,62]. Specific examples illustrating why and how timescales are calculated for a variety of such cases are provided in later sections.

1.2. Some Fundamental Concepts and Definitions

The timescale literature is replete with terms like Lagrangian, Eulerian, parcels, particles, constituents, volumes, tracers, and water types, making it difficult to avoid confusion. Therefore, before launching into the pith of this paper, we first attempt to clarify some terms and concepts in order to minimize confusion in the sections that follow. While some definitions are well-established (e.g., the basic Lagrangian and Eulerian descriptions), agreement among scientists and mathematicians is not unanimous regarding others of these concepts and terms. Thus, the following discussion of parcels, particles, volumes, etc. represents merely how we authors have chosen to define them (largely following [63]). Regardless, we intend (and hope) that some discussion of these fundamentals may help us find our way in this jungle!

1.2.1. Constituents, Particles, Parcels, and Types

Water in aquatic environments is a mixture containing a large number of dissolved and particulate constituents (e.g., pure water, dissolved gases, pollutants, nutrients, sediment, plankton cells, etc.). At any time and position, each constituent may be ascribed a "concentration", a concept associated with various definitions and, hence, various units (e.g., g/kg, kg/m^3, mol/L, cells/L, etc.). Since there is a huge number of constituents, it may be convenient to focus on groups of constituents, i.e., aggregates, whose concentrations may be seen to obey equations similar to those pertaining to individual constituents [63,64]. This is why many use the word "constituent" (or a similar term) even if the substance under consideration is actually an aggregate (e.g., salt). The water in an aquatic ecosystem is itself an "aggregate," consisting of all of its constituents. Pure water is by far its dominant constituent, making the density of the water mixture close to that of pure water. The water mixture density may be regarded as a constant in most terms of the equations to be dealt with (the "Boussinesq approximation").

A "particle" is a metaphor useful for verbal or written interpretations as well as Lagrangian calculations (see below). It is a discrete material point having zero volume and non-zero mass [63]. An individual particle, as defined herein, only contains mass of a single constituent and, depending on that constituent, may contain many ions, molecules, sediment grains, or plankton cells, all of which share the same history (Figure 1). The mass of a particle of a given constituent must be much smaller than the total mass of the constituent present in the domain of interest but may be much larger than that of a single molecule, to prevent excessive demands on computational resources (see discussion of Lagrangian approaches below). Because a particle has zero volume, it cannot have volume-normalized concentrations or densities associated with it. On the other hand, concentrations and densities are definable for "elemental volumes." An elemental volume is a control volume, delineated only by thought [63], that can contain many particles representing a variety of constituents. Its size is much smaller than the smallest resolved macroscopic processes. A fluid parcel is an elemental volume that moves with the fluid mixture velocity, i.e., the mass-weighted average of the velocities of all the molecules present in it (see Figure 1). (This is termed the "barycentric velocity" [65]). Under the Boussinesq approximation, a water parcel's volume is constant in time, but its shape is not, and its mass is also considered constant. However, the masses of its individual constituents (i.e., the precise mixture of constituent particles contained within the volume) may change over time due to diffusive transport through its boundaries (see Figure 2). Thus, a fluid parcel (or "water parcel") does not always contain the same molecules or atoms over time. Clearly, the "water parcel" concept, as defined herein, is a mathematical notion very different from that of "particle".

Depending on its origin or other differentiating factors, water at any location and time may be split into several water "types". Water types can be differentiated or "marked" by tracers, which can be measured during transport [66–69]. Tracers are constituents that are, ideally, inert (they undergo no reactions) and hydrodynamically "passive". As for any other constituent or aggregate, every water type may be viewed as being made up of (water) particles, which should not be confused with water parcels.

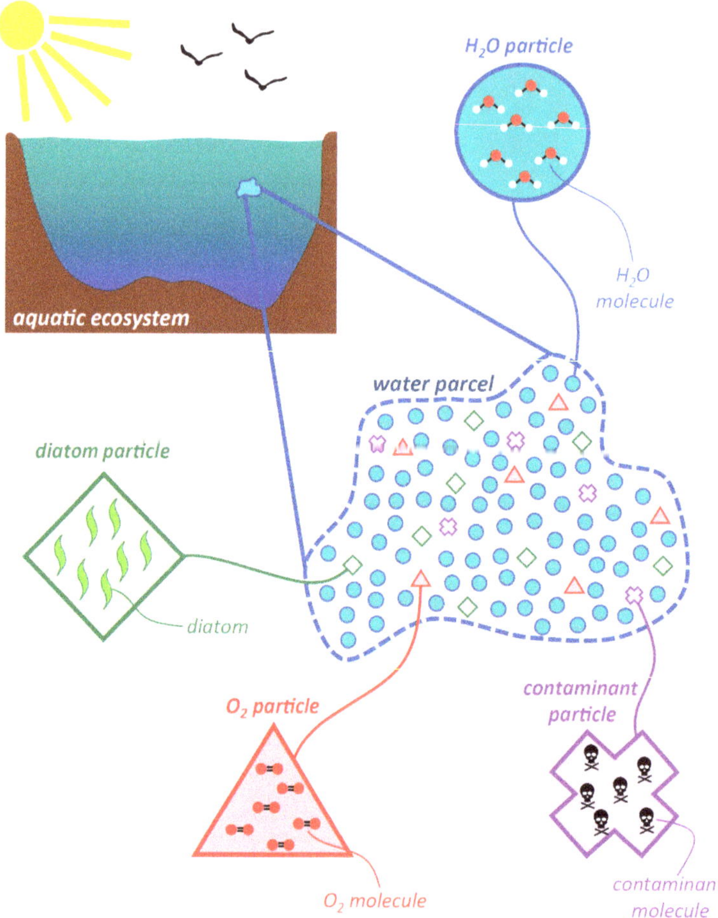

Figure 1. Cartoon depicting the relationships between water parcels, particles, and molecules, cells, etc., as defined herein. A water parcel is a mixture of particles, the most numerous of which are pure water particles. A particle is a material point at which many atoms, molecules, cells, etc., of an individual constituent or aggregate are concentrated. (Following Deleersnijder et al. [63]).

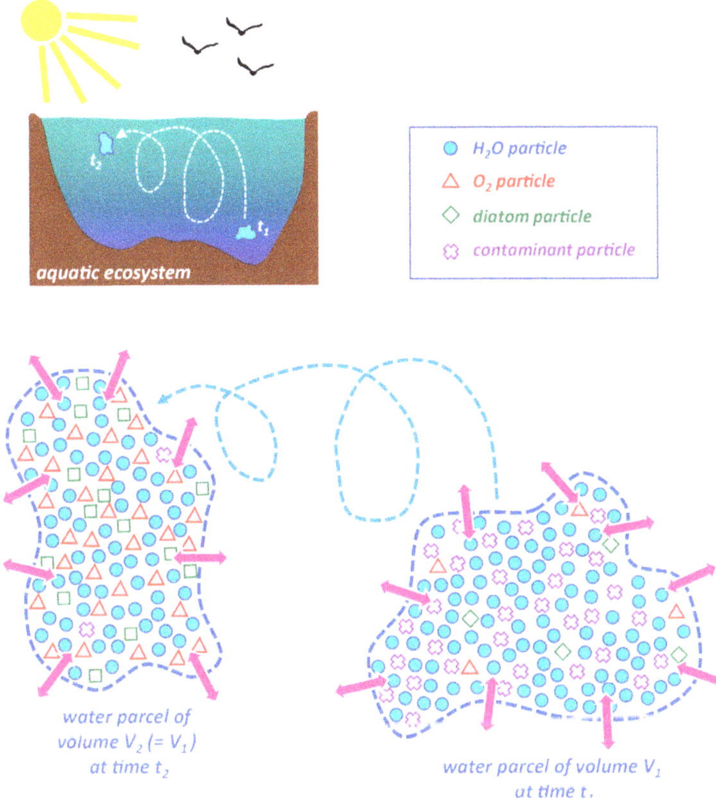

Figure 2. Cartoon depicting a water parcel as it is transported through an aquatic ecosystem between times t_1 and t_2. The water parcel's volume is constant, but its shape is not. Due to diffusion (magenta arrows), the particles contained within the parcel at t_2 are not the same as the particles contained in the parcel at t_1. Each "particle" is composed of multiple molecules, atoms, or cells of a particular constituent or aggregate. (Following Deleersnijder et al. [63]).

1.2.2. Lagrangian and Eulerian Descriptions and Approaches

The two approaches for describing fluid motion are the *Lagrangian description* (which follows the paths and histories of specific individual fluid parcels) and the *Eulerian description* (in which time-dependent variables are defined at fixed positions in space) [70,71]. One can think of these two descriptions as different reference frames for an observer of changes in some fluid property. In the Eulerian frame of reference, the observer sits at a fixed point in space, similar to a moored sensor, and observes "local" changes in fluid properties over time [71]; this stationary Eulerian observer also has the capacity to see enough of its neighborhood to evaluate local space derivatives. For an observer in the Lagrangian frame (i.e., one who jumps on a fluid parcel and rides along with it), the observed changes are a combination of "local" changes with time and "advective" changes due to transport of the parcel and observer across spatial gradients in the water property [71].

The concentration for each constituent may be obtained numerically from the solution of an appropriate reactive transport equation (RTE), i.e., a partial differential equation taking into account advection, diffusion, reactions (if any), and settling (for negatively buoyant particulate matter). The RTE is frequently solved with an Eulerian approach, treating the water and its constituents as "continuous media". Except in very idealized situations for which an analytical solution is possible, the RTE is solved numerically, and the continuous concentration field is discretised in time (timestep by timestep) and space (gridcell by gridcell). Reactions in the Eulerian method are dealt with relatively efficiently. The main challenge with Eulerian numerical solutions lies in the representation of advection, i.e., avoidance of both spurious oscillations and artificial smoothing of concentration gradients [72–74].

In the Lagrangian approach, each constituent under study is concentrated into so-called "particles" [63]. The motion of Lagrangian particles is simulated numerically by means of a time-marching procedure. During each time increment, the displacement of a particle is the sum of a deterministic drift and a stochastic component related to diffusive processes [75–77]. Lagrangian computational methods are generally superior to Eulerian methods for the representation of advection. Due to the stochasticity, however, the fate of a single particle is irrelevant when the aim is to derive a concentration [78,79]: A large number of numerical particles must be seeded into the domain of interest and tracked in order to obtain accurate concentration fields. Reactions can be taken into account, which is usually done in an Eulerian mode [80]. Well-designed Eulerian and Lagrangian schemes must result in concentration estimates converging to the exact solution as the space and time increments decrease for the former methods and as the time resolution and the number of particles increase for the latter. Therefore, discrepancies between Eulerian and Lagrangian simulation results are always due to numerical inaccuracies (or erroneous implementation) and, hence, must not be ascribed to supposedly irreconcilable differences between the two approaches. While conservative Eulerian methods are (by definition) ideally suited for the evaluation of fluxes [81], Lagrangian methods are often used for assessing connectivity [82–86].

Timescales may be employed in order to diagnose the behavior of every constituent, including a water type. They can be evaluated with either an Eulerian or a Lagrangian method. While mathematical descriptions of reactive transport (i.e., models) most often rely on the Eulerian perspective, verbal descriptions and interpretations usually take the Lagrangian perspective [79].

1.3. Transport Timescales

The most commonly used category of timescales in water science, engineering, and management are those falling under the category of "transport timescales" (e.g., residence time, flushing time, water age, transit time, etc.). These individual transport timescales each have distinct (though in some cases multiple) definitions and methods of estimation. Regrettably, in practice, the terms are often used loosely and interchangeably, with imprecise, fluid, or sometimes unexplained definitions and calculation methods [87,88]. Those implementing diagnostic timescales must be aware of such difficulties in order to avoid misunderstandings or even blatant errors. What transport timescales all have in common is they communicate approximately how long water, or a constituent transported with the water, has spent, will spend, or takes to arrive in a defined water body or subregion thereof as a result of physical transport processes. Transport timescales can be useful on their own or may be co-analyzed with other sorts of timescales to understand reactive transport [61,89–91]. Below, we define some frequently used transport timescales:

- *Residence time*—Although the term "residence time" is frequently used to mean a variety of things [88,92–94], one of the most common definitions is the time taken by a particle to leave a water body or defined region of interest [92,95–97]. Because particles originating at different locations and times within a water body may require different amounts of time to exit, residence time (according to this definition) is a function of location and time [87,92,97]. A strict interpretation of this residence time definition is the time taken to leave a water body *for the first time* (see Figure 3), an important distinction in tidal systems where oscillatory transport can cause particles to exit and then re-enter the domain of interest one or more times [26,95,98,99]. Numerical simulations currently offer the best methods for estimating time- and position-dependent timescales in realistic domains [66,97,100,101]; however, other (field-based [102–105], analytic [22,59]) methods may also provide trustworthy estimates, though with less resolution or with additional simplifying assumptions. Other residence time definitions, which are not location- and time-specific, also exist and see wide application (see "flushing time" below).

- *Age*—Age is defined as the time elapsed since a particle entered a water body or defined region [88,94,96,106]. Because the time to reach a specific location after entering will vary across the water body and over time, age (like residence time, as per our preferred definition above) is also time- and location-specific (see Figure 3). Age is seen as the complement to the location- and time-specific residence time: while age is the time taken since entering to reach location x within a water body, residence time is the time remaining within the water body after reaching location x [87,88,96,106]. Some authors have generalized the common definition for age above, arriving at the following: "the time elapsed since the parcel under consideration left the region in which its age is prescribed to be zero" [63,64].

- *Transit time*—Transit time has been defined as the total time for a particle to travel across an entire water body or defined region, from entrance to exit [93,96]. Therefore, transit time is the sum of the location- and time-specific age and residence time (see Figure 3). Some authors have taken advantage of the fact that transit time is equivalent to age computed at the downstream boundary or exit of a water body [28,107]. Travel time is similar to transit time, in that it usually references the time taken to travel between two defined points in space [28]. The transit time and location- and time-specific age and residence time are easily derived analytically for a plug flow situation (see Appendix A).

- *Exposure time*—Exposure time goes forward where the strict definition of residence time stops. While the strict, spatially and temporally variable residence time only accounts for time spent within a defined region until leaving it the first time, exposure time accounts for the *total time* spent within the domain of interest [87], including "all subsequent re-entries" [95] (see Figure 3). Thus, exposure time may be of particular relevance in systems with oscillatory tidal flows [108]. When computing exposure time with a numerical model, it is important that the computational domain be larger than the domain of interest [95], since transport processes outside the domain of interest control particle re-entry.

- *Flushing time*—"*Flushing time is a bulk or integrative parameter describing the general exchange characteristics of a waterbody without identifying detailed underlying physical processes or their spatial distribution*" ([27], adapted from [87]). There are numerous methods for defining and quantifying flushing times, many of them mathematically quite simple. For example, if advection is expected to dominate exchange between the domain of interest and an adjacent water body (as for a river reach), an advective flushing time may be estimated simply as V/Q, where V is the volume of the domain of interest, and Q is the rate of volumetric flow through it. For this situation, V/Q estimates the time for *all* water in the domain of interest to be replaced, whereas $\frac{1}{2}(V/Q)$ represents the *mean* time for replacement of the original water. Analogously, if we assume that an estuary behaves similarly to a "plug flow reactor", i.e., with perfect cross-sectional mixing but zero streamwise mixing, V/Q would represent the time needed to replace all the water initially in the estuary by water entering through its upstream boundary (Figure 4). Some variations on this approach include: (A) substitution of V with freshwater volume V_{fw} and of Q with freshwater inflow rate Q_{fw}, if one is interested in the time to replace freshwater [52,109] (this is often called the "freshwater fraction method" [58,110]); or (B) substitution of V and Q, respectively, with scalar mass M and scalar flux F (in units (mass/time)), if one is concerned with time for replacement of a scalar quantity [87]. (Incidentally, the V/Q [90,104], V_{fw}/Q_{fw} [109], and M/F [111] formulations are sometimes called "residence times".) It should be noted that the V/Q estimate depends on the (sometimes arbitrary) size of the domain of interest [112].

- *e-folding flushing time*—Another construct for quantifying time for flushing is the *e*-folding time ($\tau_{e\text{-}fold}$). This approach capitalizes on the frequently observed exponential-like decrease of constituent mass within a water body over time as it is subjected to flushing. This roughly exponential decrease is often observed in the results of coastal transport simulations [87,100,101,112–115] (see Figure 5) and tracer experiments [116,117]. Mathematically, the exponential form results from assuming a constant flow rate through a perfectly well-mixed system of constant volume, as for a CSTR (continuously stirred tank reactor) [87]. The well-mixed assumption employed here (Figure 6) is in stark contrast to the plug flow assumption above (Figure 4) and thus may be the more appropriate assumption for estuaries subject to strong (e.g., tidal) dispersive mixing. $\tau_{e\text{-}fold}$ may be obtained as (A) the reciprocal of the specific decay rate calculated from an exponential best-fit to a concentration time series [87,100,112,113] or simply as (B) the time when mass falls to $1/e$ (37%) of its initial value [114,117]. If the CSTR assumptions are perfectly met, $\tau_{e\text{-}fold} = V/Q$, but if they are not met (e.g., for basins with bidirectional, tidal exchange flow), V/Q may not accurately characterize the effective flushing time captured by methods (A) or (B) above [87]. Although the well-mixed assumption is almost never satisfied, the *e*-folding construct is nonetheless employed widely and can work well in representing the net effect of all flushing processes acting on a basin. It is important to note the quantitative difference between this flushing time approach (which characterizes flushing of only 63%, or $1\text{-}e^{-1}$, of initial mass; Figure 6) and the simple advective V/Q, V_{fw}/Q_{fw}, and M/F approaches above, whose aim is to characterize 100% replacement of initial mass or volume (Figure 4). Indeed, any perfect CSTR would never truly experience 100% replacement of initial mass, as suggested by the exponential dependency of concentration on time. Even so, for an inert constituent in a well-mixed system, the concentration *tends* to zero as time tends to infinity, resulting in a *finite* domain-averaged residence time, which is equal to the *e*-folding time [88,94,113,118].

- *Tidal prism flushing time*—Another class of flushing time approaches for estuaries—tidal prism models—prominently acknowledges tides as a flushing agent [119,120]. The most basic form for the tidal prism flushing time is $V \cdot T_{tide}/V_p$ [58], where V is estuary volume, T_{tide} is the tidal period, and V_p is the tidal prism volume (i.e., estuary volume difference between high and low tides). Applications of this general approach may vary in the way V and V_p are defined or calculated [27,110]. Moreover, authors have employed a range of assumptions and adjustments for capturing the influence of freshwater inflow or return flow at the seaward boundary [27,58,119]. Like the *e*-folding time, the tidal prism flushing (or "turnover" [58,110]) time is based on the assumption of well-mixedness [87,119].

- *Turnover time*—The V/Q [58], V_{fw}/Q_{fw} [110], M/F [88,94], *e*-folding [121], and other bulk approaches [110] are also sometimes called "turnover times." A relatively new approach for estimating bulk estuary turnover timescales is based on the total exchange flow (TEF) through a cross section at the estuary mouth; TEF is calculated using an isohaline framework [122], and the TEF timescale τ_{TEF} may be thought of as *"the ratio of the mass of salt in the estuary to the salt flux into the estuary"* [110]. (τ_{TEF} is also called a "residence time" [122].) In addition to physical processes, the term "turnover time" is frequently applied to biological or geochemical processes as well [62,90,123–125].

- *Retention time*—The term "retention time" is frequently, though not exclusively, used to refer to how long constituents (e.g., nutrients, sediment, organisms) remain within a particular aquatic environment or sub-environment [14,126]. Mechanisms influencing constituent retention can include both hydrodynamic processes (e.g., pools, eddies, and dead zones [14]; stratification and mixing [127]), sedimentation [14], biogeochemical processing [14], and motility of organisms [127]. Hydraulic "retention time" is sometimes treated interchangeably with "residence time" [128] or with expressions described herein as "flushing times" [129].

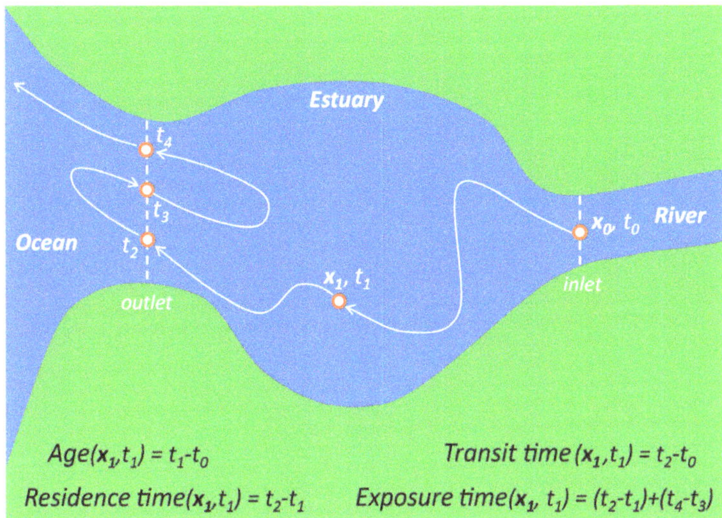

Figure 3. Schematic depicting the relationships between space- and time-dependent age, (strict) residence time, transit time, and exposure time, following Zimmerman [96], Delhez [98], Shen and Haas [121], Viero and Defina [130], Andutta et al. [22], and others. The dots represent successive locations for a single particle following a trajectory passing through locations x_i at times t_i. x_0 and t_0 are the initial location and time.

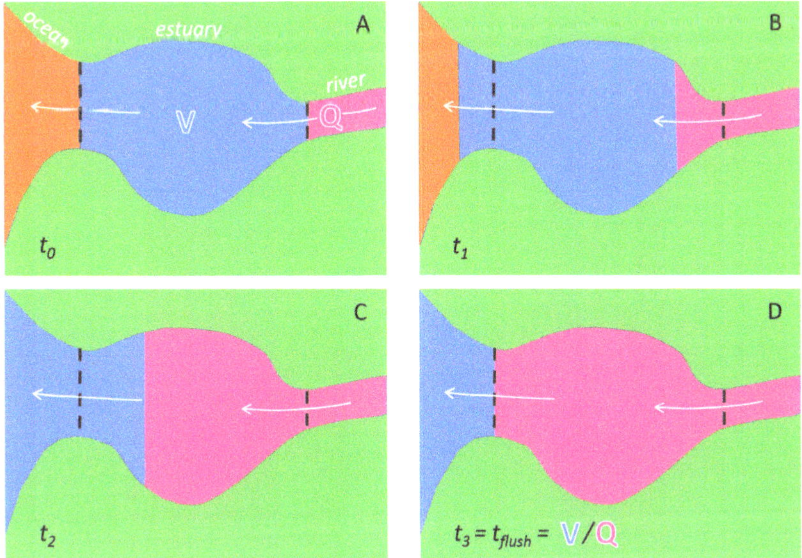

Figure 4. Simplified depiction of advective, river-driven estuarine flushing, idealized as plug flow (perfect mixing over the flow cross section, zero mixing in the streamwise direction). Panels (**A–D**) follow a progression through time of river water gradually replacing estuarine water initially present at time t_0. V is estuarine volume, and Q is river discharge. River water is depicted as magenta; original estuarine water is blue; water outside the estuary mouth is orange. Gray dashed lines represent upstream and downstream boundaries of the estuary.

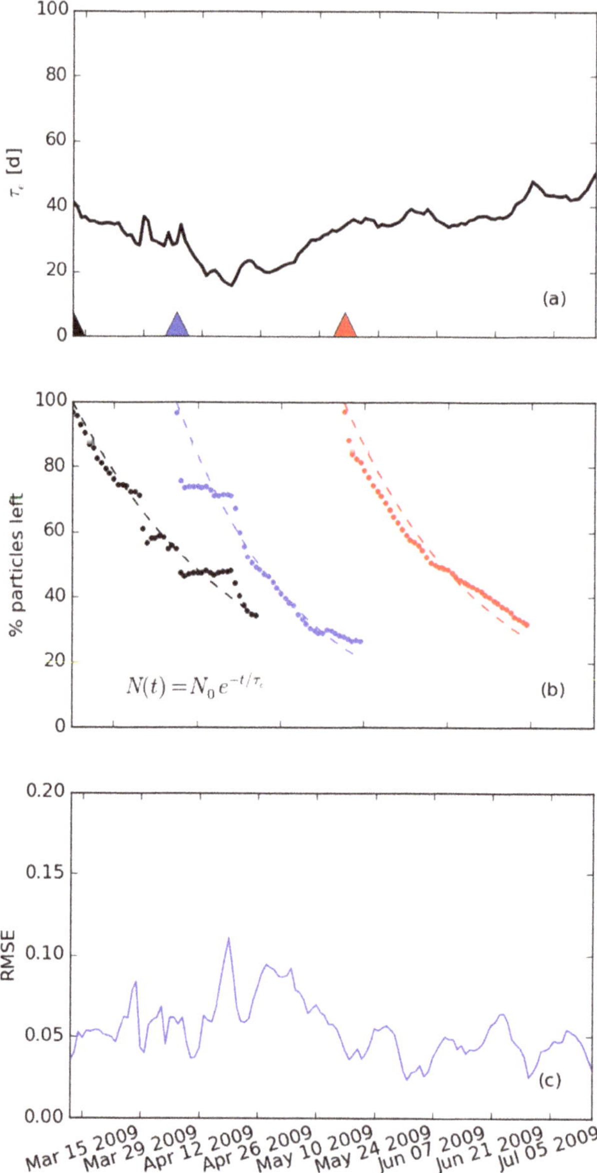

Figure 5. Based on a series of 45-day numerical particle transport simulations of Galveston Bay (TX, USA) by Rayson et al. [100]: (**a**) *e*-folding flushing times for particles initialized on each day for a period spanning mid-March to mid-July 2009. Triangles represent start times for simulations used for exponential fits shown in (**b**), with the blue triangle representing a high discharge period and the red triangle representing a low discharge period. (**b**) Example exponential fits for particle -tracking simulations with the three different start times indicated by the triangles in (**a**). Blue (red) dots and and dashed lines represent the model output and curve fit, respectively, for high (low) discharge periods. (**c**) RMSE (root mean square error) of the exponential best fit for all times modeled. (Reproduced with permission from M. Rayson, Journal of Geophysical Research: Oceans; published by Wiley, 2016).

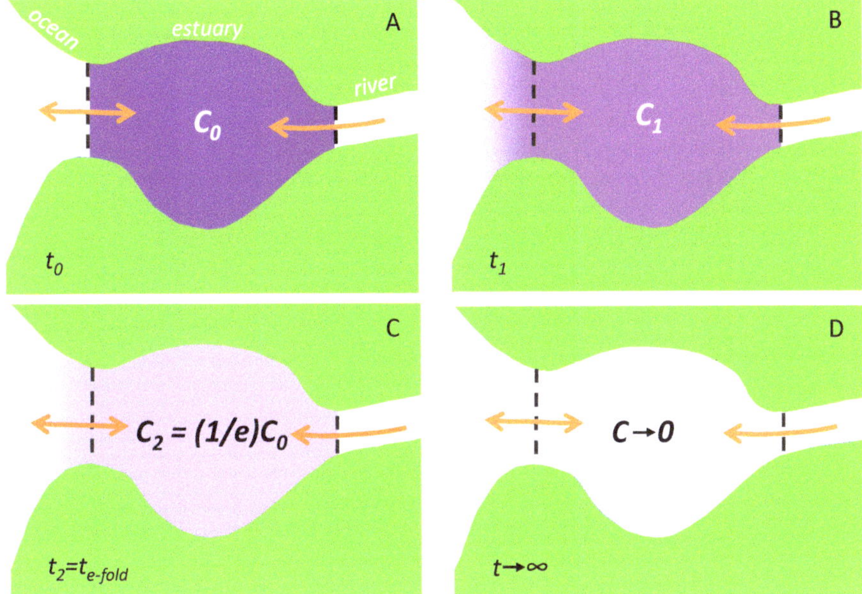

Figure 6. Simplified depiction of the *e*-folding flushing time, driven by river, tidal, and/or other flushing processes. Panels (**A**–**D**) follow a progression through time of *C*, the estuarine concentration of a tracer or other constituent. The *e*-folding mathematical construct is based on the assumption of perfect mixing within the water body of interest (in this case, the estuary). Dark gray dashed lines represent upstream and downstream boundaries of the estuary. Dark purple represents initial estuarine water. Light gray represents replacement water.

Given the variety of transport timescale definitions and estimation approaches, it can be challenging to identify the most useful timescale for addressing a particular question for a specific environment or set of conditions. Useful comparisons of various transport timescales and discussion of their assumptions and applicability can be found in [22,58,93,100,110,112,131].

1.4. What Are Timescales Good for?

There are several advantages of and uses for diagnostic timescales in assessments of water related issues. Briefly, here are some ways of implementing timescales that we expound upon in later sections:

- *A more meaningful substitute for primitive variables and native process rates:* Computed or measured primitive variables (e.g., velocity, pressure, temperature, concentration; also known as state variables) and native process rates (e.g., velocity, production, growth) are not always conducive to interpretation in their raw form [79,89]. (Here, we use the term "native process rate" to refer to the typical rate variable(s) used in connection with a particular process, e.g., velocity or discharge for water movement, or specific growth rate for biomass growth.) On the other hand, diagnostic timescales can incorporate valuable contextual information that native process rates and primitive variables do not. For that reason, timescales can serve as auxiliary variables that might better illuminate a scientific problem [79,89]. For example, the primitive variable "velocity" alone contains no additional problem-specific information that can aid the user in understanding the practical *effect* of that velocity: it is just a velocity. Whereas the advective timescale τ_{adv}—the timescale counterpart to velocity—typically conveys the time needed for a particle to traverse a specified water body or distance (e.g., the time taken by a fisherman's cooler to travel to the river mouth from the upstream location where it, sadly, fell overboard). Therefore, in comparison to a process rate or primitive variable, a timescale can in many cases take the user farther on an interpretive level by communicating what the process rate, materially, means in the context of the scientific question at hand.

- *A common currency for comparing speeds of processes:* Timescales provide a common cross-disciplinary currency by which the speed of disparate processes can be compared [23]. For example, consider the observed reduction in the concentration of a decaying pollutant in a river over the first couple days after release. Relevant process rates (e.g., decay (1/time), river discharge (volume/time)) can be transformed into timescales (τ_{decay}, τ_{flush}) that can then be directly compared. Therefore, if τ_{decay} is, for instance, 0.2 day and τ_{flush} is 30 days, the ~2 order-of-magnitude difference in timescales suggests that decay is a much faster process than river-driven flushing and is likely primarily responsible for any significant concentration reduction in the couple days following pollutant release. Since they all carry the same units, timescales can thus help bridge the gap between scientific disciplines and make quick, back-of-the-envelope assessments of dominant processes possible. Timescale ratios can represent the competition between processes; in some cases, such dimensionless numbers can serve as simple indicators of how an ecosystem might respond to a combination of different physical, biological, or geochemical processes [21,23,25,132–136].

- *Distilling numerical model outputs* [89,137]: The output files of numerical fluid flow models can be immense. Making sense of all those gigabytes, or even terabytes, of spatially and temporally detailed data is a non-trivial effort [79,137,138]. Timescales can extract the essence from such comprehensive datasets. In contrast to other analysis techniques that might provide spatially (temporally) detailed glimpses of the output at limited points in time (space), timescales can integrate across space and/or time and take advantage of most, if not all, of the results [79,138]. For this reason, timescales derived from the results of complex numerical models may be considered "holistic" [79,138]. Importantly, a model-derived timescale, such as the transit time for a particle through an estuary, may be considered holistic in a second sense: it takes into account all processes and forcings included in the model that influence the transport (e.g., river flow, tides, wind, density gradients, etc.) [139]. It is this second meaning that we refer to hereinafter.

- *Comparing systems across space or time:* An effective way of enhancing understanding of an aquatic system is through comparison with other systems or through assessing the functioning of a single system under different conditions over time. Timescales can help encapsulate the general physical or ecological state of aquatic systems across space or time, do so in a way that is relatively simple and intuitive, and allow for easy comparisons.

- *Building simple(r) models:* The partial differential equations (PDEs) governing hydrodynamics and scalar transport are complex, as they are composed of many terms describing multiple influences on momentum and mass balances. Because high-quality (i.e., stable and accurate) numerical solutions to the governing equations can be computationally costly, justifiable simplification of these PDEs is therefore a worthwhile activity. One simplification approach implements timescales of variability in combination with other (e.g., velocity, length, pressure, density) scales to estimate the relative magnitudes of individual terms in time-marching equations [2]; terms that "scale" much smaller than other terms may be justifiably neglected, with the equations reducing to the most essential terms and, hopefully, the numerical solution becoming more tractable and efficient. Another method of simplification involves quantifying the primary processes with timescales, creating dimensionless ratios with those timescales, and then substituting those ratios appropriately into a time- or space-dependent equation. The conversion of a mathematical relationship into dimensionless form can significantly reduce the complexity—and increase the solvability—of the equation [21,23]).

- *Assessing connectivity:* Transport timescales can contribute substantially to assessments of connectivity between different aquatic systems or subregions within a system [56,95,140–142].

In fact, transport timescales can form the basis for one important assessment tool—the "connectivity matrix" [95,140] (see Section 3.4).

- *In conceptual models:* Timescales are often invoked in conceptual models or qualitative descriptions of how systems work. Even if not quantified or clearly defined, well-known terms such as "residence time" capture a general meaning that a scientific or management audience can conceptually follow. Timescales are frequently used (in mental models, written descriptions, cartoons, schematics, etc.) to qualitatively explain ecological phenomena such as phytoplankton bloom development in coastal systems [6,143], legacy phosphorus across watersheds [14], coastal hypoxia [11], nutrient release from sediments in shallow lakes [144], and eutrophication in lakes [145] and coastal systems [146].

This paper focuses on timescales as diagnostic tools in the analysis of reactive transport problems in coastal waters and adjacent domains of interest. Hopefully, the information herein will be as useful to readers who have never before applied timescale methods as it will be to those who have. In the following sections, we describe various methods of diagnostic timescale estimation (Section 2); review previous studies in which diagnostic timescales have been implemented to understand, analyze, model, or explain how (primarily coastal) ecosystems function (Section 3). Throughout Sections 2 and 3, we describe the relationship between the holism of a timescale (i.e., process richness incorporated within it) and the complexity of the mathematical methods employed to derive it. In the Discussion (Section 4), we elaborate (following other authors before us) on the importance of carefully choosing, calculating, and describing timescales, as well as the concept of timescale holism. Finally, the Conclusions (Section 5) summarize the main points presented and make broad connections between the timescales discussed throughout.

2. How Are Diagnostic Timescales Estimated?

There are numerous approaches for estimating timescale magnitudes. Depending on the type of timescale, available computational resources or observational data, and the relative importance of expedience versus accuracy, there are usually rough pencil-and-paper approaches as well as more careful, calculation- or data-intensive methods that may be employed. If there are multiple feasible methods for attaching a numerical value to a timescale, then it can be useful and informative to implement them all and compare the results (e.g., [87,104,115]), as some approaches may capture underlying processes neglected by others.

2.1. Combining Process Rates with Other Scales

One relatively straightforward approach involves taking the reciprocal of a process rate and then combining with other appropriate dimensional (e.g., length, velocity, concentration) scales such that the remaining dimension is *time* (see Table 1 for examples) [52,92]. This method is often used in biological or geochemical studies and should be viable whenever characteristic values for the process rate and other needed scales are available. If the process rate or other scales are expected to exhibit a broad (e.g., more than one order of magnitude) range of values for the problem and setting under study, then it can be informative to use those ranges to provide an estimated range for the timescale [93]. Several aquatic processes, common algebraic expressions for their corresponding timescales, and their associated process rates are shown in Table 1.

Table 1. Processes operating in aquatic systems, associated native process rates and their units, and common mathematical expressions for their corresponding timescales. Scales combined with process rates to construct timescales include: L (length), L_z (vertical length), V (volume), M (integrated mass within a water body), B_p (phytoplankton biomass concentration), B_a (areal biomass concentration), DO (dissolved oxygen concentration , η (nutrient concentration). Specific growth or decay rate μ may be positive (growth) or negative (decay). Decay rate μ_{decay} is assumed positive. Unless specified otherwise, concentrations here are assumed volumetric. Timescale expressions shown here may be adjusted if available parameters or units are different from those shown.

Process	Native Process Rate	Units	Timescale	Relevant Citations
Diffusion/Dispersion/Mixing	Diffusion/Dispersion/Mixing Coefficient (K)	$length^2/time$	L^2/K	[42,49,52,59,147]
Advection	Velocity (U)	$length/time$	L/U	[23,49,59]
Flushing by river flow	Volumetric flow rate (Q)	$length^3/time$	V/Q	[87]
Flushing by scalar flux	Mass flux rate (F)	$mass/time$	M/F	[42]
[1] Growth or decay	Specific growth or decay rate (μ)	$1/time$	$1/|\mu|$	[25]
[2] Decay by one-half	Specific decay rate (μ_{decay})	$1/time$	$\ln(2)/\mu_{decay}$	[125]
[3] Growth by factor of 2	Specific growth rate (μ_{growth})	$1/time$	$\ln(2)/\mu_{growth}$	[127]
Sinking/settling	Sinking speed (w)	$length/time$	L_z/w	[59,60]
[4] Productivity	Areal Productivity (P_a)	$biomass/(length^2\text{-}time$	B_a/P_a	[125]
[4] Benthic consumption	Grazing/Filtration/Clearance rate (BG)	$length^3/(length^2\text{-}time)$	L_z/BG	[23,25,62,124]
Zooplankton grazing	Zooplankton community grazing rate (ZG)	$biomass/(length^3\text{-}time)$	B_p/ZG	[23,148]
Oxygen consumption	Net oxygen consumption rate (C_{DO})	$mass$ $O_2/(length^3\text{-}time)$	DO/C_{DO}	[21,132]
[4] Nutrient uptake	Nutrient uptake rate (v)	$mass$ $nutrient/(length^3\text{-}time)$	η/v	[90]

[1] This timescale is sometimes called an e-folding time or mean life [149] for decaying substances. [2] This timescale is typically called a "half-life." If the decay rate carries a negative sign, then the applicable expression is $\ln(0.5)/\mu_{decay}$. [3] This timescale is typically called a "doubling time." [4] These timescales are sometimes referred to as "turnover" times.

The choice of process rates and auxiliary scales should be guided by the specific question at hand and a priori knowledge of the system. For example, if we are interested in understanding whether vertical mixing is slow enough to allow for algal accumulation in the euphotic zone, then we might (1) estimate the algal growth timescale τ_{growth} as the reciprocal of a typical specific net growth rate in the euphotic zone, (2) estimate the timescale for vertical mixing τ_{mix}^{vert} as the square of the water column depth divided by K_{vert}, a typical (e.g., mean or mid-depth [25]) turbulent diffusivity for the water column, and (3) compare the two timescales. (An argument could be made to use half of the water column depth as the characteristic length scale, but since these scaling exercises are meant to be approximate, it may not matter significantly.) If τ_{mix}^{vert} is significantly shorter (i.e., at least an order of magnitude smaller) than τ_{growth}, then we would expect vertical mixing to be rapid enough to prevent an algal bloom in the euphotic layer. If, on the other hand, τ_{mix}^{vert} is significantly longer than τ_{growth}, then we would not expect vertical mixing to be strong enough to single-handedly prevent a surface bloom. If we instead wish to understand whether *longitudinal* dispersion is fast enough to limit algal accumulation within a defined water body, then (1) an algal growth timescale might be more appropriately based on a typical (e.g., mean) net growth rate over the water column, especially if vertically well-mixed, and (2) the mixing timescale would be more appropriately estimated as the square of the water body length divided by K_{long}, a longitudinal dispersion coefficient [42]. Furthermore, if transport through a water body is known to be governed primarily by advection induced by river flow as opposed to dispersive processes, then an advective timescale (e.g., water body volume V divided by river discharge Q) may be a more relevant transport timescale to compare with the algal growth timescale. Incidentally, the relative importance of advection versus dispersion (or diffusion) is a matter that itself can be illuminated using this sort of scaling approach: The well-known Peclet number (i.e., the ratio of a diffusive timescale to an advective timescale) is a dimensionless ratio implemented for this very purpose [22,59,88].

A variety of methods can be employed to obtain biogeochemical rates that can then be transformed into timescales, as in Table 1. Middelburg and Nieuwenhuize [90] performed shipboard measurements and incubations with running estuarine water to obtain nitrogen concentrations and specific uptake rates, which were manipulated to obtain absolute uptake rates and then turnover times for particulate nitrogen, ammonium, and nitrate. Phytoplankton growth timescales have been estimated as the reciprocal of specific net growth rates based on numerical models, measurements of primary production, or published relationships [23]. Middelburg et al. [125] determined algal turnover times for microphytobenthos as B:P (biomass:production) ratios based on tidal flat core samples and ^{14}C uptake experiments. Timescales for algal losses to bivalve grazing have been calculated from water depth and grazing rates based on benthic biomass samples, published temperature-dependent pumping rate relationships, and laboratory-based expressions incorporating the food-limiting effect of concentration boundary layers [23,62]. Lopez et al. [148] estimated the specific loss rate of phytoplankton to zooplankton grazing based on tow net sampling, analyses to obtain carbon weight and community grazing rate, and measurements of phytoplankton biomass; that specific loss rate was then combined with benthic grazing losses to then obtain a collective timescale for loss [23]. Shen et al. [21] estimated the timescale for biochemical oxygen consumption based on temperature, surface dissolved oxygen concentration, and net oxygen consumption rate, which was taken as the sum of sediment oxygen demand and net water column respiration and based on previously published measurements and modeling constants. Crump et al. [91] calculated estuarine bacterial community doubling times from bacterial production (based on leucine incorporation) and bacterial cell counts. A timescale for contaminant depuration was calculated as the biological half-life of trace elements in mussels fed radiolabeled diatoms in a laboratory [150]. The timescale for 50% survival for larvae of broadcast spawning corals was quantified in laboratory experiments starting with gametes collected in the field (Nozawa and Okubo 2011); these "T_{50}" values were ultimately compared with model-computed residence times to gain insight into ecological connectivity and the potential for self-seeding [135,136].

Timescales based on simple algebraic combinations of process rates and other parameters (as in Table 1) are usually low on the holism scale, in that they typically do not account for multiple major drivers or underlying processes. This is not necessarily a bad thing. Timescales that each isolate an individual process can be useful for assessing governing processes via comparisons with other process-specific timescales.

2.2. Transport Timescales Based on Observational Data

Observational data from the field can provide characteristic values for process rates and auxiliary scales (e.g., discharge, velocity, depth, concentration) for use with the method described in Section 2.1 [23,25,90,125]. Observations can also provide a strong empirical basis for more directly estimating transport timescales. Field-based approaches have the important advantage that the acquired transport information is obtained in the *actual* water body, in which all relevant processes (river flow, wind, tides, etc.) are operative, making the derived timescales holistic [139].

A significant distinction between observational strategies is whether the measurements are Lagrangian (following a water parcel through time and space) or Eulerian (observed at prescribed locations in space that are determined by humans, not hydrodynamics). Below, we describe drifter-based approaches. While the information obtained from drifters is not well suited for straightforwardly estimating fluxes, their Lagrangian nature can reveal the transport pathways and ultimate fate of solutes, particles or biota in the water, as well as their associated timescales of transport [151]. On the other hand, tracer-based approaches are generally Eulerian (e.g., those involving measurements of velocity, flow rate, concentration, etc., at set locations) and can also provide bases for transport timescale estimation but may not predict fate or specific transport trajectories [151].

2.2.1. Drifter-Based Experiments

Lagrangian drifter experiments in the field have permitted the direct measurement of residence times [102–105,152] and transit times [103] (Figure 7) within specific regions; residence times within circulation features such as currents, gyres, and eddies [153]; and travel times between defined areas [55,56,142]. This general approach can involve vessel-based [151,154] or satellite-based [56,152,153] drifter tracking (e.g., see Figure 8A), with the latter becoming increasingly more affordable given recent technological advances [103,104]. In addition, low-cost buoyant objects such as driftcards [104] or plastic "daisy-like" drifters [155], whose finding time, location, and identifying information are reported by citizen finders, can be released by the thousands [104,156]. Drifter-based methods have been deployed in the deep waters of the Adriatic Sea [152] and coastal Antarctica [153], in fjords such as the Strait of Georgia in the Salish Sea [104], and in shallower bays such as Faga'alu Bay (American Samoa) [102] and the San Francisco Bay-Delta (CA, USA) [151,154]. If tracked at high enough frequency, drifters can not only reveal overall transport timescales (e.g., how long it took a water parcel to travel from point A to point B) but also the specific travel pathways taken. Such information can be particularly valuable in tidal systems, where travel paths can be especially circuitous and unintuitive (see Figure 8B–E). Pathway or precise transport time information is likely not achievable with drift cards or other objects that are not tracked at adequately high frequency; however, if many driftcards are found in a given area, a crude estimate of transit time might be provided by the earliest driftcards found [104]. Limitations of drifter-based field approaches include the impracticality of releasing large numbers of real drifters, especially compared to the analogous number possible in numerical models [104,142,152]; grounding and potential refloating of drifters (see Figure 8D,E) [104]; for surface drifters, the *"constraint to follow the 2D surface flow"* [157], potentially diverging from a true representation of water particles, which can be mixed vertically and thereby experience a range of velocities [104]; wave and wind interactions [104,157,158]; global positioning system (GPS) inaccuracy [157]; and the finite lifetime of satellite-tracked drifters due to battery failure or other factors [103,152].

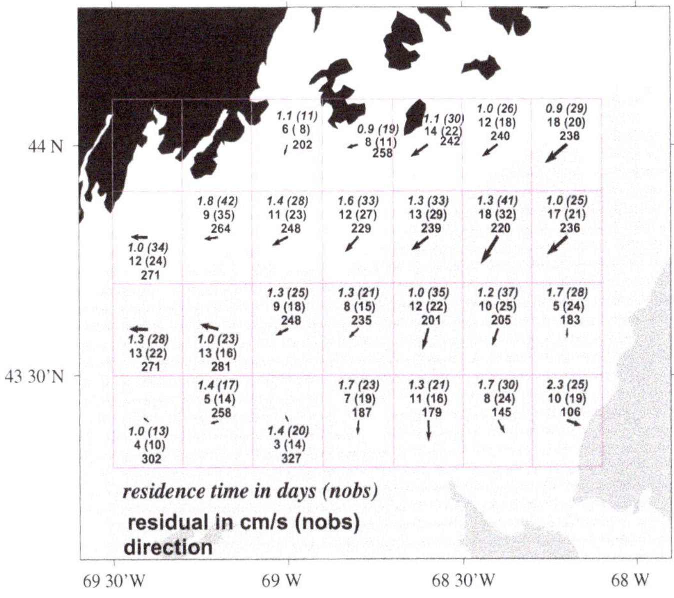

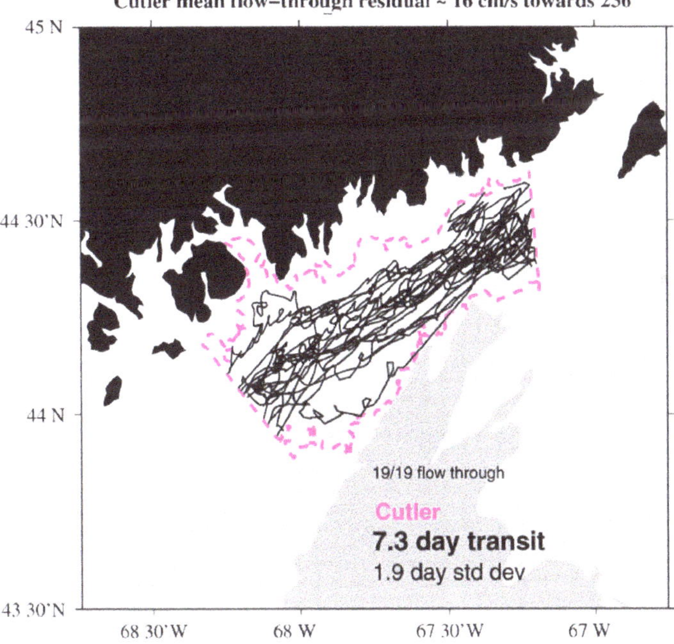

Figure 7. Results from the drifter field studies of Manning et al. [103] in the Gulf of Maine. Upper Panel: calculated residence times in days (italics), low frequency speed in cm/s, and direction in degrees True. Number of observations ("nobs") is in parentheses. Lower Panel: tracks of drifters entering waters offshore Cutler Maine from the northeast and heading southwest in the Eastern Maine Coastal Current. Transit time (7.3 d) is the mean time for drifters to traverse the region outlined in purple. (Modified from Manning et al. [103], with permission from Elsevier).

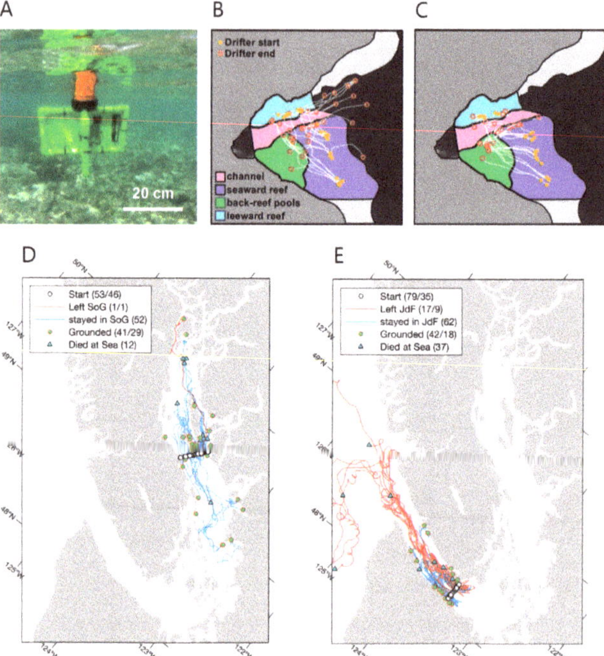

Figure 8. From the drifter studies of Storlazzi et al. [102] in Faga'alu Bay (American Samoa): (**A**) a deployed drifter; individual drifter tracks, with orange symbols representing drifter deployment locations and red circles representing drifter recovery locations for conditions of (**B**) calm and (**C**) strong winds. (Modified from Storlazzi et al. [102].) From the drifter field studies of Pawlowicz et al. [104]: tracks for drifters released in (**D**) the northern Strait of Georgia (SoG) and (**E**) Victoria Sill in the Salish Sea. Statistics in legends represent the number of tracks for each category; when two numbers are provided separated by a slash, the first is number of tracks, and the second is the number of unique drifter IDs [104]. "JdF" is "Juan de Fuca" Strait. (Modified from Pawlowicz et al. [104] and licensed under CC BY 4.0 (https://creativecommons.org/licenses/by/4.0/)).

A novel twist on the drifter approach involved the acoustic tagging of juvenile salmon to ascertain fish travel times through defined reaches and then draw linkages between travel times, river flow, routing, and fish survival [159]. These fish travel times represent an extra-holistic timescale in that they not only include the effects of processes influencing flow but also incorporate the effects of fish behavior.

2.2.2. Tracer-Based Experiments

Another class of field-based approaches for quantifying timescales involves both artificial and natural tracer studies. Artificial tracers include those released into surface waters either intentionally (e.g., rhodamine [117,160,161] or fluorescein [162] dye; NaCl in freshwater [37]; controlled radionuclide discharges from nuclear fuel reprocessing plants [163,164]) or unintentionally (e.g., radioactivity from the Fukushima Daiichi [165,166] and Chernobyl nuclear plant accidents [167] or from nuclear weapons testing [167]). Natural tracers include salinity [160,168], radioactive isotopes (e.g., Ra: [169–171]; Th and U: [172]); and stable isotopes (H and O: [173]). Field measurements of these tracers can be analyzed in a variety of ways (sometimes in combination with models) to estimate timescales such as water age [160,169–171,173,174]; travel or transit time [160,169]; residence time [160,171]; residence time in the ocean surface mixed layer [172]; flushing time [168]; or environmental half-time [165,166]. Along with extensive application in estuaries and other coastal systems [160,166,169–171,173], field tracer methods have also been implemented in streams [37], catchments [41], constructed wetlands [161], and in the open ocean [172]. Timescale estimates gleaned from these approaches have proven useful for evaluating the performance of numerical models [165,174] and have improved understanding of nutrient uptake [37,173], phytoplankton dynamics [173], trace metal export from the ocean surface mixed layer [172], the magnitude of a radioactive contaminant source [166], "biological tides" in constructed wetlands [161], seaward transport of river plumes [169], and fluid retention within seagrass [117] or macroalgae [162] canopies.

As one example of novel tracer-based approaches, Downing et al. [173] measured ratios of stable isotopes of hydrogen and oxygen in water at high-frequency aboard a high-speed boat as it wound its way along a sampling circuit through a complex tidal environment (the Cache Slough Complex in the Sacramento-San Joaquin Delta, USA; see Figure 9). Analyses of the isotope measurements permitted estimation of water age [173] along the transect. Estimated water age was co-analyzed with other parameters measured along the sampling circuit (e.g., nitrate, chlorophyll *a* fluorescence) to improve understanding of the linkages between transport time, algal production, and nutrient uptake (Figure 9A–C) [173]. Moreover, the authors used fits to an exponential relationship between change-in-nitrate versus change-in-water-age along boat tracks to obtain channel-specific estimates of whole-ecosystem net nitrate uptake rates (Figure 9D,E). As an alternative to the traditional tracer *salinity*, the authors' estimates of water age were later used to assess the skill of a numerical transport model for this environment, where characteristically low salinities can be considerably influenced by often poorly quantified agricultural return flows [174]. This same approach was used to evaluate the influence of an emergency drought barrier (installed to prevent salinity intrusion) on transport times, water quality, and ecosystem processes in a different part of that same ecosystem [62].

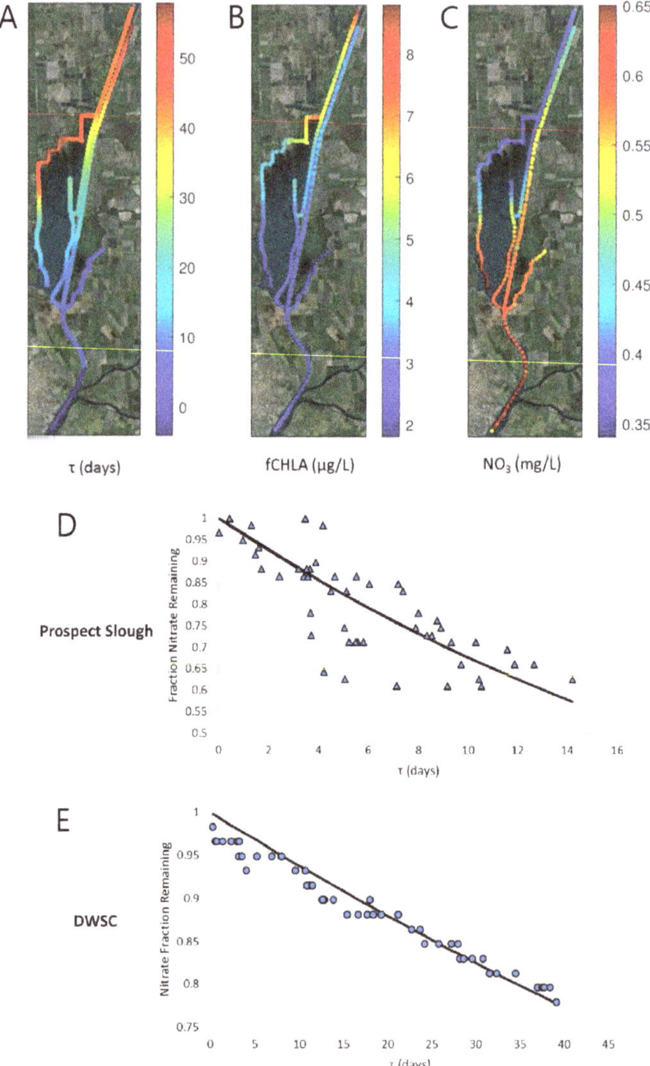

Figure 9. (**A**) Water age "τ", (**B**) chlorophyll *a* fluorescence, and (**C**) nitrate, based on concurrent mapping by Downing et al. [173] aboard a high-speed boat in the Cache Slough Complex of the Sacramento-San Joaquin Delta (USA). Low (high) fCHLA generally corresponded with small (large) τ. Nitrate had roughly the opposite pattern relative to τ. For (**D**) Prospect Slough and (**E**) the Sacramento Deep Water Ship Channel ("DWSC"), fits to an exponential relationship between change-in-nitrate versus change-in-water-age along boat tracks, used to estimate total-ecosystem net nitrate uptake rate. Estimated uptake rates were 0.039 d^{-1} in Prospect Slough and 0.006 d^{-1} in the DWSC. (Adapted from Downing et al. [173] (https://pubs.acs.org/doi/10.1021/acs.est.6b05745), with permission from American Chemical Society. This is an unofficial adaptation of an article that appeared in an ACS publication. ACS has not endorsed the content of this adaptation or the context of its use. Further permissions related to the material excerpted should be directed to the ACS).

2.3. Transport Timescales Based on Numerical Models

With the ongoing improvements in numerical methods for surface water hydrodynamics and transport, as well as continual advances in computational resources, the application of numerical models for estimating transport timescales is becoming increasingly common. There is a variety of methods for doing so, including forward and backward methods and approaches implementing numerical tracers or particles. Similar to timescales derived from field-based methods, those extracted from a numerical model can also be highly holistic [139], with the timescale holism limited by the holism of the model (i.e., all processes accounted for in the model that influence tracer or particle distribution will be accounted for in the derived timescales, but those that are missing from the model will not be "felt" by the timescales [175]). Although they may be holistic, model-based timescales, by their very design, tend to focus on the larger time and space scales of motion and filter out the smaller time and space variations.

2.3.1. Forward Methods

The most common overall model-based approach for quantifying timescales—the forward approach—is in some ways the most intuitively simple because it involves running numerical transport models for the purpose they are usually designed: marching forward in time. Numerical tracers or particles are injected into or released within a water body, and then they are transported by a hydrodynamic model's computed velocities, diffusivities, etc. The computed concentration fields or particle distributions over time are analyzed in order to extract information about how long water—or the "stuff" transported with it—has spent or will spend within a defined domain or on a trajectory to another.

Forward model-based, particle-tracking approaches have been applied in a variety of coastal environments and beyond. For example, Defne and Ganju [101] implemented hydrodynamic and Lagrangian transport models in the Barnegat Bay-Little Egg Harbor estuary (NJ, USA) to quantify spatially variable residence times, as well as whole-estuary flushing parameters. Nearly 80,000 virtual particles were released uniformly in the horizontal every hour for one day. Particles were tracked until they left the estuarine system, with residence time for each particle recorded as the time elapsed between release and exit from the system (see Figure 10). They also applied the classic *e*-folding approach and its "double-exponential" variation [175] to the totality of particles to quantify system-level flushing times. (In some cases, a double-exponential can offer an improved fit to a tracer "decay" timeseries, relative to the single exponential form described in Section 1.3 [101,175]). Moreover, Defne and Ganju [101] ran multiple simulations, turning individual forcings on and off and allowing for the identification of mechanisms most dominant in controlling flushing (Figure 10). Similar Lagrangian approaches have been applied to obtain transport timescales in: New Caledonia [112], the coastal transition zone off California (USA) [112], the Bay of Quinte (Ontario, Canada) [147], the Virginia Coast Reserve (USA) [176], the Mururoa atoll lagoon (French Polynesia) [113], the Great Barrier Reef (Australia) [45], and Galveston Bay (TX, USA) [177].

Forward-running models implementing conservative numerical tracers (an Eulerian approach) are also commonly used for assessing timescales. For example, flushing (or renewal) time can be obtained by tracking total tracer mass in a defined region and identifying the time needed for mass to decay to a prescribed level (e.g., $1/e$ [114,121] or some other fraction [178] of initial mass). Alternatively, an exponential or similar curve fit to the total-mass timeseries can allow for estimation of flushing (or turnover) time [112,114,121,175,179]. This regional approach can also be applied at the scale of a single grid cell, by fitting an exponential to the cell's concentration timeseries and obtaining a local flushing timescale as the reciprocal of the fitted decay coefficient; if this procedure is performed for all grid cells, maps of local transport time can be constructed [112,114]. Some authors have applied other constructs (e.g., Takeoka's [88] "remnant function" concept [180], or the freshwater fraction method [168]) to extract spatially variable [180] or region-wide [168] transport timescales from tracer simulations.

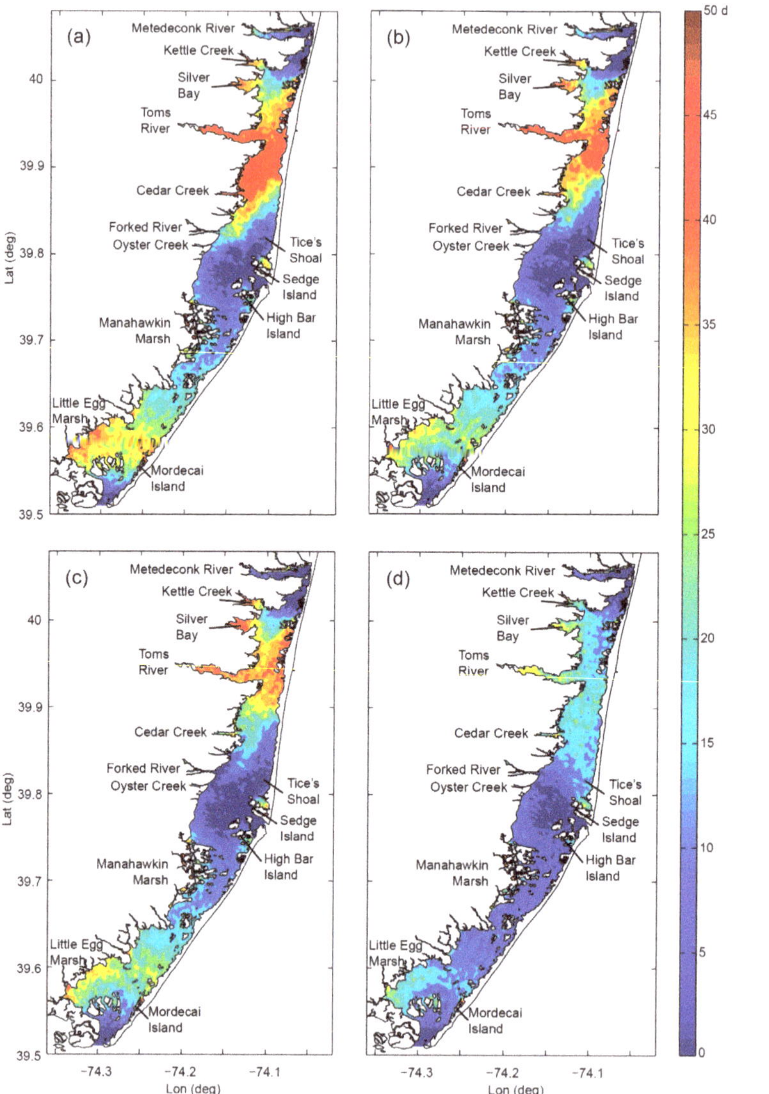

Figure 10. Spatially variable residence times computed by Defne and Ganju [101] with coupled 3D hydrodynamic and particle tracking models applied to Barnegat Bay-Little Egg Harbor (NJ, USA). The scenarios shown are (**a**) tidal forcing only, (**b**) tidal plus remote coastal forcing, (**c**) like (**b**) but with river flow added, (**d**) like (**c**) but with meteorological forcing added. Two inlets—Little Egg Inlet at the southern end and Barnegat Inlet near the center—connect the ocean and estuary (see Figure 1 in [101] for detailed site map). (Modified from Defne and Ganju [101]).

Over the past few decades, advanced theories have been developed for evaluating timescales at every time and location in the atmosphere [51,181,182], in aquifers [28,183], and in surface water bodies [63,64]. These timescales are generally derived from the solutions of partial differential equations (e.g., [64,97,140,141,182]). One such forward approach used extensively in coastal aquatic systems allows for the computation of spatially and temporally variable age of water (or of a constituent in the water) based on the solution of two forward advection-diffusion-reaction PDE's [63,64]. This approach accounts for the fact that, due to diffusion, production, and destruction, any water parcel will likely contain particles with a distribution of ages. Accordingly, the core variable is the age distribution function, which may be viewed as the histogram of the ages of the particles of the constituent (or group of constituents, including the water itself) under consideration at a given time and location. Explicitly computing this variable may be computationally demanding [184], for five independent variables (time, 3 space coordinates, and the age) are to be dealt with. However, most studies have focused on the mean age (i.e., the mass weighted age of the particles under consideration), which is the ratio of the first-order moment of the distribution function (the "age concentration") to the zeroth-order one (the concentration). Both the age concentration and the concentration satisfy coupled reactive transport equations in the time-space domain and, hence, are relatively easily computed. This approach has been applied and/or extended for the investigation of sediment transport [131,185–187], contaminants sorbed to sediment particles [107], pathways and fate of nutrients [188], interactions between ecosystem components (e.g., phytoplankton, zooplankton, nutrients) [189], connectivity [140,190], water renewal rates of semi- enclosed water bodies [66,100,174,191–196], ventilation of the deep ocean [49,197], and building reduced-complexity models that help interpret the results of complex ones [49]. A related forward method allows for the computation of average residence time for, practically, a limited number of subregions within a water body and/or start times [95,192]. Mathematically and numerically, this is an easily tractable problem for obtaining regional residence times [192] and exposure times [95], the latter having been shown useful in quantifying connectivity between subregions of a water body [95] (see Section 3.4 for more detail).

2.3.2. Backward Methods

Other advanced theories that rely on adjoint modelling, leading to backward-in-time numerical integration, have been presented over the past couple decades, also with applications to the atmosphere [182], groundwater [28], and surface waters [97–99,141]. Most relevant to the present discussion, the method of Delhez et al. [97] provides a computationally efficient means of obtaining surface water residence time *at every grid cell and time step*, not just for a limited number of locations, regions, or times as with the forward approach mentioned above [95,192]. Depending on the solution of an adjoint advection-diffusion problem, this backward-in-time approach has been extended to compute exposure times [98,99], thus allowing computation of the total cumulative time a particle spends within a defined water body, including time spent during multiple visits. This general method has been applied extensively in coastal systems including the English Channel and southern North Sea [97,99], the Scheldt Estuary [66], Brazilian estuaries [22], and the Chesapeake Bay [198]. For the lower James River (VA, USA), a tidal tributary of the Chesapeake, this approach [97] has been employed in the study of how transport processes influence the observed origins of harmful algal blooms [199] (see Section 3.4). Moreover, the theory has been generalized to the vertical dimension for computing light exposure of phytoplankton [200].

A particularly useful extension of the adjoint residence time theory allows for the calculation of partial residence times, i.e., the amounts of time a particle spends in different subregions before exiting the water body [141]. As Lin and Liu [141] point out, this application is useful for understanding connectivity between subregions of an aquatic system. Figure 11 illustrates those authors' calculation of partial residence times (PRTs) for Jiaozhou Bay (China). They divided Jiaozhou Bay (the control region, ω) into 6 subregions (ω_1–ω_6; Figure 11A), and their novel extension of the adjoint approach permitted them to compute PRTs for particles initialized at specific points in space (numbered stars in Figure 11A). For each of those seven release locations, Figure 11B shows the PRTs representing time spent in subregions ω_1–ω_6 before exiting the control region. For a given release location, the sum of all six PRTs (shaded portions of each bar in Figure 11B) equals the total residence time, i.e., the total time taken to leave the bay (top height of each bar). For pollutants discharged from a specific point location, this sort of information can quantify for resource managers how much time the pollutants spend in defined subregions on their way out of the bay [141], thereby highlighting areas potentially most impacted. PRTs are also displayed for each subregion ω_i as time spent in ω_i for particles released at every location in the domain (Figure 11C–H). These maps highlight the portions of the domain contributing particles spending the most time in a specific subregion and could, for example, provide insight into the major nutrient sources to a subregion and how much time those nutrients spend in the subregion before getting flushed out.

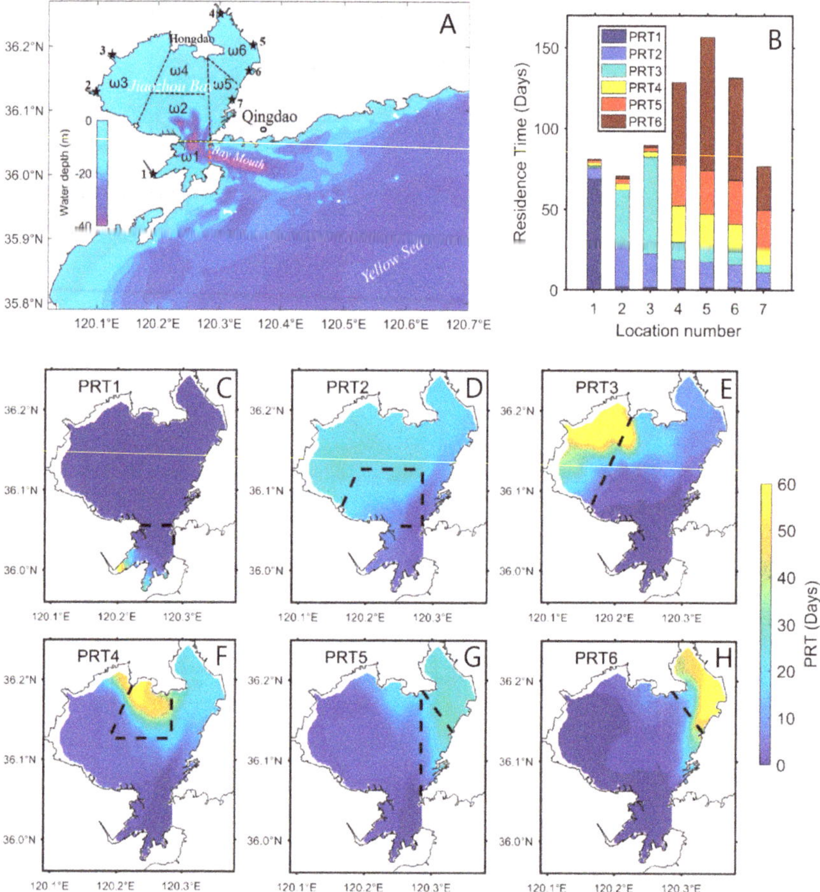

Figure 11. Lin and Liu's [141] (**A**) bathymetry map of Jiaozhou Bay (China), showing six subregions (ω_1–ω_6) in which partial residence times (PRTs) were calculated in (**B–H**), and seven release points (stars) for which PRTs in the subregions are shown in (**B**). (**B**) For particles initiated at each of seven locations, PRTs shown are time spent in each of six subregions before leaving the bay. For a given release location, the sum of the PRTs equals the total residence time within Jiaozhou Bay. (**C–H**) Spatial maps for each subregion representing time spent in the subregion for particles initiated at every location in the domain. Dashed lines represent the boundaries of each subregion. (Adapted by permission from Springer Nature Customer Service Center GmbH: Springer Nature, Ocean Dynamics, Partial residence times: determining residence time composition in different subregions, Lin and Liu, 2019. https://www.springer.com/journal/10236).

3. Timescale Applications for Explaining Ecosystem Processes and Variability in Water Quality

In this section, we describe previous studies that have referenced, estimated, and/or somehow implemented diagnostic timescales in order to help explain how aquatic ecosystems operate. We pay specific attention to biological and geochemical processes and responses of biota or water quality to (physical or other) environmental conditions. We proceed by grouping studies according to different modes of timescale use, so each type of use may include references to a variety of ecosystem variables, processes, or questions.

3.1. Timescales in Conceptual Models

Timescales are often used in a qualitative or semi-quantitative manner as components of conceptual models for helping explain how aquatic ecosystems are believed to operate. In such cases, the term "residence time" is often invoked, even though it is frequently neither defined nor quantified. Therefore, although there exist clear (albeit varied) mathematical definitions of residence time, that term is very frequently used—and understood—to refer *generally* to how long water, particles, organisms, or solutes spend in (or on their way to or from) a certain area, without specifying the details of how it might actually be calculated. "Retention" and "turnover" time are other terms often referred to in conceptual models.

Prominent (inter-related) areas in which timescales have been invoked conceptually to explain aquatic ecosystem dynamics include nutrient processing, phytoplankton dynamics, eutrophication, and hypoxia. For example, in their review of legacy phosphorus in watersheds, Sharpley et al. [14] explained that "hotspots" of phosphorus retention and cycling can occur in areas with slower flows and longer water retention times (e.g., pools, eddies, channel margins) and in areas with sharp gradients in water and sediment retention times (e.g., where rapidly flowing water meets standing water). Boyer et al. [201] explained Florida Bay's (USA) observed spatial differences in total organic nitrogen (TON), total phosphorus (TP), and phytoplankton biomass (as well as salinity and total organic carbon) as driven by differences in freshwater inputs and water residence time and, consequently, evaporation rates. In outlining his contemporary conceptual model of coastal eutrophication, Cloern [146] identified residence time as one component of the "filter" (the set of physical and biological attributes) that sets the sensitivity of individual coastal ecosystems to nutrient enrichment. Scavia et al. [143] linked climate change to estuarine phytoplankton bloom development, with residence time playing a key role: where freshwater runoff decreases, water residence time will increase, and phytoplankton production will also be expected to rise if the phytoplankton doubling time is shorter than the residence time. In such cases, susceptibility of coastal systems to eutrophication could be consequently heightened. Those authors also identified the potential role of humans in further altering residence times (e.g., by storing more freshwater within the watershed to combat drought), thereby intensifying algal production and vulnerability to eutrophication. Paerl and Huisman [202] described how massive cyanobacteria blooms have occurred when high-residence time drought periods follow intense precipitation and nutrient discharge events—a scenario that could become more prevalent with global warming. Similar to Scavia et al. [143], Paerl and Huisman [202] also suggested that human interventions intended to control flow variability (e.g., construction of dams or sluices) could further increase residence times and thereby exacerbate ecological and human health problems caused by cyanobacteria. Rabalais and Turner [203] and Rabalais et al. [11] cited long water residence time as one of the key factors (along with stratification) controlling the likelihood that a coastal system will develop hypoxia. Residence time featured prominently in Durand's [204] conceptual model of the aquatic food web of the Upper San Francisco Estuary (CA, USA), providing a linkage mechanism between physical forcings such as hydrology, tides, and water diversion and the spatial and temporal variability of nutrients, phytoplankton, and zooplankton.

Water residence time has also been identified as an important factor in conceptual models of estuarine metabolism. Hopkinson and Vallino [205] pointed to water residence time as an important influence on the autotrophic–heterotrophic nature of an estuary. They described how the relative magnitudes of the water residence (or "turnover") time and biogeochemical time constants (e.g., for organic matter decomposition or autotrophic and heterotrophic production) can determine whether decomposition or biomass accumulation are significant within an estuary. Viewing water residence time from a biogeochemical perspective, those authors saw it as a descriptor of the time for materials to be processed in a system and thereby a potential limit on whether reactions can go to completion; the *material* residence time (and thus the time for reactions to proceed) could be effectively lengthened beyond the water residence time by the settling of organic particles to the bottom [205]. Relatedly, Battin et al. [206] developed a conceptual model of organic carbon processing to help explain how terrestrial organic carbon, which had long been believed to be recalcitrant, could fuel net heterotrophy in rapidly flowing fluvial networks, as recent data had indicated. Those authors proposed that hydrological storage and retention zones along the path to the ocean (created by, for example, morphological features, rough and highly permeable streambeds, debris, floodplains, or estuarine turbidity maxima) create "geophysical opportunities" [206] for microorganisms to metabolize organic carbon. In such environments, the residence time of microorganisms may be extended beyond that of water through attachment to surfaces (e.g., as biofilms).

3.2. Implementing Timescales in Building Simple Models

Some timescales can collapse a complex process or collection of processes into a single number (hence, the holistic label referred to earlier). For example, a transport timescale, properly calculated, can simultaneously account for wind-, tide-, river-, and density-driven hydrodynamics. Similarly, a benthic grazing timescale can integrate the contributions of community composition and biomass, pumping rates of different species, concentration boundary layers, and water column depth into a single value. Some timescales are also designed to integrate over space and/or time, removing spatial or temporal detail for a "bird's eye" view of an aquatic system. Because timescales are such powerful encapsulators of complexity, they can prove useful in developing reduced-complexity mathematical models of ecosystem function.

3.2.1. Simple Models of the Physical Environment

There are several examples where timescales were used as tools to distill hydrodynamic complexity and then design simple models capturing the general physical behavior. For example, Liu et al. [207] ran multi-decadal simulations with global ocean-ice models, implementing a novel variation on an age tracer approach [208] to compute coastal residence time (CRT) worldwide (Figure 12A). The goal was to quantify a coastal retention timescale that reflects the time spent by a water parcel in the coastal zone [207]. Those authors described CRT as the *"total time a water parcel stays in any part of the global coastal ocean rather than a specified domain (i.e., a water parcel would accumulate CRT while traveling alongshore from one coastal system to another)"* [207]. Moreover, while CRT for a water parcel accumulates with time spent in the coastal zone, CRT is gradually diminished with time spent in the open ocean; a water parcel that leaves, and then re-enters, the coastal zone thus returns with a lower CRT than that with which it left. This approach allows the "coastal signature" of a water parcel to gradually increase (or decrease) depending on time spent inside (or outside) the coastal zone [209]. CRT, by this definition, is similar to "exposure time" because both metrics continue to accumulate when a water parcel is within the domain of interest, even after having left. They are different, however, in that exposure time is preserved when a water parcel is outside the domain of interest, while CRT diminishes with time outside the domain. Given latitudinal differences observed in their computational results (Figure 12A), as well as the expectation that the degree of geometric enclosure could influence CRT, the authors [207] fitted a simple algebraic model (Figure 12B) of three-dimensional (3D) model-computed CRT [d] as a function of the Coriolis frequency f [1/s] (similar to Sharples et al. [210]) and $\chi = V/S$ [m], the ratio of the total volume of a coastal system to its total open boundary area. The simple model explained 73% of the variability in simulated CRT, thus providing a convenient method for estimating CRT. Delhez [98] first identified an inherent problem with the concept of exposure time, i.e., that it (as traditionally defined) will become infinite in a computational domain limited by impermeable boundaries. As a solution to this issue, he introduced first-order decay in his calculation of the exposure time, somewhat similar to the diminishment of CRT outside the coastal zone by Liu et al. [207].

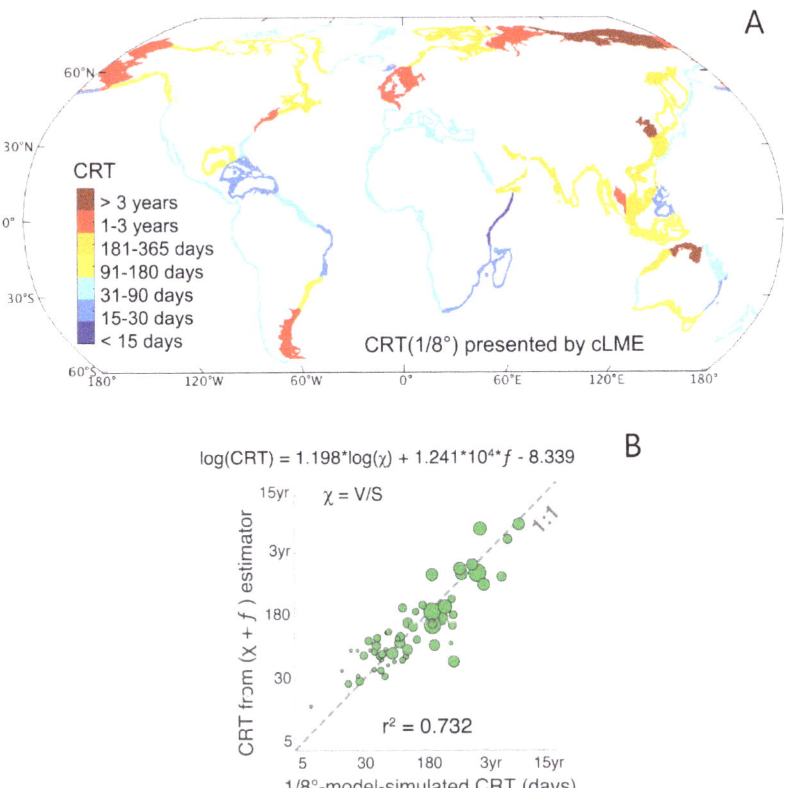

Figure 12. (**A**) A global map of coastal residence times (CRTs) simulated by Liu et al. [207] using high-resolution, coupled global ocean-ice models and a novel variation on an age tracer approach; (**B**) simple model of CRT as a function of Coriolis parameter f and a geometric parameter χ, which is the ratio of total coastal system volume to total open boundary area. (Modified with permission from Xiao Liu, Geophysical Research Letters; published by Wiley, 2019).

Other reduced-complexity models where timescales played a fundamental role include: (1) the simple but effective (R^2 = 0.74 and 0.95) regression models of Kärnä and Baptista [194] relating system-wide "renewing water age" (computed by a detailed 3D model) to observed river discharge and tidal range for the lower Columbia River Estuary (USA), thus allowing easy, quick estimates of water renewal timescales when 3D model simulations are not available; (2) the use by Mouchet and Deleersnijder [49] and [211] of mean ages and age distributions as a metric for evaluating the fidelity of the one-dimensional (1D) "leaky funnel" model to 3D models of ocean ventilation; (3) the derivation by Deleersnijder et al. [59] of simple estimates for mean residence time of sinking particles in the surface mixed layer; and (4) the development by Palazzoli et al. [179] of a simple polynomial relationship for the flushing-induced tracer decay coefficient (reciprocal of e-folding flushing time), as a function of wind speed and direction for the Virginia Coast Reserve, a complex system of interconnected shallow coastal bays and inlets on the United States east coast. Yet more examples are to be found in [22,212,213].

3.2.2. Simple Ecological Models Using Physical Timescales

A number of authors have taken advantage of the ability of transport timescales to capture the net effect of complex hydrodynamics on ecological processes. For example, Dettmann [214] derived simple algebraic models of estuarine nitrogen dynamics as a function of "freshwater residence time" (τ_{fw}). He started with an annual mass balance equation for total mass of biologically active, water-column nitrogen (m_N) in an estuary,

$$\frac{dm_N}{dt} = I - E - R \tag{1}$$

where I is the total rate of nitrogen input from upland and oceanic sources, E is the rate of export to the sea, and R is the net annual rate of within-estuary removal of water column nitrogen, assumed to be proportional to m_N. After making a number of simplifying assumptions (e.g., steady state, negligible nitrogen contribution from the ocean), Dettmann [214] arrived at the following dimensionless expression for $F_{E(I)}$, the annual net export (export to the sea minus input from the sea) expressed as a fraction of upland loading:

$$F_{E(I)} = \frac{1}{1 + \alpha\tau_{fw}} \tag{2}$$

as well as the below relationship for the annual fraction of upland loading that is denitrified:

$$F_{D(I)} = \frac{\varepsilon\alpha\tau_{fw}}{1 + \alpha\tau_{fw}} \tag{3}$$

In Equations (2) and (3), α is a first-order rate coefficient representing the net loss of nitrogen from the estuarine water column due to internal processes, and ε is the fraction of total internal losses accounted for by denitrification. The simplicity of Dettmann's [214] above expressions is impressive, especially considering how well they fit previously published estuarine data (Equation (2): $r^2 = 0.94$ with $\alpha = 0.3$ month^{-1}; Equation (3): $r^2 = 0.85$ with $\alpha = 0.3$ month^{-1}, $\varepsilon = 0.69$; Figure 13A,B). Moreover, the relationships make intuitive sense: the fraction of nitrogen input that is exported (denitrified) decreases (increases) as the transport timescale increases. This is logical because the longer nitrogen spends within an estuary, the more opportunity for it to incur denitrification and other loss processes, leaving less for export.

Transport timescales have also proven useful in the development of simple models of phytoplankton dynamics. One such model was developed in order to (1) test a common, intuitive conceptual model that was helping shape multi-billion dollar ecosystem management plans in the Sacramento-San Joaquin Delta (SSJD), CA, USA, and then to (2) communicate the findings with a clarity that ecosystem managers, engineers, and scientists alike could find useful and relevant. The conceptual model, framed by [215] as a hypothesis to be tested, was that: *"Habitats with longer transport times (slower hydrodynamics) are associated with higher phytoplankton biomass and productivity than habitats with shorter transport times (faster hydrodynamics)"*. This conceptual model was important in ecosystem restoration planning because, unlike many coastal systems that produce excessive amounts of phytoplankton biomass, the SSJD is characterized by low phytoplankton biomass. Because SSJD phytoplankton biomass was low enough to limit the growth of some zooplankton species [216,217] and experienced a long-term decrease [218] alongside similar declines in herbivorous zooplankton [219,220] and fish [221,222], low phytoplankton biomass and productivity were implicated as factors contributing to the declines of the upper trophic levels. Consequently, SSJD restoration plans included actions aimed at amplifying primary productivity [223]. The above conceptual model, which was helping guide those plans, did not account for the filtration pressure of the exotic clam *Corbicula fluminea* [224,225]. The authors [215] therefore used the following simple algebraic model for habitat averaged phytoplankton biomass as a function of transport time (defined as a "transit time", i.e., (habitat length)/velocity) to test whether the above hypothesis holds in the presence of clams:

$$B_{hab} = \frac{B_{in}}{\mu_{eff}\tau_{tran}}\left[\exp\left(\mu_{eff}\tau_{tran}\right) - 1\right]$$

(4)

B_{hab} is habitat averaged algal biomass concentration, B_{in} is algal biomass concentration flowing into the habitat, μ_{eff} is the effective phytoplankton growth rate (accounting for depth-averaged algal growth, respiration, zooplankton grazing, and clam grazing), and τ_{tran} is transport time. Operative assumptions included a vertically well-mixed water column and steady-state conditions. A similar equation was derived also for habitat averaged phytoplankton net productivity. Results from the simple models (Figure 13C,D) showed clearly that the hypothesis does not always hold: Hydrodynamically "slower" habitats can be less productive than "faster" ones if benthic grazing is strong enough to render the effective phytoplankton growth rate negative. Further, it was evident that the range of possible outcomes broadens with longer transport times. Therefore, since it is difficult to predict the response of non-native bivalves to restoration, the ultimate functioning of created habitats—especially those with long transport times—is highly uncertain. This simple model was able to clearly demonstrate that widely held intuitive, management-relevant conceptual models of phytoplankton dynamics do not always hold—and can, in fact, be reversed—in the presence of strong benthic grazing. This same lesson could have been demonstrated with more complex 1D, two-dimensional (2D) or 3D models, but the ultra-simple timescale-based form of Equation (4) isolated the salient processes and conveyed the message more effectively than more complex approaches might have.

A global view of denitrification was taken by Seitzinger et al. [54], who developed spatially distributed global-scale estimates of denitrification across system types including terrestrial soils, groundwater, lakes, reservoirs, rivers, estuaries, continental shelves, and oceanic oxygen minimum zones. One part of their analysis revealed that, when data representing estuaries, river reaches, lakes, and continental shelves were combined, "water residence time" could explain a large portion of the variability in the annual fraction of nitrogen (N) inputs that is denitrified. The empirical relationship derived from that combined data set,

$$\%\ N\ removed = 23.4(Water\ Residence\ Time)^{0.204}$$

(5)

where water residence time is in months, fits the data well ($R^2 = 0.56$). To aid in their global-scale estimates of denitrification, those authors then used this simple empirical model (Equation (5)) to estimate denitrification in lakes and reservoirs, and developed a similar estuary-specific relationship (% N removed = 16.1 (*Water Residence Time*)$^{0.30}$, $r^2 = 0.62$). In this case, "water residence time" was likely defined and calculated in more than one way, given the large number of sources contributing to the dataset [226]. Regardless, and in spite of the gross simplification of complicated and site-specific transport processes by the single parameter "water residence time", strong and useful relationships were obtained. Like Dettmann's [214] relationship (Equation (3) above), the empirical models of Seitzinger et al. [54] are also consistent with intuition: as time spent by imported nitrogen within a water body increases, the longer the time available for processing and biogeochemical removal of that nitrogen.

In their well-known work on the fate of nutrients at the land-sea margin, Nixon et al. [226] similarly compiled a collection of site-specific datasets to reveal strong linear-log empirical relationships between "residence time" and the fractional net export of nitrogen and phosphorus (P) from lakes and estuaries. Sharples et al. [210] powerfully applied simple empirical models based on the work of Seitzinger et al. [54] and Nixon et al. [226]. Their objective was to provide worldwide estimates of the N and P exported from the shelf to the open ocean. First, they developed a simple mechanistic model of how a river plume behaves after exiting an estuary, leading to straightforward relationships for estimating plume residence times on continental shelves worldwide (see Figure 14A). Combining (1) their global estimates of residence time on the shelf (Figure 14A), (2) empirical relationships between fractional nutrient export and residence time based on Seitzinger et al. [54] and Nixon et al. [226], and (3) a database of worldwide riverine nutrient loads [227], Sharples et al. [210] then produced global maps of riverine nutrient percentage (Figure 14B) and magnitude (Figure 14C) exported from shelves to the open ocean. These estimates ignore nutrient processing within estuaries, so estimated shelf-to-ocean export magnitudes are seen as an upper bound.

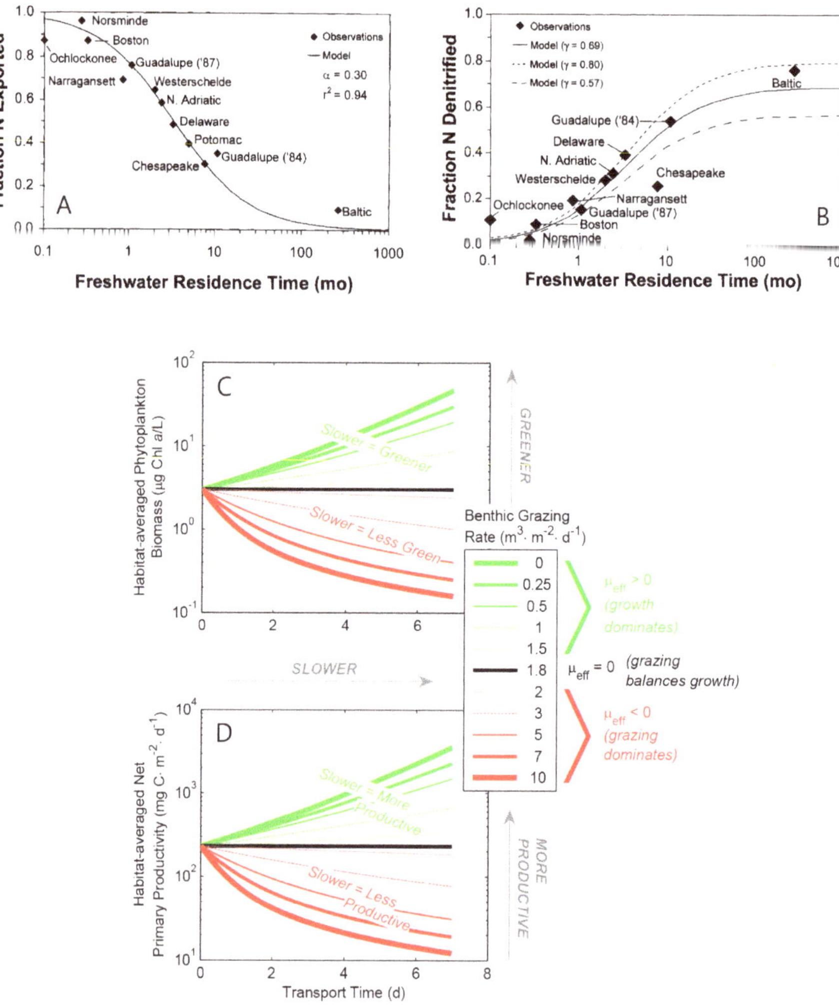

Figure 13. Dettmann's [214] simple models for fraction of upland nitrogen loading to an estuary that is (**A**) exported (Equation (2) herein) and (**B**) denitrified (Equation (3) herein), expressed as functions of "freshwater residence time" and fit to data for several estuaries. "γ" in Dettmann's [214] denitrification plot (**B**) is referred to as "ε" in Equation (3) and the text herein. (Modified from Dettmann [214].) Calculations of habitat-averaged phytoplankton (**C**) biomass and (**D**) productivity based on Lucas and Thompson's [215] simple models expressed as a function of transport time (Equation (4) herein for algal biomass). (Modified from Lucas and Thompson [215]).

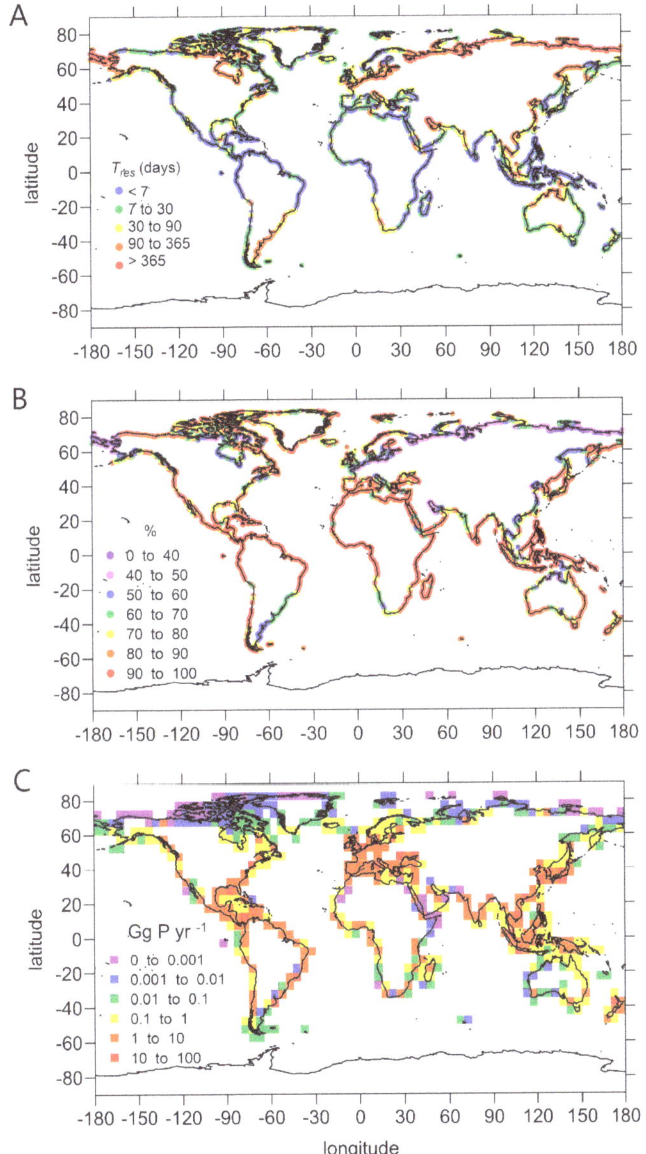

Figure 14. Based on the global scale, simple mathematical modeling of Sharples et al. [210]: (**A**) average residence time "T_{res}" on the continental shelf; (**B**) estimated proportion of riverine DIP (dissolved inorganic phosphorus) exported to the open ocean; (**C**) estimated annual DIP mass export to the open ocean. The authors performed the same calculations for dissolved inorganic nitrogen and provided uncertainty estimates (not shown here). (Modified with permission from J. Sharples, Global Biogeochemical Cycles; published by Wiley, 2017).

All of these studies exemplify how the synthesizing power of transport timescales can facilitate the development of simple, useful, and intuitive models for estimating biogeochemical responses to physical processes in coastal (and other) aquatic systems. These simple mathematical models have, in some cases, enabled large—even global—scale estimates of reactive constituent processing and delivery, an undertaking that may have been infeasible with detailed numerical transport-reaction models due to computational constraints and data limitations.

3.2.3. Simple Ecological Models Using Physical and Biogeochemical Timescales

Simplified mathematical models can incorporate ecological or geochemical timescales as well as transport timescales. For example, Lucas et al. [23] derived the following idealized model of algal transport, growth and loss in a generic vertically well-mixed aquatic system based on a common, steady-state plug flow equation:

$$B(x) = B_{out} = B_{in} \exp\left(\frac{\mu_{growth} - \mu_{loss}}{u} x\right) \tag{6}$$

where B_{in} is the phytoplankton biomass concentration entering a water body at the upstream boundary; $B(x)$ is phytoplankton biomass at distance x downstream from the inlet (if the length of the domain is x, then $B(x)$ is the same as B_{out}, the concentration exiting the domain at the downstream boundary); μ_{growth} and μ_{loss}, respectively, are the algal specific growth and combined in situ loss (e.g., grazing, senescence, sedimentation) rates (1/time); and u is the transport velocity (length/time). Substituting in timescales for advective transport ($\tau_{tran} = x/u$), growth ($\tau_{growth} = 1/\mu_{growth}$), and loss ($\tau_{loss} = 1/\mu_{loss}$), and combining timescales into ratios, they arrived at the following dimensionless relationship:

$$B^*_{out} = \frac{B_{out}}{B_{in}} = \exp\left(\left[1 - \frac{1}{\tau^*_{loss}}\right]\tau^*_{tran}\right) \tag{7}$$

where B^*_{out} is the outgoing biomass concentration normalized by the incoming biomass concentration, and τ^*_{loss} and τ^*_{tran} are, respectively, the loss and transport timescales normalized by the growth timescale. (τ^*_{tran} is comparable to the "Damköhler number", the dimensionless ratio of transport and reaction timescales used in chemical engineering [228] and in the hydrologic sciences [229] (see Section 3.4).). In Equation (6), the dependent variable (B_{out}) is a function of five parameters and variables; whereas the dependent variable in Equation (7) is a function of only two, allowing the relationship to be plotted (and, importantly, visualized) on a 2D surface (Figure 15 herein). Equation (7) and Figure 15 provide a simple tool for explaining why phytoplankton biomass can have a variety of relationships with transport time: biomass (B^*_{out}) increases with time spent in a water body (i.e., moving rightward in Figure 15) if growth is faster than in situ loss ($\tau^*_{loss} > 1$), but decreases with transport time (τ^*_{tran}) if loss is faster than growth ($\tau^*_{loss} < 1$). If growth and aggregate loss rates are similar ($\tau^*_{loss} \approx 1$), biomass does not change much while inside the water body ($B^*_{out} \approx 1$), regardless of the transport time. In summary (and contrary to the intuition of some), transport time does not determine whether phytoplankton biomass increases or decreases within an aquatic system; rather, the growth-loss balance (represented by τ^*_{loss}) does [23]. The reader is referred to a recent publication by Wang et al. [24], who developed an analytical model for downstream phytoplankton concentration in a 1D advective system, going beyond the model in Equations (6) and (7) by incorporating a non-linear reaction term (e.g., to incorporate the effects of self-shading or phytoplankton-dependent grazing). Reducing to Equation (6) above under simplified conditions, that model has two primary components—water age and accumulative growth—and agrees well with observations in the James River.

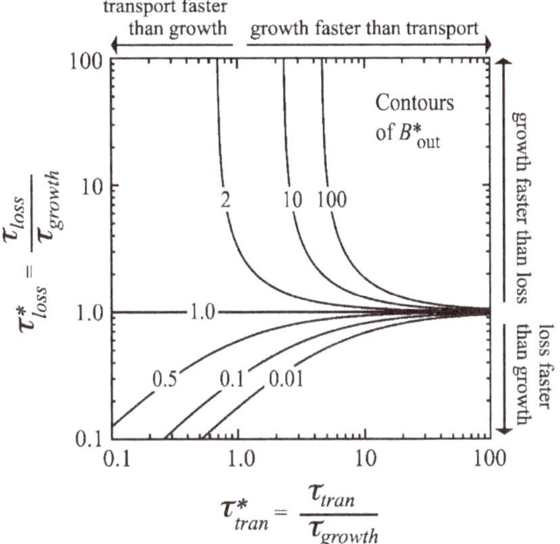

Figure 15. Contours of B^*_{out}, the ratio of outgoing algal biomass concentration to incoming concentration, as a function of two dimensionless parameters, τ^*_{loss} (the ratio of the algal loss timescale to the growth timescale) and τ^*_{tran} (the ratio of the transport timescale to the algal growth timescale). Based on the simple, timescale-based mathematical model of [23], Equation (7) herein. (From Lucas et al. [23]).

Shen et al. [21] applied a similar approach to a different problem: hypoxia in the deep waters of the Chesapeake Bay. They first derived a closed-form, steady-state 1D (along-estuary) relationship for bottom-water dissolved oxygen (DO), accounting for three dominant processes: horizontal replenishment due to gravitational circulation, vertical replenishment via exchange with the surface layer, and consumption based on the combination of sediment oxygen demand and organic carbon decay in the water column. This expression (Equation (4) in [21]; not shown herein), though mathematically straightforward, described bottom DO as a function of 8 variables and parameters. After defining a timescale for each major governing process (τ_e for longitudinal transport driven by gravitational circulation, τ_v for vertical exchange, and τ_b for consumption), creating timescale ratios, and substituting those ratios into their 1D equation, Shen et al. [21] arrived at the following predictor of bottom layer DO concentration, c:

$$\frac{c}{c_s} = 1 - \frac{1}{\tau^*_b}\left(1 - e^{-\tau^*_e}\right) \tag{8}$$

where c_s is surface DO concentration, $\tau^*_b = \frac{\tau_b}{\tau_v}$, and $\tau^*_e = \frac{\tau_e}{\tau_v}$. (Equation (8) also incorporated the assumption that bottom and surface DO were equal at the estuary mouth.) τ^*_b (τ^*_e) represents the competition between consumption (gravitational circulation) and vertical exchange processes. Equation (8) succeeded in reducing the expression for c to a problem with only three independent variables. The relationship governing dimensionless bottom DO (c/c_s) could thus be plotted in two dimensions, and the influence of the governing processes on the development (or avoidance) of hypoxia could be visualized (Figure 16). Notwithstanding the simplicity of Equation (8), estimates of bottom DO from this model compared well with observations (Figure 17), demonstrating how a complex hydrodynamic-biogeochemical problem could be broken down to a quantitatively accurate and illustrative algebraic relationship involving three timescales.

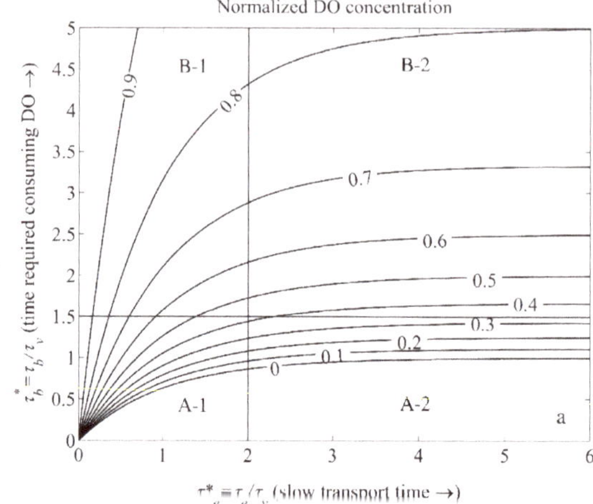

Figure 16. Contours of c:c_s (ratio of bottom layer dissolved oxygen concentration (DO) to surface layer DO), as a function of two dimensionless parameters, τ_b^* (the biochemical consumption timescale normalized by the vertical exchange timescale) and τ_e^* (the timescale for transport driven by gravitational circulation normalized by the vertical exchange timescale). Based on the simple, timescale-based mathematical model of Shen et al. [21] (Equation (8) herein). Rectangular regions delineate regimes associated with control of DO by particular processes and/or likelihood of hypoxia. (Reuse and minor adaptation from Shen et al. [21], with permission from Wiley. © 2013, by the Association for the Sciences of Limnology and Oceanography, Inc.)

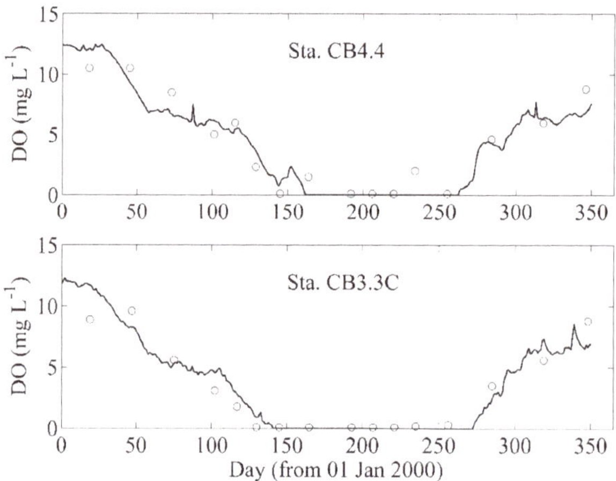

Figure 17. Comparisons of the simple Shen et al. [21] model (solid lines) and DO observations at two stations in the Chesapeake Bay. (Reuse and minor adaptation from Shen et al. [21], with permission from Wiley. © 2013, by the Association for the Sciences of Limnology and Oceanography, Inc.).

3.3. Assessing Relative Speeds or Dominance of Processes

As described briefly in Section 1.4, because timescales all carry the same units, they represent a single cross-disciplinary currency allowing for the comparison of the speeds of disparate processes, be they physical, biological, or chemical. Many authors have taken advantage of this translational characteristic of timescales to gain insight into which simultaneously acting processes exert primary control over ecosystem functions and responses. Timescale comparison can also provide a simple approach for assessing the likelihood of a particular ecosystem response. For example, in their review of coastal hypoxia, Fennel and Testa [132] defined a non-dimensional number γ as the ratio of a hypoxia timescale τ_{hyp} to water residence time. Akin to the DO consumption timescale τ_b of Shen et al. [21], τ_{hyp} was calculated as the ratio of an initial oxygen concentration to a volumetric oxygen consumption rate and represents the biogeochemically driven time to hypoxia occurrence. Residence time was taken to represent the time of restricted oxygen supply (i.e., how long biogeochemical consumption can operate uncountered by supply). The authors stated that γ *"relates the two factors contributing to hypoxia generation—net biochemical oxygen consumption and restricted supply of oxygen, which is related to water residence time"* [132]. They hypothesized that γ must be less than 1 for hypoxia to occur because, however slow oxygen consumption may be, hypoxia may still develop if hydrodynamically driven oxygen supply is impeded for an adequately long period of time. On the other hand, if oxygen consumption is rapid, hypoxia may be prevented if residence times are very short and oxygen is thus supplied on a frequent basis. Fennel and Testa [132] tested their hypothesis by estimating τ_{hyp} and residence time for nine hypoxic estuary and shelf systems (see Figure 18 herein), finding that indeed $\gamma < 1$ (biogeochemical depletion is faster than replenishment) for the majority of hypoxic systems studied. (The non-conformance of two systems—the Gulf of St. Lawrence and the Namibian shelf—was explained by an assumed, uniformly applied initial oxygen concentration that was likely too high for those two environments due to the importance of low-oxygen source waters.) The implementation of timescales thus allowed the authors to capture a great deal of the physical-biogeochemical complexity surrounding hypoxia development and distill it down to a simple ratio that performs well in describing hypoxia occurrence.

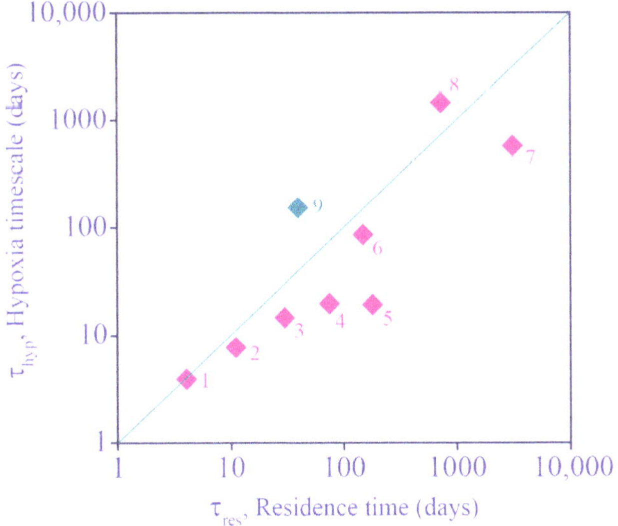

Figure 18. Hypoxia timescale versus residence time for several hypoxic estuarine and shelf systems, as estimated by Fennel and Testa [132]. Systems falling below the diagonal 1:1 line are consistent with the authors' hypothesis that γ, the ratio of the hypoxia timescale to the residence time, is less than unity for hypoxia to occur. Systems analyzed: (1) Pearl River Estuary (China); (2) East China Sea; (3) Northern Gulf of Mexico; (4) Long Island Sound (USA); (5) Chesapeake Bay (USA); (6) Northwestern Black Sea; (7) Baltic Sea; (8) Gulf of St. Lawrence (Canada); (9) Namibian Shelf. (Redrawn from Fennel and Testa [132] with the permission of K. Fennel.)

Other interdisciplinary studies that similarly used timescale comparisons to understand, explain, or predict coastal ecosystem responses include:

- *Estuarine nitrogen processing:* In their studies covering several European estuaries, Middelburg and Nieuwenhuize compared water "residence time" estimates to turnover times for particulate nitrogen, nitrate, ammonium [90], and amino acids [123], providing insight into which nutrient forms may become limiting [90] and whether individual forms will be significantly modified during transport through an estuary [90,123].

- *Hypoxia development in a tidal river:* In their study of the effect of water diversion structures on water quality in a complex, heavily managed tidal environment, Monsen et al. [230] compared 2D model-computed *e*-folding flushing times to half-lives for biological oxygen demand (BOD) [231]. They found that when a physical barrier was installed on a branch of the San Joaquin River (CA, USA), consequently forcing all flow through the mainstem, flushing times on the mainstem could decrease enough (relative to BOD half-life) to prevent the development of hypoxia, a frequent occurrence in a deep portion of the mainstem San Joaquin.

- *Nutrient processing on shelves and export to the open ocean:* Sharples et al. [210] compared their global-scale, latitudinally varying estimates of continental shelf residence times (Figure 14A herein) with nutrient processing times (assumed independent of latitude) in a discussion of which shelf regions would be expected to experience more (middle to high latitudes) or less (low latitudes) nitrate removal before exchange with the open ocean occurs.

- *Development of a unique estuarine bacterial community:* In their study of the Parker River Estuary and Plum Island Sound (MA, USA), Crump et al. [91] studied the conditions for the development of a unique community of estuarine bacterioplankton, as opposed to the advected populations of riverine or marine origin that were prevalent in the estuary. They compared water residence times and bacterial doubling times across seasons and the salinity gradient, finding that a local estuarine community developed at intermediate salinity only in the summer and fall, when water residence time was much longer than average doubling time, thus allowing the local community ample time to develop. In contrast, no local bacterial community developed in spring, when residence time was similar to average doubling time—apparently short enough to prevent the development of new estuarine bacterioplankton populations [91].

- *Benthic control of phytoplankton biomass:* Several authors have compared benthic grazing timescales to transport and/or phytoplankton growth timescales to understand controls on estuarine aquaculture potential [134] or phytoplankton biomass [25,61,124,232–234]. Extending the conceptual model of Dame [233] (who expanded that of Smaal and Prins [234]), Strayer et al. [232] presented a graphical conceptual model (Figure 19A) of phytoplankton regulation as a function of hydrologic residence time on the horizontal axis and bivalve clearance time (i.e., time for a bivalve population to clear the overlying water column of phytoplankton through their pumping) on the vertical axis. They described three regimes within that 2D timescale space, each associated with a different control on phytoplankton biomass (advective loss, bivalve grazing, or phytoplankton growth), stating that the regime boundaries would vary as a function of phytoplankton net growth rate. The Strayer et al. [232] conceptual model (Figure 19A) was used to show how bivalve clearance rates changed as a function of bivalve invasion or population decline. The Strayer et al. [232] conceptual model was later extended through (1) the generalization of the benthic grazing timescale to include potentially any in situ loss process and (2) normalization of the loss and transport timescales by the algal growth timescale (Figure 19B) [23]. The latter model was derived from the simple, dimensionless expression in Equation (7), was consistent with the Strayer model control domains, and showed that the regime boundaries are in fact defined by two timescale ratios, i.e., at $\tau^*_{loss} = 1$, $\tau^*_{tran} = 1$, and $\tau^*_{loss} = \tau^*_{tran}$ (see description in Section 3.2.3). These conceptual models, together, demonstrate the utility of timescales (and their ratios) in understanding and delineating the conditions under which an ecosystem response (e.g., algal biomass accumulation) is controlled by one of several processes.

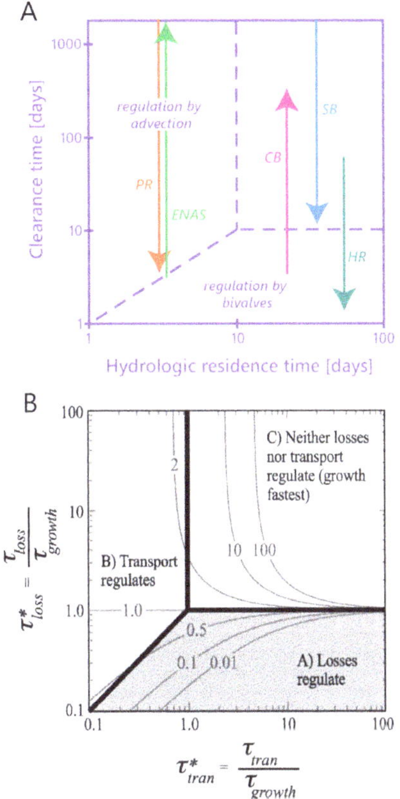

Figure 19. (**A**) The conceptual model of Strayer et al. [232], which extended that of Dame [233] and described three domains of control of phytoplankton (i.e., by advection, bivalve grazing, or phytoplankton growth). Strayer et al. [232] explained that domain boundaries may be different from those shown, depending on phytoplankton net growth rates. Arrows describe how bivalve clearance times in five estuarine, river, and stream ecosystems changed over time as a result of bivalve invasion or population decline. Ecosystems are the following: *HR*, the Hudson River (NY, USA) after the *Dreissena polymorpha* (zebra mussel) invasion; *SB*, Suisun Bay (CA, USA) after invasion by *Potamocorbula amurensis*; *CB*, the Chesapeake Bay (USA) after the decline of oyster populations; *ENAS*, a typical eastern North American stream after unionid decline; and *PR*, the freshwater tidal Potomac River (MD, USA) after the *Corbicula fluminea* invasion. (Redrawn from Strayer et al. [232] with the permission of D. Strayer.) (**B**) Reprise of Figure 15 with shaded areas added to describe domains of control on phytoplankton biomass [23], extending the conceptual model of Strayer et al. [232] in panel (**A**). Contours represent values of B^*_{out}, the ratio of outgoing algal biomass concentration to incoming concentration. (From Lucas et al. [23].)

3.4. Evaluating Connectivity

Quantification of the connectivity between aquatic ecosystems, or between sub-regions within a single ecosystem, can be critical to understanding issues such as pollutant dispersal, protection of sensitive areas, algal bloom location, and other challenges faced by resource managers. Timescale estimation can form an important foundation for performing such quantitative assessments. For example, de Brauwere et al. [95] ran a 2D tracer transport model for the Scheldt Estuary (Belgium, The Netherlands), implementing the forward approach of Gourgue et al. [192] to compute exposure times. Following [235], they divided the estuary into 13 subdomains (Figure 20A), each of which had an associated numerical tracer. Their approach permitted them to compute for each region *i* and tracer a subdomain exposure time (SET), i.e., the total time spent in subdomain *j* by water initialized in subdomain *i*, including successive visits to subdomain *j*. The SET provides *"a rough picture of where the water parcels released at different places spend most of the time on their journey out of the domain of interest"* [95]. After normalizing SET by the total time spent in the estuary, these quantitative interconnections between subdomains were visualized as connectivity matrices (Figure 20B), inspired by the dependency matrix of Braunschweig et al. [236]. The connectivity matrix allows for the identification of "preferential connections" and disconnections between subdomains, as well as regions with longer relative exposure times [95]. As pointed out by the authors [95], this sort of information can be useful in identifying which parts of the larger system will likely be affected by pollution released in a particular subregion.

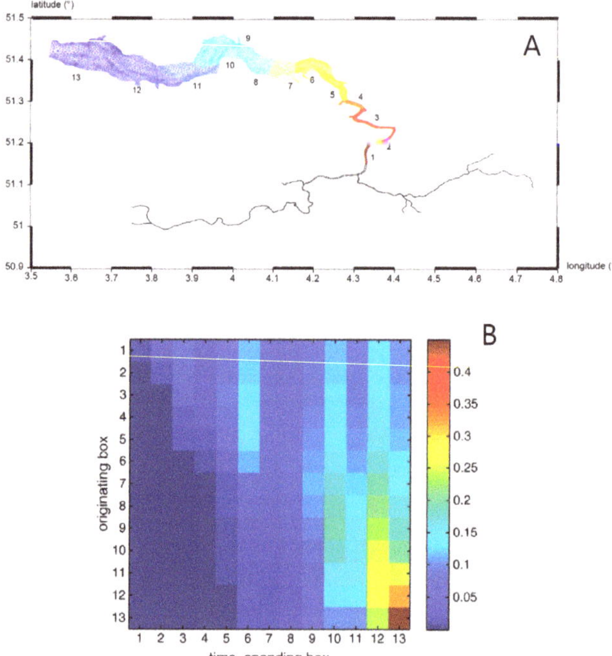

Figure 20. (**A**) Zoom-in of the computational mesh of De Brauwere et al. [95], showing subregions of the Scheldt Estuary referred to in (**B**). Subregions were based on the compartmentalization of [235]. (**B**) Connectivity matrix based on computations of "subdomain exposure times" with a 2D tracer transport model. Colors represent the relative time spent in a particular subregion numbered on the horizontal axis by tracer initialized in a subregion on the vertical axis. (Modified from De Brauwere [95], with permission from Elsevier).

Inspired by Liu et al. [190], Mouchet et al. [140] produced similar matrices of connectivity by generalizing the concept of age to "partial ages", i.e., the amounts of time spent by a particle in different subregions on its way to location *x* within a water body. Age can be conceptualized with a clock attached to a particle, the clock beginning to tick when the particle enters the water body (or at the moment of the particle's birth [140]); the age is the time noted at the instant the particle arrives at location *x*. With partial age, on the other hand, every water particle has several clocks (one for each subregion) rather than one, and only one clock is ticking at a time, depending on the subdomain in which the particle is located [140]. Unlike the traditional concept of age, which provides only time spent in the system generally before reaching *x*, partial age provides information on the histories of particles and *"some knowledge of the paths followed by the particles to reach a given region"* [140]. The authors applied this approach to the problem of ventilation of the world's deep oceans by water parcels after they touch the surface. Those authors defined subregions of the world ocean (Figure 22A) and developed connectivity matrices based on simulations with a global ocean circulation model (Figure 22B). Manning et al. [103] developed a similar connectivity matrix for the Gulf of Maine based on the analysis of real drifter tracks. The reader is also referred to the work of Lin and Liu [141], who provided a method for computing "partial residence times" (i.e., the amounts of time spent by a particle in different subregions *before leaving* a water body; see Figure 11 and Section 2.3.2). Other studies employing timescales in the investigation of connectivity include:

- *Exposure of marine protected areas (MPAs) to shipping-related pollution:* Delpeche-Ellmann et al. [56] analyzed the paths of GPS-tracked surface drifters released in the Gulf of Finland's main shipping fairway, providing insight into which MPAs on the edges of the Gulf are most likely to be affected by pollutants originating in the fairway, as well as timescales for transport to the MPAs. The transport timescales provide information for environmental managers regarding the time available to respond to pollutant spills and contain them before they reach MPAs.

- *"Material connectivity":* Oldham et al. [229] noted that, in the field of hydrology, there have been numerous efforts at characterizing *hydrological* or *hydraulic* connectivity between landscapes; whereas, to their knowledge, there had been no attempts to "*characterise connectivity in terms of the 'effectiveness' of transferring material,*" a notion which those authors termed "material connectivity." They argued that material connectivity must account for both physical transport and biological or chemical processing, since two environments may have strong hydrological connectivity between them but, if material carried by the water undergoes significant removal during transit, the material connectivity may be poor. The ratio of a transport timescale τ_{tran} to a reaction or "material processing" timescale τ_{rxn}—termed the Damköhler number (*Da*) in the chemical engineering literature and generalized by Oldham et al. [229]—was proposed to capture the conditions under which material connectivity is strong or weak. For example, when reactions remove a constituent during transit and $Da = \tau_{tran}/\tau_{rxn} \gg 1$, transport is very slow compared to in situ loss processes; the constituent material will be substantially lost during transport, resulting in material disconnectivity even under conditions of hydraulic connectivity. On the other hand, if $Da \ll 1$, transport is very fast compared to processing, the material behaves essentially conservatively, and material connectivity is therefore strong. Relatedly, Brodie et al. [237] estimated residence times for freshwater and several water quality constituents exported to the Great Barrier Reef and made the case that residence times of pollutants in that system are potentially much greater than those of the water itself, contrary to common assumptions.

- *Harmful algal bloom (HAB) initiation in geometrically complex estuaries:* Qin and Shen [199] performed both theoretical analyses and 3D numerical modeling to understand the roles of estuary geometry and hydrodynamic connectivity between estuary subregions in determining where HABs are first observed to begin. (For their species of interest, a density of 1000 cells/mL was defined as the HAB threshold). Their idealized analytical model (in which residence time was a key parameter) predicted that the location of first HAB occurrence in a hydraulically interconnected system of two water bodies (e.g., the mainstem of a tidal river and its tributary) is determined by the relative ratios of residence time to volume (τ_r/V) for the two water bodies. A HAB was predicted to be observed first in the water body with the larger τ_r/V ratio, i.e., the longer residence time and/or smaller volume. Results from numerical experiments with a 3D transport-reaction model of the lower James River (Figure 21A) were consistent with the theoretical model, demonstrating that—regardless of the initial source location of cells—flushing (represented by model-computed τ_r) and subregion volume V are indeed dominant factors determining where a HAB is first observed. Specifically, their 3D simulations were initiated with a non-zero algal concentration in the bottom layer of the lower James River mainstem (see Figure 21B), to represent cyst release in that region; initial algal concentrations were zero elsewhere, including in the tributaries. Nonetheless, only a few days were needed for concentrations in the tributaries to be higher than in the mainstem, initiated by cell transport from the mainstem driven by estuarine circulation. Simulated bloom-level densities ultimately developed first in the tributaries (Figure 21D), as predicted by the theoretical model. Both numerical and analytical results are consistent with, and help explain, first occurrences of toxic algal blooms in that system, which are frequently observed in the Lafayette River, a relatively small tributary to the James with a long residence time.

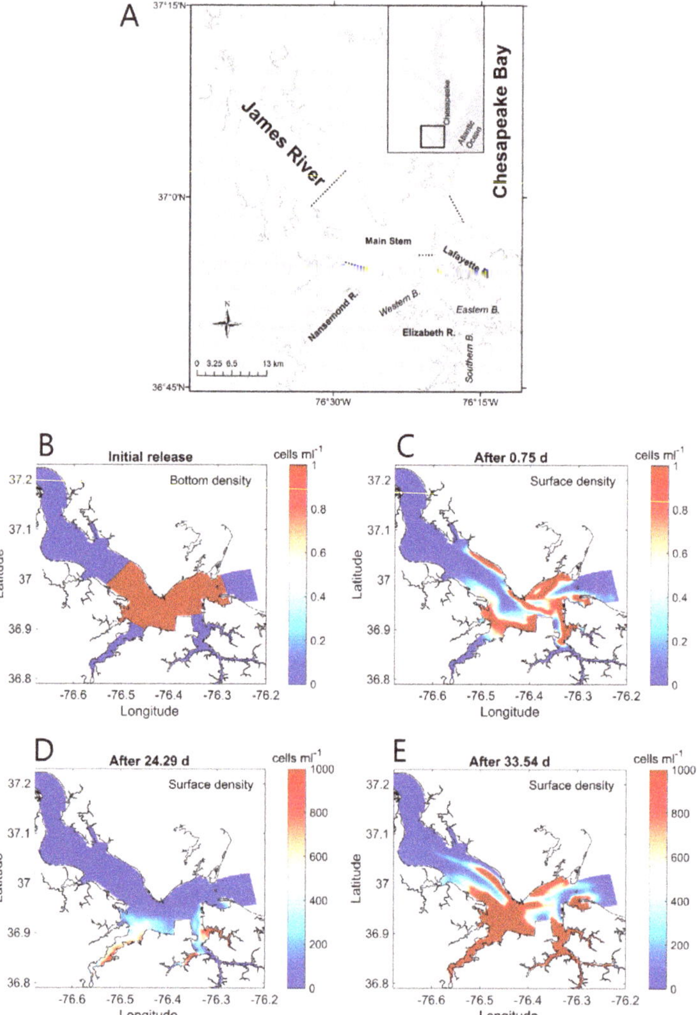

Figure 21. (**A**) Map of the lower James River (USA) and its tributaries [199]. From a 3D model simulation performed by Qin and Shen [199], algal cell densities (**B**) specified as the initial condition (non-zero cell densities initially only in the bottom layer of the lower James River mainstem), and computed cell densities (**C**) after 0.75 d; (**D**) after 24.29 d, when average surface density of the entire Lafayette River first reached bloom levels (1000 cells/mL); and (**E**) after 33.54 d, when the average surface density of the mainstem first reached bloom levels. These results are consistent with observations and with a simple theoretical model indicating that a simple parameter—the ratio of subregion residence time to its volume—can predict where harmful algal blooms are first observed [199]. (Modified from Qin and Shen [199], with permission from Elsevier).

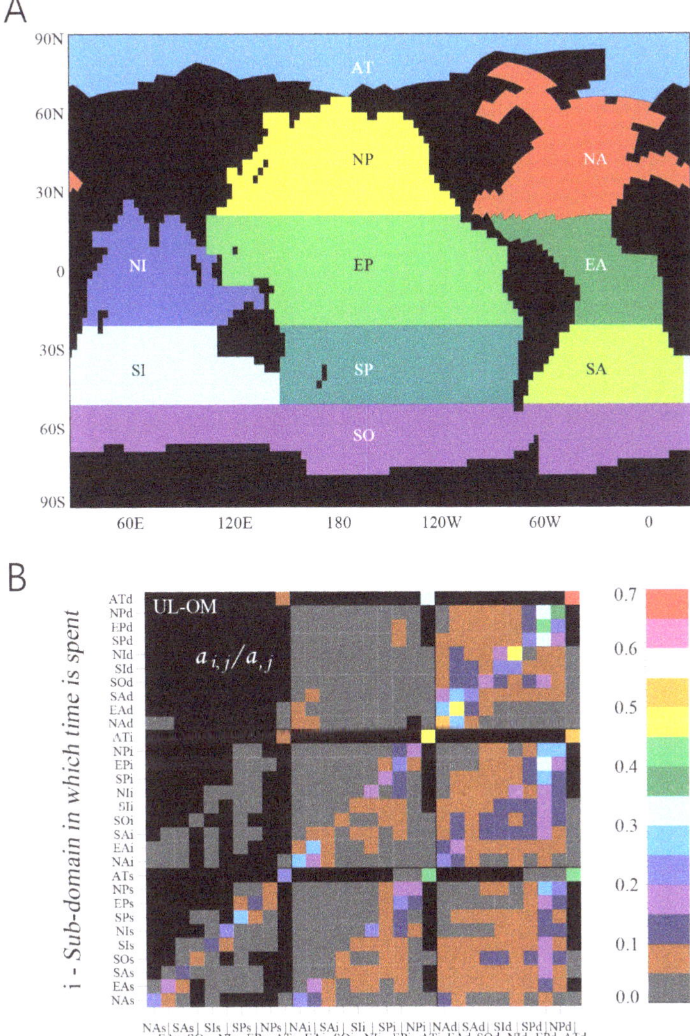

Figure 22. (**A**) Horizontal partitioning of the world ocean by Mouchet et al. [140] for use with a global ocean circulation model to evaluate connectivity between 30 different subdomains (each horizontal partition is split into three boxes in the vertical dimension, denoted by "s" for surface, "i" for intermediate, or "d" for deep in (**B**)). (**B**) Connectivity matrix showing computed "partial age" ($a_{i,j}$) for all subdomains, i.e., the mean time spent by particles in any subdomain *i* (vertical axis) before reaching the subdomain of interest *j* (horizontal axis). Partial age is normalized by the mean (total) water age in the corresponding sub-domain. (Adapted by permission from Springer Nature Customer Service Center GmbH: Springer Nature, Ocean Dynamics, Partial ages: diagnosing transport processes by means of multiple clocks, Mouchet et al., 2016. https://www.springer.com/journal/10236).

3.5. Comparing Systems across Space or Time

Timescale estimates can serve as useful metrics to explain differences in functioning between aquatic ecosystems, or within a single system as a function of space or time (or, equivalently, varying conditions). As an example of all three comparison types, Peierls et al. [57] and Hall et al. [238] analyzed sample data for phytoplankton biomass (chlorophyll *a*) and estimated flushing times for two microtidal North Carolina (USA) estuaries—the New River Estuary (NewRE; [57,238]) and the Neuse River Estuary (NRE; [57])—to understand phytoplankton dependence on hydrologic variability and other factors. Because these estuaries are river dominated, the authors implemented the "date-specific freshwater replacement method" [239] to obtain flushing times across a range of hydrologic conditions for 9 (11) contiguous estuary segments encompassing their sampling stations in the NewRE (Figure 23A) (NRE (Figure 23C)). This transport timescale represented for each estuary segment the cumulative sum of flushing times upstream of and including that segment, serving as an estimate of the freshwater age [238]. This approach collapsed two parameters—location within the estuary and flow rate—into a single parameter (an advective timescale), while also producing a larger dataset than would have resulted if they had treated the estuary as a whole [57]. Phytoplankton biomass for both rivers had a non-monotonic relationship with flushing time (see Figure 23B,D herein), displaying a positive slope for flushing times shorter than a threshold value (~10 d [57]), a negative slope for flushing times above the threshold, and peak values near the threshold. The unimodal phytoplankton–flushing time relationship was interpreted as an indicator of a changing growth-loss balance over space and time [23,57] (see also Section 3.2.3 above). Specifically, the positive phytoplankton–flushing time relationship for shorter flushing times was taken as an indicator that intrinsic growth rate in those cases was faster than losses, likely due to high riverine nutrient concentrations in upstream reaches [238]. Whereas the negative phytoplankton–flushing time relationship for flushing times larger than the threshold was seen as an indicator of in situ losses that were faster than growth, possibly due to a combination of nutrient- limited growth and enhanced zooplankton grazing at the longer flushing times. This hypothesis was bolstered by the occurrence of nitrate depletion at similar flushing times as for peak algal biomass (i.e., around the 10-day threshold) [57]. Notable was the fact that these two distinct estuaries exhibited similar phytoplankton responses to flushing time, as well as similar threshold values [57]. The authors suggested that these unimodal chlorophyll *a*-flushing time patterns may be expected in other river-dominated estuaries where primary production is driven by riverine nutrients and flushing times range from values too-short to amply-long for complete assimilation of riverine nutrient loads [57,238]. Hall et al. [238] found similar non-monotonic relationships between photopigment concentrations (indicators of phytoplankton community composition) and flushing time in the NewRE. These linked studies provide a valuable example of how a suitably defined timescale can concentrate spatial and temporal variability into a single metric, thereby bringing simplicity and shape to ecological complexity and assisting in the identification of useful patterns.

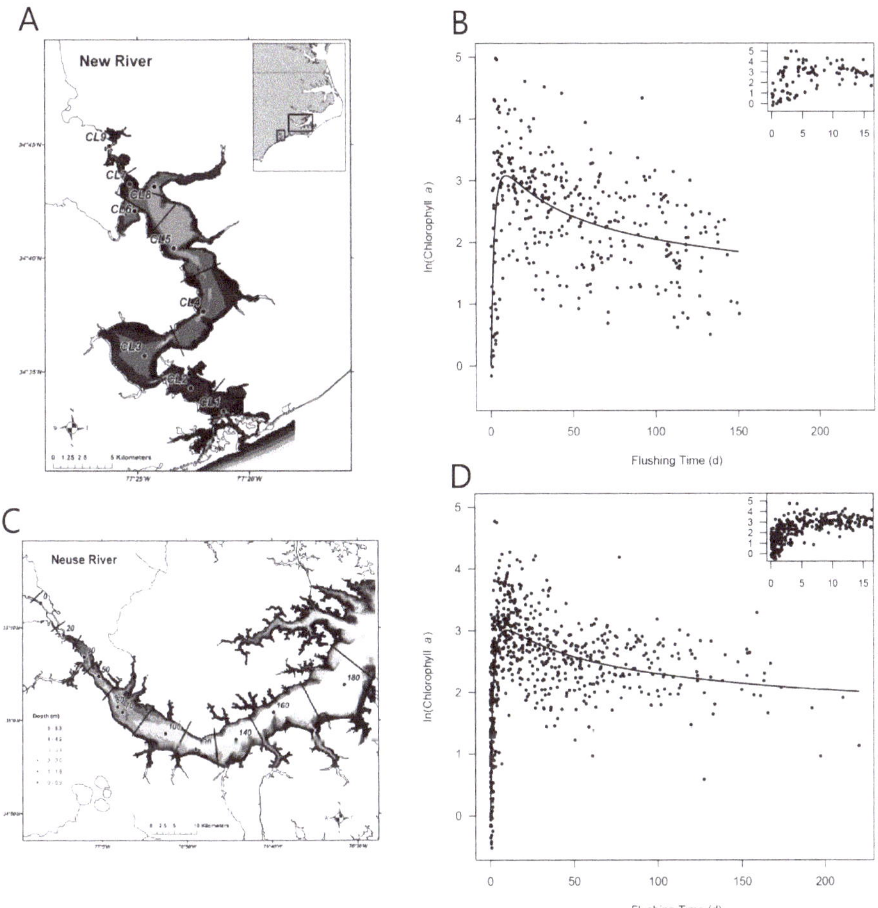

Figure 23. From Peierls et al. [57], maps of the (**A**) New River Estuary (NewRE) and (**C**) Neuse River Estuary (NRE); for the (**B**) NewRE and (**D**) NRE, observation based ln(chlorophyll *a*) versus flushing time estimated with the "date-specific freshwater replacement method" [239]. (Adapted by permission from Springer Nature Customer Service Center GmbH: Springer Nature, Estuaries and Coasts, Non-monotonic Responses of Phytoplankton Biomass Accumulation to Hydrologic Variability: A Comparison of Two Coastal Plain North Carolina Estuaries, Peierls et al., 2012. https://www.springer.com/journal/12237).

Other examples of studies in which timescales served as key diagnostics in cross-system, spatial or temporal ecosystem comparisons include [54,177,199,210,214,226,240–242], as well as the following:

- *Ecosystem responses to management actions:* To understand changes in hydrodynamics, water quality, and ecosystem processes induced by the installation of a temporary physical salinity-intrusion barrier in the Sacramento-San Joaquin Delta (CA, USA), Kimmerer et al. [62] employed high-speed boat-based isotope mapping (same approach as in [173]) to produce spatial patterns of water age with and without the barrier. Benthic grazing turnover time (i.e., time for benthic bivalve population to filter through the entire overlying water column) was also estimated as one measure of ecosystem response to related changes in salinity.

- *Variability and drivers of estuarine flushing:* In order to investigate the sensitivity of flushing in Mobile Bay (AL, USA) to river flow, wind, and baroclinic forcing, Du et al. [243] estimated both bulk (*e*-folding flushing time) and spatially variable (freshwater age) transport timescale metrics using a 3D numerical model. Deriving a simple empirical flushing time–discharge relationship based on a set of sensitivity runs and comparing to previous estimates based on a 2D depth-integrated model [244], they concluded that baroclinic processes reduce flushing times by approximately half. The spatial and temporal transport time patterns produced in these analyses (Figure 24 herein) could serve as valuable information toward interpreting variability in water quality and ecosystem processes.

- *Retention of harmful algal cells:* Ralston et al. [127] employed a 3D coupled hydrodynamic-biological model of the Nauset Estuary (MA, USA) to explore the physical and biological processes controlling recurrent blooms of the toxic alga *Alexandrium fundyense*. Implementing an *e*-folding approach to calculate *A. fundyense* residence times under a range of conditions, they explored the influence of swimming behavior, spring-neap tidal phase, wind, and stratification on retention of cells in one of the estuary's salt ponds, concluding that all four processes are major factors determining retention. Although growth and mortality were turned off in these simulations, the computed residence times are particularly holistic, in that they not only include 3D hydrodynamic processes but also organism behavior (see Figure 25 herein).

- *Ecosystem transformations by bivalves:* The graphical timescale-based conceptual model of Strayer et al. [232] (see Figure 19A and Section 3.3 above) describes the evolution of five aquatic ecosystems in response to major changes in bivalve grazer populations. The process controlling phytoplankton was shown to be capable of shifting between advection, grazing, and algal growth as a function of either bivalve invasion or population decline.

- *Hydrologic influence on zooplankton communities:* Augmenting an 18-year field dataset with calculated water residence times, Burdis and Hirsch [33] explored several potential environmental drivers of zooplankton community structure in a natural riverine lake. As hypothesized, they found that water residence time was the most important driver of zooplankton abundance and community structure. Similar to Peierls et al. [57] and Hall et al. [238], use of a transport timescale allowed these authors to collapse spatial location and temporally variable hydrology into a single variable associated with each sample.

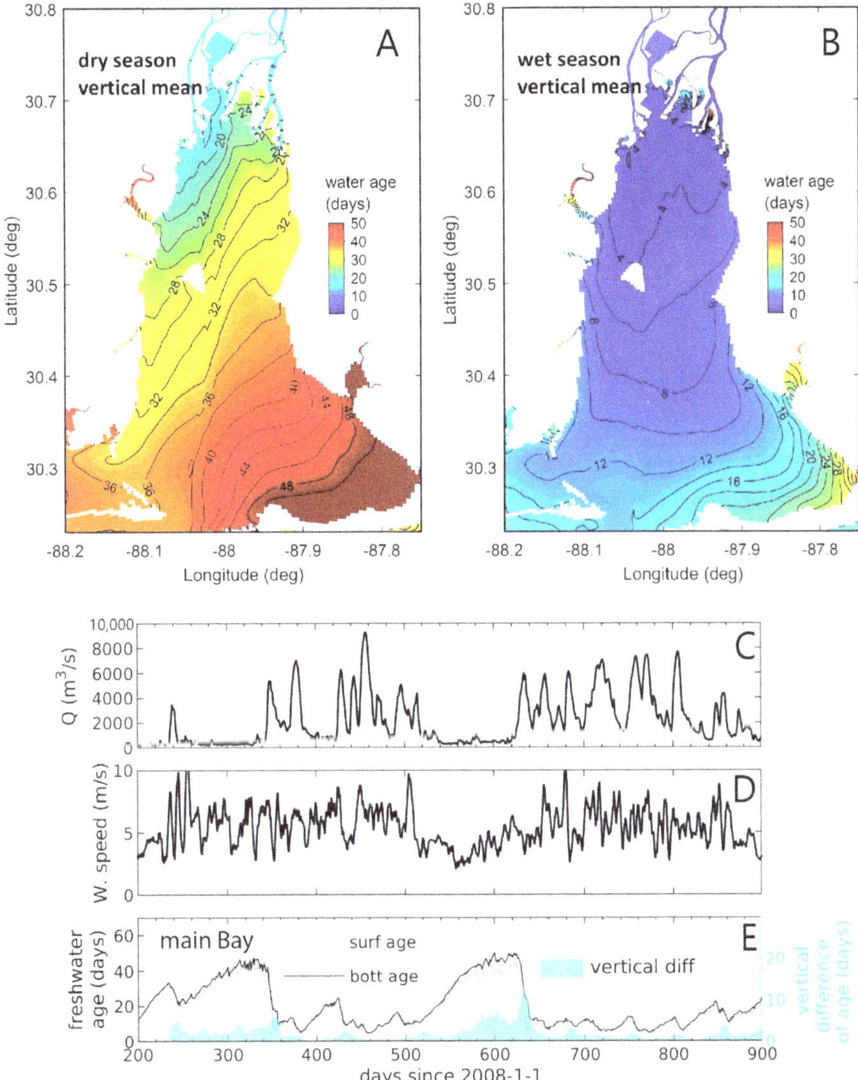

Figure 24. Maps of computed vertical mean freshwater age in Mobile Bay for (**A**) the dry season and (**B**) the wet season, based on the 3D numerical modeling of Du et al. [243]. Timeseries of (**C**) river discharge, (**D**) wind speed, and (**E**) computed freshwater age averaged over the main bay. For the age timeseries, surface water is gray, bottom water is black, and the vertical age difference is cyan [243]. (Modified with permission from J. Du, Journal of Geophysical Research: Oceans; published by Wiley, 2018).

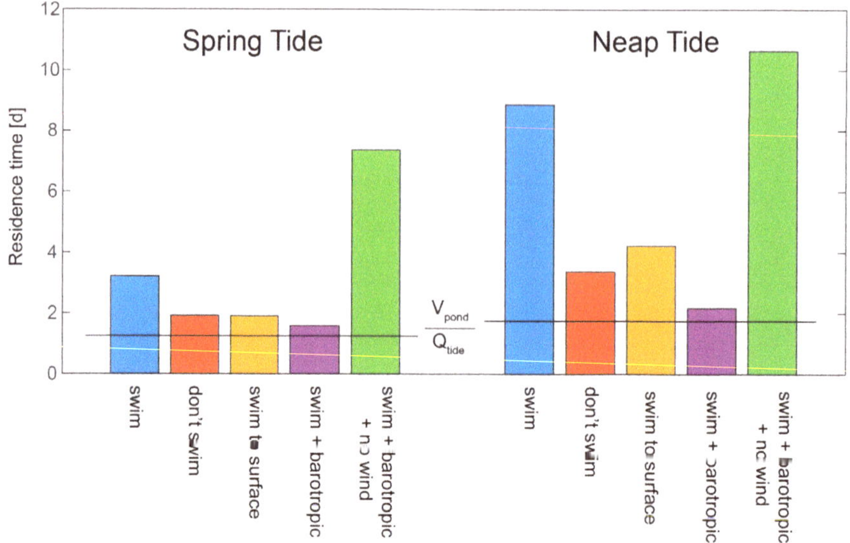

Figure 25. Computed residence times for the toxic alga *A. fundyense* based on the 3D model of Ralston et al. [127] for a pond within the Nauset Estuary (MA, USA). The different bars represent a variety of swimming and forcing cases under spring and neap tide conditions. "Swim": diel vertical migration up to $1/k_w$ depth, where k_w is the light attenuation coefficient. "Don't swim": no vertical migration. "Swim to surface": diel migration to the surface. "Swim + barotropic": diel vertical migration to $1/k_w$ with barotropic physics (uniform water density and thus no stratification). "Swim + barotropic + no wind": diel vertical migration to $1/k_w$ with barotropic physics and zero wind forcing. Horizontal lines: the residence time for tidal exchange assuming a well-mixed pond (volume of pond/tidal volume exchange), shown for reference. (Redrawn from Ralston et al. [127] with the permission of D. Ralston).

4. Discussion

4.1. The Timescale "Tower of Babel"

In their seminal 1973 article on diagnostic timescales, Bolin and Rodhe [94] stated (what should have been) the obvious: *"To avoid misunderstandings and even erroneous conclusions it is important to introduce precise definitions and to use them with care."* Surprisingly, or not, this wise piece of advice has been ignored by many [79]. (Indeed, we authors have at times committed the sins of sloppiness, ambiguity, and imprecision when using or referring to timescales in our own work.) This has led to a situation half-jokingly referred to as the "Tower of Babel" [79] by Viero and Defina [137], which we interpret as a reference to a wealth of poorly defined diagnostic timescales used rather carelessly or even timescales contradicting their very definitions, eventually causing misleading interpretations and conclusions to be produced [26,79].

The collective efforts of many scientists persist toward (1) establishing clear, consistent, and rigorous timescale definitions, (2) carefully choosing timescales and calculation methods appropriate to a scientific question, and (3) providing detail and transparency with respect to assumptions and calculation methods in presentations of studies implementing timescales. Realistically though, we may never—as an aquatic science community—converge on a universal set of timescale terms and definitions (objective (1) above). For that reason, objectives (2) and (3) are all the more important. Thus, the recommendations of Bolin and Rodhe [94] and many others [79,87,88,110] remain as relevant as ever.

For evidence of the importance of choosing timescales with care, one need only look at the numerous studies that have estimated different transport timescales and/or implemented different estimation methods for a single water body and set of conditions and then compared the results. Table 2 cites several such studies, summarizing for each one the magnitudes of the different timescale types and assessing the range of values as the ratio of the maximum transport timescale magnitude for that study to the minimum. In most cases cited in Table 2, timescale magnitudes spanned at least two orders of magnitude, demonstrating the criticality of choosing the most suitable timescale for the scientific question and setting of interest. Moreover, just as there is much to be learned from inter-comparisons between aquatic ecosystems, portions of an ecosystem, or behaviors of an ecosystem across different time periods, valuable insights can be gained from the comparison of different timescales. For example, dispersive timescales for the Bay of Quinte were on the order of 1–3 years, whereas other transport timescales were on the order of a month or two (Table 2, [147]). Oveisy et al. [147] viewed this difference as an indicator that advective transport must play an important role in flushing of that system. The reader is also referred to Andutta et al. [22], who performed an extensive comparison of several transport timescale estimates for eight different estuaries (not included herein) and found variability similar to that shown in Table 2.

It is interesting and encouraging to note that, despite the quantitative differences between different timescale types as shown in Table 2, some synthetic studies relying on transport timescale values from several sources and water bodies (and calculated using a diversity of methods) have nonetheless produced statistically (and ecologically) significant relationships. In particular, Nixon et al. [226], Dettmann [214], and Seitzinger et al. [54] all relied on diverse data sources for transport timescales to develop their simple mathematical models describing nutrient fate as a function of transport time. (Note that [54,214] drew on data from [226].) Their models performed well, especially for ecology! One can wonder whether the performance of these models would be improved further if consistent transport timescale estimation methods had been available to populate each of the authors' datasets.

Table 2. Compilation of transport timescales estimated in previous studies. Data is based on sources in "Author(s)" column. $max(\tau)/min(\tau)$ is the ratio of the maximum timescale value to the minimum value for a water body and set of conditions. Q is volumetric flow rate. V is water body volume. M is total tracer mass. $\dot{M}$ is mass loading rate. L is length. U is mean velocity. U' is average deviation from depth-mean velocity. A_o is tidal amplitude. TEF is "total exchange flow" [122], an approach for estimating a salinity turnover time. "tc" is tidal cycles. Footnotes provide methodological information. Other specifics such as temporal or spatial averaging of parameters or timescales vary between authors; please see those publications for details.

Author(s)	Water Body	Time Period/Conditions	Timescale[(approach)]	Value	$\frac{max(\tau)}{min(\tau)}$
Jouon et al. [112]	SW lagoon of New Caledonia	Constant, moderate trade wind	V/Q [1] Mean residence time [2] e-folding [3]	6.8 d 10.8 d 11.4 d	2
Lemagie and Lerczak [110]	Yaquina Bay (USA)	$Q = 10$ m^3/s, $A_o = 125$ cm	TEF [1,4] Tidal prism [4] Freshwater fraction [1,4] Transit (e-folding) [2,4] Flushing (e-folding) [2,4]	3.96 tc 1.27 tc 12.63 tc 32.6 tc 5.16 tc	26
Monsen et al. [87]	Mildred Island (USA)	June 1999 (low flow)	V/Q [5] e-folding [6] $M/\dot{M}$ [7] Mean age [6,8]	31–50 d 7.7 d 8.3–9.1 d 1.8 d	17–28
Oveisy et al. [147]	Bay of Quinte (Canada)	Summer 2004	e-folding [3,0] V/Q [1,9,11] Residence time [2,10] Dispersion [,10]	44 d 64 d 52 d 1.7 y	14
Rayson et al. [100]	Galveston Bay (USA)	Mid–late April 2009 (peak flow)	Freshwater fraction [1] TEF [1,12] Mean residence time [2,12] e-folding [2,12] Mean age [2,12]	~10 d ~20 d ~20 d ~20 d ~20 d	2
Rayson et al. [100]	Galveston Bay (USA)	Late July 2009 (low flow)	Freshwater fraction [1,12] TEF [1,12] Mean residence time [2,12] e-folding [2,12] Mean age [2,12]	~200 d ~20 d ~25 d ~50 d ~30 d	10
Tartinville et al. [113]	Mururoa atoll Lagoon (French Polynesia)	Tides, wind, hoa inflow, stratification	L/U [1] L/U' [1] Diffusion [13] e-folding [2]	8.3 d 5.3 d 5900 d 114 d	1113

[1] 3D hydrodynamic model. [2] 3D model with particle tracking. [3] 3D model with tracer(s). [4] Based on power law regression of computed timescales as a function of discharge and tidal amplitude. [5] 2D hydrodynamic model. [6] 2D model with particle tracking. [7] 2D model with tracer. [8] Mean of average ages for two locations and two time periods. [9] Observations. [10] Mean of timescales calculated for individual tributary inflows. [11] Based on total discharge from all main tributaries. [12] Estimated based on visual inspection of published figures. [13] Diffusivity based on Okubo [245].

4.2. Holism of Timescales

In Section 2, we discussed three broad categories of methods for estimating diagnostic timescales: (1) arithmetic manipulation of process rates, (2) field-based approaches implementing drifters or tracers, and (3) solution of partial differential equations with numerical models. Here, we discuss how categories (1) and (3) (primarily mathematical approaches) may be viewed as inhabiting different regions on a continuum of mathematical complexity. Further, we describe implications of that mathematical complexity for the potential holism of the resulting timescale and also discuss field-based timescale methods in this context.

Scientists for whom field-based timescale estimation approaches are not an option still have a broad range of methodological choices. We propose that method choice in that case can be reduced to two primary considerations: mathematical complexity of the calculation method and holism of the resulting timescale (i.e., the degree to which all relevant processes operating in the real system are taken into account). A holistic timescale is one that represents the net effect of a broad collection of driving processes (e.g., tides, wind, river inflow, density gradients, reactions, organism behavior) [139]. A non-holistic or "atomistic" timescale, by contrast, only represents a single process or tightly entwined set of related processes (such as the cross-sectional shear and mixing (and all the processes that influence them) that together result in the "process" of longitudinal (or shear flow [52,246]) dispersion). In cases where one wants to compare a biogeochemical process with the overall effect of physical transport, a holistic transport timescale including the effects of all major hydrodynamic influences may be particularly useful. Timescale holism is represented by the vertical axis in Figure 26.

If we consider the mathematical complexity of the timescale estimation method (horizontal axis in Figure 26), the simple arithmetic relationships in Table 1 (and Section 2.1) inhabit the left end of the schematic (Regime A). These sorts of methods were available long before realistic multi-dimensional numerical modeling was computationally feasible; thus, we refer to their results as "classical" timescales, following Deleersnijder [139]. These are generally "bulk" approaches and, as such, typically do not carry much if any resolution in space or time. As they usually describe a single process (e.g., advection or diffusion), these relationships (e.g., L/U or L^2/K, respectively) tend to be relatively atomistic (see filled circle in Figure 26). Consequently, classical timescales have proven useful in estimating the relative magnitudes of the terms in the governing equations of eco-hydrodynamics [139] or in comparing the speeds of different processes operating in an aquatic system (Section 3.3). It should be noted that while these classical algebraic timescale expressions may have the advantage of being mathematically simple, the methods to quantify the necessary parameters can be non-trivial.

Over the past couple decades, a very different set of timescale estimation approaches has emerged, involving detailed multi-dimensional numerical models that solve PDEs [22,139]. These methods (Section 2.3) reside in the middle to right side of Figure 26 and have the potential to be highly holistic (e.g., [174]; asterisk in Figure 26). The details of how the model is implemented and how much process richness is captured by the model simulation determine the level of holism associated with the resulting timescale. In fact, timescales derived in this way can be ultra-holistic, not only incorporating many hydrodynamical processes and forcings but also biological or geochemical processes, if represented in the model [107,127,188,189]. Many of these methods allow the calculation of timescales at any time or position in the computational domain (e.g., see Figures 10–12 and 24), a key difference from typical bulk approaches.

In contrast to more atomistic timescales, holistic timescales are not as well suited to understanding the relative speed (or, potentially, dominance) of individual processes (i.e., "process attribution" in Figure 26). If we take the advection versus diffusion example, the atomistic timescales L/U and L^2/K allow for the direct comparison of the two processes. Whereas, residence time derived from a realistic 3D transport model will likely incorporate both processes into it, communicating their combined effect; this is something a classical timescale usually cannot achieve, unless one process is far more dominant than all others. Thus, atomistic timescales may bear little quantitative resemblance to holistic timescales [87,113], since they exclude the subtle and complex interplay between multiple processes operating in real systems and captured by realistic models [139]. Process attribution is perhaps less easy with a numerical PDE-based method than with simple algebraic expressions, but it is not impossible. It simply requires a different approach, such as sensitivity analyses that turn individual processes on or off, or coefficients up or down (e.g., [101,127]; triangles and five-pointed stars in Figure 26).

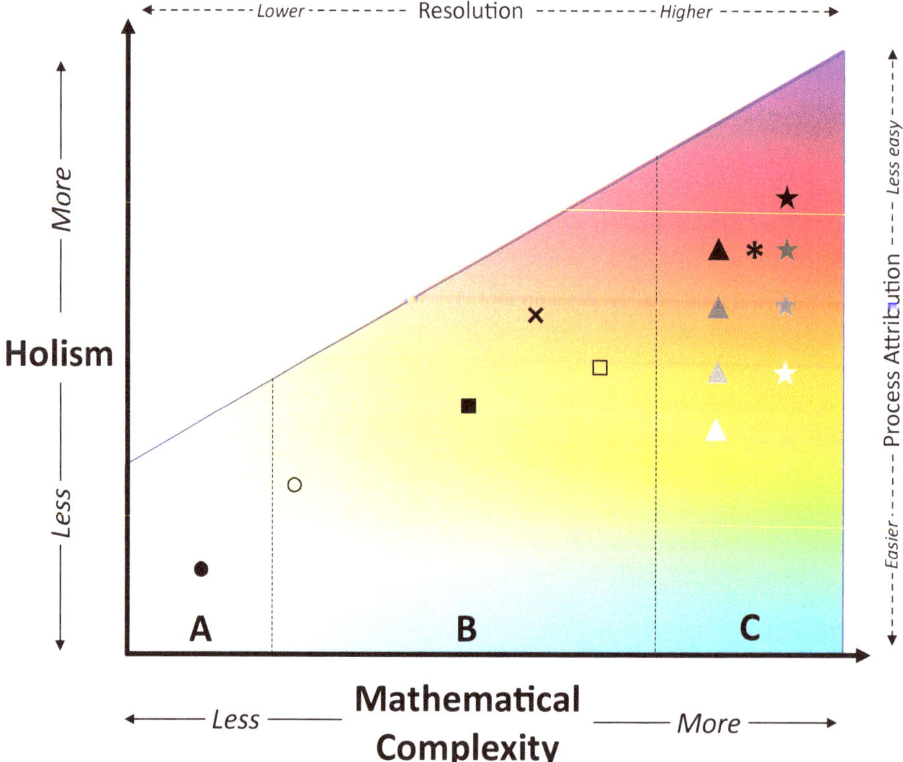

Figure 26. Schematic of diagnostic timescale holism as a function of the mathematical complexity of the calculation method (for computed timescales only). Timescales based on simple algebraic expressions tend to be less holistic, but potentially more useful for purposes of assessing dominant processes (Regime A). Complex numerical models have the potential to produce highly holistic timescales (Regime C) as well as timescale estimates at high spatial and temporal resolution. The effective level of holism depends on the process richness captured by the model simulation. More holistic timescales may be less useful for disentangling the relative speeds (and potential dominance) of individual processes ("process attribution"). Moderately holistic timescales may be derived from moderately complex numerical models or methods (Regime B) or from complex models that exclude some important processes (mid-region of Regime C). Examples: $\bullet$—$\tau_{adv} = L/U$ or $\tau_{diff} = L^2/K$. $\circ$—$\overline{\tau_{res}}$ and $\overline{\tau_{exp}}$ from Andutta et al. [22], Equations (9) and (10) herein. Timescales derived from the 1D models of Delhez and Deleersnijder [200] or Vallino and Hopkinson [160] (); the 2D depth-averaged model of Monsen et al. [87] ($\square$); the 1D hydrodynamic-biological model of Delhez et al. [189] ($\times$); the 3D hydrodynamic and transport model of Gross et al. [174] ($*$); the 3D hydrodynamic and particle tracking model described by Defne and Ganju [101], with progressively more physical processes included (starting with white triangle progressing upward to black triangle; also see Figure 10 herein); the 3D hydrodynamic-ecological model of Ralston et al. [127] with progressively more physical processes and dinoflagellate swimming behaviors (starting with white five-pointed star up to black five-pointed star; also see Figure 25 herein).

Holistic timescales obtained from the numerical solution of PDEs are quite complex. The equations may be considered indisputable, but the initial and, above all, boundary conditions are not. They have a tremendous impact on the values of the computed timescales and must be prescribed with care, in accordance with the declared objectives of the study [79]. This crucial point is sometimes overlooked. For instance, there are many published papers in which the timescale related PDEs are correctly laid out with, unfortunately, little said about boundary conditions.

There is a middle ground between the classical, atomistic timescales and those estimated from detailed numerical models. Some authors have shown that a small increase in mathematical complexity can markedly increase timescale holism. For example, based on the adjoint of the 1D advection–diffusion equation applied to V, a portion of the volume of an idealized infinite pipe, Andutta et al. [22] derived analytical expressions for domain-averaged residence time and exposure time (Equations (9) and (10) below, respectively), under the combined influence of advection and diffusion:

$$\overline{\tau_{res}} = \frac{V}{Q_R}\left(\frac{1}{2}\right) + \frac{V}{Q_R}\left(\frac{1}{e^{Pe}-1} - \frac{1}{Pe}\right) \tag{9}$$

$$\overline{\tau_{exp}} = \frac{V}{Q_R}\left(\frac{1}{2}\right) + \frac{V}{Q_R}\left(\frac{1}{Pe} - \frac{1-e^{-Pe}}{Pe^2}\right) \tag{10}$$

Pe is the dimensionless Peclet number, the ratio of the diffusive timescale to the advective timescale, and Q_R is the volumetric flow rate. Andutta et al. [22] also derived similar closed-form relationships for location-specific residence time and exposure time and for the water renewal time as well (not shown). With these expressions (see open circle in Figure 26), one can buy two processes for barely more than the calculational price of one!

Similarly, it has been shown that, for a well-mixed aquatic system subjected to steady-state hydrodynamic exchange processes with the surrounding environment, the effective residence time for a reactive tracer undergoing first-order decay is [35]:

$$\tau_{res}^* = \frac{\tau_{hydro}\tau_{decay}}{\tau_{hydro} + \tau_{decay}} \tag{11}$$

where τ_{res}^* is the mean time for particles to leave the domain by crossing an open boundary as dictated by the hydrodynamics and/or by vanishing as a result of the (e.g., radioactive, biogeochemical) decay process. τ_{decay} is the mean life of the tracer (i.e., $1/\mu_{decay}$, where μ_{decay} is the specific decay rate and is assumed positive), and τ_{hydro} is the time that would be taken by a conservative particle to leave the domain under hydrodynamic forcing only. These timescales satisfy [118]:

$$\tau_{res}^* \leq min\left(\tau_{hydro}, \tau_{decay}\right) \tag{12}$$

In other words (and unsurprisingly), the time a particle is to be taken into consideration in the domain of interest is no larger than the timescale characterizing outward transport or that related to the first-order decay. The combination of both processes causes the tracer to vanish faster than if only one of these phenomena were at work. τ_{hydro} could be estimated via classical algebraic residence time formulations, resulting in a more atomistic transport timescale (Regime A in Figure 26), by moderately complex approaches (Regime B), or by complex multi-dimensional numerical models (Regime C), potentially producing a holistic transport timescale. The latter approach would result in a hybrid expression for τ_{res}^*, i.e., one that is a function of a classical, atomistic timescale (τ_{decay}) and a holistic one (τ_{hydro}). The elegance of Equation (11) lies in the fact that a single algebraic timescale expression captures the interactions between two disparate sets of processes (transport and decay), the reciprocal of which can be implemented as an effective loss rate in the traditional exponential decay relationship [118]:

$$m(t) = m(0)e^{-(\mu^*)t} \tag{13}$$

where $m(t)$ and $m(0)$, respectively, are the tracer mass in the domain at times t and 0, and $\mu^* = 1/\tau^*_{res}$ is an effective loss rate resulting from the combination of hydrodynamic transport processes and non-transport decay processes. It is possible also to express the combined effect of decay and oscillatory transport between a domain and its adjacent environment (as captured by the exposure time) with a simple expression similar to Equation (11) above [118]. Other moderately complex mathematical methods for estimating timescales could involve simpler numerical models, such as 1D models (e.g., [160,200]; filled square in Figure 26), a 2D depth-averaged model incorporating tides, water diversions, and river flow but not wind or stratification (e.g., [87]; open square in Figure 26), or a 1D physical-biological model (e.g., [189]; "×" in Figure 26).

Figure 26 represents a first (and admittedly simplified) attempt at schematically capturing the general relationship between mathematical complexity and holism for computationally derived timescales. But what about timescales based on *field observations*, such as those involving tracers or drifters? Our expertise does not lie in field-based methods, so we will leave the development of such a diagram, if useful, to the appropriate experts. That said, we have reason to believe that such a diagram for field-based timescales would differ from Figure 26. First, field tracer- or drifter-derived timescales are inherently holistic, because observed drifter movements and tracer concentration fields are subject to the full set of physical drivers operating in the real system. These timescales are neither reliant on a modeler's realistic incorporation of all relevant processes into their model, nor are they subject to numerical inaccuracies or instabilities, although they may be subject to other limitations or errors [104,142,152], as discussed in Section 2.2. For example, timescales based on drifters that track the surface or another fixed depth may be less holistic than those based on tracers because the former would not be free to travel vertically and thus sample the range of velocities that real water parcels would [104,142]. Second, complex mathematical methods have been applied to field tracer or drifter data toward a variety of objectives such as enhancing spatial coverage [142] or revealing temporal variability [41]. Advanced mathematical treatments have been implemented to correct for disconnects between the behavior of drifters (e.g., which are subject to grounding) and that of water particles (which generally refloat after touching the shore) [104]. Such corrective methods could be viewed as enhancing the holism of the timescale. On the other hand, advanced statistical approaches have also helped in disaggregating the effects of different processes (e.g., mean advection versus eddies [142]) on transport timescales. Thus, for field-based timescales, increased mathematical complexity appears to potentially result in either increased or decreased timescale holism.

Much emphasis is placed (in this paper and in aquatic science generally) on the transport and renewal timescales of *water*. But many resource-management questions concern constituents other than pure water. It is an open question whether and to what extent water transport or renewal timescales are representative of, for example, dissolved and particulate pollutants, planktonic organisms, and suspended sediment [35,237,247]. We must often assume, given the information that is available, that water is a proxy for the other constituents carried with it. Ultra-holistic timescales, which incorporate reaction-, behavior-, or buoyancy-driven processes as well as hydrodynamic ones (e.g., [107,127,189])—and their comparison to pure water transport timescales—may help us understand how good of a proxy water is for transported constituents. This is an area ripe for future study.

5. Conclusions

In the foregoing pages, we have discussed a variety of diagnostic timescale definitions, estimation methods, and applications, with a focus on coastal aquatic systems. It is critical to realize that most, if not all, of the timescales referred to above actually belong to only two categories, namely, the timescales concerned with the *past* and those looking into the *future*. Simply put, these two categories, respectively, can be considered in terms of the two types of questions that they aim to address for a particle: (1) How much time has elapsed since *appearing* in the domain of interest? (2) How much time will pass until it *disappears* from the domain of interest?

The timescales of the first class may be called "age" in a generic manner, provided this concept is given a sufficiently general definition. Accordingly, we suggest that the age of a particle be defined as the time elapsed since it began to be taken into account, i.e., the time since it entered the domain of interest by crossing a boundary, by hitting a boundary where the age is (re-)set to zero, or by being produced by a reactive process. This description (following [189]) clearly goes beyond the traditional transport-specific definition of "age" in Section 1.3.

The timescales looking into the future may be termed "exposure time". For a given particle, it represents the time it will spend in the domain of interest until the particle ceases to be taken into consideration, either by being transported out of the domain once and for all or by being destroyed by a reaction. This broader definition also transcends the more traditional transport-oriented definition. The strict residence time is a special case of the exposure time, for in this case the particle is no longer considered at the instant it hits for the first time a boundary where the particle is assumed to be discarded.

The aforementioned timescales (age and exposure time) are useful for estimating the water renewal rate of a semi-enclosed domain. To do so, the water is split into two types, i.e., the water present in the domain at the initial instant (original water) and the water progressively replacing it (renewing water). To evaluate how fast the original water leaves the domain, its exposure time is evaluated. The age of the renewing water allows one to estimate the rate at which this water enters the domain. This generic methodology was outlined in Gourgue et al. [192] and was applied by de Brye et al. [66] and Pham Van et al. [248]. At steady state, the domain-averaged age is equal to the domain-averaged residence time [249].

We have described how diagnostic timescales, which may be estimated in countless ways, can serve as useful tools for distilling the complexity of real ecosystems or numerical model outputs down to one or a few meaningful parameters; comparing the speeds of disparate (e.g., hydrodynamic, biological, geochemical, radiological) processes; quantifying connectivity; building simple ecosystem models; comparing ecosystems, portions of an ecosystem, or behaviors of a single system over time; and conveying qualitatively or semi-quantitatively how ecosystems work in conceptual models. The methods with which timescales are estimated can determine their applicability to the above uses and their relevance for addressing a given scientific question. One of the most appealing aspects of timescales lies in the simplicity they can lend as tools in environmental problem solving. Inspired by another scientist who reduced exceptional complexity down to the elegant and seemingly simple, we now recall the wise advice commonly attributed to Albert Einstein: "*Everything should be made as simple as possible, but not simpler.*" Timescales represent one method of reaching toward that simplicity.

Author Contributions: Conceptualization, L.V.L. and E.D.; writing—original draft preparation, L.V.L. and E.D.; writing—review and editing, L.V.L. and E.D. All authors have read and agreed to the published version of the manuscript.

Funding: L.V.L. was funded by the Water Mission Area (Integrated Modeling and Prediction Division) of the U.S. Geological Survey.

Acknowledgments: The authors wish to thank their colleagues and friends who provided input and/or inspiration: Fernando P. Andutta, Pierre Archambeau, Jean-Marie Beckers, Brian Bergamaschi, Larry Brown, Jon Burau, Jim Cloern, Fabien Cornaton, Jay Cuetara, Benoît Cushman-Roisin, Anouk de Brauwere, Benjamin de Brye, Zafer Defne, Benjamin Dewals, Eric J.M. Delhez, Insaf Draoui, Katja Fennel, Neil Ganju, Olivier Gourgue, Ed Gross, Thomas W.N. Haine, Nathan Hall, Arnold W. Heemink, Jeff Koseff, Inga Koszalka, Jonathan Lambrechts, Vincent Legat, Xiao Liu, Mark Marvin-diPasquale, Stephen Monismith, Nancy Monsen, Anne Mouchet, Hans Paerl, Ben Peierls, Chien Pham Van, François Primeau, Jian Shen, Robin Stewart, Dave Strayer, Kinsey Swartz, Jan Thompson, Valentin Vallaeys, and Eric Wolanski. E.D. is an honorary research associate with the Belgian Fund for Scientific Research (F.R.S.-FNRS). Some elements of this article originated when he served as a guest lecturer, then part-time professor with the Delft University of Technology, Delft, The Netherlands. The authors wish to thank the editor, Pichayapong Srisawad, for his helpfulness and Jan Thompson and three anonymous reviewers, whose comments led to significant improvements of this article.

Conflicts of Interest: The authors declare no conflict of interest.

Appendix A

Let us consider a plug flow in a channel, i.e., velocity U is assumed cross-sectionally uniform and longitudinal mixing is zero. For a longitudinally uniform flow cross-section A and a constant volumetric flow rate Q (i.e., no tides), U is also longitudinally uniform (i.e., $U = Q/A$) and channel volume $V = AL$ (see Figure A1).

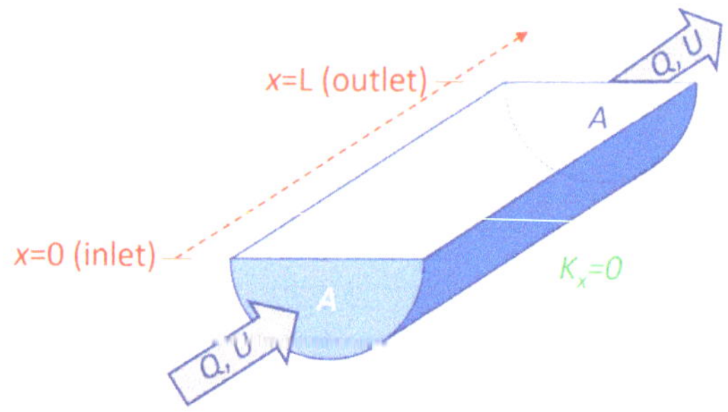

Figure A1. Depiction of plug flow (velocity U is uniform over the flow cross section; longitudinal diffusion K_x is zero) in an idealized channel, for which cross-sectional area A is longitudinally uniform, channel length is L, and volumetric flow rate Q (and therefore U) is constant and positive.

Under these assumptions for a channel of length L, some transport timescales can be easily derived analytically, as follows:

Age at location x (the time since entering the domain at $x = 0$):

$$a(x) = \frac{x}{U} \tag{A1}$$

Residence time at location x (time needed to travel to outlet at $x = L$ from starting location x):

$$\theta(x) = \frac{L - x}{U} \tag{A2}$$

Transit time (time needed to traverse the entire channel from $x = 0$ to $x = L$):

$$\varphi = a(L) = \frac{L}{U} = \frac{V}{Q} \tag{A3}$$

Domain averaged age:

$$\bar{a} = \frac{1}{L} \int_0^L a(x)dx = \frac{L}{2U} = \frac{V}{2Q} \tag{A4}$$

Domain averaged residence time:

$$\bar{\theta} = \frac{1}{L} \int_0^L \theta(x)dx = \frac{L}{2U} = \frac{V}{2Q} \tag{A5}$$

Note that the transit time (Equation (A3)) is twice the domain averaged age or residence time (Equations (A4) and (A5)). For one-dimensional analytical expressions for residence time and exposure time in the presence of both advection and longitudinal dispersion, the reader is referred to Andutta et al. [22].

References

1. *Cambridge Learner's Dictionary*; Cambridge University Press: Cambridge, UK, 2019.
2. Cushman-Roisin, B.; Beckers, J.-M. *Introduction to Geophysical Fluid Dynamics: Physical and Numerical Aspects*; Cushman-Roisin, B., Beckers, J.-M., Eds.; Academic Press: Waltham, MA, USA, 2011; Volume 101, pp. 1–828.
3. Lucas, L.V. Time scale (OR "Timescale"). In *Encyclopedia of Estuaries*; Kennish, M.J., Ed.; Springer: Dordrecht, The Netherlands, 2016.
4. Cloern, J.E. Tidal stirring and phytoplankton bloom dynamics in an estuary. *J. Mar. Res.* **1991**, *49*, 203–221. [CrossRef]
5. Litaker, W.; Duke, C.S.; Kenney, B.E.; Ramus, J. Short-term environmental variability and phytoplankton abundance in a shallow tidal estuary. II. Spring and Fall. *Mar. Ecol. Prog. Ser.* **1993**, *94*, 141–154. [CrossRef]
6. Cloern, J.E. Phytoplankton bloom dynamics in coastal ecosystems: A review with some general lessons from sustained investigation of San Francisco Bay, California. *Rev. Geophys.* **1996**, *34*, 127–168. [CrossRef]
7. Jay, D.A.; Geyer, W.R.; Montgomery, D.R. An ecological perspective on estuarine classification. In *Estuarine Science: A Synthetic Approach to Research and Practice*; Hobbie, J.E., Ed.; Island Press: Washington, DC, USA, 2000; pp. 149–176.
8. Mostofa, K.M.G.; Liu, C.Q.; Zhai, W.D.; Minella, M.; Vione, D.; Gao, K.S.; Minakata, D.; Arakaki, T.; Yoshioka, T.; Hayakawa, K.; et al. Reviews and Syntheses: Ocean acidification and its potential impacts on marine ecosystems. *Biogeosciences* **2016**, *13*, 1767–1786. [CrossRef]
9. Malamud-Roam, F.P.; Ingram, B.L.; Hughes, M.; Florsheim, J.L. Holocene paleoclimate records from a large California estuarine system and its watershed region: Linking watershed climate and bay conditions. *Quat. Sci. Rev.* **2006**, *25*, 1570–1598. [CrossRef]
10. Cronin, T.M.; Vann, C.D. The sedimentary record of climatic and anthropogenic influence on the Patuxent estuary and Chesapeake Bay ecosystems. *Estuaries* **2003**, *26*, 196–209. [CrossRef]
11. Rabalais, N.N.; Diaz, R.J.; Levin, L.A.; Turner, R.E.; Gilbert, D.; Zhang, J. Dynamics and distribution of natural and human-caused hypoxia. *Biogeosciences* **2010**, *7*, 585–619. [CrossRef]
12. Monismith, S.G.; Kimmerer, W.; Burau, J.R.; Stacey, M. Structure and flow-induced variability of the subtidal salinity field in Northern San Francisco Bay. *J. Phys. Oceanogr.* **2002**, *32*, 3003–3019. [CrossRef]
13. MacCready, P.; Geyer, W.R. Advances in Estuarine Physics. *Annu. Rev. Mar. Sci.* **2010**, *2*, 35–58. [CrossRef]
14. Sharpley, A.; Jarvie, H.P.; Buda, A.; May, L.; Spears, B.; Kleinman, P. Phosphorus legacy: Overcoming the effects of past management practices to mitigate future water quality impairment. *J. Environ. Qual.* **2013**, *42*, 1308–1326. [CrossRef]
15. Blumberg, A.F.; Khan, L.A.; St. John, J.P. Three-dimensional hydrodynamic model of New York Harbor region. *J. Hydraul. Eng.* **1999**, *125*, 799–816. [CrossRef]
16. Wang, L.; Samthein, M.; Erlenkeuser, H.; Grimalt, J.; Grootes, P.; Heilig, S.; Ivanova, E.; Kienast, M.; Pelejero, C.; Pfaumann, U. East Asian monsoon climate during the Late Pleistocene: High-resolution sediment records from the South China Sea. *Mar. Geol.* **1999**, *156*, 245–284. [CrossRef]
17. Breithaupt, J.L.; Smoak, J.M.; Byrne, R.H.; Waters, M.N.; Moyer, R.P.; Sanders, C.J. Avoiding timescale bias in assessments of coastal wetland vertical change. *Limnol. Oceanogr.* **2018**, *63*, S477–S495. [CrossRef] [PubMed]
18. Stacey, M.T.; Burau, J.R.; Monismith, S.G. Creation of residual flows in a partially stratified estuary. *J. Geophys. Res.* **2001**, *106*, 13–37. [CrossRef]
19. Chapin, F.S., III; Woodwell, G.M.; Randerson, J.T.; Rastetter, E.B.; Lovett, G.M.; Baldocchi, D.D.; Clark, D.A.; Harmon, M.E.; Schimel, D.S.; Valentini, R.; et al. Reconciling carbon-cycle concepts, terminology, and methods. *Ecosystems* **2006**, *9*, 1041–1050. [CrossRef]
20. Cohen, A.S. The past is a key to the future: Lessons paleoecological data can provide for management of the African Great Lakes. *J. Great Lakes Res.* **2018**, *44*, 1142–1153. [CrossRef]
21. Shen, J.; Hong, B.; Kuo, A.Y. Using timescales to interpret dissolved oxygen distributions in the bottom waters of Chesapeake Bay. *Limnol. Oceanogr.* **2013**, *58*, 2237–2248. [CrossRef]
22. Andutta, F.P.; Ridd, P.V.; Deleersnijder, E.; Prandle, D. Contaminant exchange rates in estuaries—New formulae accounting for advection and dispersion. *Prog. Oceanogr.* **2014**, *120*, 139–153. [CrossRef]
23. Lucas, L.V.; Thompson, J.K.; Brown, L.R. Why are diverse relationships observed between phytoplankton biomass and transport time? *Limnol. Oceanogr.* **2009**, *54*, 381–390. [CrossRef]

24. Wang, Z.G.; Wang, H.; Shen, J.; Ye, F.; Zhang, Y.L.; Chai, F.; Liu, Z.; Du, J.B. An analytical phytoplankton model and its application in the tidal freshwater James River. *Estuar. Coast. Shelf Sci.* **2019**, *224*, 228–244. [CrossRef]

25. Koseff, J.R.; Holen, J.K.; Monismith, S.G.; Cloern, J.E. Coupled effects of vertical mixing and benthic grazing on phytoplankton populations in shallow, turbid estuaries. *J. Mar. Res.* **1993**, *51*, 843–868. [CrossRef]

26. Delhez, J.M.; de Brye, B.; de Brauwere, A.; Deleersnijder, E. Residence time vs influence time. *J. Mar. Syst.* **2014**, *132*, 185–195. [CrossRef]

27. Lucas, L.V. Implications of estuarine transport for water quality. In *Contemporary Issues in Estuarine Physics*, 1st ed.; Valle-Levinson, A., Ed.; Cambridge University Press: New York, NY, USA, 2010; pp. 272–306.

28. Cornaton, F.; Perrochet, P. Groundwater age, life expectancy and transit time distributions in advective-dispersive systems: 1. Generalized reservoir theory. *Adv. Water Resour.* **2006**, *29*, 1267–1291. [CrossRef]

29. Ambrosetti, W.; Barbanti, L.; Sala, N. Residence time and physical processes in lakes. *J. Limnol.* **2003**, *62*, 1–15. [CrossRef]

30. Rueda, F.J.; Cowen, E.A. Residence time of a freshwater embayment connected to a large lake. *Limnol. Oceanogr.* **2005**, *50*, 1638–1653. [CrossRef]

31. Rueda, F.J.; Moreno-Ostos, E.; Armengol, J. The residence time of river water in reservoirs. *Ecol. Model.* **2006**, *191*, 260–274. [CrossRef]

32. McDonald, C.P.; Lathrop, R.C. Seasonal shifts in the relative importance of local versus upstream sources of phosphorus to individual lakes in a chain. *Aquat. Sci.* **2017**, *79*, 385–394. [CrossRef]

33. Burdis, R.M.; Hirsch, J.K. Crustacean zooplankton dynamics in a natural riverine lake, Upper Mississippi River. *J. Freshw. Ecol.* **2017**, *32*, 240–258. [CrossRef]

34. Zwart, J.A.; Sebestyen, S.D.; Solomon, C.T.; Jones, S.E. The Influence of Hydrologic Residence Time on Lake Carbon Cycling Dynamics Following Extreme Precipitation Events. *Ecosystems* **2017**, *20*, 1000–1014. [CrossRef]

35. Vollenweider, R.A. Input-output models: With special reference to the phosphorus loading concept in limnology. *Schweiz. Z. Für Hydrol.* **1975**, *37*, 53–84. [CrossRef]

36. Hester, E.T.; Hammond, B.; Scott, D.T. Effects of inset floodplains and hyporheic exchange induced by in-stream structures on nitrate removal in a headwater stream. *Ecol. Eng.* **2016**, *97*, 452–464. [CrossRef]

37. Seybold, E.; McGlynn, B. Hydrologic and biogeochemical drivers of dissolved organic carbon and nitrate uptake in a headwater stream network. *Biogeochemistry* **2018**, *138*, 23–48. [CrossRef]

38. Benettin, P.; Bertuzzo, E. *tran*-SAS v1.0: A numerical model to compute catchment-scale hydrologic transport using StorAge Selection function. *Geosci. Model Dev.* **2018**, *11*, 1627–1639. [CrossRef]

39. Remondi, F.; Kirchner, J.W.; Burlando, P.; Fatichi, S. Water flux tracking with a distributed hydrological model to quantify controls on the spatio-temporal variability of transit time distributions. *Water Resour. Res.* **2018**, *54*, 3081–3099. [CrossRef]

40. Rozemeijer, J.; Klein, J.; Hendriks, D.; Borren, W.; Ouboter, M.; Rip, W. Groundwater-surface water relations in regulated lowland catchments; hydrological and hydrochecmical effects of a major change in surface water level management. *Sci. Total Environ.* **2019**, *660*, 1317–1326. [CrossRef] [PubMed]

41. Onderka, M.; Chudoba, V. The wavelets show it—The transit time of water varies in time. *J. Hydrol. Hydromech.* **2018**, *64*, 295–302. [CrossRef]

42. Geyer, W.R.; Signell, R.P. A reassessment of the role of tidal dispersion in estuaries and bays. *Estuaries* **1992**, *15*, 97–108. [CrossRef]

43. Banas, N.S.; Hickey, B.M. Mapping exchange and residence time in a model of Willapa Bay, Washington, a branching, macrotidal estuary. *J. Geophys. Res.* **2005**, *110*, C11011. [CrossRef]

44. Alosairi, Y.; Pokavanich, T. Residence and transport time scales associated with Shatt Al-Arab discharges under various hydrological conditions estimated using a numerical model. *Mar. Pollut. Bull.* **2017**, *118*, 85–92. [CrossRef]

45. Andutta, F.P.; Ridd, P.V.; Wolanski, E. The age and the flushing time of the Great Barrier Reef waters. *Cont. Shelf Res.* **2013**, *53*, 11–19. [CrossRef]

46. Drake, C.W.; Jones, C.S.; Schilling, K.E.; Amado, A.A.; Weber, L.J. Estimating nitrate-nitrogen retention in a large constructed wetland using high-frequency, continuous monitoring and hydrologic modeling. *Ecol. Eng.* **2018**, *117*, 69–83. [CrossRef]

47. White, S.A. Design and season influence nitrogen dynamics in two surface flow constructed wetlands treating nursery irrigation runoff. *Water* **2018**, *10*, 8. [CrossRef]

48. Koszalka, I.M.; Haine, T.W.N.; Magaldi, M.G. Fates and travel times of Denmark Strait overflow water in the Irminger Basin. *J. Phys. Oceanogr.* **2013**, *43*, 2611–2628. [CrossRef]

49. Mouchet, A.; Deleersnijder, E. The leaky funnel model, a metaphor of the ventilation of the World Ocean as simulated in an OGCM. *Tellus* **2008**, *60*, 761–774. [CrossRef]

50. Rutherford, K.; Fennel, K. Diagnosing transit times on the northwestern North Atlantic continental shelf. *Ocean Sci.* **2018**, *14*. [CrossRef]

51. Waugh, D.W.; Hall, T.M. Age of stratospheric air: Theory, observations, and models. *Rev. Geophys.* **2002**, *40*, 1–26. [CrossRef]

52. Fischer, H.B.; List, E.J.; Koh, R.C.Y.; Imberger, J.; Brooks, N.H. *Mixing in Inland and Coastal Waters*; Academic Press: San Diego, CA, USA, 1979.

53. MacCready, P.; Banas, N.S. Residual circulation, mixing, and dispersion. In *Treatise on Estuarine and Coastal Science*; Wolansky, E., McLusky, D.S., Eds.; Academic Press: Waltham, MA, USA, 2011; Volume 2, pp. 75–89.

54. Seitzinger, S.; Harrison, J.A.; Bohlke, J.K.; Bouwman, A.F.; Lowrance, R.; Peterson, B.; Tobias, C.; Van Drecht, G. Denitrification across landscapes and waterscapes: A synthesis. *Ecol. Appl.* **2006**, *16*, 2064–2090. [CrossRef]

55. Stabeno, P.J.; Danielson, S.L.; Kachel, D.G.; Kachel, N.B.; Mordy, C.W. Currents and transport on the Eastern Bering Sea shelf: An integration of over 20 years of data. *Deep-Sea Res. Part II* **2016**, *134*, 13–29. [CrossRef]

56. Delpeche-Ellmann, N.; Torsvik, T.; Soomere, T. Tracks of surface drifters from a major fairway to marine protected areas in the Gulf of Finland. *Proc. Est. Acad. Sci.* **2015**, *64*, 226–233. [CrossRef]

57. Peierls, B.L.; Hall, N.S.; Paerl, H.W. Non-monotonic Responses of Phytoplankton Biomass Accumulation to Hydrologic Variability: A Comparison of Two Coastal Plain North Carolina Estuaries. *Estuaries Coasts* **2012**, *35*, 1376–1392. [CrossRef]

58. Sheldon, J.E.; Alber, M. The calculation of estuarine turnover times using freshwater fraction and tidal prism models: A critical evaluation. *Estuaries Coasts* **2006**, *29*, 133–146. [CrossRef]

59. Deleersnijder, E.; Beckers, J.-M.; Delhez, E.J.M. The residence time of settling particles in the surface mixed layer. *Environ. Fluid Mech.* **2006**, *6*, 25–42. [CrossRef]

60. de Brauwere, A.; Deleersnijder, E. Assessing the parameterization of the settling flux in a depth-integrated model of the fate of decaying and sinking particles, with application to fecal bacteria in the Scheldt Estuary. *Environ. Fluid Mech.* **2010**, *10*, 157–175. [CrossRef]

61. Buzzelli, C.; Parker, M.; Geiger, S.; Wan, Y.; Doering, P.; Haunert, D. Predicting system-scale impacts of oyster clearance on phytoplankton productivity in a small subtropical estuary. *Environ. Model. Assess.* **2013**, *18*, 185–198. [CrossRef]

62. Kimmerer, W.; Wilkerson, F.; Downing, B.; Dugdale, R.; Gross, E.; Kayfetz, K.; Khanna, S.; Parker, A.E.; Thompson, J.K. Effects of drought and the emergency drought barrier on the ecosystem of the California Delta. *San Franc. Estuary Watershed Sci.* **2019**, *17*, 1–28. [CrossRef]

63. Deleersnijder, E.; Campin, J.M.; Delhez, E.J.M. The concept of age in marine modelling I. Theory and preliminary model results. *J. Mar. Syst.* **2001**, *28*, 229–267. [CrossRef]

64. Delhez, E.J.M.; Campin, J.M.; Hirst, A.C.; Deleersnijder, E. Toward a general theory of the age in ocean modelling. *Ocean Model.* **1999**, *1*, 17–27. [CrossRef]

65. de Groot, S.R.; Mazur, P. *Non-Equilibrium Thermodynamics*; North-Holland Publishing Company: Amsterdam, The Netherlands, 1962; p. 510.

66. de Brye, B.; de Brauwere, A.; Gourgue, O.; Delhez, E.J.M.; Deleersnijder, E. Water renewal timescales in the Scheldt Estuary. *J. Mar. Syst.* **2012**, *94*, 74–86. [CrossRef]

67. Cox, M.D. An Idealized Model of the World Ocean. 1. The Global-Scale Water Masses. *J. Phys. Oceanogr.* **1989**, *19*, 1730–1752. [CrossRef]

68. Haine, T.W.N.; Hall, T.M. A generalized transport theory: Water-mass composition and age. *J. Phys. Oceanogr.* **2002**, *32*, 1932–1946. [CrossRef]

69. Baretta-Bekker, H.J.G.; Duursma, E.K.; Kuipers, B.R. (Eds.) Tracer. In *Encyclopedia of Marine Sciences*; Springer: Berlin, Germany, 1998; p. 357.

70. Batchelor, G.K. *An Introduction to Fluid Dynamics*, 1st ed.; Cambridge University Press: Cambridge, UK, 2000; p. 615.

71. Kundu, P.K.; Cohen, I.M. *Fluid Mechanics*, 3rd ed.; Elsevier Academic Press: San Diego, CA, USA, 2004; p. 759.

72. Thuburn, J. TVD schemes, positive schemes, and the universal limiter. *Mon. Weather Rev.* **1997**, *125*, 1990–1995. [CrossRef]

73. Gross, E.S.; Koseff, J.R.; Monismith, S.G. Evaluation of advective schemes for estuarine salinity simulations. *J. Hydr. Engrgy* **1999**, *125*, 1199–1209. [CrossRef]

74. van Slingerland, P. An Accurate and Robust Finite Volume Method for the Advection Diffusion Equation. Master's Thesis, Delft University of Technology, Delft, The Netherlands, 2007.

75. Hunter, J.R.; Craig, P.D.; Phillips, H.E. On the Use of Random-Walk Models with Spatially-Variable Diffusivity. *J. Comput. Phys.* **1993**, *106*, 366–376. [CrossRef]

76. Visser, A.W. Using random walk models to simulate the vertical distribution of particles in a turbulent water column. *Mar. Ecol. Prog. Ser.* **1997**, *158*, 275–281. [CrossRef]

77. Visser, A.W. Lagrangian modelling of plankton motion: From deceptively simple random walks to Fokker-Planck and back again. *J. Mar. Syst.* **2008**, *70*, 287–299. [CrossRef]

78. van Sebille, E.; Griffies, S.M.; Abernathey, R.; Adams, T.P.; Berloff, P.; Biastoch, A.; Blanke, B.; Chassignet, E.P.; Cheng, Y.; Cotter, C.J.; et al. Lagrangian ocean analysis: Fundamentals and practices. *Ocean Model.* **2018**, *121*, 49–75. [CrossRef]

79. Deleersnijder, E.; Draoui, I.; Lambrechts, J.; Legat, V.; Mouchet, A. Consistent boundary conditions for age calculations. *Water* **2020**, *12*, 1274. [CrossRef]

80. Alessandrini, S.; Ferrero, E. A hybrid Lagrangian-Eulerian particle model for reacting pollutant dispersion in non-homogeneous non-isotropic turbulence. *Phys. A* **2009**, *388*, 1375–1387. [CrossRef]

81. Peyret, R.; Taylor, T.D. *Computational Methods for Fluid Flow*; Springer: New York, NY, USA, 1983.

82. Condie, S.A.; Andrewartha, J.R. Circulation and connectivity on the Australian North West Shelf. *Cont. Shelf Res.* **2008**, *28*, 1724–1739. [CrossRef]

83. Thomas, C.J.; Lambrechts, J.; Wolanski, E.; Traag, V.A.; Blondel, V.D.; Deleersnijder, E.; Hanert, E. Numerical modelling and graph theory tools to study ecological connectivity in the Great Barrier Reef. *Ecol. Model.* **2014**, *272*, 160–174. [CrossRef]

84. Frys, C.; St-Amand, A.; Le Henaff, M.; Figueiredo, J.; Kuba, A.; Walker, B.; Lambrechts, J.; Vallaeys, V.; Vincent, D.; Hanert, E. Fine-Scale Coral Connectivity Pathways in the Florida Reef Tract: Implications for Conservation and Restoration. *Front. Mar. Sci.* **2020**, *7*. [CrossRef]

85. Grech, A.; Hanert, E.; McKenzie, L.; Rasheed, M.; Thomas, C.; Tol, S.; Wang, M.Z.; Waycott, M.; Wolter, J.; Coles, R. Predicting the cumulative effect of multiple disturbances on seagrass connectivity. *Glob. Chang. Biol.* **2018**, *24*, 3093–3104. [CrossRef] [PubMed]

86. Paris, C.B.; Helgers, J.; van Sebille, E.; Srinivasan, A. Connectivity Modeling System: A probabilistic modeling tool for the multi-scale tracking of biotic and abiotic variability in the ocean. *Environ. Model. Softw.* **2013**, *42*, 47–54. [CrossRef]

87. Monsen, N.E.; Cloern, J.E.; Lucas, L.V.; Monismith, S.G. A comment on the use of flushing time, residence time, and age as transport time scales. *Limnol. Oceanogr.* **2002**, *47*, 1545–1553. [CrossRef]

88. Takeoka, H. Fundamental concepts of exchange and transport time scales in a coastal sea. *Cont. Shelf Res.* **1984**, *3*, 311–326. [CrossRef]

89. Deleersnijder, E.; Mouchet, A.; Delhez, E.J.M. *Diagnostic Timescales in Fluid Flows: From the Tower of Babel to Partial Differential Problems*; Université Catholique de Louvain: Louvain-la-Neuve, Belgium, 2018; pp. 1–26. Available online: http://hdl.handle.net/2078.1/196273 (accessed on 10 March 2018).

90. Middelburg, J.J.; Nieuwenhuize, J. Uptake of dissolved inorganic nitrogen in turbid, tidal estuaries. *Mar. Ecol. Prog. Ser.* **2000**, *192*, 79–88. [CrossRef]

91. Crump, B.C.; Hopkinson, C.S.; Sogin, M.L.; Hobbie, J.E. Microbial biogeography along an estuarine salinity gradient: Combined influences of bacterial growth and residence time. *Appl. Environ. Microb.* **2004**, *70*, 1494–1505. [CrossRef]

92. Lucas, L.V. Residence time. In *Encyclopedia of Estuaries*; Kennish, M.J., Ed.; Springer: Dordrecht, The Netherlands, 2016.

93. Sheldon, J.E.; Alber, M. A comparison of residence time calculations using simple compartment models of the Altamaha River Estuary, Georgia. *Estuaries* **2002**, *25*, 1304–1317. [CrossRef]

94. Bolin, B.; Rodhe, H. A note on the concepts of age distribution and transit time in natural reservoirs. *Tellus* **1973**, *25*, 58–62. [CrossRef]

95. de Brauwere, A.; de Brye, B.; Blaise, S.; Deleersnijder, E. Residence time, exposure time, and connectivity in the Scheldt Estuary. *J. Mar. Syst.* **2011**, *84*, 85–95. [CrossRef]

96. Zimmerman, J.T.F. Mixing and flushing of tidal embayments in the western Dutch Wadden Sea. Part I: Distribution of salinity and calculation of mixing time scales. *Neth. J. Sea Res.* **1976**, *10*, 149–191. [CrossRef]

97. Delhez, E.J.M.; Heemink, A.W.; Deleersnijder, E. Residence time in a semi-enclosed domain from the solution of an adjoint problem. *Estuar. Coast. Shelf Sci.* **2004**, *61*, 691–702. [CrossRef]

98. Delhez, E.J.M. On the concept of exposure time. *Cont. Shelf Res.* **2013**, *71*, 27–36. [CrossRef]

99. Delhez, E.J.M. Transient residence and exposure times. *Ocean Sci.* **2006**, *2*, 1–9. [CrossRef]

100. Rayson, M.D.; Gross, E.S.; Hetland, R.D.; Fringer, O.B. Time scales in Galveston Bay: An unsteady estuary. *J. Geophys. Res. Oceans* **2016**, *121*, 2268–2285. [CrossRef]

101. Defne, Z.; Ganju, N.K. Quantifying the Residence Time and Flushing Characteristics of a Shallow, Back-Barrier Estuary: Application of Hydrodynamic and Particle Tracking Models. *Estuaries Coasts* **2015**, *38*, 1719–1734. [CrossRef]

102. Storlazzi, C.D.; Cheriton, O.M.; Messina, A.M.; Biggs, T.W. Meteorologic, oceanographic, and geomorphic controls on circulation and residence time in a coral reef-lined embayment: Faga'alu Bay, American Samoa. *Coral Reefs* **2018**, *37*, 457–469. [CrossRef]

103. Manning, J.P.; McGillicuddy, D.J.; Pettigrew, N.R.; Churchill, J.H.; Incze, L.S. Drifter Observations of the Gulf of Maine Coastal Current. *Cont. Shelf Res.* **2009**, *29*, 835–845. [CrossRef]

104. Pawlowicz, R.; Hannah, C.; Rosenberger, A. Lagrangian observations of estuarine residence times, dispersion, and trapping in the Salish Sea. *Estuar. Coast. Shelf Sci.* **2019**, *225*, 1–16. [CrossRef]

105. Corcoran, A.A.; Reifel, K.M.; Jones, B.H.; Shipe, R.F. Spatiotemporal development of physical, chemical, and biological characteristics of stormwater plumes in Santa Monica Bay, California (USA). *J. Sea Res.* **2010**, *63*, 129–142. [CrossRef]

106. Lucas, L.V. Age. In *Encyclopedia of Estuaries*; Kennish, M.J., Ed.; Springer: Dordrecht, The Netherlands, 2016.

107. Delhez, E.J.M.; Wolk, F. Diagnosis of the transport of adsorbed material in the Scheldt estuary: A proof of concept. *J. Mar. Syst.* **2013**, *128*, 17–26. [CrossRef]

108. Camacho, R.A.; Martin, J.L. Hydrodynamic Modeling of First-Order Transport Timescales in the St. Louis Bay Estuary, Mississippi. *J. Environ. Eng.* **2013**, *139*, 317–331. [CrossRef]

109. Dyer, K.R. *Estuaries: A Physical Introduction*, 2nd ed.; John Wiley & Sons Ltd.: Chichester, UK, 1997; p. 195.

110. Lemagie, E.P.; Lerczak, J.A. A Comparison of Bulk Estuarine Turnover Timescales to Particle Tracking Timescales Using a Model of the Yaquina Bay Estuary. *Estuaries Coasts* **2015**, *38*, 1797–1814. [CrossRef]

111. Geyer, W.R.; Morris, J.T.; Pahl, F.G.; Jay, D.A. Interaction between physical processes and ecosystem structure: A comparative approach. In *Estuarine Science: A Synthetic Approach to Research and Practice*; Hobbie, J.E., Ed.; Island Press: Washington, DC, USA, 2000; pp. 177–206.

112. Jouon, A.; Douillet, P.; Ouillon, S.; Fraunie, P. Calculations of hydrodynamic time parameters in a semi-opened coastal zone using a 3D hydrodynamic model. *Cont. Shelf Res.* **2006**, *26*, 1395–1415. [CrossRef]

113. Tartinville, B.; Deleersnijder, E.; Rancher, J. The water residence time in the Mururoa atoll lagoon: Sensitivity analysis of a three-dimensional model. *Coral Reefs* **1997**, *16*, 193–203. [CrossRef]

114. Hinrichs, C.; Flagg, C.N.; Wilson, R.E. Great South Bay After Sandy: Changes in Circulation and Flushing due to New Inlet. *Estuaries Coasts* **2018**, *41*, 2172–2190. [CrossRef]

115. Huguet, J.-R.; Brenon, I.; Coulombier, T. Characterisation of the water renewal in a macro-tidal marina using several transport timescales. *Water* **2019**, *11*, 2050. [CrossRef]

116. Wolanski, E.; Mazda, Y.; King, B.; Gay, S. Dynamics, Flushing and Trapping in Hinchinbrook Channel, a Giant Mangrove Swamp, Australia. *Estuar. Coast. Shelf Sci.* **1990**, *31*, 555–579. [CrossRef]

117. Hansen, J.C.R.; Reidenbach, M.A. Turbulent mixing and fluid transport within Florida Bay seagrass meadows. *Adv. Water Resour.* **2017**, *108*, 205–215. [CrossRef]

118. Deleersnijder, E. *A Quick Refresher on Exposure and Residence Times in a Well-Mixed Domain*; Université catholique de Louvain: Louvain-la-Neuve, Belgium, 2020; pp. 1–4. Available online: http://hdl.handle.net/2078.1/229804 (accessed on 15 May 2020).

119. Sanford, L.P.; Boicourt, W.C.; Rives, S.R. Model for estimating tidal flushing of small embayments. *J. Waterw. Port Coast. Ocean Eng.* **1992**, *118*, 635–654. [CrossRef]

120. van de Kreeke, J. Residence time: Application to small boat basins. *J. Waterw. Port Coast. Ocean Eng* **1983**, *109*, 416–428. [CrossRef]

121. Shen, J.; Haas, L. Calculating age and residence time in the tidal York River using three-dimensional model experiments. *Estuar. Coast. Shelf Sci.* **2004**, *61*, 449–461. [CrossRef]

122. MacCready, P. Calculating Estuarine Exchange Flow Using Isohaline Coordinates. *J. Phys. Oceanogr.* **2011**, *41*, 1116–1124. [CrossRef]

123. Middelburg, J.J.; Nieuwenhuize, J. Nitrogen uptake by heterotrophic bacteria and phytoplankton in the nitrate-rich Thames estuary. *Mar. Ecol. Prog. Ser.* **2000**, *203*, 13–21. [CrossRef]

124. Caraco, N.F.; Cole, J.J.; Raymond, P.A.; Strayer, D.L.; Pace, M.L.; Findlay, S.; Fischer, D.T. Zebra mussel invasion in a large, turbid river: Phytoplankton response to increased grazing. *Ecology* **1997**, *78*, 588–602. [CrossRef]

125. Middelburg, J.J.; Barranguet, C.; Boschker, H.T.S.; Herman, P.M.J.; Moens, T.; Heip, C.H.R. The fate of intertidal microphytobenthos carbon: An in situ C-13-labeling study. *Limnol. Oceanogr.* **2000**, *45*, 1224–1234. [CrossRef]

126. Hong, B.; Liu, Z.H.; Shen, J.; Wu, H.; Gong, W.P.; Xu, H.Z.; Wang, D.X. Potential physical impacts of sea-level rise on the Pearl River Estuary, China. *J. Mar. Syst.* **2020**, *201*. [CrossRef]

127. Ralston, D.K.; Brosnahan, M.L.; Fox, S.E.; Lee, K.D.; Anderson, D.M. Temperature and Residence Time Controls on an Estuarine Harmful Algal Bloom: Modeling Hydrodynamics and Alexandrium fundyense in Nauset Estuary. *Estuaries Coasts* **2015**, *38*, 2240–2258. [CrossRef]

128. Larson, M.; Nunes, A.; Tanaka, H. Semi-analytic model of tidal-induced inlet flow and morphological evolution. *Coast. Eng.* **2020**, *155*. [CrossRef]

129. Chen, Y.H.; Yu, K.F.; Hassan, M.; Xu, C.; Zhang, B.; Gin, K.Y.H.; He, Y.L. Occurrence, distribution and risk assessment of pesticides in a river-reservoir system. *Ecotoxicol. Environ. Saf.* **2018**, *166*, 320–327. [CrossRef]

130. Viero, D.P.; Defina, A. Water age, exposure time, and local flushing time in semi-enclosed, tidal basins with negligible freshwater inflow. *J. Mar. Syst.* **2016**, *156*, 16–29. [CrossRef]

131. Ralston, D.K.; Geyer, W.R. Sediment Transport Time Scales and Trapping Efficiency in a Tidal River. *J. Geophys. Res. Earth* **2017**, *122*, 2042–2063. [CrossRef]

132. Fennel, K.; Testa, J.M. Biogeochemical Controls on Coastal Hypoxia. *Annu. Rev. Mar. Sci.* **2019**, *11*, 105–130. [CrossRef] [PubMed]

133. Gross, E.S.; Koseff, J.R.; Monismith, S.G. Three-dimensional salinity simulations of South San Francisco Bay. *J. Hydraul. Eng.* **1999**, *125*, 32–46. [CrossRef]

134. Kim, J.H.; Hong, S.J.; Lee, W.C.; Kim, H.C.; Eom, K.H.; Jung, W.S.; Kim, D.M. Estimation of the Effect of Flushing Time on Oyster Aquaculture Potential in Jaran Bay. *Ocean Sci. J.* **2019**, *54*, 559–571. [CrossRef]

135. Wolanski, E. Bounded and unbounded boundaries—Untangling mechanisms for estuarine-marine ecological connectivity: Scales of m to 10,000 km—A review. *Estuar. Coast. Shelf Sci.* **2017**, *198*, 378–392. [CrossRef]

136. Golbuu, Y.; Gouezo, M.; Kurihara, H.; Rehm, L.; Wolanski, E. Long-term isolation and local adaptation in Palau's Nikko Bay help corals thrive in acidic waters. *Coral Reefs* **2016**, *35*, 909–918. [CrossRef]

137. Viero, D.P.; Defina, A. Renewal time scales in tidal basins: Climbing the Tower of Babel. In *Sustainable Hydraulics in the Era of Global Change: Advances in Water Engineering and Research*, 1st ed.; Erpicum, S., Dewals, B., Archambeau, P., Pirotton, M., Eds.; CRC Press: London, UK, 2016; pp. 338–345.

138. Deleersnijder, E.; Delhez, E.J.M. Timescale- and tracer-based methods for understanding the results of complex marine models. *Estuar. Coast. Shelf Sci.* **2007**, *74*. [CrossRef]

139. Deleersnijder, E. *Classical vs. Holistic Timescales: The Mururoa Atoll Lagoon Case Study*; Université Catholique de Louvain: Louvain-la-Neuve, Belgium, 2019; pp. 1–12. Available online: http://hdl.handle.net/2078.1/224391 (accessed on 18 April 2020).

140. Mouchet, A.; Cornaton, F.; Deleersnijder, E.; Delhez, E.J.M. Partial ages: Diagnosing transport processes by means of multiple clocks. *Ocean Dyn.* **2016**, *66*, 367–386. [CrossRef]

141. Lin, L.; Liu, Z. Partial residence times: Determining residence time composition in different subregions. *Ocean Dyn.* **2019**, *69*, 1023–1036. [CrossRef]

142. Rypina, I.I.; Fertitta, D.; Macdonald, A.; Yoshida, S.; Jayne, S. Multi-iteration approach to studying tracer spreading using drifter data. *J. Phys. Oceanogr.* **2017**, *47*, 339–351. [CrossRef]

143. Scavia, D.; Field, J.C.; Boesch, D.F.; Buddemeier, R.W.; Burkett, V.; Cayan, D.R.; Fogarty, M.; Harwell, M.A.; Howarth, R.W.; Mason, C.; et al. Climate change impacts on U.S. Coastal and Marine Ecosystems. *Estuaries* **2002**, *25*, 149–164. [CrossRef]

144. Beklioglu, M.; Romo, S.; Kagalou, I.; Quintana, X.; Bécares, E. State of the art in the functioning of shallow Mediterranean lakes: Workshop conclusions. *Hydrobiologia* **2007**, *584*, 317–326. [CrossRef]

145. Schindler, D.W. Recent advances in the understanding and management of eutrophication. *Limnol. Oceanogr.* **2006**, *51*, 356–363. [CrossRef]

146. Cloern, J.E. Our evolving conceptual model of the coastal eutrophication problem. *Mar. Ecol. Prog. Ser.* **2001**, *210*, 223–253. [CrossRef]

147. Oveisy, A.; Boegman, L.; Rao, Y.R. A model of the three-dimensional hydrodynamics, transport and flushing in the Bay of Quinte. *J. Great Lakes Res.* **2015**, *41*, 536–548. [CrossRef]

148. Lopez, C.B.; Cloern, J.E.; Schraga, T.S.; Little, A.J.; Lucas, L.V.; Thompson, J.K.; Burau, J.R. Ecological values of shallow-water habitats: Implications for restoration of disturbed ecosystems. *Ecosystems* **2006**, *9*, 422–440. [CrossRef]

149. Cochran, J.K.; Miquel, J.C.; Armstrong, R.; Fowler, S.W.; Masque, P.; Gasser, B.; Hirschberg, D.; Szlosek, J.; Baena, A.M.R.Y.; Verdeny, E.; et al. Time-series measurements of Th-234 in water column and sediment trap samples from the northwestern Mediterranean Sea. *Deep-Sea Res. Partt II* **2009**, *56*, 1487–1501. [CrossRef]

150. Wang, W.X.; Fisher, N.S.; Luoma, S.N. Kinetic determinations of trace element bioaccumulation in the mussel Mytilus edulis. *Mar. Ecol. Prog. Ser.* **1996**, *140*, 91–113. [CrossRef]

151. Cuetara, J.I.; Burau, J.R. Drifter Studies in Open Shallow Water Habitats of the San Francisco Bay and Delta. Available online: https://archive.usgs.gov/archive/sites/sfbay.wr.usgs.gov/watershed/drifter_studies/index.html (accessed on 11 February 2020).

152. Falco, P.; Griffa, A.; Poulain, P.-M.; Zambianchi, E. Transport Properties in the Adriatic Sea as Deduced from Drifter Data. *J. Phys. Oceanogr.* **2000**, *30*, 2055–2071. [CrossRef]

153. Zhou, M.; Niiler, P.P.; Hu, J.-H. Surface currents in the Bransfield and Gerlache Straits, Antarctica. *Deep-Sea Res. Part II* **2002**, *49*, 267–280. [CrossRef]

154. Lacy, J.R. Circulation and Transport in a Semi-Enclosed Estuarine Subembayment. Ph.D. Thesis, Stanford University, Stanford, CA, USA, 2000.

155. Conomos, T.J.; Peterson, D.H.; Carlson, P.R.; McCulloch, D.S. *Movement of Seabed Drifters in the San Francisco Bay Estuary and the Adjacent Pacific Ocean: A Preliminary Report;* Geological Survey Circular 637-B; United States Geological Survey: Washington, DC, USA, 1970; pp. 1–8.

156. Conomos, T.J. Movement of spilled oil as predicted by estuarine nontidal drift. *Limnol. Oceanogr.* **1975**, *20*, 159–173. [CrossRef]

157. Rypina, I.I.; Kirincich, A.; Lentz, S.; Sundermeyer, M. Investigating the Eddy Diffusivity Concept in the Coastal Ocean. *J. Phys. Oceanogr.* **2016**, *46*, 2201–2218. [CrossRef]

158. Poulain, P.M.; Gerin, R.; Mauri, E.; Pennel, R. Wind Effects on Drogued and Undrogued Drifters in the Eastern Mediterranean. *J. Atmos. Ocean. Tech.* **2009**, *26*, 1144–1156. [CrossRef]

159. Perry, R.W.; Pope, A.C.; Romine, J.G.; Brandes, P.L.; Burau, J.R.; Blake, A.R.; Ammann, A.J.; Michel, C.J. Flow-mediated effects on travel time, routing, and survival of juvenile Chinook salmon in a spatially complex, tidally forced river delta. *Can. J. Fish. Aquat. Sci.* **2018**, *75*, 1886–1901. [CrossRef]

160. Vallino, J.J.; Hopkinson, C.S., Jr. Estimation of dispersion and characteristic mixing times in Plum Island Sound Estuary. *Estuar. Coast. Shelf Sci.* **1998**, *46*, 333–350. [CrossRef]

161. Bois, P.; Childers, D.L.; Corlouer, T.; Laurent, J.; Massicot, A.; Sanchez, C.A.; Wanko, A. Confirming a plant-mediated "Biological Tide" in an aridland constructed treatment wetland. *Ecosphere* **2017**, *8*, 1–16. [CrossRef]

162. Nishihara, G.N.; Terada, R.; Shimabukuro, H. Effects of wave energy on the residence times of a fluorescent tracer in the canopy of the intertidal marine macroalgae, Sargassum fusiforme (Phaeophyceae). *Phycol. Res.* **2011**, *59*, 24–33. [CrossRef]

163. Dahlgaard, H.; Herrmann, J.; Salomon, J.C. A Tracer Study of the Transport of Coastal Water from the English-Channel through the German-Bight to the Kattegat. *J. Mar. Syst.* **1995**, *6*, 415–425. [CrossRef]

164. Bailly du Bois, P.; Dumas, F.; Solier, L.; Voiseux, C. In-situ database toolbox for short-term dispersion model validation in macro-tidal seas, application for 2D-model. *Cont. Shelf Res.* **2012**, *36*, 63–82. [CrossRef]

165. Bailly du Bois, P.; Garreau, P.; Laguionie, P.; Korsakissok, I. Comparison between modelling and measurement of marine dispersion, environmental half-time and 137Cs inventories after the Fukushima Daiichi accident. *Ocean Dyn.* **2014**, *64*, 361–383. [CrossRef]

166. Bailly du Bois, P.; Laguionie, P.; Boust, D.; Korsakissok, I.; Didier, D.; Fiévet, B. Estimation of marine source-term following Fukushima Dai-ichi accident. *J. Environ. Radioact.* **2012**, *114*, 2–9. [CrossRef] [PubMed]

167. Bressac, M.; Levy, I.; Chamizo, E.; La Rosa, J.J.; Povinec, P.P.; Gastaud, J.; Oregioni, B. Temporal evolution of ^{137}Cs, ^{237}Np, and $^{239+240}$Pu and estimated vertical $^{239+240}$Pu export in the northwestern Mediterranean Sea. *Sci. Total Environ.* **2017**, *595*, 178–190. [CrossRef] [PubMed]

168. Uncles, R.J.; Torres, R. Estimating dispersion and flushing time-scales in a coastal zone: Application to the Plymouth area. *Ocean Coast Manag.* **2013**, *72*, 3–12. [CrossRef]

169. Souza, T.A.; Godoy, J.M.; Godoy, M.L.D.P.; Moreira, I.; Carvalho, Z.L.; Salomao, M.S.M.B.; Rezende, C.E. Use of multitracers for the study of water mixing in the Paraiba do Sul River estuary. *J. Environ. Radioact.* **2010**, *101*, 564–570. [CrossRef]

170. Dulaiova, H.; Burnett, W.C. Evaluation of the flushing rates of Apalachicola Bay, Florida via natural geochemical tracers. *Mar. Chem.* **2008**, *109*, 395–408. [CrossRef]

171. Eller, K.T.; Burnett, W.C.; Fitzhugh, L.M.; Chanton, J.P. Radium Sampling Methods and Residence Times in St. Andrew Bay, Florida. *Estuaries Coasts* **2014**, *37*, 94–103. [CrossRef]

172. Black, E.E.; Lam, P.J.; Lee, J.M.; Buesseler, K.O. Insights From the U-238-Th-234 Method Into the Coupling of Biological Export and the Cycling of Cadmium, Cobalt, and Manganese in the Southeast Pacific Ocean. *Glob. Biogeochem. Cycles* **2019**, *33*, 15–36. [CrossRef]

173. Downing, B.D.; Bergamaschi, B.A.; Kendall, C.; Kraus, T.E.C.; Dennis, K.J.; Carter, J.A.; Von Dessonneck, T.S. Using Continuous Underway Isotope Measurements To Map Water Residence Time in Hydrodynamically Complex Tidal Environments. *Env. Sci. Technol.* **2016**, *50*, 13387–13396. [CrossRef]

174. Gross, E.; Andrews, S.; Bergamaschi, B.; Downing, B.; Holleman, R.; Burdick, S.; Durand, J. The Use of Stable Isotope-Based Water Age to Evaluate a Hydrodynamic Model. *Water* **2019**, *11*, 2207. [CrossRef]

175. Choi, K.W.; Lee, J.H.W. Numerical determination of flushing time for stratified water bodies. *J. Mar. Syst.* **2004**, *50*, 263–281. [CrossRef]

176. Safak, I.; Wiberg, P.L.; Richardson, D.L.; Kurum, M.O. Controls on residence time and exchange in a system of shallow coastal bays. *Cont. Shelf Res.* **2015**, *97*, 7–20. [CrossRef]

177. Du, J.B.; Park, K.; Yu, X.; Zhang, Y.L.J.; Ye, F. Massive pollutants released to Galveston Bay during Hurricane Harvey: Understanding their retention and pathway using Lagrangian numerical simulations. *Sci. Total Environ.* **2020**, *704*. [CrossRef] [PubMed]

178. Guillou, N.; Thiebot, J.; Chapalain, G. Turbines' effects on water renewal within a marine tidal stream energy site. *Energy* **2019**, *189*. [CrossRef]

179. Palazzoli, I.; Leonardi, N.; Jimenez-Robles, A.M.; Fagherazzi, S. Velocity skew controls the flushing of a tracer in a system of shallow bays with multiple inlets. *Cont. Shelf Res.* **2020**, *192*. [CrossRef]

180. Miguel, L.L.A.J.; Castro, J.W.A.; Machava, S.F.A. Dynamics of water exchange and salt flux in the Macuse Estuary, central Mozambique, southern Africa. *Afr. J. Mar. Sci.* **2019**, *41*, 203–219. [CrossRef]

181. Hall, T.M.; Plumb, R.A. Age as a diagnostic of stratospheric transport. *J. Geophys. Res.* **1994**, *99*, 1059–1070. [CrossRef]

182. Holzer, M.; Hall, T.M. Transit-time and Tracer-age Distributions in Geophysical Flows. *J. Atmos. Sci.* **2000**, *57*, 3539–3558. [CrossRef]

183. Cornaton, F.; Perrochet, P. Groundwater age, life expectancy and transit time distributions in advective-dispersive systems; 2. Reservoir theory for sub-drainage basins. *Adv. Water Resour.* **2006**, *29*, 1292–1305. [CrossRef]

184. Delhez, E.J.M.; Deleersnijder, E. The concept of age in marine modelling II. Concentration distribution function in the English Channel and the North Sea. *J. Mar. Syst.* **2002**, *31*, 279–297. [CrossRef]

185. Mercier, C.; Delhez, E.J.M. Diagnosis of the sediment transport in the Belgian Coastal Zone. *Estuar. Coast. Shelf Sci.* **2007**, *74*, 670–683. [CrossRef]

186. Gong, W.P.; Shen, J. A model diagnostic study of age of river-borne sediment transport in the tidal York River Estuary. *Environ. Fluid Mech.* **2010**, *10*, 177–196. [CrossRef]

187. Zhu, L.; Gong, W.; Zhang, H.; Huang, W.; Zhang, R. Numerical study of sediment transport time scales in an ebb-dominated waterway. *J. Hydrol.* **2020**, *591*, 125299. [CrossRef]

188. Radtke, H.; Neumann, T.; Voss, M.; Fennel, W. Modeling pathways of riverine nitrogen and phosphorus in the Baltic Sea. *J. Geophys. Res. Ocean.* **2012**, *117*. [CrossRef]

189. Delhez, E.J.M.; Lacroix, G.; Deleersnijder, E. The age as a diagnostic of the dynamics of marine ecosystem models. *Ocean Dyn.* **2004**, *54*, 221–231. [CrossRef]

190. Liu, Z.; Wang, H.Y.; Guo, X.Y.; Wang, Q.; Gao, H.W. The age of Yellow River water in the Bohai Sea. *J. Geophys. Res. Ocean.* **2012**, *117*. [CrossRef]

191. Andrejev, O.; Myrberg, K.; Lundberg, P.A. Age and renewal time of water masses in a semi-enclosed basin—Application to the Gulf of Finland. *Tellus A* **2004**, *56*, 548–558. [CrossRef]

192. Gourgue, O.; Deleersnijder, E.; White, L. Toward a generic method for studying water renewal, with application to the epilimnion of Lake Tanganyika. *Estuar. Coast. Shelf Sci.* **2007**, *74*, 628–640. [CrossRef]

193. Meier, H.E.M. Modeling the pathways and ages of inflowing salt- and freshwater in the Baltic Sea. *Estuar. Coast. Shelf Sci.* **2007**, *74*, 610–627. [CrossRef]

194. Kärnä, T.; Baptista, A.M. Water age in the Columbia River estuary. *Estuar. Coast. Shelf Sci.* **2016**, *183*, 249–259. [CrossRef]

195. Li, Y.Y.; Feng, H.; Zhang, H.W.; Su, J.; Yuan, D.K.; Guo, L.; Nie, J.; Du, J.L. Hydrodynamics and water circulation in the New York/New Jersey Harbor: A study from the perspective of water age. *J. Mar. Syst.* **2019**, *199*. [CrossRef]

196. Shang, J.C.; Sun, J.; Tao, L.; Li, Y.Y.; Nie, Z.H.; Liu, H.Y.; Chen, R.; Yuan, D.K. Combined Effect of Tides and Wind on Water Exchange in a Semi-Enclosed Shallow Sea. *Water* **2019**, *11*, 1762. [CrossRef]

197. Bendtsen, J.; Gustafsson, K.E.; Soderkvist, J.; Hansen, J.L.S. Ventilation of bottom water in the North Sea-Baltic Sea transition zone. *J. Mar. Syst.* **2009**, *75*, 138–149. [CrossRef]

198. Du, J.B.; Shen, J. Water residence time in Chesapeake Bay for 1980–2012. *J. Mar. Syst.* **2016**, *164*, 101–111. [CrossRef]

199. Qin, Q.B.; Shen, J. Physical transport processes affect the origins of harmful algal blooms in estuaries. *Harmful Algae* **2019**, *84*, 210–221. [CrossRef]

200. Delhez, E.J.M.; Deleersnijder, E. Residence time and exposure time of sinking phytoplankton in the euphotic layer. *J. Biol.* **2010**, *262*, 505–516. [CrossRef]

201. Boyer, J.N.; Fourqurean, J.W.; Jones, R.D. Spatial characterization of water quality in Florida Bay and Whitewater Bay by multivariate analyses: Zones of similar influence. *Estuaries* **1997**, *20*, 743–758. [CrossRef]

202. Paerl, H.W.; Huisman, J. Blooms like it hot. *Science* **2008**, *320*, 57–58. [CrossRef]

203. Rabalais, N.N.; Turner, R.E. Oxygen depletion in the Gulf of Mexico adjacent to the Mississippi River. In *Past and Present Water Column Anoxia*; Neretin, L.N., Ed.; Springer: Dordrecht, The Netherlands, 2006; pp. 225–245.

204. Durand, J. A conceptual model of the aquatic food web of the Upper San Francisco Estuary. *San Franc. Estuary Watershed Sci.* **2015**, *13*, 1–37. [CrossRef]

205. Hopkinson, C.S.; Vallino, J.J. The Relationships among Mans Activities in Watersheds and Estuaries—A Model of Runoff Effects on Patterns of Estuarine Community Metabolism. *Estuaries* **1995**, *18*, 598–621. [CrossRef]

206. Battin, T.J.; Kaplan, L.A.; Findlay, S.; Hopkinson, C.S.; Marti, E.; Packman, A.I.; Newbold, J.D.; Sabater, F. Biophysical controls on organic carbon fluxes in fluvial networks. *Nat. Geosci.* **2008**, *1*, 95–100. [CrossRef]

207. Liu, X.; Dunne, J.P.; Stock, C.A.; Harrison, M.J.; Adcroft, A.; Resplandy, L. Simulating Water Residence Time in the Coastal Ocean: A Global Perspective. *Geophys. Res. Lett.* **2019**, *46*, 13910–13919. [CrossRef]

208. England, M.H. The Age of Water and Ventilation Timescales in a Global Ocean Model. *J. Phys. Oceanogr.* **1995**, *25*, 2756–2777. [CrossRef]

209. Liu, X.; (IMSG at NOAA/NWS/NCEP/EMC, College Park, Maryland, USA). Personal communication, 2020.

210. Sharples, J.; Middelburg, J.J.; Fennel, K.; Jickells, T.D. What proportion of riverine nutrients reaches the open ocean? *Glob. Biogeochem. Cycles* **2017**, *31*, 39–58. [CrossRef]

211. Mouchet, A.; Deleersnijder, E.; Primeau, F. The leaky funnel model revisited. *Tellus A* **2012**, *64*. [CrossRef]

212. Deleersnijder, E.; Wang, J.; Mooers, C.N.K. A two-compartment model for understanding the simulated three-dimensional circulation in Prince William Sound, Alaska. *Cont. Shelf Res.* **1998**, *18*, 279–287. [CrossRef]

213. Abdelrhman, M.A. Simplified modeling of flushing and residence times in 42 embayments in New England, USA, with special attention to Greenwich Bay, Rhode Island. *Estuar. Coast. Shelf Sci.* **2005**, *62*, 339–351. [CrossRef]

214. Dettmann, E.H. Effect of water residence time on annual export and denitrification of nitrogen in estuaries: A model analysis. *Estuaries* **2001**, *24*, 481–490. [CrossRef]

215. Lucas, L.V.; Thompson, J.K. Changing restoration rules: Exotic bivalves interact with residence time and depth to control phytoplankton productivity. *Ecosphere* **2012**, *3*, 117. [CrossRef]

216. Jassby, A.D.; Cloern, J.E.; Cole, B.E. Annual primary production: Patterns and mechanisms of change in a nutrient-rich tidal ecosystem. *Limnol. Oceanogr.* **2002**, *47*, 698–712. [CrossRef]

217. Müller-Solger, A.B.; Jassby, A.D.; Müller-Navarra, D.C. Nutritional quality of food resources for zooplankton (Daphnia) in a tidal freshwater system (Sacramento-San Joaquin River Delta). *Limnol. Oceanogr.* **2002**, *47*, 1468–1476. [CrossRef]

218. Jassby, A.D. Phytoplankton in the Upper San Francisco Estuary: Recent biomass trends, their causes and their trophic significance. *San Franc. Estuary Watershed Sci.* **2008**, *6*, 1–24.

219. Kimmerer, W.J.; Orsi, J.J. Changes in the zooplankton of the San Francisco Bay Estuary since the introduction of the clam *Potamocorbula amurensis*. In *San Francisco Bay: The Ecosystem*; Hollibaugh, J.T., Ed.; Pacific Division of the American Association for the Advancement of Science: San Francisco, CA, USA, 1996; pp. 403–424.

220. Winder, M.; Jassby, A.D. Shifts in zooplankton community structure: Implications for food web processes in the Upper San Francisco Estuary. *Estuaries Coasts* **2011**, *34*, 675–690. [CrossRef]

221. Bennett, A.W.; Moyle, P.B. Where have all the fishes gone? Interactive factors producing fish declines in the Sacramento-San Joaquin Estuary. In *San Francisco Bay: The Ecosystem*; Hollibaugh, J.T., Ed.; Pacific Division of the American Association for the Advancement of Science: San Francisco, CA, USA, 1996; pp. 519–542.

222. MacNally, R.; Thomson, J.R.; Kimmerer, W.J.; Feyrer, F.; Newman, K.B.; Sih, A.; Bennett, W.A.; Brown, L.; Fleishman, E.; Culberson, S.D.; et al. Analysis of pelagic species decline in the upper San Francisco Estuary using multivariate autoregressive modeling (MAR). *Ecol. Appl.* **2010**, *20*, 1417–1430. [CrossRef] [PubMed]

223. ICF International. *Chapter 3.1 and 3.2—Conservation Strategy. Administrative Draft. Bay Delta Conservation Plan*; ICF 00610.10; ICF International: Sacramento, CA, USA, 2012.

224. ICF International. *Chapter 5—Effects Analysis. Administrative Draft. Bay Delta Conservation Plan*; ICF 00282.11; ICF International: Sacramento, CA, USA, 2012.

225. ICF International. *Appendix F—Ecological Effects. Working Draft. Bay Delta Conservation Plan*; ICF 00282.11; ICF International: Sacramento, CA, USA, 2011.

226. Nixon, S.W.; Ammerman, J.W.; Atkinson, L.P.; Berounsky, V.M.; Billen, G.; Boicourt, W.C.; Boynton, W.R.; Church, T.M.; Di Toro, D.M.; Elmgren, R.; et al. The fate of nitrogen and phosphorus at the land-sea margin of the North Atlantic Ocean. *Biogeochemistry* **1996**, *35*, 141–180. [CrossRef]

227. Mayorga, E.; Seitzinger, S.P.; Harrison, J.A.; Dumont, E.; Beusen, A.H.W.; Bouwman, A.F.; Fekete, B.M.; Kroeze, C.; Van Drecht, G. Global Nutrient Export from WaterSheds 2 (NEWS 2): Model development and implementation. *Environ. Model. Softw.* **2010**, *25*, 837–853. [CrossRef]

228. Isaac, B.J.; Parente, A.; Galletti, C.; Thornock, J.N.; Smith, P.J.; Tognotti, L. A Novel Methodology for Chemical Time Scale Evaluation with Detailed Chemical Reaction Kinetics. *Energy Fuel* **2013**, *27*, 2255–2265. [CrossRef]

229. Oldham, C.E.; Farrow, D.E.; Peiffer, S. A generalized Damkohler number for classifying material processing in hydrological systems. *Hydrol. Earth Syst. Sci.* **2013**, *17*, 1133–1148. [CrossRef]

230. Monsen, N.E.; Cloern, J.E.; Burau, J.R. Effects of flow diversions on water and habitat quality: Examples from California's highly manipulated Sacramento-San Joaquin Delta. *San Franc. Estuary Watershed Sci.* **2007**, *5*, 1–16. [CrossRef]

231. Volkmar, E.C.; Dahlgren, R.A. Biological oxygen demand dynamics in the Lower San Joaquin River, California. *Environ. Sci. Technol.* **2006**, *40*, 5653–5660. [CrossRef]

232. Strayer, D.L.; Caraco, N.F.; Cole, J.J.; Findlay, S.; Pace, M.L. Transformation of freshwater ecosystem by bivalves. *Bioscience* **1999**, *49*, 19–27. [CrossRef]

233. Dame, R.F. *Ecology of Marine Bivalves: An Ecosystem Approach*; CRC Press: Boca Raton, FL, USA, 1996.

234. Smaal, A.C.; Prins, T.C. The uptake of organic matter and the release of inorganic nutrients by bivalve suspension feeder beds. In *Bivalve Filter Feeders in Estuarine and Coastal Ecosystem Processes*; Dame, R.F., Ed.; Springer: Berlin, Germany, 1993; Volume 33, pp. 271–298.

235. Soetaert, K.; Herman, P.M.J. Estimating Estuarine Residence Times in the Westerschelde (the Netherlands) Using a Box Model with Fixed Dispersion Coefficients. *Hydrobiologia* **1995**, *311*, 215–224. [CrossRef]

236. Braunschweig, F.; Martins, F.; Chambel, P.; Neves, R. A methodology to estimate renewal time scales in estuaries: The Tagus Estuary case. *Ocean Dyn.* **2003**, *53*, 137–145. [CrossRef]

237. Brodie, J.; Wolanski, E.; Lewis, S.; Bainbridge, Z. An assessment of residence times of land-sourced contaminants in the Great Barrier Reef lagoon and the implications for management and reef recovery. *Mar. Pollut. Bull.* **2012**, *65*, 267–279. [CrossRef]

238. Hall, N.S.; Paerl, H.W.; Peierls, B.L.; Whipple, A.C.; Rossignol, K.L. Effects of climatic variability on phytoplankton community structure and bloom development in the eutrophic, microtidal, New River Estuary, North Carolina, USA. *Estuar. Coast. Shelf Sci.* **2013**, *117*, 70–82. [CrossRef]

239. Alber, M.; Sheldon, J.E. Use of a date-specific method to examine variability in the flushing times of Georgia estuaries. *Estuar. Coast. Shelf Sci.* **1999**, *49*, 469–482. [CrossRef]

240. Wheat, E.E.; Banas, N.S.; Ruesink, J.L. Multi-day water residence time as a mechanism for physical and biological gradients across intertidal flats. *Estuar. Coast. Shelf Sci.* **2019**, *227*. [CrossRef]

241. Wan, Y.S.; Qiu, C.; Doering, P.; Ashton, M.; Sun, D.T.; Coley, T. Modeling residence time with a three-dimensional hydrodynamic model: Linkage with chlorophyll a in a subtropical estuary. *Ecol. Model.* **2013**, *268*, 93–102. [CrossRef]

242. Delesalle, B.; Sournia, A. Residence Time of Water and Phytoplankton Biomass in Coral-Reef Lagoons. *Cont. Shelf Res.* **1992**, *12*, 939–949. [CrossRef]

243. Du, J.; Park, K.; Shen, J.; Dzwonkowski, B.; Yu, X.; Yoon, B.I. Role of Baroclinic Processes on Flushing Characteristics in a Highly Stratified Estuarine System, Mobile Bay, Alabama. *J. Geophys. Res. Oceans* **2018**, *123*, 4518–4537. [CrossRef]

244. Webb, B.M.; Marr, C. Spatial variability of hydrodynamic timescales in a broad and shallow estuary: Mobile Bay, Alabama. *J. Coast. Res.* **2016**, *32*, 1374–1389.

245. Okubo, A. *Diffusion and Ecological Problems: Mathematical Models*; Springer: Berlin/Heidelberg, Germany; New York, NY, USA, 1980.

246. Geyer, W.R.; Chant, R.; Houghton, R. Tidal and spring-neap variations in horizontal dispersion in a partially mixed estuary. *J. Geophys. Res. Ocean.* **2008**, *113*. [CrossRef]

247. Hong, B.; Wang, G.; Xu, H.; Wang, D. Study on the transport of terrestrial dissolved substances in the Pearl River Estuary using passive tracers. *Water* **2020**, *12*, 1235. [CrossRef]

248. Pham Van, C.; de Brye, B.; de Brauwere, A.; Hoitink, A.J.F.T.; Soares-Frazao, S.; Deleersnijder, E. Numerical Simulation of Water Renewal Timescales in the Mahakam Delta, Indonesia. *Water* **2020**, *12*, 1017. [CrossRef]

249. Deleersnijder, E. *Toward a Generic Method to Estimate the Water Renewal Rate of a Semi-Enclosed Domain*; Université Catholique de Louvain: Louvain-la-Neuve, Belgium, 2011; pp. 1–46. Available online: http://hdl.handle.net/2078.1/155414 (accessed on 21 September 2020).

Article

Characterisation of the Water Renewal in a Macro-Tidal Marina Using Several Transport Timescales

Jean-Rémy Huguet *, Isabelle Brenon and Thibault Coulombier

LIENSs, La Rochelle University, 2 rue Olympe de Gouges, 17000 la Rochelle, France; isabelle.brenon@univ-lr.fr (I.B.); thibault.coulombier@univ-lr.fr (T.C.)
* Correspondence: jean-remy.huguet@univ-lr.fr; Tel.: +33-0516-496534

Received: 11 June 2019; Accepted: 26 September 2019; Published: 30 September 2019

Abstract: In this paper, we investigate the water renewal of a highly populated marina, located in the south-west of France, and subjected to a macro-tidal regime. With the use of a 3D-numerical model (TELEMAC-3D), three water transport timescales were studied and compared to provide a fully detailed description of the physical processes occurring in the marina. Integrated Flushing times (IFT) were computed through a Eulerian way while a Lagrangian method allowed to estimate Residence Times (RT) and Exposure Times (ET). From these timescales, the return-flow (the fraction of water that re-enters the marina at flood after leaving the domain at ebb) was quantified via the Return-flow Factor (RFF) and the Return Coefficient (RC) parameters. The intrinsic information contained in these parameters is thoroughly analysed, and their relevance is discussed. A wide range of weather-marine conditions was tested to provide the most exhaustive information about the processes occurring in the marina. The results highlight the significant influence of the tide and the wind as well as the smaller influence of the Floating Structures (FS) on the renewal. Besides, this study provides the first investigation of the water exchange processes of La Rochelle marina. It offers some content that interest researchers and environmental managers in the monitoring of pollutants as well as biological/ecological applications.

Keywords: marina; water renewal; transport timescales; return-flow; macro-tidal; wind influence; floating structures

---

1. Introduction

Over the years, the increasing development of coastal areas has modified the quality of water and sediments as well as marine habitats. Ports, which are the main interfaces between cities and the sea, are primarily subjected to a multi-source of contamination due to intense anthropogenic activities. Their complex geometry and infrastructure (e.g., quays, channels, and docks) induce low circulation and stagnant waters which tend to enhance and to control the fate of contaminants. Given the tendency of pollutants to remain confined and settle on the bottom, the pollution generated within the ports is of grave concern. Then, improving water and sediment quality is of vital importance for the sustainable development of coastal waters.

In recent decades, managers have been under increasing pressure to demonstrate the environmental skills of the port they manage. Some studies focused on the effect of diffuse pollution originating from urban water runoffs and boat repair activities [1,2] as well as accidental oil spills [3,4]. Metal concentrations in water and sediments were also monitored and investigated [5,6] but characterising the water quality of such areas is still a challenge as it requires many parameters.

Although ports are considered as low-energy systems, the hydrological pattern established within the port basin cannot be neglected in any study [7]. The rate of renewal of a basin is useful information that provides a first-order description of its dynamics. Numerous transport timescales have been defined through the literature to quantify this renewal [8–12]. With the growing use of numerical modelling, these water transport timescales are useful parameters to condense the considerable amount of data in intelligible and quantitative information [13]. The most commonly used, "flushing time", "residence time", "age", and "exposure time" have been applied in a wide range of studies all over the world [14–17]. From the environmental port management point of view, these transport timescales offer an interesting indication of the spatial and temporal variability of the dynamic and of the susceptibility to pollution. However, such timescale descriptors need to be carefully employed because there is no real consensus on their application [11].

After an expansion in 2014 (corresponding to the NE basin in Figure 1), La Rochelle Marina, located in the southwestern part of France, is currently considered as the biggest marina on the European Atlantic coast. Despite the environmental policy and ecological awareness of the marina, equipment and maritime activities may be a source of the pollution [18]. The main objective of this study is, therefore, to characterise the water renewal of La Rochelle Marina due to the horizontal and vertical variability of its currents. This contribution can be considered as a first scientific investigation because, even if the importance of such timescales is evident, no references or estimates were available in the literature for a similar marina.

To describe water renewal mechanisms accurately, we performed a large number of simulations with a 3D hydrodynamic model calibrated and validated in a previous study [19]. The specificity of the model is to take into consideration the considerable number of structures floating in the marina (e.g., docks, boats). In this study, three timescales are compared (flushing time, residence time and exposure time) to find the most relevant parameter to describe water renewal of the domain. Besides, we estimated the return-flow in different ways, the fraction of water that leaves the marina at ebb tide before re-entering it at the next flood tide [20], and its effect on the tidal flushing of the marina was analysed. The above-mentioned timescales and quantities were computed for different scenarios to characterise the influence of wind, tide and floating structures on the water renewal. In the next section, study site and numerical computations will be presented before introducing definitions and concepts of chosen timescales and quantities. In Section 3, a Lagrangian validation is carried out while timescale results are shown in Section 4 and discussed in Section 5.

2. Materials and Methods

2.1. Study Site

2.1.1. La Rochelle Marina

The city of La Rochelle, the provincial administrative centre of the department, has a land area of 28.43 km^2 and a population of 80,000 inhabitants. Its marina, created in 1972, has been the biggest marina (50 ha) along the Atlantic coast, since its expansion in 2014. This 900 m-long and 820 m-wide semi-enclosed area is divided in three basins totalling 4500 moorings, distributed along 15 km of floating docks. The southeastern (SE) basin is the more prominent, with 22 ha, while the western (W) and the northeastern (NE) basin, present respectively 17 and 15 ha. At sea, the marina is accessible by a 110 m wide main entrance, while the NE basin offers two openings: 150 m wide to the northeast and a 64 m wide to the southwest which connects the NE basin to the W basin (Figure 1). The marina is not spared by siltation and has to spend 10 per cent of its total budget to dredge around 200 000 m^3 of cohesive sediment each year. The annual sediment deposition can overpass 50 cm in some of its basins (Pers. Comm. La Rochelle Marina), which requires recurring dredging of the basins, 8 months a year.

The environmental policy of the marina led to an ISO 14001 certification, an international reference in sustainable development. Despite their effort, water quality and marine biodiversity are still impacted by intensive anthropogenic inputs. Several potential sources of contamination have been identified in the marina (symbolized in Figure 1): the rainwater outlet, where runoff waters from streets and roadways can flow abundantly during storm events; the fairing area where boats are maintained (antifouled, painted and sanded); the fuel station where dripping fuel can be discharged. We can also consider diffuse pollution from boat activities.

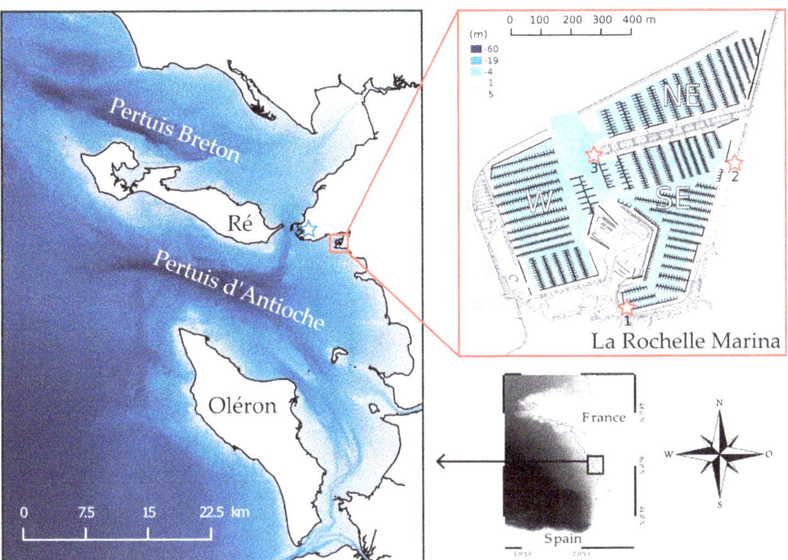

Figure 1. Bathymetry/topography map of the modelling domain (left) and the La Rochelle marina (right). Depths are given with respect to mean-sea-level, and the straight, bold black line indicates the shoreline in the left figure while La Pallice weather station is symbolised by a blue bordered white star. At right, the floating docks are signified by a black line, and the maritime infrastructures are indicated by a grey line. The red-bordered white stars numbered 1, 2, and 3, represent the location of the rainwater outlet, the fairing area, and the fuel station, respectively. Western, southeastern and northeastern basins are denoted by white characters W, SE and NE, respectively.

2.1.2. Geomorphology and Hydrodynamics of the Coastal Area

La Rochelle Marina is located in the northern landward part of the Pertuis d'Antioche embayment along the French Atlantic Coast, in the central part of the Bay of Biscay. This shallow water coastal area is protected from the Atlantic Ocean by Ré and Oléron islands and connected to the Pertuis Breton embayment through a narrow inlet. The bathymetry is characterised by silty to sandy-silty bottoms, with a 44 m deep trench and many tidal flats. The coastal area is considered as a mixed, wave and tide-dominated estuary [21]. The tidal regime is semidiurnal and tidal range varies from 2 m during neap tides to more than 6 m during spring tides. This macro-tidal environment is dominated by M2, and its amplitude grows to more than 1.8 m in the inner part of the estuaries due to resonance and shoaling [22]. Because of resonance occurring on the Bay of Biscay shelf, the quarter-diurnal tidal constituents (M4, MS4 and MN4) are strongly amplified shoreward [23]. In the embayments, average freshwater inflows are about 2 orders of magnitude less than tidal flows [24].

2.1.3. Meteorological Context

The study area is subjected to seasonal climate variations. Summer presents a weak low-pressure system activity resulting in weak northeasterly winds while northwest thermic breezes mainly dominate littoral. Low-pressure systems that cross the Atlantic Ocean from autumn are the most active during winter, generating extreme west and northwest winds. North Atlantic Oscillation (NAO) partly controls the inter-annual variability of the wind regime in the whole Bay of Biscay [25]. Weather-marine conditions were collected by La Pallice weather station (blue-bordered white star in Figure 1) over the temporal interval 2015–2018 (Figure 2). Data analysis reveals a predominance of four winds over the area of study: north-western (22% mean yearly occurrence), western (21% mean yearly occurrence), north-eastern (19% mean yearly occurrence) and southern (14% mean yearly occurrence).

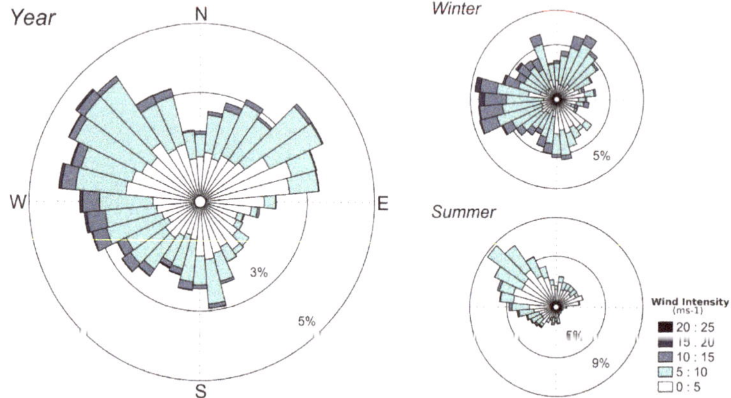

Figure 2. The wind regime in La Pallice station over the period 2015–2018. Winter and summer periods are also distinguished to characterise their variability in terms of direction and intensity. The legend box at right indicates the intensity of wind velocity in ms^{-1}.

2.2. Numerical Model Implementation

To calculate water transport timescales, we used the TELEMAC-3D model [26], a three-dimensional hydrodynamic model adapted to free-surface flow. In this study, three-dimensional Navier-Stokes equations were solved in non-hydrostatic mode. Bottom stress was computed through a Chézy parametrisation over an unstructured grid. The semi-implicit Galerkin finite element method is used to solve continuity and momentum equations and a Lagrangian-Eulerian treatment of advective terms and a semi-implicit way ensure numerical stability. More insight and details about the equations are presented in [19]. TELEMAC-3D provides the possibility of taking into account passive tracers in the model domain. The model solves the advection and diffusion equation of a conservative tracer.

The modelled area is 35 km wide and 100 km long and is discretised on a 41,000 nodes unstructured grid, with resolution from 2 km offshore to nearly 5 m in the marina [19]. The size of the computational domain (4757.9 km^2) is considered sufficiently large so that the prescribed boundary conditions negligibly affect the tracer mass in the marina (0.5 km^2) as requested by [27]. The model has eight vertical sigma levels, which are treated with the Arbitrary Lagrangian-Eulerian method [28] and lead to a total of 320,000 nodes. Our bathymetry originates from French Navy (hereafter SHOM) [29] and single beam surveys acquired in the marina. Then, the topography of intertidal areas is determined using a LiDAR survey, acquired in 2010 (LITTO3D, French National Geographic Institute and SHOM). Along its open boundary, the model is forced by 34 astronomical tidal constituents obtained by linear interpolation from the global tide model FES2014 [30]. Atmospheric forcing is set over the whole domain with hourly sea-level atmospheric pressure and 10 m wind speed and direction originating from the Climate Forecast System Reanalysis (CFSR) provided by the National Center for Environmental Prediction (NCEP). All the simulations used a time step set to 5 s after sensitivity analysis. Wave effects are not simulated because the marina is considered sufficiently sheltered from ocean waves. Water fluxes across the sea surface (precipitation-evaporation) and rivers input are also neglected. Floating docks and moorings that occupy more than a third of the marina surface were implemented in the model by adding head losses at the surface. Their implementation and their effect on circulation are presented in [19] as well as validation results in terms of currents and water levels.

2.3. Water Transport Timescales

Flow exchanges between inner water and the open sea control the water quality of the domain. Numerous transport timescales have been tested to assess water renewal in semi-enclosed areas. This study focuses on three transport timescales: flushing, residence, and exposure time. They provide different space and time-dependent quantitative measures of the water renewal. They can help to determine water mass dynamics in aquatic systems and its influence on both geochemical and biological processes. Their definitions and numerical computation are explained in this section.

2.3.1. Flushing Time

Flushing time characterises the general exchange characteristics of the waterbody and has been described as the ratio of the mass of a scalar in a semi-enclosed domain to the rate of renewal of the scalar [11]. It reflects the mean time spent by any pollutant discharged in the marina and measures the effectiveness of flushing to remove any pollutant from the water body of the marina.

Here, FT calculation is based on the study of the evolution of a conservative tracer initially introduced at several locations in La Rochelle Marina. Flushing time was primarily defined as the time required to reduce the mass of the conservative tracer to 37% (e-1) of its initial value in the water body [31–33]. This parameter was sometimes referred to as "e-folding flushing time" [11,34]. In this study, we used 37% as a percentage reference, but it should be noted that it varies according to the authors (e.g., 2% in [35] and 50% in [36]). Computation of flushing time is based on the assumption that the marina behaves as a continuously stirred tank reactor (CSTR). The main assumption for a CSTR is that any introduction of mass is wholly mixed throughout the domain, so the concentration of a constituent exiting the system is equal to the concentration everywhere inside the CSTR [11,37]. Assuming that a known quantity of conservative tracer has been introduced only at $t = 0$ and that the water entering the marina can mix totally with the existent water of the marina, the total mass of the tracer is:

$$M(t) = M_0 e^{-t/FT}, \tag{1}$$

$M(t)$ is the mass of conservative tracer that remains in the marina, M_0 the initial mass of conservative tracer introduced in the marina and t the time since the tracer has been discharged from a specific location. To spatially determine the flushing time of the marina, we computed the Integrated Flushing Time (hereafter *IFT*) at several locations of the marina. This timescale, defined by [33] is the time for the mean mass of the tracer over the whole marina to fall below 37% of the initial mass discharged at a specific location in the marina. *IM* and IM_0 correspond to the spatial integration of the tracer mass over the entire domain, under stable conditions. By releasing tracer at a hundred source points evenly distributed in the marina domain, we obtained spatially-varying *IFT*. Building on Equation (1), for the same previous assumptions, *IFT* was computed through the following equation:

$$IM(t) = IM_0 e^{-t/IFT}, \tag{2}$$

where $IM(t)$ is the mass of conservative tracer over the whole marina for a release at a specific location, IM_0 the initial mass of conservative tracer released at a specific location. The use of a 3D-model has permitted to compute *IFT* at each layer of the water column.

2.3.2. Residence and Exposure time

Residence time (hereafter RT) is a fundamental concept defined as the time until a water parcel, or particle at a specified location within the water body, leaves the system [8,38,39]. Similarly to the Lagrangian Water Transport Time (or Water Transit Time) defined by [40], RT is a property of the water parcel that is carried in and out of the marina by the hydrodynamic processes. A vast number of particles is necessary to capture the diffusive processes generated by the small scale turbulence [41,42]. For the computation of RT, 400 particles were released at arbitrary locations, and each particle was representative of a 1000 m^2 area. Then, the RT was computed for each particle, from their release to the first time they reached the boundaries (symbolised by the two main entrances of the marina). No random walk diffusion was implemented in the particle tracking method which will give us an interesting comparison axis with the IFT.

Previous residence time studies underestimated the total time that particles spend in the domain because of the possibility for the particles to re-enter the domain [11,43]. This conceptual drawback of residence time timescale led to unrealistic timescales in tidal systems where water parcels can leave and re-enter the domain many times, especially close to the boundaries. Reference [11] then introduced the concept of exposure time to take into account returning water parcels or particles. By computing the total time that a particle spends in the domain, exposure time offers an interesting alternative to residence time and is particularly suited for macro-tidal seas. The two timescales are very similar in terms of region of interest and numerical computation, but exposure time requires additional assumptions and post-processing treatment. Indeed, the latter dwells on the hydrodynamic processes that occur not only within but also outside the marina. The computational domain must be much larger than the domain of interest, and the open boundaries need to be located far enough from the area of interest [11–16]. For this study, we supposed that both open boundaries and model domain extension have a negligible influence on the computed exposure times. Then, we defined Exposure Time (hereafter ET) to get a spatial distribution of exposure time for each particle. While the RT was computed by detecting the first departure of the particles from the marina, we computed ET by considering the total time spent by the particles inside the marina before their last departure.

2.3.3. Estimation of the Return-Flow

Comparing Residence Time (RT) and Exposure Time (ET) provides information about the contribution of returning water at each tidal cycle. For the same numerical computation, RT equal to ET indicates that the fraction of water flowing out of the domain during ebb tide is totally lost into the open sea, and does not return into the domain on the next flood tide. Conversely, ET much higher than RT indicates that a significant fraction of water that flowed out the area during ebb tide has returned into the area on the next flood tide. This fraction of effluent water that returns to the domain is called "Return-flow Factor (RFF)". It represents the fate of the water once it is outside the domain, and was introduced by [20] to figure the movements of water parcels in semi-enclosed areas where the tide is significant. The RFF depends on the phase and strength difference between the flow along the coast and the flow in the connecting channel, and the mixing that occurs between open seas and water masses flowing out of the domain [20].

The timescales mentioned above can help us to compute RFF with different approaches. RFF was firstly introduced via the tidal prism method, which is a classical approach to easily estimate flushing time in tidal systems. This method is appropriate to small and well-mixed embayments with low river inputs compared to tidal flows, and sufficiently large enough receiving water to dilute water exiting the system [11,14,20]. The flushing time T_f is then given by:

$$T_f = \frac{TV}{(1-b)P},$$ (3)

where T is the average tidal period, V the basin volume, P the intertidal volume and b, that lies in the interval [0, 1], represents the return-flow factor via the expression:

$$b = 1 - \frac{TV}{PT_f},$$ (4)

If $b = 0$, no water, previously ejected, returns into the embayment and the situation $b = 1$ corresponds to a situation where all water returns to the embayment. Knowing the average flushing time of the marina (FT_{av}) thanks to the tracer simulations described in Section 2.3.1, the average return-flow factor b can be obtained with:

$$RFF = 1 - \frac{TV}{FT_{av}P},$$ (5)

The return-flow can also be described through the concept of return coefficient developed by [44]. It represents the relative difference between exposure time and residence time and is equal to:

$$r = \frac{E - R}{E},$$ (6)

With E the exposure time and R the residence time. The coefficient r also lies in the interval [0, 1], and the first limit ($r = 0$) corresponds to a situation where the exposure time is equal to residence time, while $r = 1$ is reached when the exposure time is much higher than the residence time of the domain. The Lagrangian Particle Tracking methodology developed in Section 2.3.2 permits to compute average residence (RT_{av}) and exposure times (ET_{av}) of the marina which transforms Equation (6) in:

$$RC = \frac{ET_{av} - RT_{av}}{ET_{av}},$$ (7)

These two expressions of the return-flow factor were computed for each scenario described in the next section.

2.4. General Simulation Set Up

Several simulations were carried out to calculate water transport timescales over La Rochelle marina, under meteorological and tidal forcing. Every scenario tested and analysed during this study are resumed in Table 1. The theoretical steady winds applied on the model domain correspond to the prevailing winds of the area: west, north-east, and south winds. Model results with north-west winds are not presented here because they are similar to model results with westerly winds. Five atmospheric conditions were tested: one without wind, three with an averaged 7.5 ms^{-1} wind from several directions and one 15 ms^{-1} west wind, typical of winter events (Table 1).

Table 1. Summary of the scenarios tested and analysed in the study.

Conditions		Scenarios											
		a	c	e	g	i	k	b	d	f	h	j	l
Tides	Neap	●	●	●	●	●	●						
	Spring							●	●	●	●	●	●
Wind	West 7.5 ms^{-1}					●				●			
	North-East 7.5 ms^{-1}				●						●		
	South 7.5 ms^{-1}						●					●	
	West 15 ms^{-1}							●					●
With floating structures (FS)		●	●	●	●	●		●	●	●	●	●

Every simulation was computed for both spring tides (tidal range about ±6 m) and neap tide conditions (tidal range about ±2 m). The tidal conditions correspond to periods from 17 March 2017 to 3 April 2017. Furthermore, to analyse the influence of floating structures (hereafter FS) on the marina hydrodynamics, simulations without FS were investigated with the tide only. Considering that the tidal phase at the moment of release is considered as one of the main factors influencing water transport timescales [40], four releases were implemented for each simulation: at low tide, rising tide, high tide and ebb tide. As TELEMAC-3D offers the possibility to release passive Eulerian tracer and Lagrangian particles in the same simulation, it led to a total of 48 simulations. The duration of each simulation was

10 days, with 1 spin-up day. The hypothesised scenarios tested do not reproduce the real situation, but they provide an overview of the typical hydrodynamic processes that affect the water renewal in La Rochelle Marina.

3. Lagrangian Validation

In a previous study [19], the validation of the model was done in term of water levels and currents intensity, at numerous locations, inside and outside the marina. Here, the simulated velocities and trajectories of particles are compared with the observational data collected by the drifting buoys. The analysis offered in this section gives a spatially continuous distribution of model skill in the upper layer of La Rochelle harbour and marina entrance. The Lagrangian time series of surface velocity allow evaluating the performance of the model to reproduce dispersion trends of surface waters, which differs from the Eulerian assessment commonly used in validation methodology.

In the framework of the study, several drifting buoys were released for a wide range of temporal and spatial scales. The drifters, manufactured by Pacific Gyres (Oceanside, CA, USA), are composed of a surface float with a diameter of 0.3 m and a drogue dimension of 1.2 m length. The buoy uses Iridium to send float positions with 5-min temporal resolution. First experiments involved their deployment at the marina entrance with less than 1-h transport while the following occurred more offshore for timescales reaching more than one day. The range of hydrodynamics and atmospheric conditions tested, and the settings of the experiments were resumed in Table 2 while the mean trajectories of each drifting buoy experiment were resumed in Figure 3a. Because of maritime infrastructure and boat navigation, more release of drifting buoys was needed to track sufficient currents patterns inside the marina.

Table 2. The drifting buoys deployments and their corresponding settings.

Deployment Date	Number of Drifters per Release	Number of Releases	Location	Duration	Weather-Marine Conditions
9 May 2017	3	5	46°08′54.1″ N 1°10′08.0″ W	Over one flood tide	Spring tides with strong south winds
17 May 2017	3	5	46°08′54.1″ N 1°10′08.0″ W	Over one flood tide	Neap tides with calm weather
14 June 2017	5	1	46°08′16.2″ N 1°10′46.4″ W	One ebb tide	Neap tides with calm weather
26 June 2017	5	1	46°08′15.8″ N 1°10′46.6″ W	One ebb tide	Spring tides with calm weather
15 January 2019	3	1	46°08′39.2″ N 1°10′52.7″ W	24 h	Neap tides with mixed winds

To facilitate the comparison, the real-time positions of the drifting buoys were averaged for each release while the positions of 10 simulated drifters were averaged for each corresponding release. Comparison is visible in Figure 3 and shows fair agreement between the observed and simulated circulation of the drifters. While the distance between simulated and observed drifters can reach 600 m after one day of release, the comparison indicates a good correlation in term of velocities and a 6 cms^{-1} Root-Mean-Squared-Discrepancy (hereafter RMSD). The meridional (V) and zonal (U) components of velocity also display a fair agreement but with less accuracy and consistency (Table 3). Globally, drifter buoy trajectories are better reproduced in the bay than in the marina, in particular during calm weather conditions where RMSD reaches 0.01 ms^{-1}. The tide in the bay rapidly and homogeneously advected drifter buoys while complex currents and micro-scale structures at the western marina entrance caused the buoys to diverge from each other. The model is less effective at reproducing the currents at the marina entrance, but the RMSD and R^2 results show a good reproducibility in particular concerning the meridional component of the velocity (Table 3). First and last deployments are characterised by stronger winds and the presence of waves in the bay. As the waves and their effects (Stokes drifts, interaction wave-currents) are not implemented in this modelling study, the quality of predictions of the model is slightly decreased (Table 3).

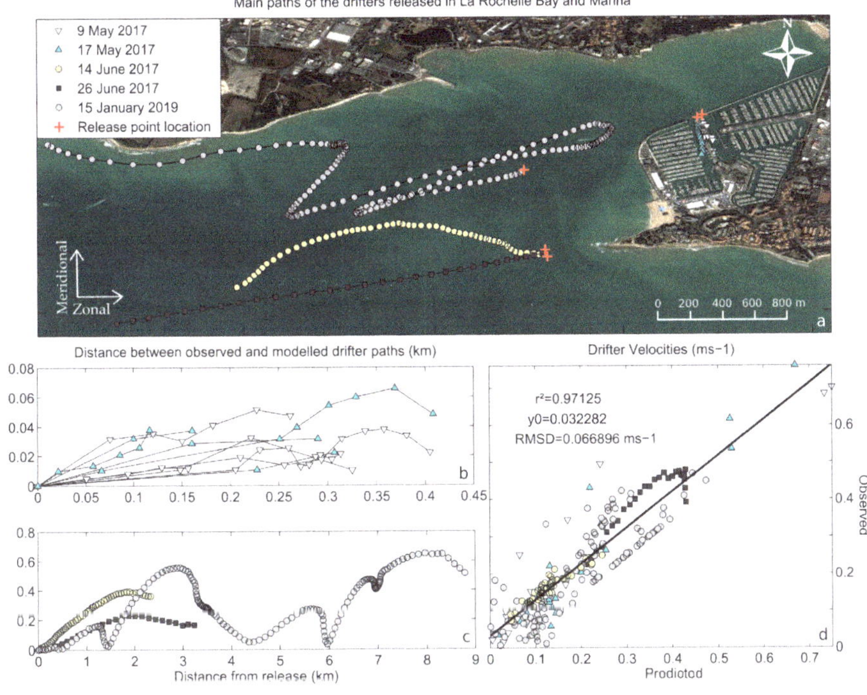

Figure 3. (**a**) The mean paths of drifters released in La Rochelle Bay and marina entrance; (**b**) and (**c**) The distance between observed and modelled drifter paths is displayed for every deployment; (**d**) Correlation of predicted-observed velocities (in ms^{-1}).

Table 3. Statistical analysis between the observed and predicted circulation of drifters. RMSD and linear regression analysis R^2 were computed for zonal (U) and meridional (V) components of the velocity.

Deployment Date	RMSD (ms^{-1})		R^2	
	U	V	U	V
9 May 2017	0.19	0.15	0.69	0.75
17 May 2017	0.17	0.14	0.71	0.82
14 June 2017	0.01	0.04	0.91	0.80
26 June 2017	0.04	0.06	0.88	0.69
15 January 2019	0.11	0.17	0.76	0.72

4. Results

4.1. Integrated Flushing Time

Figure 4 shows the spatial distribution of vertically-averaged Integrated Flushing Time (IFT) within La Rochelle marina for two configurations (with and without Floating Structures (FS)) associated with several combinations of wind and tide. While the first scenario considers only the tide without FS, scenarios with the wind were computed with the FS. To quantify differences between each case, we also averaged IFT (spatially and across the tidal phases) and estimated the standard deviation of its spatial variability (Spatial Standard Deviation, hereafter SSD) over the marina. We computed the standard deviation of IFT between each release (corresponding to different phases of the tide) and named it Tidal Phase Standard Deviation (hereafter TPSD).

Statistical results are visible in Table 4 while Figure 4 displays the spatial distribution of IFT. Spatially, the NE basin generally presents the lower values while the SE basin generally displays the higher values and the higher variability of IFT. IFT generally increases from the entrances to the most sheltered areas and in particular the southern part of the SE basin. The neap tide conditions exhibit 2 to 4 times larger IFT and SSD than spring tides (Figure 4, Table 4). The presence of FS increases IFT in the southern part of the SE basin during spring tides (Figure 4b–d), while it decreases IFT in the NE basin during neap tides (Figure 4a–c). While their presence slightly increases the mean IFT (23.6/22.5 h at spring tides and 89.3/88.9 h at neap tides), it considerably increases SSD during spring tides (11.1/6.2 h) and neap tides (30.2/22.7 h) (Table 4).

The wind significantly reduces IFT and SSD (on average 2 to 3 times lower) compared with situations with tide only (Figure 4c,d), in particular for the west and south directions. Although all wind cases decrease IFT in all basins relative to the tide-only case, the IFT patterns and directionality of IFT gradients vary with wind direction. For example, on neap tides the highest IFT tends to be in the southern part of the SE basin for north east wind (Figure 1g)) in the northern part of the SE basin for south wind (Figure 4i); and in the southern parts of the W and SE basins for west wind (Figure 4e). West wind is the most impacting on neap tides, and its effect increases with its magnitude. TPSD is significantly lower than SSD and is decreased by wind action and also by the presence of floating structures (Table 4). Here, we only present vertically-averaged results because the surface and bottom layers were shown to be comparable in terms of IFT (data not shown), at the scale of the marina. The relative homogeneity of the water column was found to be slightly affected by the wind.

Table 4. Mean, Spatial Standard Deviation (SSD) and Tidal Phase Standard Deviation (TPSD) calculated for IFT, RT, ET, RFF and RC for the weather-marine scenarios (letters) defined in Table 1. Statistics results are in hours for IFT, RT and ET and without units for RFF and RC.

Parameters		Weather–Marine Scenarios											
		Neap Tides						Spring Tides					
		a	c	e	g	i	k	b	d	f	h	j	l
IFT	Mean	88.9	89.3	27.3	43.6	35.4	25.6	22.5	23.6	13.5	19.4	12.5	10.1
	SSD	22.7	30.2	12.1	19.1	13.5	13	6.2	11.1	3.7	4.9	3.1	5.6
	TPSD	6.8	4.3	3.6	1.6	3.9	3.3	4.2	3.7	2.8	2.6	2	2.9
RT	Mean	31.1	37	24.9	33.9	26.4	23.9	13.3	15.4	9.3	10.3	8.1	9.6
	SSD	28.3	42.9	26.8	31.2	26.7	27.9	12.7	20.6	7.5	10.6	5.1	15.4
	TPSD	4.6	7.1	3.1	2.5	4.1	3.8	4.3	5.3	3.6	1.6	2.5	2.8
ET	Mean	124.1	129.3	35.5	69.9	41.3	27.5	34.5	34.8	19.3	19.5	14.2	11.8
	SSD	24.1	25.7	27.2	36.2	27.7	27.6	17.4	21.2	14.9	13.1	9.1	16.7
	TPSD	5	6.4	3.4	1.3	6.4	6	6.1	8	3.9	0.3	2.3	5.1
RFF	Mean	0.72	0.73	0.12	0.45	0.32	0.06	0.64	0.66	0.41	0.59	0.36	0.19
	SSD	0.02	0.01	0.11	0.02	0.08	0.10	0.07	0.05	0.11	0.10	0.10	0.11
	TPSD	0.05	0.03	0.03	0.01	0.01	0.01	0.04	0.03	0.01	0.01	0.02	0.02
RC	Mean	0.75	0.73	0.19	0.51	0.36	0.09	0.61	0.56	0.47	0.50	0.43	0.08
	SSD	0.03	0.04	0.05	0.04	0.05	0.11	0.09	0.06	0.13	0.06	0.08	0.1
	TPSD	0.06	0.07	0.04	0.03	0.01	0.01	0.06	0.05	0.02	0.01	0.03	0.03

4.2. Residence Time

The vertically-averaged Residence Time (RT) of particles within La Rochelle marina is visible in Figure 5 for the same combinations of tide and wind than Figure 4. IFT and RT share many similarities, but they are quantitatively different. Mean RT is up to 3 times lower than IFT, especially for neap tides, but its spatial variability (SSD) is significantly higher.

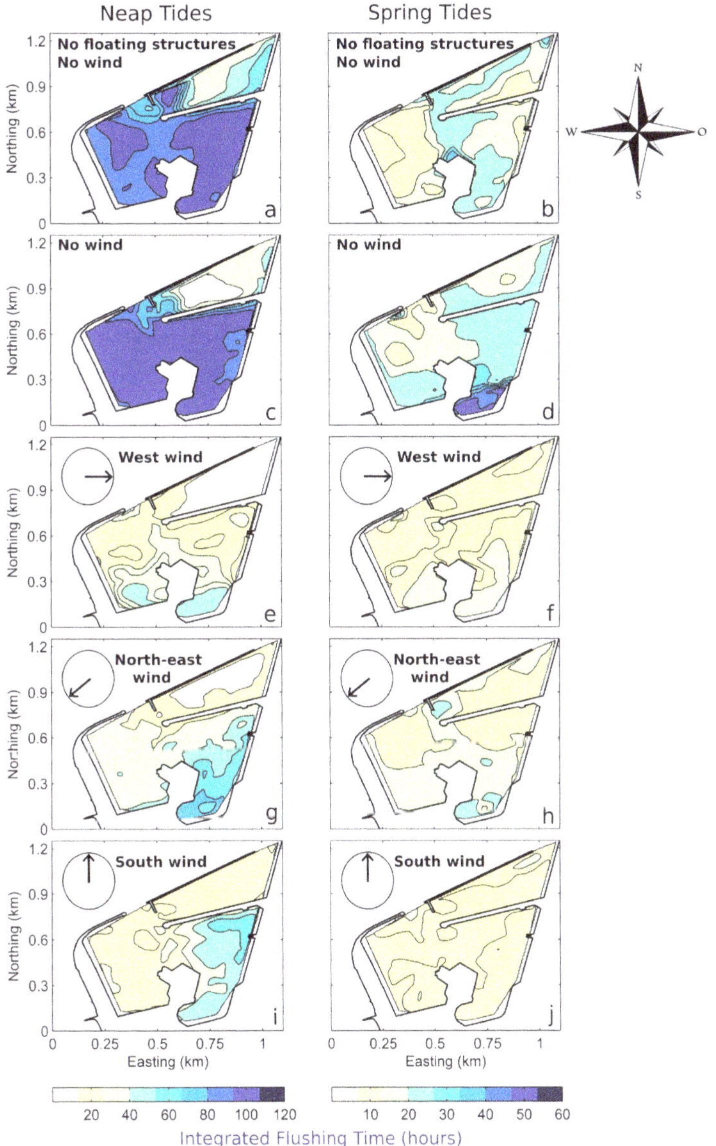

Figure 4. Spatial distribution of vertically-averaged Integrated Flushing Time (expressed in hours) within the La Rochelle marina computed for the scenario (letters) defined in Table 1. Rows characterise the wind, and FS configurations and columns correspond to the tidal regime.

According to Table 4, the presence of FS increases the RT (15.4/13.3 h at spring tides and 37/31.1 h at neap tides) and almost doubles SSD (20.6/12.7 h at spring tides and 42.9 h/28.3 h at neap tides). The wind has less influence on RT than on IFT and mainly reorganises RT spatially (Figure 5e–j), without substantially affecting the mean RT (Table 4). West and south winds stay the most impacting winds during neap and spring tides, respectively (Table 4). On neap tides, the lowest residence times are reached with intense 15 ms^{-1} west wind while on spring tides this wind configuration slightly enhances residences times compared with a 7.5 ms^{-1} west wind.

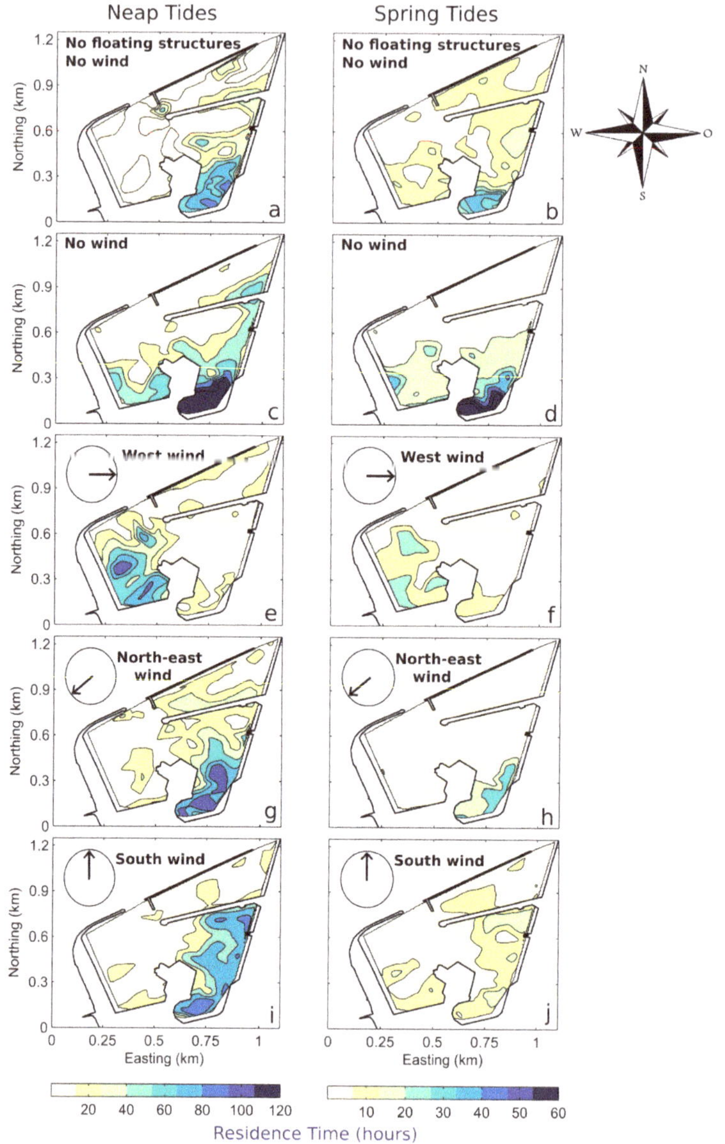

Figure 5. Spatial distribution of vertically-averaged Residence Time (expressed in hours) within the La Rochelle marina computed for each scenario (letters) defined in Table 1. Rows characterise the wind and FS configurations and columns correspond to the tidal regime.

4.3. Exposure Time

The last timescale studied is ET and represents the total time that particles spend in the domain. Its computation was derived from the release of particles used to characterise RT. ET is necessarily equal to or greater than RT and is not only a function of hydrodynamic processes within the marina but also of the return-flow. Figure 6 presents the spatial distribution of the vertically-averaged exposure times computed and the statistics results related are visible in Table 4. Their behaviour share, though, some similarities with the pattern of RT and IFT but globally their magnitudes display larger values.

Mean ET can significantly increase during neap tides, by doubling or almost quadrupling compared with spring tides (Table 4).

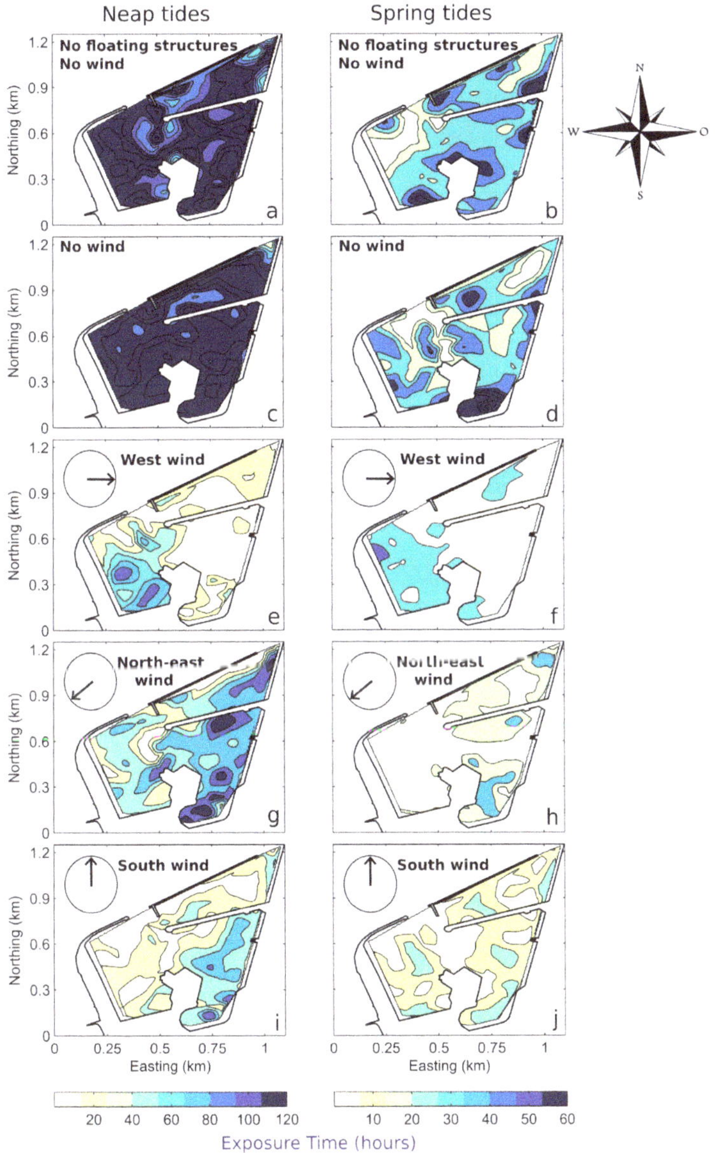

Figure 6. Spatial distribution of vertically-averaged Exposure Time (expressed in hours) within the La Rochelle marina computed for each scenario (letters) defined in Table 1. Rows characterise the wind and FS configurations and columns correspond to the tidal regime.

Similarly to IFT, the presence of FS does not increase significantly ET (34.8/34.5 h at spring tides and 129.3/124.1 h at neap tides). The wind has a more significant effect on exposure times than on residence times, and its impact is comparable for the behaviour of flushing times. It decreases ET by 2 to almost 5, particularly during neap tides with west winds (Table 4). Northeast winds have less influence than the other winds and mainly increase the exposure time of particles coming from both the southern part of the NE basin (on neap tides) and the entire SE basin, relative to the other wind cases. TPSD is still lower than SSD but is generally decreased by the presence of wind. During this study, we also observed that the particles returned preferentially to the marina through the western entrance (Figure 1) regardless of weather-marine conditions.

4.4. Characterisation of the Return-Flow

From the previous timescales, we assessed the importance of the return-flow in the marina through the expressions established in Section 2.3.3. The Return-flow Factor (RFF) was obtained through Equation (5) by using IFT values while the Return Coefficient (RC) was found via the Equation (7) by using RT and ET values. RFF and RC are inherently different as they were estimated, respectively, in a Eulerian and a Lagrangian way. These factors can provide relevant points of comparison that need to be analysed with caution. Results for each scenario were averaged and displayed in Table 4.

A large variety of return-flow values were found depending on the weather-marine conditions, but RFF globally displays the same behaviour than RC. The similarities mainly concern the influence of the wind and the tide. Neap tides show the higher variability of RC and RFF, with the lower and higher values respectively obtained with and without wind (Table 4). Without wind, the return-flow is always greater than 0.5, but its magnitude can drastically decrease during neap tides with the presence of wind, and particularly with the west wind (0.12 ± 0.11 and 0.19 ± 0.05 for RFF and RC, respectively). This decrease is also visible for spring tides but with less intensity (Table 4). These parameters can approach zero during intense 15 ms-1 western events. The maximum values are also obtained during neap tides without wind, with and without the presence of FS for RFF (0.73 ± 0.01) and for RC (0.75 ± 0.03), respectively. In general terms, the macro-tidal influence and the configuration of the area generate a high return-flow that is weakened under the action of the wind but not substantially affected by the presence of FS. The last result confirms the weak impact of floating structures on the circulation outside the marina.

5. Discussion

5.1. Assessment of the Main Drivers of the Water Renewal

Although the tide is a primary driver of the hydrodynamics, [19] has shown that the wind, but also infrastructures, had a substantial impact on the physical processes in the marina. The results, visible in Section 4, provide a robust view of their influence in the macro-tidal marina of La Rochelle. Here, we gather the information collected through the study of the three transport timescales to discuss the importance of the tide, the wind and the Floating Structures (FS) in the marina renewal.

5.1.1. Tidal Influence

Flushing, residence and exposure times (Figures 4–6) reveal a strong spatial heterogeneity of the renewal. The NE and the SE basins generally present the lowest and largest timescale values in the marina, respectively, but the behaviour of the timescales significantly vary depending on the conditions. Close to the entrances, the water parcels are rapidly advected because of the continuous effect of the tidal currents, which decreases renewal times. On the opposite, the innermost parts of the marina are more sheltered from the flows and display the longest renewal times. This marked horizontal spatial variability has been found in estuaries and coastal embayments with macro-tidal influence [14,33,47] but also with micro-tidal influence [16,48,49].

The renewal time and its spatial variability increase as the tidal range decreases in our area: the neap tide is a very limiting factor in the renewing of waters. Indeed, tidal currents that are 3 to 4 times lower in intensity within the marina during neap tides increase the renewal time from a factor of 2–3 (for RT) to in some cases almost a factor of 4 (for IFT and ET). This result shows that the advective coastal tidal movement is a main driver of the water flow in the marina is as it was demonstrated in previous studies [40,50,51].

Other European ports investigated the behaviour of their water masses, and the results displayed a broad range of renewal timescale values. [52,53] found global flushing times of 288 and 120 h, for a Sicilian micro-tidal and an Australian macro-tidal port, respectively. In Spain, the local flushing times reached about 125 and 163 h for the meso-tidal port of Bilbao and the micro-tidal port of Barcelona, respectively [32,54]. Thus, the renewal timescales are not strictly related to the tidal regime. The shape and dimension of ports also need to be considered.

5.1.2. Wind Influence

Even if the marina is tidally dominated, the wind also has a substantial impact on the water renewal because of the presence of shallow depths. Its influence is exerted differently depending on its direction and its intensity, but the wind globally enhances the water exchange, up to a factor of almost 5 during neap tides. This significant effect on renewal timescales values has already been reported for several coastal environments [34,55–58]. The wind has a more significant influence on exposure and flushing times than on residence times, especially during neap tides (Table 4). Contrary to RT, ET and IFT take into account the return-flow into the marina. This suggests that, even if the wind has a significant influence on processes inside the marina, its impact is larger on outside processes and their interaction with processes inside the marina. It mainly pushes waters out from the marina influence and facilitates their renewal in the local environment. Among the moderate wind cases, South and west winds are the most impacting winds during spring tides and neap tides, respectively (Table 4). Water exchanges are indeed significantly enhanced by the west wind action, as reported in [19]. The effect of the south wind was less noticeable on marina flows [19], but we could explain its impact on renewal by the marina configuration. Its two main entrances (Figure 1) are almost perpendicular to the direction of propagation of southern winds which tends to evacuate water bodies more rapidly while hindering their return to the marina. The mechanisms involved during the west and south wind action are different and they are not strictly proportional to the tidal range. The funnel configuration of the bay linked to the complex architecture of the marina and the numerous tidal flats in the area could explain the non-linearity of the processes involved from neap tides to spring tides. Finally, as the variability depends mainly on the tidal forcing and the dominant wind regime, the winter which is characterised by tougher western winds, should display a faster renewal as found in [33,47,59].

The transport timescales were computed at the bottom layers, and a slight vertical variability was found in the water renewal (data not shown). With tide only, surface layers always experience the highest renewal even if the difference with bottom layers does not overpass 2 h. Even if it is less clear for north-east and south winds, west wind generally enhances the variability of the renewal between the surface and bottom layers. This result differs from the vertical heterogeneity found by some authors [32,60] that demonstrated a straight vertical variability of the renewal depending on the wind regime. In the marina, the presence of shallow depths associated with the macro-tidal forcing of the area ensure a quasi-homogeneity of the water column that is slightly affected by the action of winds.

5.1.3. Influence of Floating Structures

Floating structures (FS) have a lower influence than the wind on the total water renewal of the marina. They increase the spatial variability of the renewal by reducing the water renewal in the most sheltered parts and slightly enhancing the hydrodynamics close to the entrances [19]. With two entrances, the NE basin is particularly exposed to this enhancement. The presence of FS is also involved in the increase of the water renewal by channelling the circulation and reducing the flood-related micro-scale eddies [19]. These eddies, generated by the tide-topography interaction, develop during the flood and affect the residual circulation within the marina. Their existence can interfere with the flushing processes and subsequently generate an increase in the residence time of the particles [61]. The broad coverage of floating bodies in the marina reduces the eddy activity by decreasing velocities in the inner parts the marina and focusing it along the entrances. Although FS reduces eddies, the general decrease of currents induced is sufficient to lower the renewal of water bodies in a significant portion of the marina.

5.1.4. Influence of the Tidal Phase

Given the macro-tidal regime of the area, the tidal phase release is considered as a possible factor influencing the water transport timescales. While many studies worked with only two opposite tidal phases [44,46,57], we proposed to investigate the renewal properties of the mid-ebb, the mid-flood, the high tide and the low tide. Figure 7 enables us to investigate the influence of the tidal phase of release in the computation of the three timescale values in the marina (IFT, RT and ET).

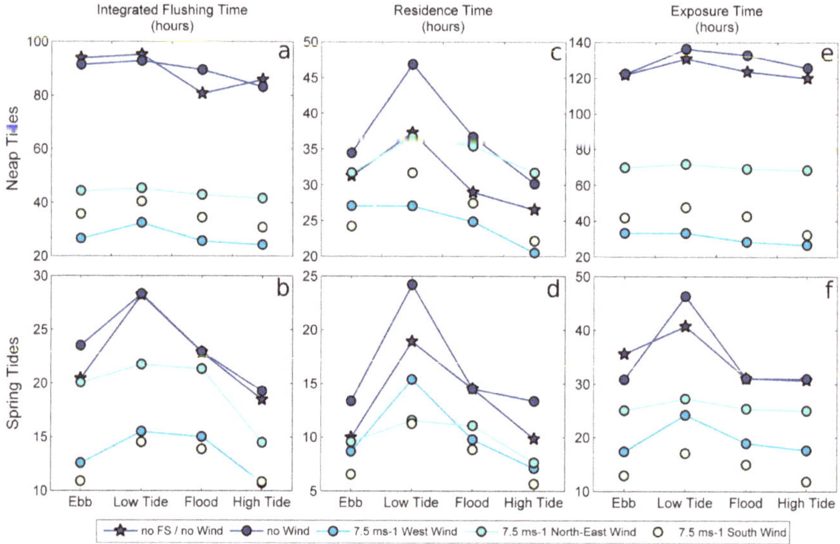

Figure 7. Spatially-averaged timescales values obtained for each release (at Ebb, Low Tide, Flood and High Tide) for each scenario.

Maximum timescale values are usually found when the release of tracer and particles occurs at low tide while the best moment to obtain lowest timescales is generally high tide (Figure 7). Generally, mid-ebb and mid-flood are quite similar in terms of renewal, even if spatial variabilities exist particularly close to the entrances. The Spatial Standard Deviation (SSD) was found to be always larger than the Tidal Phase Standard Deviation (TPSD) according to Table 4. It indicates that the variability of the renewing is more spatially than tidally dependent in the marina. Moreover, tidal phase variability of time scales generally decreased during spring tides and in the presence of wind (Table 4). This dependence on the tidal phase was already found and discussed in many studies [40,44,62].

5.2. Comparison of Water Transport Timescales

Based on the similarities shared by the several transport timescales, we highlighted general conclusions in the behaviour of the water renewal in the previous section. Here, we assess their differences to find the most relevant to describe the water transport processes occurring within the marina. The Eulerian (Integrated Flushing Time) and the Lagrangian transport timescales (Residence Time and Exposure Time) are inherently different as they follow advection-diffusion and only advection processes, respectively. The two approaches are complementary as their comparison provides exhaustive information about the motion features in the marina. As discussed in [62], their good agreement could be explained in tide-dominated environments where advection is the main driver of the circulation. Conversely, their differences in their spatial distribution characterise diffusion dominated areas. The cross-correlation of these approaches indicates that the advection mostly dominates in the marina, even if diffusion processes occur in the most sheltered parts of the marina.

The spatial distribution of Lagrangian timescales (RT and ET) displays more chaotic variability and relatively less monotonic gradients than Eulerian timescale (IFT). These chaotic patterns are the result of turbulent Lagrangian flows that generate a chaotic and non-monotonic dispersion of particles. This concept of "Lagrangian chaos", demonstrated experimentally [63], and numerically [64], has also been experienced in [62]. It mainly contributes to greater spatial variability of Lagrangian timescales.

The Eulerian timescale IFT generally presents approximately average values between RT and ET. A sensitivity test of the « e-folding flushing time » parameter used to compute IFT (data not shown) demonstrated that the chosen percent reduction was a leading factor in the magnitude of IFT. As it was set to 63% in this study, it could help explain why IFT presents approximately average values between RT and ET. Although IFT and ET are different in terms of magnitude, they globally follow the same trend. Without wind, the transition from spring to neap tides increases IFT and ET much more than RT. It means that the combination of processes outside and inside the marina is more sensitive to the tidal range than processes occurring only in the marina. Wind influences RT less than the two other timescales. It implies that hydrodynamics within the marina are less sensitive to the wind than in the system bay-marina. RT is more vulnerable to the presence of FS than the two other timescales, which confirms that the presence of FS primarily modifies the circulation within the marina. The tidal phase release can significantly alter the time required for a water parcel to leave the marina, which noticeably affects the RT results. ET and IFT are less dependent on the tidal phase release (Table 4; compare TPSD/Mean). Considering only the tide, physical processes occurring in the marina are more sensitive to the tidal phase, while marina-bay exchanges are more tidal range dependent.

Some authors recommend to use Eulerian approaches or to take into consideration the return of particles in the system when using Lagrangian methods [10,65]. In such a macro-tidal environment, the importance of the return-flow on the water renewal requires that dynamic processes outside the area of study be considered. Taking into account the processes occurring outside the domain of study is of crucial importance to represent the water renewal of semi-enclosed tidal basins accurately. However, the computation of RT alone does not make it possible to fully characterise the behaviour of water bodies in a semi-enclosed domain with such tidal influence. In the framework of this study, the computation of IFT appears to be sufficient to represent the intra-marina exchanges as well as the exchanges between the bay and the marina. IFT and ET are also less sensitive to the initial moment of calculation, which ensures an easier solution for the calculation of currents and associated phenomena. Nevertheless, the information provided by RT was necessary to characterise the fate of waters leaving the marina via the Return Coefficient formulation. In that sense, the comparison of the three timescales allowed us to have a complete picture of the physical processes leading to the water renewal in the marina. As seen in the literature [11], the relevance of each time scale depends on the initial question and the physical, biological, and geochemical characteristics of the study area. Thus, the question of the usefulness or relevance of each timescale depends on the characteristics of the scientific study proposed.

5.3. Impact of the Return-Flow in the Water Renewal

The Return-flow Factor (RFF) and the Return Coefficient (RC) display the same behaviour while they are computed in two different ways. This consistency validates the several methods employed to characterise the water renewal in the marina. Although the three transport timescales (IFT, RT, and ET) show quantitative differences, their information allows computing the general return-flow occurring in the marina. The knowledge of the return-flow magnitude in a basin is also a straightforward way to represent the renewal by taking into account the dynamics outside the embayment. Its information and variability is a powerful indication of the water masses confinement in semi-enclosed tidal bays depending on the weather-marine conditions.

In the past, tidal prism flushing models have often set the return-flow to 0.5 or ignore it [66,67], but as suggested by [20], it should be characterised in any study. Many authors described the return-flow in different embayments throughout the world. For example, reference [68] found a maximum value of 0.15 for a micro-tidal creek in the US; reference [69] computed 0.2 to 0.9 from the entrance to the inner parts of a meso-tidal lagoon in China; reference [16] estimated a mean return-flow of 0.66 for a micro-tidal lagoon in Italy; an average return-flow of 0.32 was found by [70] in a meso-tidal Scottish Fjord; a maximum return-flow of 0.95 was reached in a meso to macro-tidal lagoon in France [33]. As experienced in La Rochelle marina and other embayments in the world, the return-flow is not correlated strictly to the tidal influence in the domain.

Here, the return flow is higher during neap tides than during spring tides in the absence of wind. The influence of the tide is particularly visible in the bay, and the back and forth movement generates significant trapping in the bay which extends to the marina. The funnel configuration of the bay and its sheltered position (Figure 1) combined with the tidal dominance of the area could explain the significant trapping of water masses, responsible for a high return-flow in the marina. This trapping is increased during neap tides when the excursions of water parcels are reduced. The study has shown that when water particles left the bay, they were rapidly ejected and their return to the system was hindered, especially in the marina. This mechanism is also described in [68], where the oscillatory flood and ebb processes caused the water masses to be trapped, particularly during neap tides. Conversely, [44] computed the larger return coefficient during spring tides, but the estuarine environment studied was much more influenced by the freshwater inputs.

The comparison of the weather-marine scenarios highlights the leading role of the wind on the reduction of the return-flow (Table 4). Neap tide conditions are particularly concerned by this decrease, which is directly correlated to lower activity of the tidal currents. The influence of the wind can overpass the tide, especially during intense 15 ms-1 west events when the return-flow can approach zero (Table 4). When water parcels leave the marina during ebb, the wind can rapidly flush them out of the bay where they are exposed to the high renewal rates of the local embayments [71].

6. Conclusions and Perspectives

The purpose of the present study was to characterise the water renewal of La Rochelle marina under a complete range of weather-marine conditions. We mainly focused on the physical mechanisms allowing to describe the renewal, but the results obtained might provide sufficient material for future studies related to the monitoring of pollutants and biological/ecological applications. Pollution is one of the major threats to water quality in coastal areas and understanding the physical behaviour of such environments is the first step toward efficient management of the problem. The results provided in this paper enable us to identify the most vulnerable zones concerning accidental pollution, but they can also be useful for undertaking protection and management policies. Results emphasise the substantial variability of the water renewal depending on the weather-marine conditions and the location in the marina.

Two different approaches helped us determining the temporal and spatial renewal of waters in the marina. The computation of three water transport timescales (IFT, RT and ET) led to the estimation of the return-flow in the marina via the RFF and RC formulations. Based on their study and comparison, the main findings concerning the water renewal in the marina are:

- The marina displays a strong horizontal variability of the renewal. The most confined waters are located in the south of the SE basin, while the NE basin is generally the most renewed.
- Both the tide and wind substantially impact the water renewal of the marina. The transition from spring to neap tides significantly decreases the water renewal while the presence of the wind enhances it, in particular for west and south directions.
- In the marina, physical processes responsible for the water renewal are generally more affected by the tidal phase than in its surrounding bay.
- The influence of the wind is less significant in the marina than in the system bay-marina.

- The water renewal is relatively homogeneous over the water column even with the effect of the wind (data not shown).

- As shown by the study of RT, the Floating Structures (FS) particularly decreased the water renewal in the most sheltered parts (SE basin), but they also generally increase it in the most exposed parts of the marina (NE basin). The study of both ET and IFT showed that they do not alter the circulation significantly outside the marina and thus, slightly affect the return-flow in the marina.

- The return-flow has been consistently represented both in a Lagrangian and Eulerian way. Its information is a relevant indication of the circulation processes occurring outside the marina. Without wind, return-flow in the marina is amplified by neap tides that generate significant trapping of the water masses at the scale of the bay. Both ET and IFT results showed that return-flow was a key parameter in the dynamics of water renewal in the marina.

- The return-flow is very sensitive to the wind action and the trapping processes happening in the bay can be drastically reduced. Wind enhances the dispersion of water masses away from the bay and thus decreases the return-flow in the marina. Powerful west winds (15 ms^{-1}), typical of winter conditions, can generate a near-zero return-flow.

- Without wind, return-flow is amplified by neap tides that generates significant trapping.

Given the importance of the return-flow in the region of interest, we should point out that tracer mass and particles dispersion directly depend on the size of the computational domain [27,72]. We assumed the latter to be sufficiently large enough not to affect the tracer mass and the behaviour of the particles used in the definition of the water transport timescales.

The use of only one transport timescale (IFT) enabled to thoroughly describe the renewal processes occurring in the marina but also between the marina and its local environment. This parameter is less sensitive to the tidal phase release moment than RT, displays less chaotic patterns than ET and RT and offers the possibility to compute the return-flow efficiently through the tidal prism method. However, the comparison with Lagrangian timescales allowed us to detail the physical processes responsible for the renewal and to assess the consistency of the results. It underlines the need to cross-reference complementary methods and also puts forward some methods employed to address questions in a wide range of scientific domains.

Future works should characterise the influence of freshwater inflow and waves. Indeed, many authors suggested the contribution of the freshwater to the enhancement of the renewal [47,57,59,73,74]. Even if river influence is negligible in our area, intense rainy events can significantly affect the renewal times [75]. Waves that can be energetic offshore can also modify the exchange efficiency [20,76–78]. In the next study, the influence of dredging maintenance of the marina on the water renewal will be investigated. Indeed, the water depth was found to increase the water exchange [79], and it could be interesting to identify the optimal depth to improve the water quality of the marina.

Author Contributions: J.-R.H. and I.B. designed the study. T.C. and J.-R.H. collected and analysed the surface velocities with drifting buoys. J.-R.H. performed numerical simulations to calculate the different indicators studied. J.-R.H. analysed the results and wrote the paper. I.B. and T.C. revised the paper.

Funding: This research was funded by Région Nouvelle-Aquitaine, the direction of La Rochelle marina and the CPER ("Contrat Plan Etat Région") DYPOMAR.

Acknowledgments: The authors wish to thank the team of La Rochelle Marina and in particular Adeline Thomassin, that provided technical support for the deployment of observing systems. The developing team of TELEMAC-MASCARET system is also acknowledged for making its code available. We also thank CNRS, Région Nouvelle-Aquitaine, and La Rochelle Université for their support and Météo France for providing us with meteorological data. The authors finally want to sincerely thank editor and reviewers chosen by Water Editorial Office, for their very constructive reviews that improved this paper greatly.

Conflicts of Interest: The authors declare no conflict of interest.

References

1. Ondiviela, B.; Gómez, A.G.; Puente, A.; Juanes, J.J. A pragmatic approach to define the ecological potential of water bodies heavily modified by the presence of ports. *Envion Sci. Policy* **2013**, *33*, 320–331. [CrossRef]
2. Gómez, A.G.; Ondiviela, B.; Fernández, M.; Juanes, J.J. Atlas of susceptibility to pollution in marinas. Application to the Spanish coast. *Mar. Pollut. Bull.* **2017**, *114*, 239–246. [CrossRef] [PubMed]
3. Mestres, M.; Sierra, J.P.; Mösso, C.; Sánchez-Arcilla, A. Sources of contamination and modelled pollutant trajectories in a Mediterranean harbour (Tarragona, Spain). *Mar. Pollut. Bull.* **2010**, *60*, 898–907. [CrossRef] [PubMed]
4. Grifoll, M.; Jordà, G.; Borja, Á.; Espino, M. A new risk assessment method for water quality degradation in harbour domains, using hydrodynamic models. *Mar. Pollut. Bull.* **2010**, *60*, 69–78. [CrossRef] [PubMed]
5. Fatoki, O.S.; Mathabatha, S. An assessment of heavy metal pollution in the East London and Port Elizabeth harbours. *Water SA* **2001**, *27*, 233–240. [CrossRef]
6. Bonamano, S.; Madonia, A.; Piazzolla, D.; Paladini de Mendoza, F.; Piermattei, V.; Scanu, S.; Marcelli, M. Development of a Predictive Tool to Support Environmentally Sustainable Management in Port Basins. *Water* **2017**, *9*, 898. [CrossRef]
7. Mali, M.; De Serio, F.; Dell'Anna, M.M.; Mastrorilli, P.; Damiani, L.; Mossa, M. Enhancing the performance of hazard indexes in assessing hot spots of harbour areas by considering hydrodynamic processes. *Ecol. Indic.* **2017**, *73*, 38–45. [CrossRef]
8. Bolin, B.; Rodhe, H. A note on the concepts of age distribution and transit time in natural reservoirs. *Tellus* **1973**, *25*, 58–62. [CrossRef]
9. Takeoka, H. Fundamental concepts of exchange and transport time scales in a coastal sea. *Cont. Shelf Res.* **1984**, *3*, 311–326. [CrossRef]
10. Zimmerman, J.T.F. Estuarine Residence Times. In *Hydrodynamics of Estuaries*, 1st ed.; Kjerfve, B., Ed.; CRC Press: Boca Raton, FL, USA, 1988; p. 171.
11. Monsen, N.E.; Cloern, J.E.; Lucas, L.V.; Monismith, S.G. A comment on the use of flushing time, residence time, and age as transport time scales. *Limnol. Oceanogr.* **2002**, *47*, 1545–1553. [CrossRef]
12. Delhez, E.J.M.; Heeming, A.W.; Deleersnijder, E. Residence time in a semi-enclosed domain from the solution of an adjoint problem. *Estuar. Coast. Shelf Sci.* **2004**, *61*, 691–702. [CrossRef]
13. Deleersnijder, E.; Delhez, E. Timescale and tracer-based methods for understanding the results of complex marine models. *Estuar. Coast. Shelf Sci.* **2007**, *74*, 5–7. [CrossRef]
14. Oliveira, A.; Baptista, A.M. Diagnostic modeling of residence times in estuaries. *Water Resour. Res.* **1997**, *3*, 1935–1946. [CrossRef]
15. Abdelrhman, M.A. Simplified modeling of flushing and residence times in 42 embayments in New England, USA, with special attention to Greenwich Bay, Rhode Island. *Estuar. Coast. Shelf Sci.* **2005**, *62*, 339–351. [CrossRef]
16. Cucco, A.; Umgiesser, G. Modeling the Venice Lagoon residence time. *Ecol. Model.* **2006**, *193*, 34–51. [CrossRef]
17. Sánchez-Arcilla, A.; Espino, M.; Grifoll, M.; Mösso, C.; Sierra, J.P.; Mestres, M.; Spyroupoulu, S.; Hernáez, M.; Ojanguren, A.; Sotillo, M.G.; et al. Quay design and operational oceanography. The case of bilbao harbour. *Coast. Eng. Proc.* **2011**, *1*, 51. [CrossRef]
18. Breitwieser, M.; Dubillot, E.; Barbarin, M.; Churlaud, C.; Huet, V.; Muttin, F.; Thomas, H. Assessment of the biological quality of port areas: A case study on the three harbours of La Rochelle: The marina, the fishing harbour and the seaport. *PLoS ONE* **2018**, *13*, e0198255. [CrossRef] [PubMed]
19. Huguet, J.R.; Brenon, I.; Coulombier, T. Influence of floating structures on the tide and wind-driven hydrodynamics of a highly populated marina. *J. Waterw. Port Coast. Ocean Eng.* **2019**. In press.
20. Sanford, L.P.; Boicourt, W.C.; Rives, S.R. Model for Estimating Tidal Flushing of Small Embayments. *J. Waterw. Port Coast. Ocean Eng.* **1992**, *118*, 635–654. [CrossRef]
21. Chaumillon, E.; Weber, N. Spatial variability of modern incised valleys on the French Atlantic coast: Comparison between the Charente (Pertuis d'Antioche) and the Lay-Sèvre (Pertuis Breton) incised valleys. In *Incised Valleys in Time and Space*, 1st ed.; Dalrymple, R.A., Leckie, D.A., Tillman, R.A., Eds.; SEPM Special Publications: Tarsa, OK, USA, 2006; pp. 57–85.

22. Bertin, X.; Bruneau, N.; Breilh, J.F.; Fortunato, A.B.; Karpytchev, M. Importance of wave age and resonance in storm surges; the case Xynthia, Bay of Biscay. *Ocean Model.* **2012**, *42*, 16–30. [CrossRef]

23. Le Cann, B. Barotropic tidal dynamics of the Bay of Biscay shelf: Observations, numerical modelling and physical interpretation. *Cont. Shelf Res.* **1990**, *10*, 723–758. [CrossRef]

24. Bertin, X. Morphodynamique Séculaire, Architecture Interne et Modélisation d'un Système Baie/Embouchure tidale: Le Pertuis de Maumusson et la Baie de Marennes-Oléron. Ph.D. Thesis, La Rochelle University, La Rochelle, France, 2005.

25. Dodet, G.; Bertin, X.; Tabord, R. Wave climate variability in the North-East Atlantic Ocean over the last six decades. *Ocean Model.* **2010**, *31*, 120–131. [CrossRef]

26. Hervouet, J.M. *Hydrodynamics of Free Surface Flows: Modelling with the Finite Element Method*; John Wiley and Sons: Hobken, NJ, USA, 2007; p. 360.

27. Viero, P.D.; Defina, A. Renewal time scales in tidal basins: Climbing the Tower of Babel. In *Sustainable Hydraulics in the Era of Global Chang*, 1st ed.; Taylor & Francis Group: London, UK, 2016; p. 216.

28. Donea, J.; Giuliani, S.; Halleux, J.P. An arbitrary Lagrangian-Eulerian finite element method for transient dynamic fluid-structure interactions. *Comput. Methods Appl. Mech. Eng.* **1982**, *33*, 689–723. [CrossRef]

29. SHOM. Information Géographique Maritime et Littorale de Référence. Available online: https://data.shom.fr/ (accessed on 2 July 2019).

30. AVISO+ Satellite Altimetry Data. Available online: https://www.aviso.altimetry.fr/en/data/products/auxiliary-products/global-tide-fes.html (accessed on 2 July 2019).

31. Choi, K.W.; Lee, J.H.W. Numerical Determination of Flushing Time for Stratified Water Bodies. *J. Mar. Syst.* **2004**, *50*, 263–281. [CrossRef]

32. Grifoll, M.; Del Campo, A.; Espino, M.; Mader, J.; González, M.; Borja, Á. Water renewal and risk assessment of water pollution in semi-enclosed domains: Application to Bilbao Harbour (Bay of Biscay). *J. Mar. Syst.* **2013**, *109–110*, 241–251. [CrossRef]

33. Plus, M.; Dumas, F.; Stanisière, J.Y.; Maurer, D. Hydrodynamic characterization of the Archachon Bay, using model-derived descriptors. *Cont. Shelf Res.* **2009**, *29*, 1008–1013. [CrossRef]

34. Wang, C.F.; Hsu, M.H.; Kuo, A.Y. Residence time of the Danshuei River estuary, Taiwan. *Estuar. Coast. Shelf Sci.* **2004**, *60*, 381–393. [CrossRef]

35. Andrejev, O.; Myrberg, K.; Lundberg, P.A. Age and renewal time of water masses in a semi-enclosed basin—application to the Gulf of Finland. *Tellus A Dyn. Meteorol. Oceanogr.* **2004**, *56*, 548–558.

36. Fugate, D.C. Estimation of Residence Time in a Shallow Back Barrier Lagoon, Hog Island Bay, Virginia, USA. In Proceedings of the Ninth International Conference on Estuarine and Coastal Modelling, Charleston, SC, USA, 31 October–2 November 2005.

37. Thomann, R.V.; Mueller, J.A. *Principles of Surface Water Quality Modeling and Control*; Harper & Row, Publishers: New York, NY, USA, 1987; p. 656.

38. Zimmerman, J.T.F. Mixing and flushing of tidal embayments in the Western Dutch wadden Sea, Part I: Distribution of salinity and calculation of mixing time scales. *Neth. J. Sea Res.* **1976**, *10*, 149–191. [CrossRef]

39. Dronkers, J.; Zimmerman, J.T.F. Some principles of mixing in tidal lagoons. *Oceanol. Acta* **1982**, *4*, 107–118.

40. Cucco, A.; Umgiesser, G.; Ferrarin, C.; Perilli, A.; Canu, D.M.; Solidoro, C. Eulerian and lagrangian transport timescales of a tidal active coastal basin. *Ecol. Model.* **2009**, *220*, 913–922. [CrossRef]

41. Spivakovskaya, D.; Heemink, A.W.; Deleersnijder, E. Lagrangian modelling of multi-dimensional advection-diffusion with space-varying diffusivities: Theory and idealized test cases. *Ocean Dyn.* **2007**, *57*, 189–203. [CrossRef]

42. Delhez, E.J.M.; Brye, B.; Brauwere, A.; Deleersnijder, E. Residence time vs influence time. *J. Mar. Syst.* **2014**, *132*, 185–195. [CrossRef]

43. Wolanski, E. *Estuarine Ecohydrology*, 1st ed.; Elsevier Science: Atlanta, GA, USA, 2007; p. 168.

44. Brauwere, A.; Brye, B.; Blaise, S.; Deleersnijder, E. Residence time, exposure time and connectivity in the Scheldt Estuary. *J. Mar. Syst.* **2011**, *84*, 85–95. [CrossRef]

45. Delhez, E. On the concept of exposure time. *Cont. Shelf Res.* **2013**, *71*, 27–36. [CrossRef]

46. Andutta, F.P.; Helfer, F.; Miranda, L.B.; Deleersnjider, E.; Thomas, C.; Lemckert, C. An assessment of transport timescales and return coefficient in adjacent tropical estuaries. *Cont. Shelf Res.* **2013**, *124*, 49–62. [CrossRef]

47. Barcena, J.F.; Garcià, A.; Gómez, A.G.; Àlvarez, C.; Juanes, J.A.; Revilla, A. Spatial and temporal flushing time approach in estuaries influenced by river and tide. An application in Suances Estuary (Northern Spain). *Estuar. Coast. Shelf Sci.* **2012**, *112*, 40–51. [CrossRef]

48. Wijeratne, S.; Rydberg, L. Modelling and observations of tidal wave propagation circulation and residence times in Puttalam Lagoon, Sri Lanka. *Estuar. Coast. Shelf Sci.* **2007**, *74*, 697–708. [CrossRef]

49. Ferrarin, C.; Zaggia, L.; Paschini, E.; Scirocco, T.; Lorenzetti, G.; Bajo, M.; Penna, P.; Francavilla, M.; D'Adamo, R.; Guerzoni, S. Hydrological Regime and Renewal Capacity of the Micro-tidal Lesina Lagoon, Italy. *Estuar. Coasts* **2013**, *37*, 79–93. [CrossRef]

50. Yuan, D.; Lin, B.; Falconer, R.A. A modelling study of residence time in a macro-tidal estuary. *Estuar. Coast. Shelf Sci.* **2007**, *71*, 401–411. [CrossRef]

51. Cavalcante, G.H.; Kjverfve, B.; Feary, D.A. Examination of residence time and its relevance to water quality whithin a coastal mega-structure: The Palm Jumeirah Lagoon. *J. Hydrol.* **2012**, *468–469*, 111–119. [CrossRef]

52. Lisi, I.; Taramelli, A.; Di Risio, M.; Cappucci, S.; Gabellini, M. Flushing efficiency of Augusta Harbour (Italy). *J. Coast. Res.* **2009**, *56*, 841–845.

53. Schwartz, R.A.; Imberger, J. Flushing behaviour of a coastal marina. *Coast. Eng. Proc.* **1988**, 2626–2640. [CrossRef]

54. Grifoll, M., Espino, M., Gonzàlez, M.; Ferrer, L.; Sánchez-Arcilla, A. Spatial residence time description for water discharges in harbours. In Proceedings of the 4th Internation Conference on Marine Waste Water Disposal and Marine Environment, Antalya, Turkey, 6–10 November 2006.

55. Orfila, A.; Jordi, A.; Basterretxea, G.; Vizoso, G.; Marbà, N.; Duarte, C.M.; Werner, F.J.; Tintoré, J. Residence time and Posidonia oceanica in Cabrera Archipelago National park, Spain. *Cont. Shelf Res.* **2005**, *25*, 1339–1352. [CrossRef]

56. Geyer, W.R. Influence of Wind on Dynamics and Flushing of Shallow Estuaries. *Estuar. Coast. Shelf Sci.* **1997**, *44*, 713–722. [CrossRef]

57. Canu, D.M.; Solidoro, C.; Umgiesser, G.; Cucco, A.; Ferrarin, C. Assessing confinement in coastal lagoons. *Mar. Pollut. Bull.* **2012**, *64*, 2391–2398. [CrossRef]

58. Du, J.; Shen, J. Water residence time in Chesapeake Bay for 1982–2012. *J. Mar. Syst.* **2016**, *164*, 101–111. [CrossRef]

59. Umgiesser, G.; Zemlys, P.; Erturk, A.; Razinkova-Baziukas, A.; Mežine, J.; Ferrarin, C. Seasonal renewal time variability in the Curonian Lagoon caused by atmospheric and hydrographical forcing. *Ocean Sci.* **2016**, *12*, 391–402. [CrossRef]

60. Warner, J.C.; Geyer, R.W.; Arango, H.G. Using a composite grid approach in a complex coastal domain to estimate estuarine residence time. *Comput. Geosci.* **2010**, *36*, 921–935. [CrossRef]

61. Babu, M.T.; Vethamony, P.; Desa, E. Modelling tide-driven currents and residual eddies in the Gulf of Kachchh and their seasonal variability: A marine environmental planning perspective. *Ecol. Model.* **2005**, *184*, 299–312. [CrossRef]

62. Ridderkinhof, H.; Zimmerman, J.T.F.; Philippart, M.E. Tidal exchange between the north sea and Dutch Wadden Sea and mixing timescales of the tidal basins. *Neth. J. Sea Res.* **1990**, *25*, 331–350. [CrossRef]

63. Aref, H. Stirring by chaotic advection. *J. Fluid Mech.* **1984**, *143*, 1–21. [CrossRef]

64. Arega, F.; Badr, A.W. Numerical age and residence-time mapping for a small tidal creek: Case study. *J. Waterw. Port Coast. Ocean Eng.* **2010**, *136*, 226–237. [CrossRef]

65. Arega, F. Hydrodynamic modeling and characterizing of Lagrangian flows in the West Scott Creek wetlands system, South Carolina. *J. Hydro Environ. Res.* **2013**, *7*, 50–60. [CrossRef]

66. Dyer, K.R. *Estuaries: A Physical Introduction*; John Wiley and Sons: New York, NY, USA, 1973.

67. Callaway, R.J. Flushing study of South Beach Marina, Oregon. *J. Waterw. Port Coast. Ocean Eng.* **1981**, *102*, 47–58.

68. Arega, F.; Armstrong, S.; Badr, A.W. Modeling of residence time in the East Scott Creek Estuary, South Carolina, USA. *J. Hydro Environ. Res.* **2008**, *2*, 99–108. [CrossRef]

69. Zhang, C. Water renewal timescales in an ecological reconstructed lagoon in China. *J. Hydroinform.* **2013**, *15*, 991–1001.

70. Gillibrand, P.A. Calculating Exchange Times in a Scottish Fjord Using a Two-dimensional, Laterally-integrated Numerical model. *Estuar. Coast. Shelf Sci.* **2001**, *53*, 437–449. [CrossRef]

71. Ascione Kenov, I.; Muttin, F.; Campbell, R.; Fernandes, R.; Campuzano, F.; Machado, F.; Franz, G.; Neves, R. Water fluxes and renewal rates at Pertuis d'Antioche/Marennes-Oléron Bay, France. *Estuar. Coast. Shelf Sci.* **2015**, *167*, 32–44. [CrossRef]
72. Bialik, R.J.; Karpiński, M. On the effect of the window size on the assessment of particle diffusion. *J. Hydraul. Res.* **2018**, *56*, 560–566. [CrossRef]
73. Huang, W.; Spaulding, M. Modelling residence-time response to freshwater input in Apalachicola Bay, Florida, USA. *Hydrol. Process.* **2002**, *16*, 3051–3064. [CrossRef]
74. Asselin, S.; Spaulding, M.L. Flushing Times for Providence River Based on Tracer Experiments. *Estuaries* **1993**, *16*, 830–839. [CrossRef]
75. Malhadas, M.S.; Neves, R.J.; Leitão, P.C.; Silva, A. Influence of tide and waves on water renewal in Obidos Lagoon, Portugal. *Ocean Dyn.* **2010**, *60*, 41–55. [CrossRef]
76. Azevedo, A.; Oliveira, A.; Fortunato, A.B.; Bertin, X. Application of an Eulerian-Lagrangian oil spill modeling system to the Prestige accident: Trajectory analysis. *J. Coast. Res.* **2009**, *56*, 777–781.
77. Delpey, M.T.; Ardhuin, F.; Otherguy, P.; Jouon, A. Effects of waves on coastal water dispersion in a small estuarine bay. *J. Geophys. Res.* **2014**, *119*, 70–86. [CrossRef]
78. Guo, W.; Wu, G.; Liang, B.; Xu, T.; Chen, X.; Yang, Z.; Xie, M.; Jiang, M. The influence of surface wave on water exchange in the Bohai Sea. *Cont. Shelf Res.* **2016**, *118*, 128–142. [CrossRef]
79. Malhadas, M.S.; Silva, A.; Leitão, P.C.; Neves, R. Effect of the Bathymetric Changes on the Hydrodynamic and Residence Time in Obidos Lagoon (Portugal). *J. Coast. Res.* **2009**, *56*, 549–553.

Article

Wind Effects on the Water Age in a Large Shallow Lake

Sien Liu [1], Qinghua Ye [1,2,*], Shiqiang Wu [3] and Marcel J. F. Stive [1]

[1]   Department of Hydraulic Engineering, Delft University of Technology, 1, Stevinweg,
     2628 CN Delft, The Netherlands; s.liu@tudelft.nl (S.L.); m.j.f.stive@tudelft.nl (M.J.F.S.)
[2]   Deltares, Boussinesqweg 1, 2629 HV Delft, The Netherlands
[3]   State Key Laboratory of Hydrology-Water Resources and Hydraulic Engineering,
     Nanjing Hydraulic Research Institute, Nanjing 210029, China; sqwu@nhri.cn
*   Correspondence: qinghua.ye@deltares.nl

Received: 27 February 2020; Accepted: 24 April 2020; Published: 27 April 2020

Abstract: As the third largest fresh water lake in China, Taihu Lake is suffering from serious eutrophication, where nutrient loading from tributary and surrounding river networks is one of the main contributors. In this study, water age is used to investigate the impacts of tributary discharge and wind influence on nutrient status in Taihu Lake, quantitatively. On the base of sub-basins of upstream catchments and boundary conditions of the lake, multiple inflow tributaries are categorized into three groups. For each group, the water age has been computed accordingly. A well-calibrated and validated three-dimensional Delft3D model is used to investigate both spatial and temporal heterogeneity of water age. Changes in wind direction lead to changes in both the average value and spatial pattern of water age, while the impact of wind speed differs in each tributary group. Water age decreases with higher inflow discharge from tributaries; however, discharge effects are less significant than that of wind. Wind speed decline, such as that induced by climate change, has negative effects on both internal and external nutrient source release, and results in water quality deterioration. Water age is proved to be an effective indicator of water exchange efficiency, which may help decision-makers to carry out integrated water management at a complex basin scale.

Keywords: shallow lake; water age; meteorological influence; sub-basins; Delft3D

---

1. Introduction

Located in the southeastern part of China, Taihu Lake is the third largest fresh water lake in China. Like most large shallow lakes all over the world and as a typical shallow lake around the middle and lower reaches of the Yangtze River, Taihu Lake is suffering from the threat of eutrophication [1–3]. Considering the multi-functionality of Taihu Lake, both as an economic resource (such as supplying drinking water, and providing flood control, irrigation, water transport, and recreation) and as a valuable ecological resource, the water quality issue and consequential algae bloom problem have caused, and are still causing huge losses to this regional industrial and economic center, ever since 1987 [4–7].

The formation of the algae bloom happens when the concentration of algae is high at a particular spot. The growth of the algae population requires proper temperature, light availability, and essential nutrients, such as Nitrogen (N) and Phosphorus (P) [8]. The local government has made many attempts to mitigate the algae bloom, including wetland restoration, water transfer from the Yangtze River, and environmental dredging [9–11]. However, these treatments have not achieved the expected goal. Research has suggested applying a nutrient input reduction strategy for Taihu Lake's water quality and ecology restoration [12–16].

Nutrients are released into the lake's water body from two types of sources—namely, internal sources from sediment resuspension, and external sources from the connected river network. The previously mentioned studies of input nutrient reduction considered the external sources of nutrient release, which are significantly correlated to the local urbanization around the lake [17]. The lake-connected river networks accumulate nutrients from various sources (such as the diffuse sources of agriculture runoff, atmospheric deposition, or point sources of industry waste water and domestic sewage) and transport these into the lake [18–21]. Nutrient distribution in Taihu Lake varies both spatially and temporally, partially because the nutrients from the inflow tributaries vary due to local economic structure differences [22,23]. In turn, the uneven distribution has caused an inter-annual difference in the location of algae bloom in the lake. During the summer, algae bloom happens in the northern part of the lake, while in the early summer, autumn, and sometimes early winter, the algae bloom happens along the southwestern lake region[24]. However, these studies usually consider the lake as a whole black-box model or as several separate regions for a roughly spatially averaged evaluation, without considering the hydrodynamics. Thus, little attention has been paid to the spatial and temporal variations of nutrient distribution inside the waterbody of the lake resulting from the influence of the difference in both external input and physical factors, such as meteorological conditions, and the lake's intrinsic characters before and after the input reduction. Besides, based on hydrodynamic studies, more attention should be paid to the effectiveness of stand-alone input nutrient reduction in different sub-basins upstream of Taihu Lake.

To quantitatively study the influence of the external nutrient input, a time scale is considered valuable, since a certain time is required for the nutrient to transport to a given location, and this amount of time is correlated to the hydrodynamics of the lake system [25]. Considering the nutrient transport from the connected river tributaries into Taihu Lake as point sources, the concept of water age is introduced here as an index to describe the time taken for the transport of nutrients within the lake. Water age is widely used in marine and fresh water systems to effectively reflect the nutrient transport and mixing process of nutrients, and to provide the spatial and temporal heterogeneity of these processes [26–28]. Studies have shown a strong correlation of in situ measured Chl-a concentration and water age distribution of external discharge into shallow lakes, which proves the significance of the existing hydrodynamics [29].

In this study, a three-dimensional Delft3D numerical model has been set up and used to investigate the transport and mixing of dissolved nutrients in Taihu Lake using the concept of water age. The purpose of this study was to answer the following questions: (1) Whether it is possible to quantitatively compare the nutrient load from different parts of the catchment river networks to Taihu Lake using the concept of water age; and (2) how the meteorological factor wind, which is the largest influencing factor for lake hydrodynamics, impacts the transport of nutrients from all over the catchment inside the lake.

This chapter is organized as follows. Section 2 is the theoretical background describing the water age theory. In the methodology Section 3, a detailed description of the geographical area, and Delft3D numerical modelling is introduced. In Section 4, the hydrodynamic results and water age distribution in various scenarios are provided. Then, in Section 5, our discussion and further extensions of this study are shown. Section 6 is the conclusion section.

2. Theoretical Background

Several transport time scales are frequently utilized in hydrodynamic, biological, and water environmental studies for multiple topics, like pollution transport tracking and water mass renewal [30]. These time scales include, for example, water age, flushing time, residence time or transit time, turnover time, and exposure time. Each of the transport time scales works within the scope of a certain application [31]. In this study, considering the complicated hydrodynamic condition and spatial heterogeneity, the concept of water age (WA) has been chosen.

The most common definition of water age is given as "the time that has elapsed since the particle under consideration left the region in which its age is prescribed as being zero" [32,33]. Thus, particularly in this study, WA is defined as the time elapsed since the tributary water with dissolved nutrients entered the lake water body, with WA being equal to zero at the boundary between the connected river network and the lake water.

Research on WA includes theoretical study, field observation, and numerical modelling [34–39]. Considering the applicability and accuracy, a numerical modelling study of WA is suitable under realistic bathymetry and hydrodynamic conditions [26,40]. There are two widely used numerical approaches for WA calculation in numerical models, namely, the Particle-Tracking Method (PTM) [41], and the Constituent-oriented Age and Residence time Theory (CART) [34,42,43]. PTM is based on a Lagrangian approach by releasing a large amount of numerical particle tracers and calculating WA from the concentration spectrum. The disadvantage is the high computational cost. While CART is based on the Eulerian method, and thus no numerical particle tracer is required, the actual transport trajectory is not provided in the model result [44]. The governing equation for the evolution of the WA concentration distribution function in CART is

$$\frac{\partial c_i}{\partial t} = p_i - d_i - \nabla \cdot (\mathbf{u}c_i - \mathbf{K} \cdot \nabla c_i) - \frac{\partial c_i}{\partial \tau}, \tag{1}$$

where c_i is concentration, t is time, $\mathbf{u}$ is velocity, τ is water age number, p_i and d_i are source and sink terms, $\mathbf{K}$ is the eddy diffusivity tensor, and $-\mathbf{K} \cdot \nabla c_i$ is the diffusive flux. Thus, the mean age at a given location $\mathbf{x}$ is calculated with Equation (2) based on the assumption that the mean age of a set of water parcels is mass-weighted, and thus, arithmetically averaged.

$$a_i(t, \mathbf{x}) = \frac{\int_0^\infty \tau c_i(t, \mathbf{x}, \tau) \, d\tau}{\int_0^\infty c_i(t, \mathbf{x}, \tau) \, d\tau}. \tag{2}$$

A numerical WA simulation based on PTM usually combines the hydrodynamic model with a random-walk model of horizontal eddy diffusion for the statistical treatment of turbulent mixing.

The position of a random particle is described by the following function:

$$\mathbf{x}(t + \Delta t) = \mathbf{x}(t) + \mathbf{u}\Delta t + \mathbf{z_n}\sqrt{2K\Delta t}, \tag{3}$$

where $\mathbf{x}(t)$ is the particle's position at time t, Δt is the time step, $\mathbf{u}$ is the velocity vector, $\mathbf{K}$ is the eddy diffusion tensor from a hydrodynamic model, and $\mathbf{z_n}$ is the normally distributed random vector with unit standard deviation and zero average value.

3. Methodology

3.1. Study Area

Taihu Lake is located in the lower part of Yangtze River Delta in the southeastern part of China, between 30°05′ N and 32°08′ N and between 119°08′ E and 122°55′ E. [27]. As a typical large shallow lake, Taihu Lake has an average depth of 1.9 m and maximum depth of no more than 3 m, while the total surface area is 2338 km². The Taihu Lake Basin, with an area of 36,900 km², has a typical subtropical monsoon climate, with a mean annual precipitation around 1200 mm, concentrated mainly in the monsoon season between May and September. The dominant prevailing wind direction is southeasterly in summer and reverses in winter, with the average wind speed ranging from 3.5 m/s to 5 m/s. [45]

Around Taihu Lake, there are over 150 tributaries connecting to the adjacent river networks, some of which are very seasonal. The altitude to the northwest of Taihu Lake is higher than the south and the east, thus water normally flows from the northwest to the southeast. However, since Taihu Lake is located at a very developed area, many artificial hydraulic structures have been constructed. Thus, the discharge and flow direction near the river inlet have been remarkably altered by human interventions.

3.2. Numerical Model Description

Delft3D, an integrated open-source modelling software developed by Deltares (Delft, The Netherlands), is used to simulate the hydrodynamics of Taihu Lake and the temporal and spatial varying water age (WA) distribution in this study. In particular, its hydrodynamic (FLOW) and water quality (WAQ) modules are applied. Delft3D-WAQ is a multi-dimensional water quality model framework, which solves the advection-diffusion-reaction equation on a predefined computational grid for a wide range of model sub-stances. Delft3D-WAQ simulation includes large numbers of substances and processes. Applications of Delft3D-WAQ include, amongst others, the eutrophication of lakes and reservoirs, dissolved oxygen depletion in stratified systems, the impact of a sewage outfall on nutrient concentrations and primary production, transport of heavy metals through an estuary, accumulation of organic micro-pollutants in fresh water basins, and the emission of greenhouse gases from reservoirs. Hydrodynamic information has been derived from the Delft3D-FLOW model [46].

The mass balance equation in Delft3D-WAQ is:

$$M_i^{t+\Delta t} = M_i^t + \Delta t \times \left(\frac{\Delta M}{\Delta t} \right)_{Tr} + \Delta t \times \left(\frac{\Delta M}{\Delta t} \right)_{P} + \Delta t \times \left(\frac{\Delta M}{\Delta t} \right)_{S}, \tag{4}$$

where M_i^t is the mass at the beginning of time step t; $\left(\frac{\Delta M}{\Delta t} \right)_{Tr}$ represents the mass changes by transport, including both advective and dispersive transport; $\left(\frac{\Delta M}{\Delta t} \right)_{P}$ represents the mass changes by physical, (bio)chemical, or biological processes; and $\left(\frac{\Delta M}{\Delta t} \right)_{S}$ represents the mass changes by sources (e.g., waste loads, river discharges).

Mass transport by advection and dispersion in Delft3D-WAQ is:

$$\frac{\partial C}{\partial t} = D_x \frac{\partial^2 C}{\partial x^2} - v_x \frac{\partial C}{\partial x} + D_y \frac{\partial^2 C}{\partial y^2} - v_y \frac{\partial C}{\partial y} + D_z \frac{\partial^2 C}{\partial z^2} - v_z \frac{\partial C}{\partial z}, \tag{5}$$

where $\frac{\partial C}{\partial t}$ is the concentration gradient, D_x is the dispersion coefficient in the x direction, and v_x is the velocity in the x direction.

3.3. Age Calculation in Delft3D

In the Delft3D model, WA calculation is similar to CART based on the mass concentration ratio of two kinds of tracers, namely, the conservative tracer and decayable tracer. The mass of the conservative tracer remains the same amount as at the released time, while the mass of the decayable tracer will decay with time at a given decay rate. The decayable tracers do not need to necessarily exist or be released into the real world—they are a reference to compute the water age in numerical modelling. The mechanism is that the two kinds of tracers will be released at the same time, and since they participate in the advection and diffusion process at the same time, the ratio of their concentration will remain the same at a fixed time. With the decay rate, the water age can be easily achieved. For the conservative tracer, the time derivative of concentration is

$$\frac{\partial c}{\partial t} = advection + dispersion + source, \tag{6}$$

while for the decayable tracer, it is

$$\frac{\partial c}{\partial t} = advection + dispersion + source - Kc, \tag{7}$$

where K is the decay rate. The formulation used to calculate water age in this study is:

$$ageTr_i = \frac{\ln\left(\frac{dTr_i}{cTr_i}\right)}{RcDecTr_i} \tag{8}$$

$$dDecTr_i = RcDecTr_i \times dTr_i,$$

where $ageTr_i$ is the age of the tracer $i[d]$; cTr_i is the concentration of the conservative tracer $i[gm - 3]$; dTr_i is the concentration of the decayable tracer $i[gm - 3]$; $RcDecTr_i$ is the first-order decay rate constant for the decayable tracer $i[d - 1]$; and $dDecTr_i$ is flux for the decayable tracer $i[gm - 3d - 1]$.

3.4. Model Setup

The hydrodynamic section of the Taihu Lake model was developed by [45] with Delft3D. This model uses a rectangular grid with a grid resolution of 1000 m horizontally, and five vertical sigma layers uniformly defined in depth. The model is driven by tributary discharge boundaries and meteorological conditions, like surface wind, evaporation, and precipitation. Over 150 tributaries have been arranged into 21 groups for simplicity. The simulation time is during the entire year of 2008, with a time step of 10 min. Detailed physical and numerical parameter sets are listed in the paper [45].

Based on upstream catchment sub-basins [22] and the boundary condition of the hydrodynamic model [45], inflow tributary boundaries have been categorized into three groups for the WA simulation (Figure 1). For each group of boundaries, a set of conservative and decayable tracers are continuously released, with the decay rate of the decayable tracers set to be 0.01/day. The northern and northwestern boundaries mainly represent the discharges from Jiangsu province (WA1), while the southern and southwestern boundaries include mountainous river discharges and tributaries from Zhejiang province (WA2). Northeastern boundaries mainly account for water transfer from the Yangtze River (WA3). The WA simulation time step is the same as in the hydrodynamic model, and the model result is recorded at every 6 h of model time. Note, both inflow and outflow occur in the above-mentioned boundaries throughout the year. Tracers released with the outflow boundary condition will be transported out of the model domain and not be included in the WA calculation. Based on Taihu Lake's geometry and hydrological features, the lake is divided into seven sub-basins. For each sub-basin, an observation point is set to monitor the WA distribution (Figure 1).

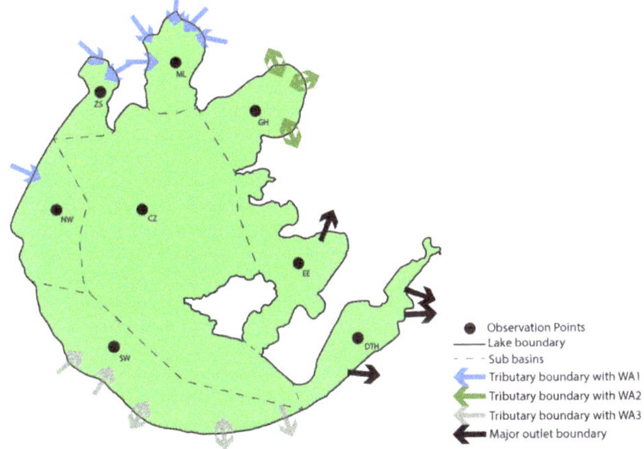

Figure 1. Tributary discharge, observation points, and sub-basins of Taihu Lake. Directions of arrows imply the major inflow/outflow direction of tributary discharge. Abbreviations stand for names of sub-basins, namely, ML: Meiliang Bay; GH: Gonghu Bay; EE: East Epigeal; DTH: Dongtaihu Bay; SW: Southwest Zone; NW: Northwest Zone; ZS: Zhushan Bay; CZ: Central Zone.

3.5. Scenarios

The calibrated hydrodynamic model was used to quantitatively investigate the influence from the surrounding river network and wind, focusing on tributary discharges, wind direction, and wind speed. A series of numerical scenarios were conducted (Table 1). A reference scenario was set up to represent the real tributary discharge and the wind record in 2008 (Scenario 1). With the actual boundary discharge conditions and wind directions, wind speed in 2008 downscaled with a ratio 0.5 and 0 (Scenario 2, 3). The other cases were designed to investigate the influence of the prevailing wind and the magnitude of inflow boundary discharge (Scenarios 4–11). Further, the influence of all wind directions was studied (Scenarios 12–18). For the constant tributary discharge case, the same amount of water is flowing out of the model domain through the major outlet boundary as the inflow of water.

Table 1. Scenarios.

Scenario	Wind		Discharge for Each WA Inlet (m^3/s)		
	Direction	Speed (m/s)	WA1	WA2	WA3
1	2008 data	2008 data	2008 data	2008 data	2008 data
2	2008 data	half 2008 data	2008 data	2008 data	2008 data
3	No wind	no wind	2008 data	2008 data	2008 data
4	SE	3.5	10	10	10
5	SE	5	10	10	10
6	SE	3.5	20	20	20
7	SE	5	20	20	20
8	NW	3.5	10	10	10
9	NW	5	10	10	10
10	NW	3.5	20	20	20
11	NW	5	20	20	20
12	No wind	/	10	10	10
13	S	3.5	10	10	10
14	SW	3.5	10	10	10
15	W	3.5	10	10	10
16	NW	3.5	10	10	10
17	N	3.5	10	10	10
18	NE	3.5	10	10	10

4. Results

4.1. Spatial and Temporal Distribution of WA

Influenced by the time-varying wind field and hydrodynamics, WA distribution for all three groups of tracers changed both spatially and temporally. Bottom and surface WA1 distribution at the last time-step of Scenario 1 is shown in Figure 2. Little WA1 vertical difference (~0.01 day) between the surface and bottom layer is observed from the model result, as well as in WA2 and WA3. The consistency in vertical WA distribution implies that, although surface and bottom horizontal flow velocity fields differ hugely [45], WA is fully mixed in vertically within each time step.

WA distribution varies spatially for WA1; the highest WA1 is over 200 days in Dongtaihu Bay, while the lowest WA1 is less than 30 days in Zhushan Bay. The phenomenon could be explained by the difference in distance from the inflow WA1 tributaries. WA1 values near the northern and western part of Taihu Lake (~30–90 days at Zhushan Bay, Meiliang Bay, Gonghu Bay, and the northern part of the Central Zone) are significantly smaller than the values of the southern and eastern part of the lake (~120–200 days at the Northwestern Zone and Dongtaihu Bay), since the WA1 tributary boundary is located in the northern and western part of the lake.

High temporal heterogeneity is also observed in the model results for WA1 distribution after each quarter of the year in Scenario 1 (Figure 3). As time passes since the model's starting time, the maximum WA1 increases. By definition, this value will not exceed the model time passed, but the location with the value varies. At the end of Q1, the maximum value (~90 days) is located at the eastern part of the lake; one quarter later, the maximum value (~120 days) moves to the northeastern part in Gonghu Bay, and the northwestern part at Dongtaihu Bay; after another quarter, the maximum value (~180 days) occurs only in Gonghu Bay in the northeast, and at the end of year, the peak value (~200 days) lies in Dongtaihu Bay. In contrast, the lowest WA1 value occurs near the WA1 tributary boundary throughout the whole year.

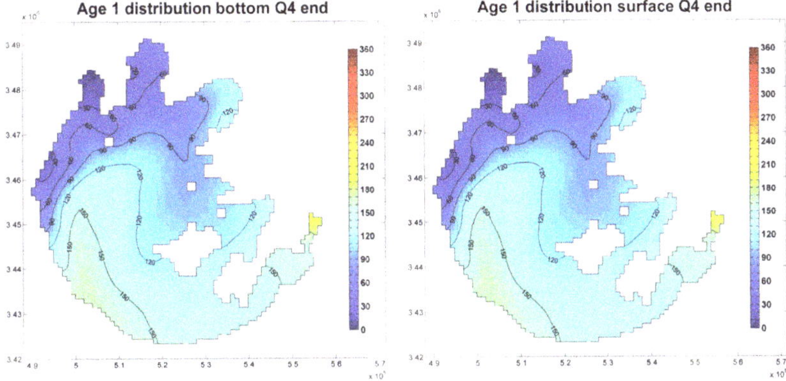

Figure 2. WA1 distribution of Scenario 1 at the last time step. (Unit: days).

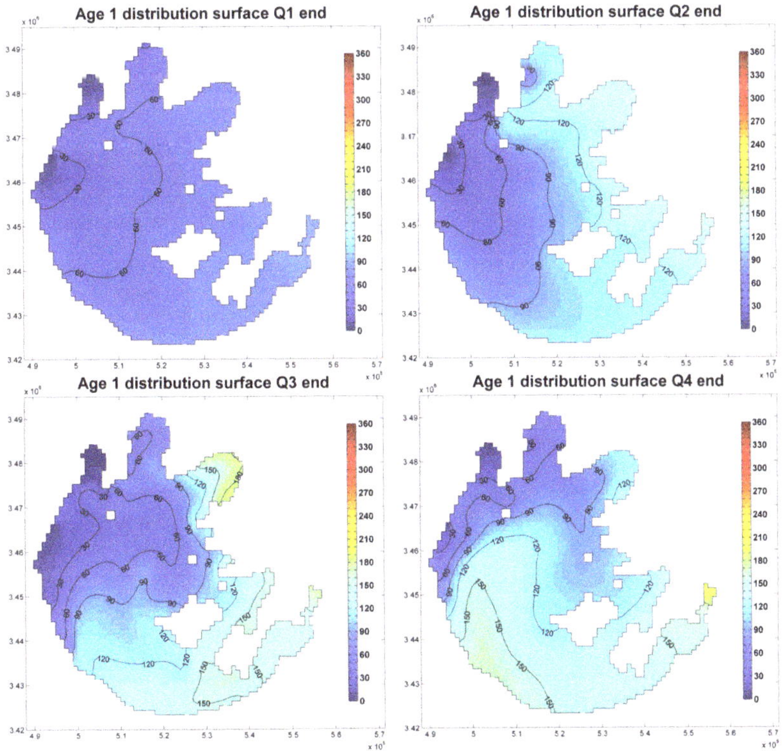

Figure 3. WA1 distribution at the end of each quarter. (Unit: days).

Since the tributary boundaries for each WA are located at different locations around Taihu Lake, the distribution of each WA also varies, both spatially and temporally. WA distributions of the final time step in Scenario 1 are shown in Figure 4. For WA1, the northern sub-basins (Zhushan Bay, Meiliang Bay, Gonghu Bay, and the northern part of the Central Zone and East Epigeal) have a lower WA1, while the Southwestern Zone and Dongtaihu Bay have a higher WA1. For WA2, WA in the western half of the lake is higher than 240days, implying that hardly any water from the WA2 boundary has reached this part of lake; while for WA3, small WA occurs near the western margin of the lake.

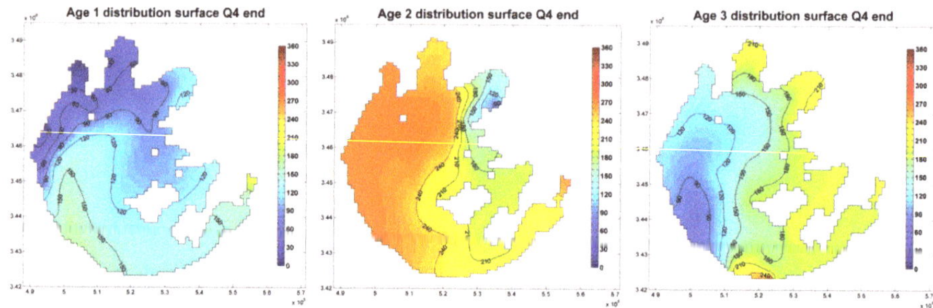

Figure 4. Water age (WA) distribution for Scenario 1 at the end of the year. (Unit: days).

Besides the distance to the tributary boundaries, the variance in WA distribution could be explained by the value of total discharge through the tributary boundary for each WA group (Figure 5).

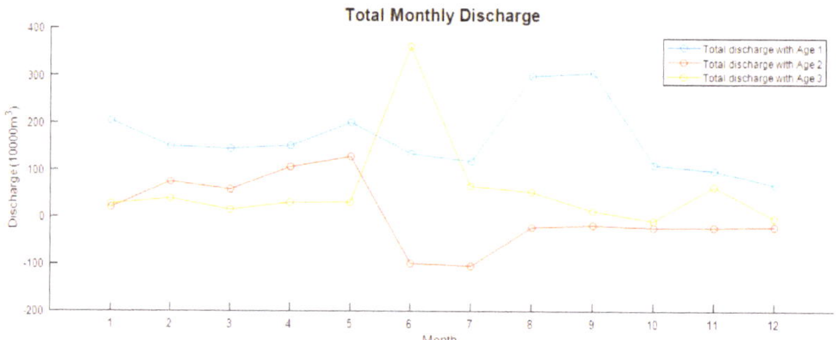

Figure 5. Total discharge for each WA discharge.

Total discharge for WA2 peaks in June then becomes almost zero for the next 5 months, which could explain the extreme high WA2 in the western half of the lake, since water from WA2 barely enters this area. While total discharge for WA1 is always positive and larger than around 100 m^3/s, lower WA1 occurs in around half the area of the lake. For WA3, discharge remains negative since June, but since two inflow boundaries among WA3 boundaries still have positive discharge from the mountainous area, the area near the western margin of the lake still has a smaller WA3. However, for all three WA, larger values occur in Dongtaihu Bay, suggesting that less inflow tributary water enters this sub-basin. This is possibly due to the narrow entrance and elongated geometry of Dongtaihu Bay.

In general, under the influence of time-varying inflow discharge, WA distributions show heterogeneity both spatially and temporally.

4.2. Wind Speed and Direction Effects

Wind influence on WA is studied by comparing the WA value at the last time step of each steady wind scenario with tributary discharge at 10 m^3/s. The WA of steady wind for each observation point differs at both the average and range values. This difference could be explained by the influence from hydrodynamics. With steady wind, the horizontal circulation patterns of Taihu Lake differ with the wind direction, which in turn influence the advection and mixing processes of inflow tributary discharge. Thus, the corresponding range and average of the WA value varies. For example, WA1 distribution within the Southwestern Zone ranges from the highest in an east wind condition (∼190 days) to the lowest with a south wind condition (∼120 days) with the average WA1 being around 140 days; while for WA3, the situation is different, and the highest WA3 is with a southwestern wind (∼120 days) and the lowest WA is with a north wind (∼70 days), with the lowest WA being around 100 days (Figure 6).

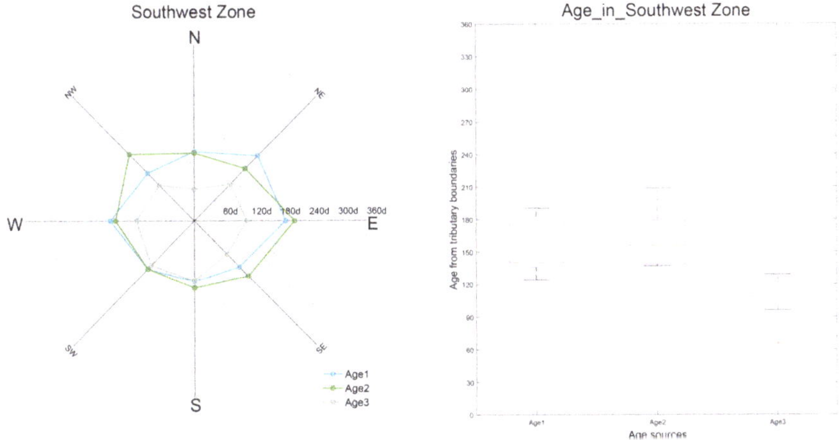

Figure 6. WA for the Southwestern Zone observation point with eight wind scenarios.

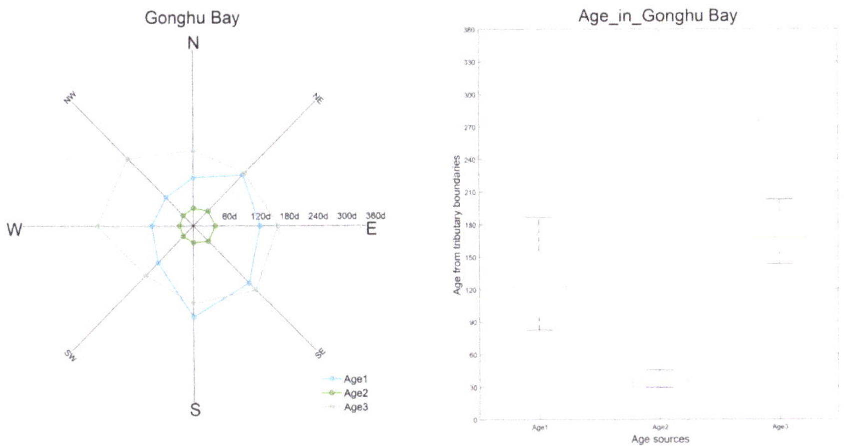

Figure 7. WA for the Gonghu Bay observation point with eight wind scenarios.

Moreover, in some sub-basins where inflow tributary discharge is nearby, the wind influence is relatively low, such as in Gonghu Bay, where WA2 boundaries are adjacent. With eight wind directions, the range of WA2 is less than 30 days, while at the same location the difference between the maximum

WA1 and lowest WA1 is larger than 100 days (Figure 7). The bias in WA implies that WA2 tributary inflow contributes more to the water retention in Gonghu Bay than WA1 and WA3 tributary discharges.

Beside wind directions, wind speed is also an important factor influencing WA distribution. In Scenario 2, the wind speed of the entire simulation is half the wind speed in scenario, which is from the real 2008 wind data. As illustrated in the comparison of the model result (Figure 8), differences occur for all three WA distributions. However, the increase or decrease of WA is site-specific and WA-specific. For the Northwestern Zone, with half wind, WA1 decreases while WA2 and WA3 increases. However, for Gonghu Bay, with half wind, all three WA values increase.

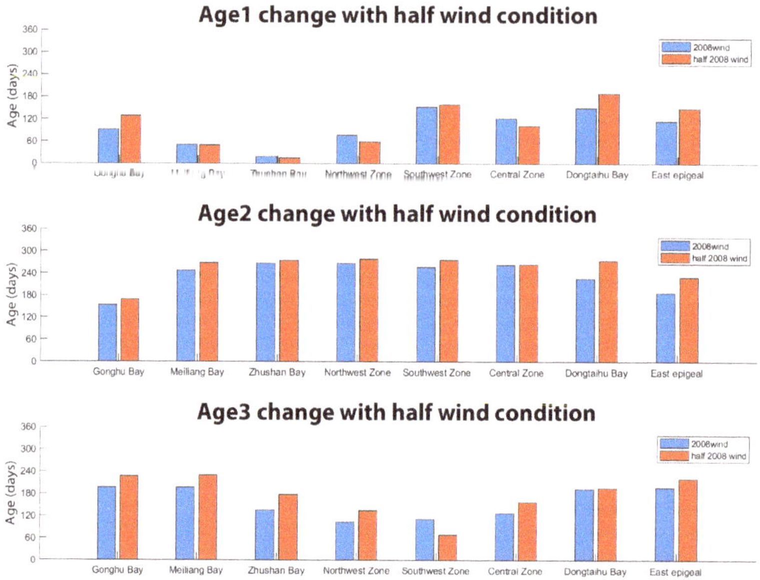

Figure 8. Water age comparison between the 2008 real wind data scenario and half wind speed scenario.

In general, wind direction and wind speed do influence WA distribution over the whole Taihu Lake. The influence is spatially heterogeneous. Change of wind direction would lead to a change in both the average value and the range of WA, while wind speed difference induces a site-specific and age-specific change of the WA value.

4.3. Discharge Effects

Discharge influence is studied by comparing scenarios with the same wind condition, but using a different inflow discharge rate. Flow discharge in Scenario 4 and Scenario 6 are 10 m^3/s and 20 m^3/s in each WA inflow tributary, respectively, while the outflow discharge at the remaining boundaries are calculated to ensure a mass balance between inflow and outflow. For both scenarios, a steady southeastern wind with 3.5 m/s wind speed is set.

The WA for all three WAs of all observation points decreases with a rising tributary inflow discharge (Figure 9), which could be explained by the enhancing hydrodynamics due to more momentum input through the tributary boundaries. The largest WA difference is the WA2 change in East Epigeal (45 days), while the smallest WA difference is the WA1 change in Zhushan Bay (13 days). Again, changes in WA show spatial heterogeneity, and that it is WA-specific.

Comparing with the impact of wind speed and wind direction change, the influence of discharge on WA is smaller, partially because it is easier to dampen the increasing momentum from tributary discharge when it has penetrated farther into the lake, while wind momentum input through surface shear stress is continuous all over the lake.

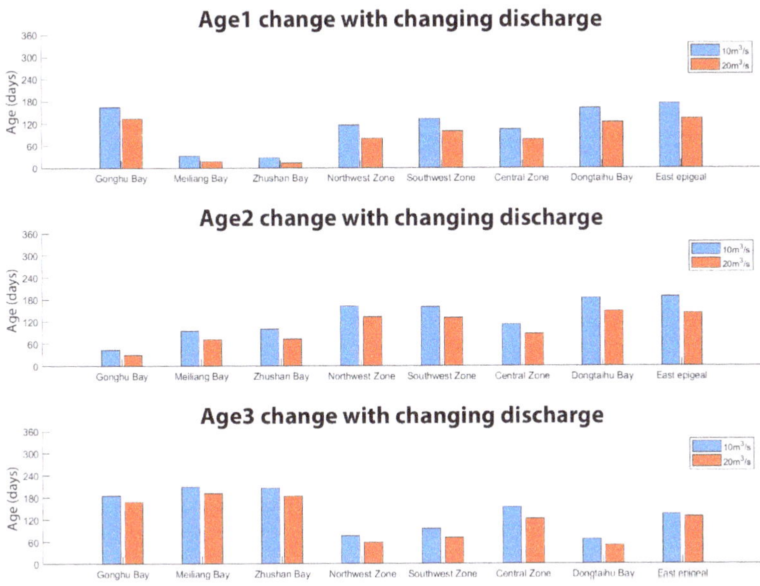

Figure 9. Water age comparison between the 2008 real wind data scenario and half wind speed scenario.

5. Discussion

5.1. Various Transport Time Scales

Transport time scales are frequently adopted in describing the hydrodynamic processes, which transport water and the constituents. Water age is one of the most favorable transport time scales, while the usage of flushing time and residence time is also very common. To extensively adopt transport time scales in large shallow lake studies, understanding the definition and limitations of these time scales is crucial.

Residence time is the time spent by a water parcel or a pollutant to leave the given water body [47]. By definition, resident time is a location-specific value, such as water age, and serves as a complement to water age, since water age is the duration for a water parcel from the inflow boundary to a given spot, while residence time is the duration from this particular location to the outflow boundary [30]. Residence time is commonly used to evaluate the inflow nutrient's further influence inside the water body [48].

Flushing time, on the other hand, is a bulk parameter to describe the exchange of water body. Flushing time is defined as the time it takes to replace all the water in a basin [32]. Flushing time could be seen as the sum of water age and residence time. This concept is frequently used in estuary and lagoon research, and early studies can be traced back to the 1950s [49]. The original method to get flushing time is the "tidal prism method"—that is, flushing time is calculated as the ratio of the mass of a scalar to the rate of renewal of the scalar, with the assumption of an instantaneous release of inflow and thorough mixing inside the water body. Another approach considering salt balance has also been adopted in previous studies [30,50]. Further studies have improved the "tidal prism method" to mitigate underestimation due to the idealized assumptions; however, the thoroughly mixing assumption is still adopted [51].

Beside residence time and flushing time, some terminologies like transit time and turn-over time are also found in literature. By definition, turn-over time is identical to flushing time, and transit time is the average residence time [32].

Based on the definition of these transport time scales, the application of these time scales are purpose-oriented. Water age is more suitable when considering the spatial distribution of influence from tributary discharge into the large shallow lake as in this study, while residence time could help to study the dilution of pollution already inside the lake. Flushing time, as the sum of water age and residence time, could be used to indicate the temporal extent and the efficiency of diluted fresh water from the external waterbody.

5.2. Radio-Age

In this study, water age is calculated with a concentration of both conservative and decayable tracers, as described in Section 3.1. In previous studies, it is usually referred to as "radio-age" [52]. However, theoretically, the value is between the water age of passive tracers (or water parcels) and of radioactive tracers. A 10% bias with water age larger than 7 years has also been reported [52]. Choice of decay rate is crucial for the modelled radio age value, and with a smaller decay rate, the difference between radio-age and water age of passive tracers is small (Equation (9)).

$$\lim_{\gamma \to 0} \tilde{a}(t, \mathbf{x}, \gamma) = a(t, \mathbf{x}, 0),\tag{9}$$

where γ is the decay rate, $\tilde{a}(t, \mathbf{x}, \gamma)$ is the calculated radio- age, and $a(t, \mathbf{x}, 0)$ is the age of the passive tracers.

To verify this further, the modelled water age averaged over the whole lake after 1 year of simulation was investigated with additional numerical tests using five decay rates, ranging from $10^{-6}/d$ to $10^{-2}/d$ (Figure 10). The age value is almost identical with a decay rate less than $10^{-3}/d$ and the difference is less than days, while the largest difference in mean water age with $10^{-2}/d$ and $10^{-6}/d$ is around 20 days.

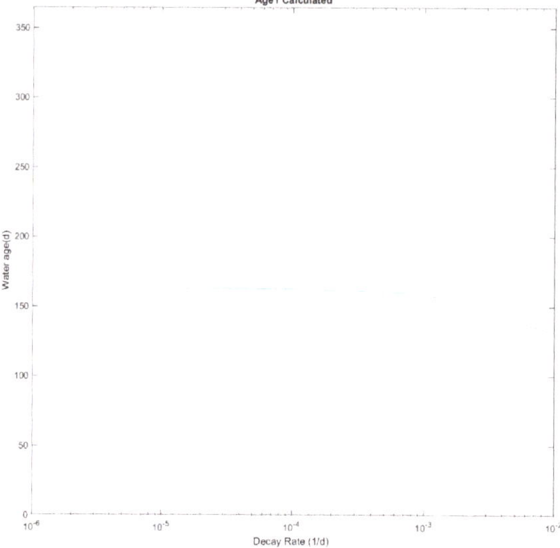

Figure 10. Water age modelled with various decay rates.

This model's results are similar to that in Figure 1 by Delhez et al. (2003) [52], where the curve of radio-age is asymptotic to the line representing the passive tracers. With a small decay rate, the value of radio-age is close to the age of passive tracers. Thus, we believe that with a smaller decay rate (say, less than $10^{-3}/d$), the radio-age distribution gives a similar indication on water horizontal circulation as the age of radioactive tracers do.

Radio age results provide diagnoses consistent with the model results interpretation, that the relative importance of water injection from tributaries in this study could be compared. Furthermore, the advantage of the Eulerian approach is that numerical results are easier to obtain than in the Lagrangian formalism [52].

5.3. Wind Change Due to Climate Change

Terrestrial near-surface wind speed has been reported to decrease due to climate change. During the last 30 years, 73% of terrestrial stations record average wind speed decline across most of the northern mid-latitudes [53–55]. For large shallow lakes like Taihu Lake, where wind influences not only the hydrodynamic conditions [45] but also the ecological statues [6,7], climate change-induced wind condition variation and its consequences should be granted more attention.

Nutrient loads in shallow water mainly come from two sources, namely, the internal sources and the external sources. Albeit low wind speed causes low waves and low corresponding bottom shear stresses, hampering sediment resuspension which is crucial for the release of internal nutrient sources, it promotes hypoxia in the bottom layer of the water column, and in turn, enhances nutrient release from sediment and counteracts the effect of declining resuspension, stimulating algae growth and finally leading to more severe eutrophication. More effort should be spent on this with numerical models.

The combination of summer water level increase and wind speed decrease, [14,56], is expected to change the distribution of nutrients from external sources in shallow lakes. To illustrate the actual influence of a change in wind speed, a comparison of model results with a steady southeastern wind but changing wind speed (Scenario 4 and Scenario 5) is illustrated (Figure 11).

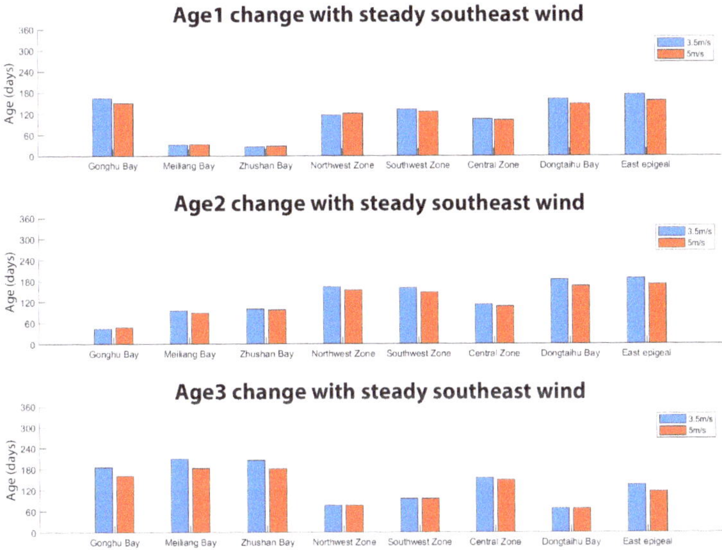

Figure 11. WA change with 3.5 m/s and 5 m/s southeastern wind.

For most observation points, wind speed decline causes an increase in WA values for all three WAs. Since less wind speed weakens the wind-induced hydrodynamics in most parts of the lake, the advection and mixing processes of incoming water is attenuated. Moreover, increased WA means that external nutrient input would stay longer in the lake, increasing the chances for cyanobacteria to capture more nutrients and to form an algae bloom [57].

Thus, less wind speed will encourage the release of inner sources and cause a longer duration of outer source nutrient inflow, both of which will induce more severe algae bloom and deterioration of water quality.

5.4. Implication of Water Age on Shallow Lake Management

The initial water age study mainly has two applications: (a) To assess the ventilation rate of semi-closed basins; (b) to infer the horizontal circulation, which focuses more on estuaries and lagoons [34]. Further studies in broader aqua systems have paid more attention to water quality issues, such as the efficiency of water transfer [27]. With climate change and a more complicated nutrient control policy, water age is able to provide more assistance in integrated water quality management of shallow lakes, such as Taihu Lake.

Firstly, water age analysis would provide help when a critical toxin leakage condition happens. With numerical models and meteorological forecasts, water age distribution maps could be generated in minutes. Thus, the spatial and temporal spread information of inflow toxins could be provided, and further measures could be decided based on that.

Secondly, water age analysis would provide information on the tributary discharge influence of critical spots, assisting the governance of water quality in a complicated management condition. In this study, three water age groups have been chosen corresponding to inflow tributary discharge from three municipalities. Situations become complicated when nutrient control and wastewater treatment involves more stakeholders. Water age analysis could also provide essential information to divide the responsibility and help improve the master plan of Taihu Basin's water quality management.

6. Conclusions

The impact of tributary discharge inflow from river discharge around Taihu Lake, the third largest fresh water lake in China, has been investigated using the concept of water age. The main purposes of this study were to provide quantitative comparison of nutrient loads from different parts of the catchment river networks and to investigate the meteorological influences on the advection and mixing process of nutrients from tributary discharge inside the lake body. In this study, the inflow tributaries were divided into three groups based on upstream catchment sub-basins and the boundary condition of the hydrodynamic model. Water age was computed using the three-dimensional Delft3D model with FLOW and WAQ module.

Model results show both spatial and temporal heterogeneity occurred in all three water age groups, which was influenced by both the distance to the tributary boundaries and total discharge through tributary boundaries for each water-age group. The influence of wind on water age was also analyzed. Change of wind direction would lead to changes in both the average value and range of water age, while wind speed difference would induce site-specific and group-specific changes of WA value. Water age decreases with rising of tributaries inflow discharge; however, the influence of discharge is less significant than that of a change of wind.

Various time scales, such as residence time and flushing time, have been discussed for clearer understanding. Wind speed decline induced by climate change was analyzed on the effect on both internal and external nutrient source release, and influences on both sources would cause water quality to be deteriorated. Lastly, further application of water age is suggested for more complicated integrated water management on a lake basin scale.

Author Contributions: Funding acquisition, S.W.; Methodology, S.L. and Q.Y.; Writing—original draft, S.L.; Writing—review and editing, Q.Y., S.W. and M.J.F.S. All authors have read and agreed to the published version of the manuscript.

Funding: This research was funded by Chinese national key research and development program (2018YFC0407200), China Scholarship Council (CSC) (201407720008) and Het Lamminga Fonds.

Acknowledgments: The authors would like to thank TBA for providing tributary discharge data.

Conflicts of Interest: The authors declare no conflict of interest.

References

1. Chen, Y.; Fan, C.; Teubner, K.; Dokulil, M. Changes of nutrients and phytoplankton chlorophyll-a in a large shallow lake, Taihu, China: An 8-year investigation. *Hydrobiologia* **2003**, *506–509*, 273–279.:HYDR.0000008604.09751.01. [CrossRef]

2. Janssen, A.B.; Teurlincx, S.; An, S.; Janse, J.H.; Paerl, H.W.; Mooij, W.M. Alternative stable states in large shallow lakes? *J. Great Lakes Res.* **2014**, *40*, 813–826. [CrossRef]

3. Jin, X. Analysis of eutrophication state and trend for lakes in China. *J. Limnol.* **2003**, *62*, 60. [CrossRef]

4. Duan, H.; Ma, R.; Xu, X.; Kong, F.; Zhang, S.; Kong, W.; Hao, J.; Shang, L. Two-Decade Reconstruction of Algal Blooms in China's Lake Taihu. *Environ. Sci. Technol.* **2009**, *43*, 3522–3528. [CrossRef] [PubMed]

5. Guo, L. ECOLOGY: Doing Battle With the Green Monster of Taihu Lake. *Science* **2007**, *317*, 1166. [CrossRef] [PubMed]

6. Paerl, H.W.; Hall, N.S.; Calandrino, E.S. Controlling harmful cyanobacterial blooms in a world experiencing anthropogenic and climatic-induced change. *Sci. Total Environ.* **2011**, *409*, 1739–1745. [CrossRef] [PubMed]

7. Qin, B.; Zhu, G.; Gao, G.; Zhang, Y.; Li, W.; Paerl, H.W.; Carmichael, W.W. A Drinking Water Crisis in Lake Taihu, China: Linkage to Climatic Variability and Lake Management. *Environ. Manag.* **2010**, *45*, 105–112. [CrossRef]

8. McGowan, S. Algal Blooms. In *Biological and Environmental Hazards, Risks, and Disasters*; Elsevier: Amsterdam, The Netherlands, 2016; pp. 5–43. [CrossRef]

9. He, W.; Shang, J., Lu, X.; Fan, C. Effects of sludge dredging on the prevention and control of algae-caused black bloom in Taihu Lake, China. *J. Environ. Sci.* **2013**, *25*, 430–440. [CrossRef]

10. Li, Y.; Tang, C.; Wang, C.; Anim, D.O.; Yu, Z.; Acharya, K. Improved Yangtze River Diversions: Are they helping to solve algal bloom problems in Lake Taihu, China? *Ecol. Eng.* **2013**, *51*, 104–116. [CrossRef]

11. Sun, X.; Xiong, S.; Zhu, X.; Zhu, X.; Li, Y.; Li, B.L. A new indices system for evaluating ecological-economic-social performances of wetland restorations and its application to Taihu Lake Basin, China. *Ecol. Model.* **2015**, *295*, 216–226. [CrossRef]

12. Ding, Y.; Xu, H.; Deng, J.; Qin, B.; He, Y. Impact of nutrient loading on phytoplankton: A mesocosm experiment in the eutrophic Lake Taihu, China. *Hydrobiologia* **2019**, *829*, 167–187. [CrossRef]

13. Janssen, A.B.; de Jager, V.C.; Janse, J.H.; Kong, X.; Liu, S.; Ye, Q.; Mooij, W.M. Spatial identification of critical nutrient loads of large shallow lakes: Implications for Lake Taihu (China). *Water Res.* **2017**, *119*, 276–287. [CrossRef] [PubMed]

14. Ke, Z.; Xie, P.; Guo, L. Ecological restoration and factors regulating phytoplankton community in a hypertrophic shallow lake, Lake Taihu, China. *Acta Ecol. Sin.* **2018**, *39*, 81–88. [CrossRef]

15. Paerl, H.W.; Xu, H.; McCarthy, M.J.; Zhu, G.; Qin, B.; Li, Y.; Gardner, W.S. Controlling harmful cyanobacterial blooms in a hyper-eutrophic lake (Lake Taihu, China): The need for a dual nutrient (N & P) management strategy. *Water Res.* **2011**, *45*, 1973–1983. [CrossRef]

16. Xu, X.; Li, W.; Fujibayashi, M.; Nomura, M.; Nishimura, O.; Li, X. Asymmetric response of sedimentary pool to surface water in organics from a shallow hypereutrophic lake: The role of animal consumption and microbial utilization. *Ecol. Indic.* **2015**, *58*, 346–355. [CrossRef]

17. Deng, X.; Xu, Y.; Han, L.; Song, S.; Yang, L.; Li, G.; Wang, Y. Impacts of Urbanization on River Systems in the Taihu Region, China. *Water* **2015**, *7*, 1340–1358. [CrossRef]

18. Bozelli, R.L.; Caliman, A.; Guariento, R.D.; Carneiro, L.S.; Santangelo, J.M.; Figueiredo-Barros, M.P.; Leal, J.J.; Rocha, A.M.; Quesado, L.B.; Lopes, P.M.; et al. Interactive effects of environmental variability and human impacts on the long-term dynamics of an Amazonian floodplain lake and a South Atlantic coastal lagoon. *Limnologica* **2009**, *39*, 306–313. [CrossRef]

19. Chen, C.; Zhong, J.C.; Yu, J.H.; Shen, Q.S.; Fan, C.X.; Kong, F.X. Optimum dredging time for inhibition and prevention of algae-induced black blooms in Lake Taihu, China. *Environ. Sci. Pollut. Res.* **2016**, *23*, 14636–14645. [CrossRef]

20. Huang, K.; Guo, H.; Liu, Y.; Zhou, F.; Yu, Y.; Wang, Z. Water environmental planning and management at the watershed scale: A case study of Lake Qilu, China. *Front. Environ. Sci. Eng. China* **2008**, *2*, 157–162. [CrossRef]

21. Xu, J.; Chen, Y.; Zheng, L.; Liu, B.; Liu, J.; Wang, X. Assessment of Heavy Metal Pollution in the Sediment of the Main Tributaries of Dongting Lake, China. *Water* **2018**, *10*, 1060. [CrossRef]

22. Wang, M.; Strokal, M.; Burek, P.; Kroeze, C.; Ma, L.; Janssen, A.B. Excess nutrient loads to Lake Taihu: Opportunities for nutrient reduction. *Sci. Total Environ.* **2019**, *664*, 865–873. [CrossRef] [PubMed]

23. Yu, G.; Xue, B.; Lai, G.; Gui, F.; Liu, X. A 200-year historical modeling of catchment nutrient changes in Taihu basin, China. *Hydrobiologia* **2007**, *581*, 79–87. [CrossRef]

24. Zhang, Y.; Lin, S.; Qian, X.; Wang, Q.; Qian, Y.; Liu, J.; Ge, Y. Temporal and spatial variability of chlorophyll a concentration in Lake Taihu using MODIS time-series data. *Hydrobiologia* **2011**, *661*, 235–250. [CrossRef]

25. Shen, J.; Yuan, H.; Liu, E.; Wang, J.; Wang, Y. Spatial distribution and stratigraphic characteristics of surface sediments in Taihu Lake, China. *Chin. Sci. Bull.* **2011**, *56*, 179–187. [CrossRef]

26. de Brye, B.; de Brauwere, A.; Gourgue, O.; Delhez, E.J.; Deleersnijder, E. Reprint of Water renewal timescales in the Scheldt Estuary. *J. Mar. Syst.* **2013**, *128*, 3–16. [CrossRef]

27. Li, Y.; Acharya, K.; Yu, Z. Modeling impacts of Yangtze River water transfer on water ages in Lake Taihu, China. *Ecol. Eng.* **2011**, *37*, 325–334. [CrossRef]

28. Qi, H.; Lu, J.; Chen, X.; Sauvage, S.; Sanchez-Pérez, J.M. Water age prediction and its potential impacts on water quality using a hydrodynamic model for Poyang Lake, China. *Environ. Sci. Pollut. Res.* **2016**, *23*, 13327–13341. [CrossRef]

29. Wu, Z.; Lai, X.; Zhang, L.; Cai, Y.; Chen, Y. Phytoplankton chlorophyll a in Lake Poyang and its tributaries during dry, mid-dry and wet seasons: A 4-year study. *Knowl. Manag. Aquat. Ecosyst.* **2014**, *412*, 6. [CrossRef]

30. Monsen, N.E.; Cloern, J.E.; Lucas, L.V.; Monismith, S.G. A comment on the use of flushing time, residence time, and age as transport time scales. *Limnol. Oceanogr.* **2002**, *47*, 1545–1553. [CrossRef]

31. Delhez, É.J.M.; de Brye, B.; de Brauwere, A.; Deleersnijder, É. Residence time vs influence time. *J. Mar. Syst.* **2014**, *132*, 185–195. [CrossRef]

32. Bolin, B.; Rodhe, H. A note on the concepts of age distribution and transit time in natural reservoirs. *Tellus* **1973**, *25*, 58–62. [CrossRef]

33. Delhez, E.J.; Campin, J.M.; Hirst, A.C.; Deleersnijder, E. Toward a general theory of the age in ocean modelling. *Ocean Model.* **1999**, *1*, 17–27. [CrossRef]

34. Deleersnijder, E.; Campin, J.M.; Delhez, E.J. The concept of age in marine modelling I. Theory and preliminary model results. *J. Mar. Syst.* **2001**, *28*, 229–267. [CrossRef]

35. Jenkins, W.; Clarke, W. The distribution of 3He in the western Atlantic ocean. *Deep Sea Res. Oceanogr. Abstr.* **1976**, *23*, 481–494. [CrossRef]

36. Johnston, C.; Cook, P.; Frape, S.; Plummer, L.; Busenberg, E.; Blackport, R. Ground Water Age and Nitrate Distribution Within a Glacial Aquifer Beneath a Thick Unsaturated Zone. *Ground Water* **1998**, *36*, 171–180. [CrossRef]

37. Karstensen, J.; Tomczak, M. Age determination of mixed water masses using CFC and oxygen data. *J. Geophys. Res. Oceans* **1998**, *103*, 18599–18609. [CrossRef]

38. Pangle, L.A.; Klaus, J.; Berman, E.S.F.; Gupta, M.; McDonnell, J.J. A new multisource and high-frequency approach to measuring δ 2 H and δ 18 O in hydrological field studies. *Water Resour. Res.* **2013**, *49*, 7797–7803. [CrossRef]

39. Wunsch, C. Oceanic age and transient tracers: Analytical and numerical solutions. *J. Geophys. Res.* **2002**, *107*, 3048. [CrossRef]

40. Li, Y.; Tang, C.; Wang, C.; Tian, W.; Pan, B.; Hua, L.; Lau, J.; Yu, Z.; Acharya, K. Assessing and modeling impacts of different inter-basin water transfer routes on Lake Taihu and the Yangtze River, China. *Ecol. Eng.* **2013**, *60*, 399–413. [CrossRef]

41. Zhang, X.Y. Ocean Outfall Modeling–Interfacing Near and Far Field Models with Particle Tracking Method. Ph.D. Thesis, Massachusetts Institute of Technology, Cambridge, MA, USA, 1995.

42. Chen, X. A laterally averaged two-dimensional trajectory model for estimating transport time scales in the Alafia River estuary, Florida. *Estuar. Coast. Shelf Sci.* **2007**, *75*, 358–370. [CrossRef]
43. Liu, W.C.; Chen, W.B.; Hsu, M.H. Using a three-dimensional particle-tracking model to estimate the residence time and age of water in a tidal estuary. *Comput. Geosci.* **2011**, *37*, 1148–1161. [CrossRef]
44. Wang, H.; Guo, X.; Liu, Z.; Gao, H. A comparative study of CART and PTM for modelling water age. *J. Ocean Univ. China* **2015**, *14*, 47–58. [CrossRef]
45. Liu, S.; Ye, Q.; Wu, S.; Stive, M. Horizontal Circulation Patterns in a Large Shallow Lake: Taihu Lake, China. *Water* **2018**, *10*, 792. [CrossRef]
46. Deltares. *Delft3D-WAQ Users Manual*; Technical Report; Deltares: Delft, The Netherlands, 2005.
47. Delhez, É.J.; Heemink, A.W.; Deleersnijder, É. Residence time in a semi-enclosed domain from the solution of an adjoint problem. *Estuar. Coast. Shelf Sci.* **2004**, *61*, 691–702. [CrossRef]
48. Rueda, F.J.; Cowen, E.A. Residence time of a freshwater embayment connected to a large lake. *Limnol. Oceanogr.* **2005**, *50*, 1638–1653. [CrossRef]
49. Choi, K.W.; Lee, J.H. Numerical determination of flushing time for stratified water bodies. *J. Mar. Syst.* **2004**, *50*, 263–281. [CrossRef]
50. Miller, R.L.; McPherson, B.F. Estimating estuarine flushing and residence times in Charlotte Harbor, Florida. via salt balance and a box model. *Limnol. Oceanogr.* **1991**, *36*, 602–612. [CrossRef]
51. Luketina, D. Simple Tidal Prism Models Revisited. *Estuar. Coast. Shelf Sci.* **1998**, *46*, 77–84. [CrossRef]
52. Delhez, É.J.; Deleersnijder, É.; Mouchet, A.; Beckers, J.M. A note on the age of radioactive tracers. *J. Mar. Syst.* **2003**, *38*, 277–286. [CrossRef]
53. McVicar, T.R.; Roderick, M.L.; Donohue, R.J.; Li, L.T.; Van Niel, T.G.; Thomas, A.; Grieser, J.; Jhajharia, D.; Himri, Y.; Mahowald, N.M.; et al. Global review and synthesis of trends in observed terrestrial near-surface wind speeds: Implications for evaporation. *J. Hydrol.* **2012**, *416–417*, 182–205. [CrossRef]
54. Stocker, T.F.; Qin, D.; Plattner, G.K.; Tignor, M.; Allen, S.K.; Boschung, J.; Nauels, A.; Xia, Y.; Bex, V.; Midgley, P.M.; et al. *Climate Change 2013—The Physical Science Basis*; Cambridge University Press: Cambridge, UK, 2014. [CrossRef]
55. Vautard, R.; Cattiaux, J.; Yiou, P.; Thépaut, J.N.; Ciais, P. Northern Hemisphere atmospheric stilling partly attributed to an increase in surface roughness. *Nat. Geosci.* **2010**, *3*, 756–761. [CrossRef]
56. Deng, J.; Paerl, H.W.; Qin, B.; Zhang, Y.; Zhu, G.; Jeppesen, E.; Cai, Y.; Xu, H. Climatically-modulated decline in wind speed may strongly affect eutrophication in shallow lakes. *Sci. Total Environ.* **2018**, *645*, 1361–1370. [CrossRef] [PubMed]
57. Ji, Z.G. *Hydrodynamics and Water Quality*; John Wiley & Sons, Inc.: Hoboken, NJ, USA, 2017. [CrossRef]

 *water*

Article

Consistent Boundary Conditions for Age Calculations

Eric Deleersnijder [1], **Insaf Draoui** [2,*], **Jonathan Lambrechts** [2], **Vincent Legat** [2]
and Anne Mouchet [3]

[1] Institute of Mechanics, Materials and Civil Engineering (IMMC) & Earth and Life Institute (ELI) L4.05.02, Université catholique de Louvain, B-1348 Louvain-la-Neuve, Belgium; eric.deleersnijder@uclouvain.be
[2] Institute of Mechanics, Materials and Civil Engineering (IMMC), Université catholique de Louvain, L4.05.02, B-1348 Louvain-la-Neuve, Belgium; jonathan.lambrechts@uclouvain.be (J.L.); vincent.legat@uclouvain.be (V.L.)
[3] Freshwater and OCeanic science Unit of reSearch (FOCUS), Sart-Tilman B5a, Université de Liège, B-4000 Liège, Belgium; A.Mouchet@uliege.be
* Correspondence: insaf.draoui@uclouvain.be; Tel.: +32-10-47-22-16

Received: 18 March 2020; Accepted: 27 April 2020; Published: 30 April 2020

Abstract: Age can be evaluated at any time and position to understand transport processes taking place in the aquatic environment, including for reactive tracers. In the framework of the Constituent-oriented Age and Residence time Theory (CART), the age of a constituent or an aggregate of constituents, including the water itself, is usually defined as the time elapsed since leaving the boundary where the age is set or reset to zero. The age is evaluated as the ratio of the age concentration to the concentration, which are the solution of partial differential equations. The boundary conditions for the concentration and age concentration cannot be prescribed independently of each other. Instead, they must be derived from boundary conditions designed beforehand for the age distribution function (the histogram of the ages, the age theory core variable), even when this variable is not calculated explicitly. Consistent boundary conditions are established for insulating, departure and arrival boundaries. Gas exchanges through the water–air interface are also considered. Age fields ensuing from consistent boundary conditions and, occasionally, non-consistent ones are discussed, suggesting that the methodology advocated herein can be utilized by most age calculations, be they used for diagnosing the results of idealised models or realistic ones.

Keywords: partial differential equations; boundary conditions; geophysical and environmental fluid flows; reactive transport; interpretation methods; diagnostic timescales; CART; age; age distribution function

1. Introduction

Today's numerical models of geophysical and environmental fluid flows and the related (reactive) transport processes produce huge output files. Making sense of all these real numbers (that is, identifying key processes and establishing causal relationships between them) is no trivial task. Analysing primitive variables (e.g., velocity, pressure, temperature, concentrations) is not always conducive to the most fruitful interpretations. Evaluating auxiliary variables introduced for diagnostic purposes is an option worth considering. In this respect, diagnostic timescales (e.g., age, transit time, residence or exposure time) have been of use in the modelling of the atmosphere [1–4], various types of water bodies [5–19], and sediment transport [39,50–52]. These diagnostic timescales paint a simplified picture of the impact that the phenomena under study have on the largest time and space scales of (reactive) transport.

The residence or exposure time [16,53] and the age are of fundamentally different natures, as may be seen, for instance, in Figure 13 of [54]. The former type of timescale looks into the future, whilst the age is concerned with the past evolution. The present study deals with age calculations.

Age and age-related variables were introduced for zero-dimensional (or reservoir) modelling by Bolin and Rodhe [55]. Then, Zimmerman [54] and Takeoka [56] paved the way for theories allowing the age to be calculated numerically at every time and location from numerical model results [57–60]. With this approach, most, if not all of the numerical results are taken into account, which is why the ensuing timescale fields may be considered holistic.

Deleersnijder [61] made an attempt at building the most general definition of the age. The latter is as follows: the age of a particle of a constituent of seawater is the time that has elapsed since it began to be taken into consideration. In many cases, particles begin to be taken into consideration at the instant they enter the domain of interest, in which case the age is the time elapsed since entering the domain (e.g., [54,56,62]). When reactions are present, it may be found to be appropriate to transfer the age from one constituent to another [63]. Other age calculation strategies exist, in which, for instance, the age is evaluated as the time elapsed since leaving the sea bottom or surface [57,58,64–67]. Where and how particles cease to be considered must also be specified. All of these considerations clearly point to the importance of the boundary conditions.

The fate of a single particle is rather irrelevant [68]—a sufficiently large number of particles must be taken into consideration. As a consequence, the mean age of a set of particles must be introduced. In accordance with the age-averaging hypothesis [69], the mean age is defined as follows [61]: the mean age of a collection of particles is the mass-weighted average of their individual ages.

The mean age, which, for simplicity, will be called "age" hereinafter, may be computed at every time and location in the Lagrangian framework or in the Eulerian one. The Constituent-oriented Age and Residence time Theory (CART, www.climate.be/cart) is an Eulerian approach to the calculation of the age of any constituent of the water, or groups of constituents (that is, aggregates), including the water itself [59,69,70]. The age is obtained as the ratio of the age concentration to the concentration of the constituent or aggregate under consideration. These variables are the solutions of coupled reactive transport equations. The initial and boundary conditions must be prescribed in accordance with the declared objectives of the diagnostic strategy. So far, insufficient attention has been paid to the formulation of the boundary conditions. Specifically, one has yet to make it clear that the boundary conditions for the concentration and age concentration cannot be built independently of each other and draw the relevant consequences. Filling this gap is the objective of the present study.

In Section 2, we recall the definition of the age concentration function and evaluate its first two moments, the concentration and age concentration, so as to obtain the mean age. No novel concept is introduced in this Section, but developments of the past two decades are taken into account, hopefully yielding a line of reasoning that is easy to comprehend. This should facilitate the understanding of the strategy for building boundary conditions that is set out in Section 3, in which various types of boundaries are considered, that is, insulating, arrival, and departure boundaries, as well as semi-permeable boundaries allowing gas exchanges between water and air. In Section 4, the results of the preceding Section are put into perspective by tackling a simple ventilation assessment problem. Conclusions are drawn in Section 5.

2. The Age Distribution Function and Its First Two Moments

Although it is possible to apply CART to compressible fluid flows, this conceptual toolbox has been used thus far exclusively in the framework of the Boussinesq approximation. Therefore, the density of the fluid (that is, a mixture of pure water and many other constituents), ρ, is assumed to be constant. Let $\mathbf{x} = (x, y, z)$ denote the position vector, where x, y, and z are Cartesian coordinates. In accordance with the continuous media approach, the relevant variables are introduced in relation to elemental control volumes. We denote $\delta\Omega(\mathbf{x})$ and $\delta V(\mathbf{x})$ as the elemental control domain located at $\mathbf{x}$ and the value of its volume, respectively.

As far as the age is concerned, CART's core variable is the age distribution function $c(t, \mathbf{x}, \tau)$ [59,71], where independent variables t and τ denote the time and the age, respectively. The latter is generally assumed to be positive definite, that is, $\tau \in [0, \infty[$. Function $c(t, \mathbf{x}, \tau)$ is defined as follows: in the abovementioned control volume, at time t, the mass of the particles of the constituent under study whose age lies in the interval $[\tau, \tau + \delta\tau]$ tends to $\rho c(t, \mathbf{x}, \tau)\delta V(\mathbf{x})\delta\tau$ as $\delta V, \delta\tau \to 0$. Clearly, the physical dimension of the age distribution function is time^{-1}, and this function may be viewed as the histogram of the particle ages at time t and location $\mathbf{x}$. Advection and diffusion proceed independently of the age of the particles being transported. Then, by having recourse to mass conservation considerations alone, Delhez et al. [59] established the equation governing the age distribution function, which reads:

$$\frac{\partial c}{\partial t} = -\nabla \cdot (c\mathbf{v} - \mathbf{K} \cdot \nabla c) - \frac{\partial c}{\partial \tau},$$ (1)

where ∇, $\mathbf{v}(t, \mathbf{x})$, and $\mathbf{K}(t, \mathbf{x})$ denote the del operator, the fluid velocity, and the diffusivity tensor, respectively. Under the Boussinesq approximation, the velocity is divergence-free, that is, $\nabla \cdot \mathbf{v} = 0$. As was argued in Appendix A of Deleersnijder et al. [69], the diffusivity tensor must be symmetric and positive-definite. Equation (1) holds valid for a passive or inert constituent (also termed tracer). By adding suitable production-destruction terms, it can be extended to take reactions into account [59]. Doing so is, however, not necessary to serve the purpose of the present study. The last term in Equation (1) is related to ageing. It may be seen as an advection term related to a unit velocity, representing the fact that the age tends to increase at the same pace as time progresses.

The mass of the constituent under study that is present in elemental control volume $\delta\Omega(\mathbf{x})$ is obtained by taking the sum over all age categories, that is,

$$\delta M(t, \mathbf{x}) = \lim_{\delta\tau \to 0} \sum_{\tau=0}^{\tau=\infty} \rho c(t, \mathbf{x}, \tau)\delta\tau\delta V(\mathbf{x}) = \rho\delta V(\mathbf{x}) \int_0^\infty c(t, \mathbf{x}, \tau)d\tau,$$ (2)

where $\rho\delta V(\mathbf{x})$ is the mass of the fluid present in $\delta\Omega(\mathbf{x})$ under the Boussinesq approximation. Then, the concentration of the constituent under consideration, defined as a mass fraction (that is, a dimensionless variable), reads:

$$C(t, \mathbf{x}) = \frac{\delta M(t, \mathbf{x})}{\rho\delta V(\mathbf{x})} = \int_0^\infty c(t, \mathbf{x}, \tau)d\tau,$$ (3)

which means that the concentration is the 0th order moment of the age distribution function. By integrating (1) over the age, Delhez et al. [59] obtained the well-known advection-diffusion equation governing the evolution of the concentration:

$$\frac{\partial C}{\partial t} = -\nabla \cdot (C\mathbf{v} - \mathbf{K} \cdot \nabla C).$$ (4)

The age content of a particle [69] is defined as the product of its age and mass. Like mass, this quantity is of an additive nature. As a consequence, the age content of the particles of the constituent under study that are present in $\delta\Omega(\mathbf{x})$ is:

$$\delta A(t, \mathbf{x}) = \lim_{\delta\tau \to 0} \sum_{\tau=0}^{\tau=\infty} \rho\tau c(t, \mathbf{x}, \tau)\delta\tau\delta V(\mathbf{x}) = \rho\delta V(\mathbf{x}) \int_0^\infty \tau c(t, \mathbf{x}, \tau)d\tau.$$ (5)

This points to the importance of the first-order moment of the age distribution function,

$$\alpha(t, \mathbf{x}) = \int_0^\infty \tau c(t, \mathbf{x}, \tau)d\tau,$$ (6)

which, in the CART-related literature, is termed age concentration. Delhez et al. [59] established the equation obeyed by $\alpha(t, \mathbf{x})$ by multiplying (1) by the age and integrating over the age, eventually yielding:

$$\frac{\partial \alpha}{\partial t} = -\nabla \cdot (\alpha\mathbf{v} - \mathbf{K} \cdot \nabla\alpha) + C.$$ (7)

Owing to its relation to the age content, the age concentration is an extensive variable. This is why it is no surprise that it satisfies a reactive transport equation. The last term in the right-hand side of (7) is related to ageing. It is through this term that the equations for the concentration and age concentration are coupled.

Concentration and age concentration Equations (4) and (7) are of a parabolic nature [72]. Therefore, to solve each of them, the initial value of their solution must be prescribed and one boundary condition has to be enforced at every point of the surface delimiting the domain of interest.

According to the abovementioned age-averaging hypothesis [69], the mean age of the particles under study that are present in $\delta\Omega(x)$ is:

$$a(t,x) = \frac{\delta A(t,x)}{\delta M(t,x)} = \frac{\rho \delta V(x)\alpha(t,x)}{\rho \delta V(x)C(t,x)} = \frac{\alpha(t,x)}{C(t,x)}. \tag{8}$$

This type of averaging is not the only one that can be conceived. The choice that has been made is the only arbitrary ingredient in the developments leading to CART's age. As opposed to the concentration and age concentration, the age is an intensive variable, rather than an extensive one. In contrast with the equations for C and α, the equation obeyed by the age, which is obtained by combining (1), (7) and (8),

$$\frac{\partial a}{\partial t} = 1 - (\mathbf{v} - 2C^{-1}\nabla C \cdot \mathbf{K}) \cdot \nabla a - \nabla \cdot (-\mathbf{K} \cdot \nabla a), \tag{9}$$

cannot be cast into a conservative form. This is why in most, if not all numerical studies, the mean age has been computed as the ratio of the age concentration to the concentration rather than by solving the age Equation (9) [39,47,65,69,73–88]. This equation may, however, prove to be useful in theoretical studies [89–91].

In the right-hand side of relation (9), the first term is associated with ageing, whilst the second one bears some similarity with an advection term. However, the expression that may be regarded as the velocity, $\mathbf{v} - 2C^{-1}\nabla C \cdot \mathbf{K}$, is not divergence-free. Its behaviour is at the root of the intriguing symmetry that the age field occasionally exhibits [89,91–93].

A number of studies focused on seawater, which may be split into several water types, components, or masses (though the word "type" is likely to be more appropriate in this instance), which can be treated as inert tracers [15,60,62,69,94,95]. Obviously, the diagnostic strategy must be designed in such a way that the sum of all the water type concentrations must be equal to a constant at any time and location. With no loss of generality, this constant may be assumed to be equal to unity. For a tracer with unit concentration, age Equation (9) simplifies to the equation solved by England [58], and the corresponding age is sometimes called ventilation, or ideal age.

3. Consistent Insulating, Departure, and Arrival Boundary Conditions

Numerically solving the equation for the distribution function presents several challenges. First, there is an additional independent variable, namely, the age (τ). For example, if the three space dimensions are taken into account, Equation (1) must be discretised in a five-dimensional space, the corresponding independent variables of which are t, x, y, z, and τ. The necessary numerical developments are not trivial and the added computational cost cannot go unnoticed [71]. Furthermore, when the distribution function exhibits a long tail, Cornaton [96] argued that standard numerical techniques are no longer appropriate, which is why he developed a method involving the Laplace transform and classical space-time discretisations. This technique is computationally efficient, but its implementation is not straightforward.

Most authors did not find it necessary to compute the age distribution function and, instead, focused on the (mean) age, obtained from the solution of the concentration and age concentration equations. These equations are to be solved under boundary conditions that must be consistent with each other. They should be derived from the boundary conditions that the age distribution function would obey if the equation governing this function was to be solved explicitly.

Accordingly, we will derive a number of consistent boundary conditions, illustrate their impact on relevant problems and, when appropriate, show that opting for inconsistent boundary conditions would have a detrimental impact on the age field. Schematically, we will address insulating, departure, and arrival boundaries. We will also consider gas exchanges through the water–air interface.

Though the illustrations below essentially deal with passive constituents, the developments leading to consistent boundary conditions apply to any type of constituent, which includes constituents undergoing reactions.

3.1. Insulating Boundary

Consider surface Γ, whose outward unit normal is denoted n. Assume that this surface is impermeable, thereby insulating the neighbouring part of the domain of interest from its environment. As a result, the velocity satisfies boundary condition

$$[\mathbf{v} \cdot \mathbf{n}]_{\mathbf{x} \in \Gamma} = 0. \tag{10}$$

Since this boundary is impermeable, no particles of the constituent under consideration, irrespective of their age, cross it, leading to

$$[(c\mathbf{v} - \mathbf{K} \cdot \nabla c) \cdot \mathbf{n}]_{\mathbf{x} \in \Gamma} = 0. \tag{11}$$

Combining (10) and (11) yields

$$[(-\mathbf{K} \cdot \nabla c) \cdot \mathbf{n}]_{\mathbf{x} \in \Gamma} = 0. \tag{12}$$

To derive the boundary condition for the concentration, we integrate (12) over the age and use relation (3):

$$\left[\left(-\mathbf{K} \cdot \nabla \int_0^\infty c d\tau \right) \cdot \mathbf{n} \right]_{\mathbf{x} \in \Gamma} = [(-\mathbf{K} \cdot \nabla C) \cdot \mathbf{n}]_{\mathbf{x} \in \Gamma} = 0. \tag{13}$$

Multiplying (12) by τ, integrating over τ and using definition (6) of the age concentration, we obtain

$$\left[\left(-\mathbf{K} \cdot \nabla \int_0^\infty \tau c d\tau \right) \cdot \mathbf{n} \right]_{\mathbf{x} \in \Gamma} = [(-\mathbf{K} \cdot \nabla \alpha) \cdot \mathbf{n}]_{\mathbf{x} \in \Gamma} = 0. \tag{14}$$

Unsurprisingly, $C(t, \mathbf{x})$ and $\alpha(t, \mathbf{x})$ obey zero normal diffusive flux boundary conditions.

Boundary conditions (12)–(14) may be cast into generic form $[(-\mathbf{K} \cdot \nabla \varsigma) \cdot \mathbf{n}]_{\mathbf{x} \in \Gamma} = 0$, where $\varsigma(t, \mathbf{x})$ is the variable whose diffusive flux is zero through the boundary. If $\mathbf{n}$ is parallel to one of the principal axes of the diffusivity tensor, then such a boundary condition simplifies to $[\nabla \varsigma \cdot \mathbf{n}]_{\mathbf{x} \in \Gamma} = 0$. For instance, assume that the diffusivity tensor takes the widely-used form of $\mathbf{K} = \mathrm{diag}(\kappa_h, \kappa_h, \kappa_v)$, where κ_h and κ_v are the horizontal diffusivity and the vertical one, respectively. If the boundary under consideration is horizontal, then the aforementioned zero normal flux boundary condition degenerates into $[\partial \varsigma / \partial z]_{\mathbf{x} \in \Gamma} = 0$, where z denotes the vertical coordinate.

Now, for illustration purposes, assume that the domain of interest, Ω, is completely insulated from its environment, implying that the abovementioned boundary conditions are to be enforced on the whole boundary (Figure 1). Further assume that the age of all the particles of a passive tracer is zero at the initial time, that is, $c(0, \mathbf{x}, \tau) = C^0(\mathbf{x})\delta(\tau)$, where $C^0(\mathbf{x})$ and $\delta(\tau)$ are the initial concentration field and the Dirac delta function, respectively. The corresponding initial age concentration is readily seen to be $\alpha(0, \mathbf{x}) = 0$. Then, at any point of Ω and time $t \geq 0$, the age distribution function is $c(t, \mathbf{x}, \tau) = C(t, \mathbf{x})\delta(\tau - t)$, implying that the age concentration and the age are $\alpha(t, \mathbf{x}) = tC(t, \mathbf{x})$ and $a(t, \mathbf{x}) = \alpha(t, \mathbf{x})/C(t, \mathbf{x}) = t$. This result is readily understood. No particles enter or leave the domain. As all particles age at the same pace as time progresses, their age is equal to the elapsed time. The aforementioned results, $\alpha(t, \mathbf{x}) = tC(t, \mathbf{x})$ and $a(t, \mathbf{x}) = t$, can also be obtained without computing the age distribution function and, instead, by solving only the concentration and age concentration equations under boundary conditions (13) and (14). This is readily seen.

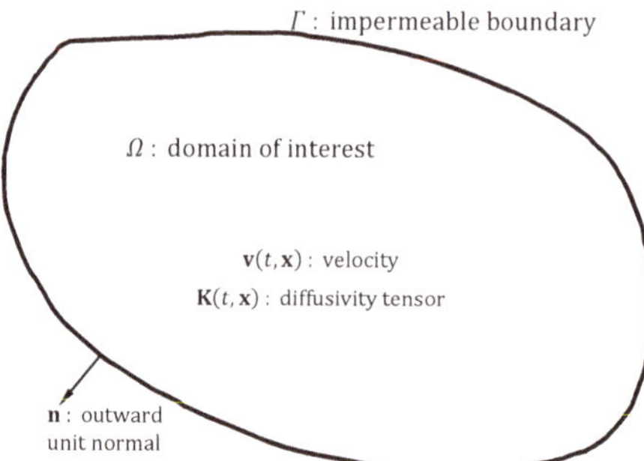

Figure 1. Schematic representation of domain of interest Ω delimited by insulating (or impermeable) boundary Γ, whose outward unit normal is vector $\mathbf{n}$, with $|\mathbf{n}| = 1$. The velocity and the diffusivity tensor are time- and position-dependent.

In this case study, that the age is equal to the elapsed time is of little diagnostic value. However, it may be regarded as a piece of information supporting the well-foundedness of boundary conditions (13) and (14). In other words, these boundary conditions allow us to obtain the expected result. Opting for another set of boundary conditions would lead to an unacceptable age field.

We should note in passing that, had we studied the evolution of a tracer undergoing a first-order decay process associated with constant mean life λ^{-1} (and, hence, half-life $\log2\lambda^{-1} \approx 0.7\lambda^{-1}$), we would have obtained the following result: $c_d(t,\mathbf{x},\tau) = e^{-\lambda t}c(t,\mathbf{x},\tau)$, $C_d(t,\mathbf{x},\tau) = e^{-\lambda t}C(t,\mathbf{x},\tau)$ and $\alpha_d(t,\mathbf{x},\tau) = e^{-\lambda t}\alpha(t,\mathbf{x},\tau)$, where subscript "$d$" identifies the fields related to the decaying tracer. As a consequence, the age would have been unchanged: $a_d(t,\mathbf{x}) = a(t,\mathbf{x})$. This is chiefly because first-order decay proceeds at the same rate at every time and position [19,97].

3.2. Departure Boundary

If the age is defined as the time elapsed since leaving a given boundary, then the age of all the particles under consideration must be set or reset to zero at the moment they touch this boundary, leading to

$$[c(t,\mathbf{x},\tau)]_{\mathbf{x}\in\Gamma} = C^{\Gamma}(t,\mathbf{x})\delta(\tau), \tag{15}$$

where, in this subsection, Γ refers to the departure boundary (that is, the boundary where the age is prescribed to be zero), which is usually a fraction of the domain boundary; $C^{\Gamma}(t,\mathbf{x})$ is the tracer concentration on departure boundary Γ, which is assumed to be known (Dirichlet boundary condition). The boundary condition for the age concentration is derived from (15) as follows:

$$[\alpha(t,\mathbf{x})]_{\mathbf{x}\in\Gamma} = \left[\int_0^{\infty} \tau c(t,\mathbf{x},\tau)d\tau\right]_{\mathbf{x}\in\Gamma} = C^{\Gamma}(t,\mathbf{x})\int_0^{\infty} \tau\delta(\tau)d\tau = 0. \tag{16}$$

Thus, the concentration and age concentration obey Dirichlet boundary conditions.

To help understand the role of these boundary conditions, consider a one-dimensional flow, in semi-infinite domain $x \in [0,\infty]$ (Figure 2). The age of a passive tracer particle is defined as the time elapsed since leaving the departure boundary ($x = 0$). At the initial instant ($t = 0$), there is no tracer in the domain and the tracer concentration is prescribed to be equal to constant C^{Γ} on the departure boundary. If positive constants U and K represent the velocity and the diffusivity, the equation obeyed by age distribution function $c(t,x,\tau)$ is

$$\frac{\partial c}{\partial t} = -U\frac{\partial c}{\partial x} + K\frac{\partial^2 c}{\partial t^2} - \frac{\partial c}{\partial \tau}. \tag{17}$$

This equation is to be solved under the following initial and boundary conditions

$$c(0, x, \tau) = 0, \quad c(t, 0, \tau) = C^{\Gamma}\delta(\tau), \quad c(t, x, 0) = 0, \quad c(t, \infty, \tau) < \infty. \tag{18}$$

Without any loss of generality, C^{Γ} may be assumed to be equal to unity. Accordingly, the solution reads (Figure 3)

$$c(t, x, \tau) = \frac{x}{\sqrt{4\pi K\tau^3}}\exp\left[-\frac{(x - U\tau)^2}{4K\tau}\right]Y(t - \tau), \tag{19}$$

where Y is the Heaviside step function, that is, function $Y(t - \tau)$ is equal to unity (resp. zero) according to whether $t > \tau$ (resp. $t < \tau$).

The concentration of the tracer is

$$C(t, x) = \int_0^{\infty} c(t, x, \tau)d\tau = \int_0^{t} \frac{x}{\sqrt{4\pi K\tau^3}}\exp\left[-\frac{(x - U\tau)^2}{4K\tau}\right]d\tau, \tag{20}$$

whilst its age concentration reads

$$\alpha(t, x) = \int_0^{\infty} \tau c(t, x, \tau)d\tau = \int_0^{t} \frac{\tau x}{\sqrt{4\pi K\tau^3}}\exp\left[-\frac{(x - U\tau)^2}{4K\tau}\right]d\tau. \tag{21}$$

The related age, $a(t, x) = \alpha(t, x)/C(t, x)$, may be seen to be smaller than the elapsed time, as it should be.

It is noteworthy that the (mean) age could have been obtained without explicitly calculating the age distribution function. The tracer concentration and age concentration are governed by the following partial differential problems:

$$\begin{cases} \dfrac{\partial C}{\partial t} = -U\dfrac{\partial C}{\partial x} + K\dfrac{\partial^2 C}{\partial t^2} \\ C(t, 0) = 1, \quad C(0, x) = 0, \quad C(t, \infty) < \infty \end{cases} \tag{22}$$

and

$$\begin{cases} \dfrac{\partial \alpha}{\partial t} = -U\dfrac{\partial \alpha}{\partial x} + K\dfrac{\partial^2 \alpha}{\partial t^2} + C \\ \alpha(t, 0) = 0, \quad \alpha(0, x) = 0, \quad \lim_{x \to \infty}\dfrac{\alpha(t, x)}{x} < \infty, \end{cases} \tag{23}$$

where the initial and boundary conditions are derived from (18). Then, lengthy manipulations would allow us to show that the solutions to (22) and (23) are (20) and (21), as expected.

At first glance, it is not obvious that concentration (20) and age concentration (21) satisfy boundary conditions $C(t, 0) = 1$ and $\alpha(t, 0) = 0$. To remove doubts, we prove in Appendix A that these boundary conditions are actually obeyed.

In the limit $t \to \infty$, the age distribution function, the concentration, the age concentration, and the mean age are

$$c_{\infty}(x, \tau) = \lim_{t \to \infty} c(t, x, \tau) = \frac{x}{\sqrt{4\pi K\tau^3}}\exp\left[-\frac{(x - U\tau)^2}{4K\tau}\right] \tag{24}$$

$$C_{\infty}(x) = \lim_{t \to \infty} C(t, x) = 1, \quad \alpha_{\infty}(x) = \lim_{t \to \infty} \alpha(t, x) = \frac{x}{U}, \quad a_{\infty}(x) = \lim_{t \to \infty} a(t, x) = \frac{x}{U}. \tag{25}$$

Thus, in the steady-state limit, the age is the time needed to travel distance x at speed U. Unfortunately, this simple result obscures the fact that, because of diffusion, the time taken for a given particle to travel from the entrance of the domain to a point located at distance x to the inlet is generally not equal to x/U, as is illustrated in Figure 3.

If diffusivity K is zero, then the age distribution function is

$$c(t, x, \tau) = \delta(\tau - x/U)Y(t - \tau), \tag{26}$$

implying that the concentration and age concentration are

$$C(t, x) = Y(t - x/U), \quad \alpha(t, x) = \frac{x}{U}Y(t - x/U). \tag{27}$$

Therefore, the age is x/U for $t > x/U$, and is undefined otherwise. These expressions can also be obtained by setting $K = 0$ in (22) and (23) and dealing with the resulting equations in the sense of distributions.

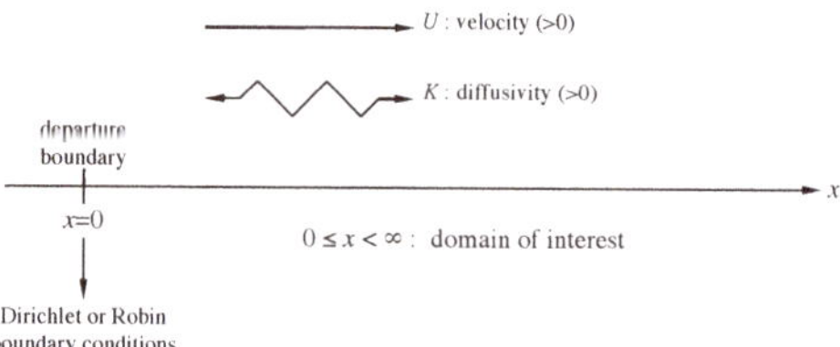

Figure 2. Schematic representation of a semi-infinite domain ($x \in [0, \infty[$) with a departure boundary at $x = 0$. On this boundary, we can either prescribe that all particles of the tracer under study have zero age (Dirichlet boundary condition) or that the age of the incoming particles is zero, which leads to a Robin boundary condition, as seen in Section 3.3.

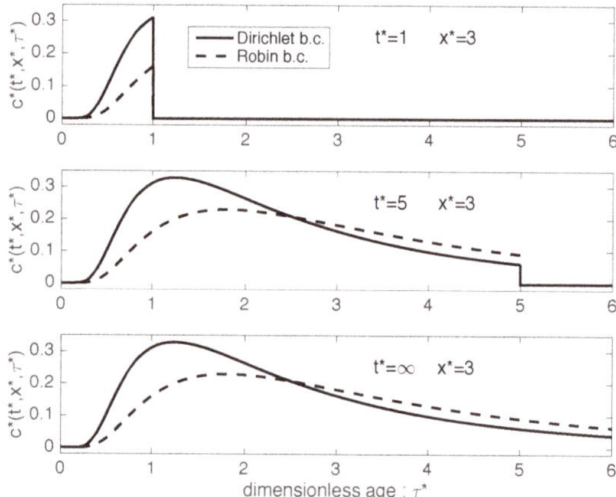

Figure 3. Illustration of age distribution functions (19) (solid line) and (34) (dashed line), which are obtained under Dirichlet and Robin boundary conditions at the inlet ($x = 0$). Dimensionless variables are used. They are identified by asterisks and are defined as follows: $t^* = U^2 t/K$, $x^* = Ux/K$, $\tau^* = U^2 \tau/K$ and $c^* = Kc/U^2$ (see Appendix B). The dimensionless age distribution functions are plotted at $x^* = 3$ as functions of the age at different instants.

Calculating the age from the concentration and age concentration by solving the relevant equations under consistent initial and boundary conditions is generally much easier than evaluating the (mean) age from the age distribution function, which is why many studies relying on CART simply did so [39,47,65,69,73–88]. However, we must bear in mind that this approach may veil some of the subtleties of the transport phenomena under study.

3.3. Departure Boundary: An Alternative Approach

Having recourse to Dirichlet boundary conditions is not the only option to deal with a departure boundary. An alternative approach consists in imposing the incoming tracer flux and prescribing that the age of the particles entering the domain is zero. Accordingly, on boundary Γ with outward unit normal n, the age distribution function must satisfy

$$[(c\mathbf{v} - \mathbf{K} \cdot \nabla c) \cdot n]_{x \in \Gamma} = -\Phi(t, x)\delta(\tau), \tag{28}$$

where $\Phi(t, x)$ is the incoming tracer flux (ms^{-1}), which we assumed to be known. In this case, the age is the time elapsed since entering the domain rather than the time elapsed since leaving boundary Γ.

Integrating (28) over the age, we obtain

$$\left[\left(\int_0^\infty cd\tau \mathbf{v} - \mathbf{K} \cdot \nabla \int_0^\infty cd\tau\right) \cdot n\right]_{x \in \Gamma} = -\Phi(t, x)\int_0^\infty \delta(\tau)d\tau, \tag{29}$$

which simplifies to the boundary condition for the concentration, that is,

$$[(C\mathbf{v} - \mathbf{K} \cdot \nabla C) \cdot n]_{x \in \Gamma} = -\Phi(t, x). \tag{30}$$

As for the age concentration, we integrate the product of the age and relation (28), yielding

$$\left[\left(\int_0^\infty \tau cd\tau \mathbf{v} - \mathbf{K} \cdot \nabla \int_0^\infty \tau cd\tau\right) \cdot n\right]_{x \in \Gamma} = -\Phi(t, x)\int_0^\infty \tau\delta(\tau)d\tau. \tag{31}$$

Then, on Γ, the age concentration satisfies:

$$[(\alpha\mathbf{v} - \mathbf{K} \cdot \nabla\alpha) \cdot n]_{x \in \Gamma} = 0. \tag{32}$$

The above relations (30) and (32) are Robin boundary conditions. With the latter, the (mean) age on boundary Γ is unlikely to be zero. This is because, on Γ, there are particles that are entering the domain (their age is zero) and also particles that have been in the domain for some time (their age is positive).

We now revisit the illustration of the preceding subsection. The only modifications to be made are related to the boundary conditions at $x = 0$, which must be transformed to

$$\left[cU - K\frac{\partial c}{\partial x}\right]_{x=0} = \Phi\delta(\tau), \quad \left[CU - K\frac{\partial C}{\partial x}\right]_{x=0} = \Phi, \quad \left[\alpha U - K\frac{\partial \alpha}{\partial x}\right]_{x=0} = 0. \tag{33}$$

We set $\Phi = U$ so that the steady-state concentration is equal to unity as in the previous illustration. Doing so entails little loss of generality. Then, the age distribution function is

$$c(t, x, \tau) = \left\{\frac{U}{\sqrt{\pi K\tau}}\exp\left[-\frac{(x - U\tau)^2}{4K\tau}\right] - \frac{U^2 e^{Ux/K}}{2K}\,\text{erfc}\left[\sqrt{\frac{U^2\tau}{4K}} + \sqrt{\frac{x^2}{4K\tau}}\right]\right\}Y(t - \tau). \tag{34}$$

The present age distribution function has a longer tail than that associated with a Dirichlet boundary condition at the incoming boundary of the domain (Figure 3). The steady-state concentration and age concentration are

$$C_\infty(x) = \lim_{t \to \infty}\int_0^\infty c(t, x, \tau)d\tau = 1 \tag{35}$$

and

$$\alpha_\infty(x) = \lim_{t \to \infty}\int_0^\infty \tau c(t, x, \tau)d\tau = \frac{K + Ux}{U^2}. \tag{36}$$

As a consequence, the steady-state age reads

$$a_\infty(x) = \frac{\alpha_\infty(x)}{C_\infty(x)} = \frac{K}{U^2} + \frac{x}{U}. \tag{37}$$

The difference between the age ensuing from the Robin boundary conditions and that obtained under Dirichlet boundary conditions is equal to a constant, namely, K/U^2, which, unsurprisingly, is an increasing (resp. decreasing) function of the diffusivity (resp. velocity).

Needless to say, the same concentration and age concentration fields could have been obtained by directly solving the equations for the concentration and age concentration under the abovementioned initial and boundary conditions.

3.4. Arrival Boundary

The particles of the constituent under study cease to be taken into account (that is, they are discarded) at the moment they touch an arrival boundary. If Γ is a boundary of this type, then the age distribution function must satisfy

$$[c(t, x, \tau)]_{x \in \Gamma} = 0 \tag{38}$$

Therefore, on Γ, the concentration and age concentration must be zero (Dirichlet boundary conditions):

$$[C(t, x)]_{x \in \Gamma} = \left[\int_0^\infty c(t, x, \tau) d\tau \right]_{x \in \Gamma} = 0 \tag{39}$$

and

$$[\alpha(t, x)]_{x \in \Gamma} = \left[\int_0^\infty \tau c(t, x, \tau) d\tau \right]_{x \in \Gamma} = 0. \tag{40}$$

Some may be left unconvinced by the above reasoning, mainly because it leads to the requirement that the concentration be zero on the boundary, causing uncertainties as to how the mean age is to be evaluated on the boundary. In Section 3 of Delhez and Deleersnijder (2006) [98], detailed Lagrangian and Eulerian developments led to a similar result. Though these calculations were made in a different context, they may be found to be helpful to grasp the issue at hand.

On the arrival boundary, the mean age appears as the ratio of two functions, the age concentration and the concentration, that are both zero. However, the age is unlikely to be arbitrarily large, for it is precisely on this boundary that the particles under study cease to be taken into consideration. In addition, the age gradient may be seen to satisfy a property that is of use for graphical representations. First, we rewrite the equation governing the (mean) age, that is, relation (9), as follows:

$$\nabla C \cdot K \cdot \nabla a = \frac{C}{2} \left[\frac{\partial a}{\partial t} - 1 + \nabla \cdot (a v - K \cdot \nabla a) \right]. \tag{41}$$

Thus, on the boundary under consideration, where concentration C is prescribed to be zero, this equation simplifies to $\nabla C \cdot K \cdot \nabla a = 0$. Since the concentration is zero on the boundary, its gradient must be normal to it and, hence, must be parallel to unit normal vector n, leading to $[n \cdot K \cdot a]_{x \in \Gamma} = 0$. Then, it is readily seen that this expression is equivalent to a zero normal diffusive age flux boundary condition

$$[(-K \cdot \nabla a) \cdot n]_{x \in \Gamma} = 0, \tag{42}$$

which, as pointed out in Section 3.1, simplifies to $[\nabla a \cdot n]_{x \in \Gamma} = 0$ if one of the principal axes of the diffusivity tensor is parallel to n. Clearly, (42) does not contradict the hypothesis that the age has a finite value on Γ.

Departure and arrival boundary conditions (15), (16) and (38)–(40) have been derived without consideration of the direction the velocity on the boundary. However, it is likely that a diagnostic strategy will be built in such a way that a departure (resp. arrival) boundary will be an incoming (resp. outgoing) boundary, that is, a boundary on which $\mathbf{v} \cdot \mathbf{n} \leq 0$ (resp. $\mathbf{v} \cdot \mathbf{n} \geq 0$).

For illustration purposes, we revisit the one-dimensional problem dealt with in Section 3.2. We keep the departure boundary with Dirichlet boundary conditions at $x = 0$ and introduce an arrival boundary at $x = L$ so that the domain of interest now has a finite length ($0 \leq x \leq L$) (Figure 4). For the sake of simplicity, we focus on steady-state solutions. Accordingly, concentration $C(x)$ and age concentration $\alpha(x)$ obey

$$\begin{cases} 0 = -U\frac{dC}{dx} + K\frac{d^2C}{dx^2} \\ C(0) = 1, \quad C(L) = 0 \end{cases} \tag{43}$$

and

$$\begin{cases} 0 = -U\frac{d\alpha}{dx} + K\frac{d^2\alpha}{dx^2} + C \\ \alpha(0) = 0, \quad \alpha(L) = 0. \end{cases} \tag{44}$$

Assuming that the concentration is equal to unity on the incoming boundary entails no loss of generality. This is readily understood. Furthermore, dealing with similar idealised, one-dimensional problems have been found to be of use in previous studies [38,62].

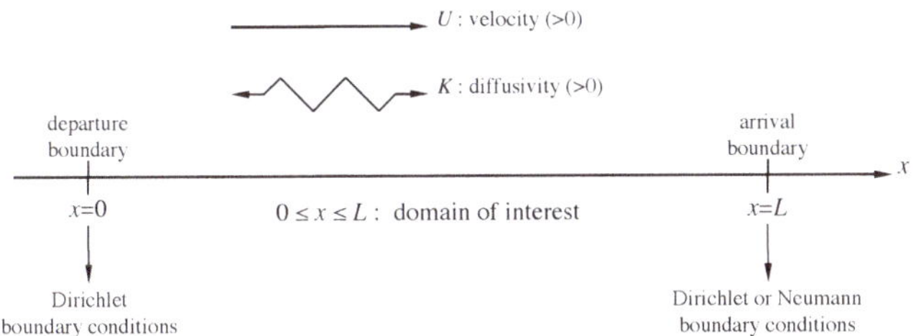

Figure 4. Schematic representation of a finite-sized domain ($x \in [0, L]$) with a departure boundary at $x = 0$ and an arrival one at $x = L$. Dirichlet boundary conditions are prescribed on the boundaries. An alternative treatment of the arrival boundary leads to the implementation of Neumann boundary conditions at $x = L$.

The concentration and age concentration are (Figure 5):

$$C(x) = \frac{e^{Pe} - e^{Ux/K}}{e^{Pe} - 1} \tag{45}$$

and

$$\alpha(x) = \frac{e^{Pe} + e^{Ux/K}}{e^{Pe} - 1}\frac{x}{U} - \frac{2e^{Pe}\left(e^{Ux/K} - 1\right)}{(e^{Pe} - 1)^2}\frac{L}{U}, \tag{46}$$

where dimensionless parameter $Pe = UL/K$ is the Peclet number, that is, the ratio of the timescale characterising diffusion (L^2/K) and that associated with advection (L/U). In the vicinity of the departure boundary ($x = 0$), the age, $a(x) = \alpha(x)/C(x)$, admits asymptotic expansion

$$a(x) \sim \frac{e^{2Pe} - 2Pe\,e^{Pe} - 1}{(e^{Pe} - 1)^2}\frac{x}{U}, \quad x \to 0, \tag{47}$$

which, unsurprisingly, simplifies to $a(x) \sim x/U$ in the limit $Pe \to \infty$. As for the arrival boundary ($x = L$), the age tends, as expected, to a finite value with a zero gradient,

$$a(x) \sim \underbrace{\frac{Pe(e^{Pe}+1)-2(e^{Pe}-1)}{e^{Pe}-1}\frac{K}{U^2}}_{a(L)} - \frac{(L-x)^2}{6K}, \quad x \to L, \tag{48}$$

with $a(L) \to L/U$ as $Pe \to \infty$. The larger the Peclet number, the closer the solutions are to their zero diffusion counterparts, that is, a unit value of concentration, with the age concentration and age equal to x/U. Such solutions cannot satisfy the boundary conditions prescribed at $x = L$. This is why the correct concentration and age concentration exhibit a boundary layer adjacent to the outgoing boundary.

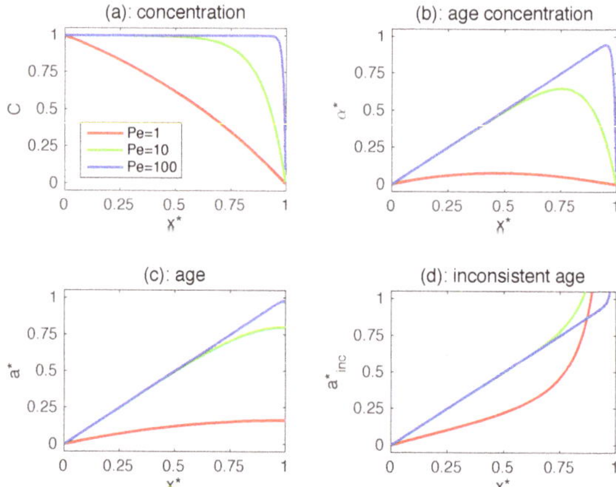

Figure 5. Illustration of (**a**) concentration (45), (**b**) age concentration (46), and (**c**) the associated age. Panel (**d**) depicts the inconsistent age that is obtained as the ratio of inconsistent age concentration (49) and correct concentration (49). Dimensionless variables are represented. They are identified by asterisks and are $x^* = x/L$, $\alpha^* = U\alpha/L$, $a^* = Ua/L$ and $a^*_{inc} = Ua^*_{inc}/L$.

To document the impact of inconsistent boundary conditions, we replace for a moment the Dirichlet boundary condition for the age concentration at $x = L$ by Neumann condition $[-K\, d\,\alpha_{inc}/dx]_{x=L} = 0$, where subscript "*inc*" refers to the inconsistent solution. This relation is the simplest type of radiation condition, which may be found in Table 1 of Bendsten et al. [65] in conjunction with a Dirichlet boundary condition for the concentration. Obviously, the concentration is not changed. The inconsistent age concentration reads

$$\alpha_{inc}(x) = \frac{e^{Pe} + e^{Ux/K}}{e^{Pe}-1}\frac{x}{U} - \frac{(2+Pe)(e^{Ux/K}-1)}{e^{Pe}-1}\frac{K}{U^2}. \tag{49}$$

The modified age is slightly different from the correct one in the neighbourhood of the incoming boundary,

$$a_{inc}(x) \sim \frac{e^{Pe}-Pe-1}{e^{Pe}-1}\frac{x}{U}, \quad x \to 0, \tag{50}$$

but has no finite limit on the outgoing boundary,

$$a_{inc}(x) \sim [Pe(1+e^{-Pe})+2(1-e^{-Pe})]\frac{K^2}{U^3(L-x)}, \quad x \to L. \tag{51}$$

In other words, the age resulting from the imposition of inconsistent boundary conditions on the downstream boundary is such that $a_{inc} \to \infty$ as $x \to L$, which is unjustifiable in a finite-sized domain of interest with an open boundary meant to allow particles to leave the domain. What is worse, the inconsistent boundary conditions impact a significant fraction of the domain of interest (Figure 5). Undoubtedly, such behaviour is unacceptable.

3.5. Arrival Boundary: An Alternative Approach

Imposing, as suggested in the previous Section, that the concentration be zero on the arrival boundary implies that the advective mass flux through this boundary is zero. Thus, the outgoing mass flux crossing the boundary is purely diffusive. An alternative approach consists in prescribing that the diffusive flux through the boundary be zero so that the outgoing flux is entirely of an advective nature. This requires boundary conditions equivalent to (11), (13), and (14) to be enforced on the arrival boundary.

To illustrate the impact of these boundary conditions, the idealised problem of the previous Section is revisited. The only modification pertains to the outgoing boundary, where the diffusive fluxes are prescribed to be zero, yielding

$$C(x) = 1, \quad \alpha(x) = \frac{x}{U} - \frac{e^{Ux/K} - 1}{e^{Pe}} \frac{K}{U^2} = a(x). \tag{52}$$

This age is larger than that obtained by imposing Dirichlet boundary conditions at the departure boundary (Figure 6). This can be proven rigorously for the present one-dimensional flow. Remarkably, a similar inequality also holds valid for a much wider class of problems [99,100].

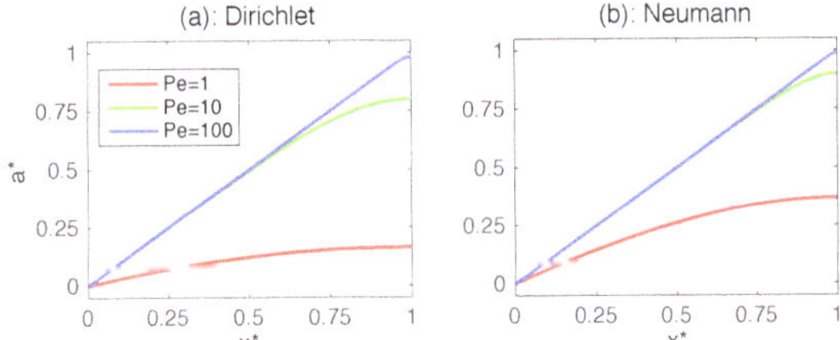

Figure 6. Illustration of the ages obtained from the solution of the one-dimensional problem outlined in Figure 4. Panel (**a**) depicts the age ensuing from Dirichlet boundary conditions imposed at the departure boundary ($x = L$) and, hence, is the same age as that represented in panel (**c**) of Figure 5. The age obtained by prescribing Neumann boundary conditions at $x = L$, that is, age (52), is illustrated in panel (**b**). Dimensionless variables are represented. They are identified by asterisks and are defined as follows: $x^* = x/L$ and $a^* = aU/L$.

3.6. Gas Exchanges through the Water–air Interface

The gas flux at the water–air interface is usually parameterised by having recourse to the concept of piston velocity [101–105]. Accordingly, the net outgoing (that is, from the water body to the atmosphere) mass flux (kg m^{-2} s^{-1}) through surface Γ of a gas whose concentration in the water is $C(t, x)$ reads

$$[(-\rho K \cdot \nabla C) \cdot n]_{x \in \Gamma} = [\rho \omega (C - C^s)]_{x \in \Gamma}, \tag{53}$$

where ω is the piston velocity (ms^{-1}), whilst C^s is the saturation concentration, that is, the water surface concentration in equilibrium with the atmosphere. In general, ω and C^s depend on time and position. If the surface concentration is greater (resp. smaller) than the saturation concentration, the gas flux is directed from the water body to the atmosphere (resp. from the air to the water). This is why the boundary condition (53) may be viewed as a relaxation boundary condition—the air-water mass flux tends to nudge the surface concentration toward C^s.

Formula (53) provides an estimate of the net flux at the water–air interface. No other information about the water–air interface is needed in order to model the concentration in the water (or in the atmosphere) of the gas under consideration. For age calculations, however, it is necessary to realise that the right-hand side of (53) actually represents the difference between the upward mass flux $(\rho\phi_C^\uparrow = [\rho\omega C]_{x\in\Gamma})$ and the downward one $(\rho\phi_C^\downarrow = [\rho\omega C^s]_{x\in\Gamma})$ (Figure 7).

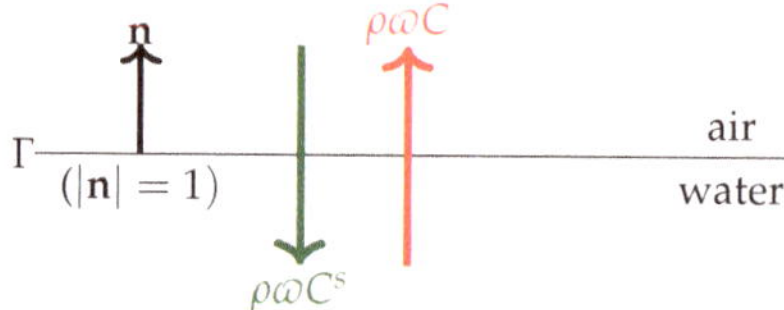

Figure 7. Illustration of the gas fluxes (kg m^{-2} s^{-1}) involved in boundary condition (53). The net gas (mass) flux through the water–air interface, which is positive in the upward direction (that is, from water to air), is the difference between the outgoing flux, $\rho\phi_C^\downarrow = [\rho\omega C]_{x\in\Gamma}$, and the incoming one, $\rho\phi_C^\downarrow = [\rho\omega C^s]_{x\in\Gamma}$, where C^s is the surface concentration in equilibrium with the atmosphere.

This piece of information is essential for building the boundary condition for the age distribution function. The outgoing (resp. incoming) mass flux of the gas particles whose age lies in the interval $[\tau, \tau + \delta\tau]$ tends to $\rho\phi_c^\uparrow \delta\tau$ (resp. $\rho\phi_c^\downarrow \delta\tau$) in the limit $\delta\tau \to 0$, with $\phi_c^\uparrow = [\omega c]_{x\in\Gamma}$ and $\phi_c^\downarrow = [\omega c^a]_{x\in\Gamma}$, where c^a is the age distribution in the atmosphere. Thus, the boundary condition for the age distribution function reads

$$\left[(-K \cdot \nabla c) \cdot n\right]_{x\in\Gamma} = \phi_c^\uparrow - \phi_c^\downarrow = \left[\omega(c - c^a)\right]_{x\in\Gamma}. \tag{54}$$

Under the assumption that the integral of c^a over the age τ is C^s, the boundary conditions for the concentration and age concentration are obtained by taking the 0th- and first-order moment of (54) (Table 1):

$$\left[(-K \cdot \nabla C) \cdot n\right]_{x\in\Gamma} = \int_0^\infty [\omega(c - c^a)]_{x\in\Gamma} d\tau = \left[\omega(C - C^s)\right]_{x\in\Gamma} \tag{55}$$

and

$$\left[(-K \cdot \nabla\alpha) \cdot n\right]_{x\in\Gamma} = \int_0^\infty [\omega(\tau c - \tau c^a)]_{x\in\Gamma} d\tau = \left[\omega(\alpha - C^s a^a)\right]_{x\in\Gamma}. \tag{56}$$

Obviously, relations (53) and (55) are equivalent. Equations (54)–(56) are usually referred to as Robin boundary conditions.

It is often assumed that all the gas particles entering the water column have the same age, $a^a(t, x)$, that is, the gas age at the lower boundary of the atmosphere. As a result, their age distribution function is $c^a = C^s \delta(\tau - a^a)$. Therefore, in the water body, the age of the dissolved gas is the sum of atmospheric age $a^a(t, x)$ and the time spent in the water body since entering it through the water–air. For ventilation studies, it may be appropriate to assume that the atmospheric age is zero (Table 1).

Table 1. Outgoing and incoming specific fluxes (ratio of a flux to the water density) at the water–air interface for the age distribution function ($\phi_c^{\uparrow,\downarrow}$), the concentration ($\phi_C^{\uparrow,\downarrow}$), and the age concentration ($\phi_\alpha^{\uparrow,\downarrow}$). As for the downward flux, the general expression and two simplified ones are taken into consideration, which consists in assuming that all the incoming gas particles have the same age, a^a (fourth column), and that this age is zero (fifth column).

Variable	Outgoing (Upward)	Incoming (Downward) Specific Flux		
	Specific Flux	General Expression	$c^a = C^s\delta(\tau - a^a)$	$c^a = C^s\delta(\tau)$
age distribution function	$\phi_c^\uparrow = \omega c$	$\phi_c^\downarrow = \omega c^a$	$\phi_c^\downarrow = \omega C^s \delta(\tau - a^a)$	$\phi_c^\downarrow = \omega C^s \delta(\tau)$
concentration	$\phi_C^\uparrow = \omega C^s$	$\phi_C^\downarrow = \omega C^s$	$\phi_C^\downarrow = \omega C^s$	$\phi_C^\downarrow = \omega C^s$
age concentration	$\phi_\alpha^\uparrow = \omega\alpha$	$\phi_\alpha^\downarrow = \omega\alpha^a$	$\phi_\alpha^\downarrow = \omega C^s a^a$	0

If the piston velocity is small, then the boundary conditions derived above tend to simplify to no-flux expressions relevant to an insulating boundary. If, on the other hand, the piston velocity is large, (54)–(56) tend to degenerate into Dirichlet boundary conditions. To assess the impact of the piston velocity, a dimensionless parameter should be derived. This can be achieved with the help of a steady-state water column model (Figure 8a). The domain of interest is defined by inequalities $-h \leq z \leq 0$, where z is the vertical coordinate and h is the height of the water column, whose water–air boundary is located at $z = 0$. As is customary in water column modelling, the lower boundary of the domain is considered to be impermeable. For the sake of simplicity, it is assumed that vertical diffusion, represented by means of constant diffusivity K, is the only process to be taken into account. Accordingly, the concentration and age concentration obey the following differential problems:

$$\begin{cases} 0 = K\dfrac{d^2C}{dx^2} \\[2mm] \left[-K\dfrac{dC}{dz}\right]_{z=-h} = 0, \quad \left[-K\dfrac{dC}{dz}\right]_{z=0} = \omega\left[C(0) - C^s\right] \end{cases} \tag{57}$$

and

$$\begin{cases} 0 = K\dfrac{d^2\alpha}{dx^2} + C \\[2mm] \left[-K\dfrac{d\alpha}{dz}\right]_{z=-h} = 0, \quad \left[-K\dfrac{d\alpha}{dz}\right]_{z=0} = \omega\left[\alpha(0) - C^s a^a\right]. \end{cases} \tag{58}$$

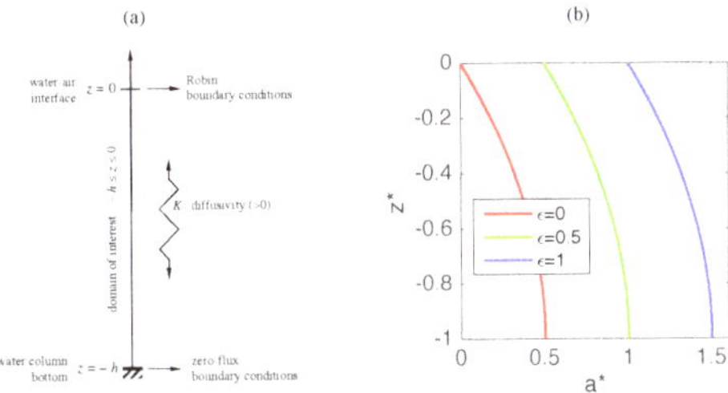

Figure 8. Schematic representation of the vertical, one-dimensional domain of interest dealt with in Section 3.6 ($-h \leq z \leq 0$) (panel (**a**)), and age (60) of the dissolved gas originating from the atmosphere for various values of dimensionless parameter ϵ (panel (**b**)). The lower boundary ($z = -h$) is impermeable. Robin boundary conditions, related to gas exchanges, are prescribed at the water–air interface ($z = 0$). Dimensionless variables are represented in panel (**b**). They are identified by asterisks and are defined as follows: $z^* = z/h$ and $a^* = K(a - a^a)/h^2$.

Unsurprisingly, the concentration is a constant: $C(z) = C^s$. The age concentration and age are readily seen to be

$$\alpha(z) = \left[a^a + \frac{h}{\varpi} - \frac{(2h+z)z}{2K} \right] C^s \tag{59}$$

and

$$a(z) = a^a + \underbrace{\left[\epsilon - \frac{(2h+z)z}{2h} \right] \frac{h^2}{K}}_{=a^0(z)}, \tag{60}$$

where ϵ is the sought-after dimensionless parameter (Figure 8). The latter is the ratio of the timescale associated with the piston velocity, h/ϖ, and the classical diffusive timescale, h^2/K, that is,

$$\epsilon = \frac{h/\varpi}{h^2/K} = \frac{K}{h\varpi}. \tag{61}$$

The larger the piston velocity, the smaller this dimensionless parameter. In the limit $\epsilon \to 0$, the age tends to the function of z that would be obtained under Dirichlet boundary conditions. The age is the sum of the atmospheric age and $a^0(z)$, which is the age ensuing from the assumption that the atmospheric age a^a is zero. Interestingly, this age shift property is satisfied in a wide class of multi-dimensional, time-dependent age calculation problems [106].

In reality, as opposed to what is represented in the above highly idealised water column model, the impact of vertical turbulent diffusion due to the surface forcing (chiefly surface wind stress) does not always extend to the bottom. Therefore, the relevant vertical length scale is not necessarily the height of the water column. It could be significantly smaller. A plausible option consists in selecting the thickness of the surface mixed layer, when such a hydro-dydnamic feature can be identified. Furthermore, the vertical eddy diffusivity and the piston velocity are likely to be time- and position-dependent, implying that their typical order of magnitude should be evaluated and subsequently introduced into (61). Accordingly, the final formulation of dimensionless parameter (61) is

$$\epsilon = \frac{\mathcal{K}}{\mathcal{HW}}, \tag{62}$$

where $\mathcal{K}$, $\mathcal{H}$, and $\mathcal{W}$ denote a typical value of the vertical eddy diffusivity near the water–air interface, the relevant vertical length scale, and the order of magnitude of the piston velocity, respectively.

For advective and diffusive transport problems, Haine [104] studied the relationship between solutions obtained under Dirichlet boundary conditions and those ensuing from Robin conditions. The obtained theoretical results are rather general, but are beyond the scope of the present study.

There is no denying that the developments related to gas exchanges through the water–air interface are somewhat cumbersome. If obtaining diagnostic quantities for such phenomena were to prove too laborious, we may wonder if it would be worth the effort, suggesting that we should perhaps restrict ourselves to the simplest approach, which consists in assuming that the atmospheric age is zero, as laid out in the rightmost column of Table 1.

4. A Simple Ventilation Assessment Problem

Each of the illustrations included in the preceding Section was meant to help comprehend the role of a single type of boundary condition. To gain further insight into the role of boundary conditions in age calculations, it may be desirable to consider a slightly more sophisticated situation. Seeking inspiration in [65,107], we will tackle a relatively simple, two-dimensional, "horizontal-vertical" ventilation assessment study with constant hydro-dynamic parameters.

Let x and z denote the horizontal coordinate and the vertical one, respectively. The domain of interest is defined by inequalities $0 \leq x \leq L$ and $-h \leq z \leq 0$ (Figure 9). Water flows in the direction of increasing x with constant (horizontal) velocity U. Horizontal and vertical diffusion is taken into account with the help of constant diffusivities K_x and K_z. The bottom ($z = -h$) is an insulating boundary. The vertical boundaries located at $x = 0$ and $x = L$ are open. The latter is an arrival boundary and the former is a departure one, and so is the water–air interface. Therefore, the age is the time elapsed since leaving the incoming boundary ($x = 0$) or the water–air interface ($z = 0$).

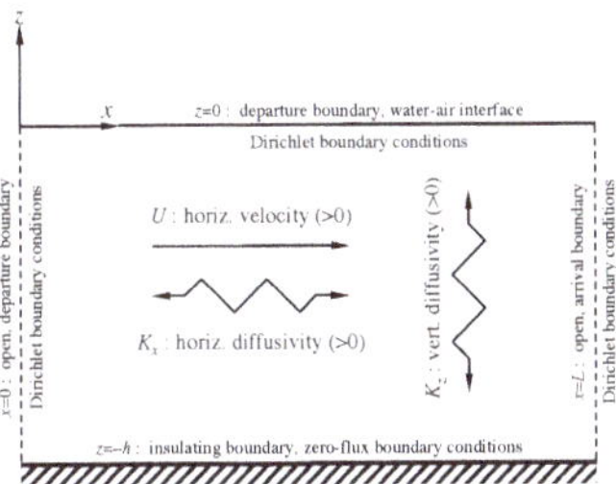

Figure 9. Schematic representation of the two-dimensional, "horizontal-vertical" domain of interest for the simple ventilation study dealt with in Section 4. The nature of the boundaries and the related boundary conditions are also indicated.

For the sake of simplicity, only steady-state solutions will be considered. Accordingly, the concentration of the ventilation tracer, $C(x, z)$, is the solution of the following partial differential problem:

$$\begin{cases} 0 = -U\frac{\partial C}{\partial x} + K_x\frac{\partial^2 C}{\partial x^2} + K_z\frac{\partial^2 C}{\partial z^2} \\ C(0, z) = 1 = C(x, 0), \quad [-K_z\frac{\partial C}{\partial z}]_{z=-h} = 0, \quad C(L, z) = 0. \end{cases} \tag{63}$$

Then, the associated age concentration, $\alpha(x, z)$, obeys

$$\begin{cases} 0 = -U\frac{\partial \alpha}{\partial x} + K_x\frac{\partial^2 \alpha}{\partial x^2} + K_z\frac{\partial^2 \alpha}{\partial z^2} + C \\ \alpha(0, z) = 0 = \alpha(x, 0), \quad [-K_z\frac{\partial \alpha}{\partial z}]_{z=-h} = 0, \quad \alpha(L, z) = 0. \end{cases} \tag{64}$$

Finally, the age is $a(x, z) = \alpha(x, z)/C(x, z)$.

To the best of our knowledge, there exists no simple analytical solution to the differential problem (63)–(64). This is why we built a numerical solution for it by having recourse to the tracer transport module of the finite-element, discontinuous Galerkin model, SLIM (www.slim-ocean.be) [108–110]. The computations are based on a fine mesh consisting of 39,204 rectangular elements. The resolution is increased near the domain boundaries in such a way that the discrete solution is believed to be very close to the exact one.

There are two crucial dimensionless numbers associated with the present transport problem. The horizontal (resp. vertical) Peclet number is the ratio of the horizontal (resp. vertical) diffusion timescale L^2/K_x (resp. h^2/K_z) to the advective timescale L/U, yielding $Pe_x = UL/K_x$ (resp. $Pe_z = h^2U/(LK_z)$). Since the vertical velocity is zero, introducing a vertical Peclet number may seem to be somewhat questionable. It is argued in Appendix C that the aforementioned vertical Peclet number is, roughly speaking, in line with common practice.

In shallow domains, such as rivers, estuaries, or coastal regions, vertical (turbulent) diffusion plays an important role. Therefore, assuming that the vertical Peclet number ranges from 1 to 100 with a typical order of magnitude of 10 would presumably be acceptable. In the aforementioned domains of interest, the aspect ratio (that is, the ratio of the vertical length scale to the horizontal one) is significantly smaller than unity, which is why the horizontal Peclet number should probably be somewhat larger than the vertical one. The concentration, age concentration, and age are displayed in Figure 10 for $(Pe_x, Pe_z) = (10, 10)$ and $(Pe_x, Pe_z) = (100, 10)$. The boundary conditions at the water–air interface mostly impact the solution in the upper part of the domain, so that the maximum of the age is always attained at $(x, z) = (L, -h)$, that is, at the lower-right corner of the domain of interest. The smaller the horizontal Peclet number, the larger the impact of the boundary conditions prescribed at the upper boundary of the domain. Clearly, the present age distribution is more complex than that obtained in Section 3.4, though there, the horizontal processes taken into account are similar. Unsurprisingly, in the vicinity of the lower boundary of the domain $(z = -h)$, with relatively large values of Pe_z, the age tends to be similar to that derived from (45)–(46).

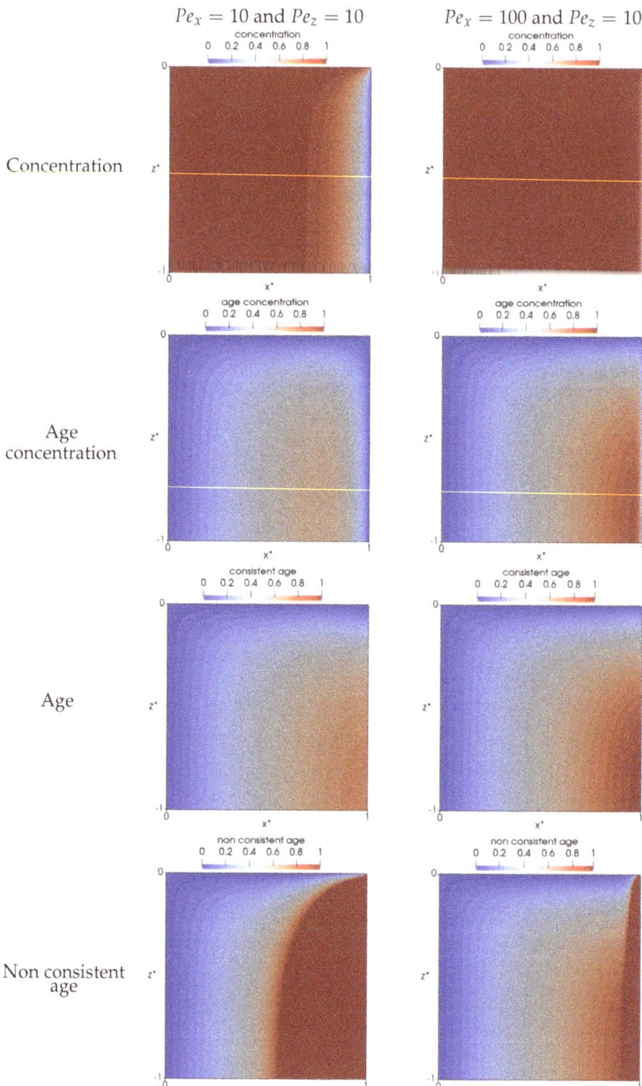

Figure 10. Illustration of the concentration, age concentration, and age from the solution of the partial differential problem (63) and (64) for $(Pe_x, Pe_z) = (10, 10)$ and $(Pe_x, Pe_z) = (100, 10)$. Dimensionless variables are represented. They are identified by asterisks and are defined as follows: $x^* = x/L$, $z^* = z/h$, $\alpha^* = U\alpha/L$, and $a^* = Ua/L$. The non-consistent age ensuing from inappropriate boundary conditions for the age concentration on the departure boundary is represented in the lowermost row of the graph.

Finally, on the departure boundary ($x = L$), we introduce an inconsistent boundary condition for the age concentration similar to that of Section 3.4, that is, $[-K \partial \alpha_{inc}/\partial x]_{x=L} = 0$. This causes the age to be infinite on the departure boundary. Nonetheless, as opposed to what was observed in the one-dimensional solution of the abovementioned Section, in the two-dimensional problem, the error does not affect a large fraction of the domain, because of the impact of the correct boundary conditions at the water–air interface. This is why we suspect that the studies that relied on non-consistent boundary conditions of the type referred to here are plagued by errors that, though non-negligible, are not catastrophic.

As illustrated by Figure 11, the error due to the use of non-consistent boundary conditions occurs in a region adjacent to the arrival boundary, whose width increases as the distance to the water–air interface increases. For the purpose of a sensitivity analysis, it is convenient to evaluate the maximum width of the region where the error is significant. A simple measure (Λ) is defined as follows: $a_{inc}(\Lambda, -h) = a(L, -h)$. This width, which is represented in Figure 12, is a decreasing (resp. increasing) function of the horizontal (resp. vertical) Peclet number. This is in agreement with elementary physical intuition.

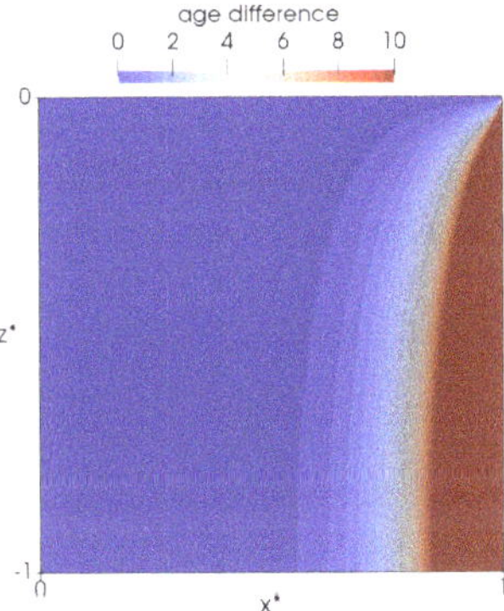

Figure 11. Difference between the non-consistent age and the correct one for $(Pe_x, Pe_z) = (10, 10)$. Dimensionless variables similar to those of Figure 10 are represented. On the departure boundary, the age difference is infinite, but the colour is saturated at a value of 10.

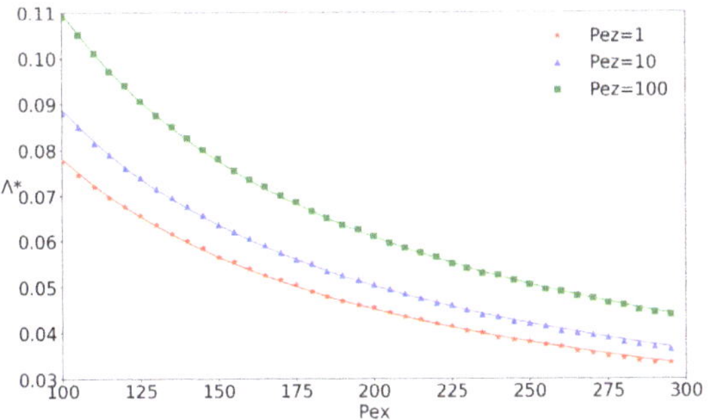

Figure 12. Dimensionless width ($\Lambda^* = \Lambda/L$) of the region where the error due to the inconsistent boundary conditions on the departure boundary is significant as a function of the horizontal Peclet number (Pe_x) for various values of the vertical Peclet number, that is, $Pe_z = 110, 100$.

5. Discussion and Conclusions

The (mean) age of a constituent of the water, or a group of constituents, including the water itself (that is, the aggregate of all the constituents), is a diagnostic timescale depending on time and position that can be obtained from the solution of a system of partial differential equations. The general form of them has been well-known since the turn of the century [59,69]. Over the past two decades, relatively little attention has been devoted to the construction of the initial and boundary conditions under which the age-related equations are to be solved. This is, however, a critical ingredient of an age-based diagnostic strategy—for the solutions of age-related (or any other) parabolic partial differential equations to be unambiguously determined, the initial and boundary conditions must be precisely defined. In this regard, casualness must be ruled out, as has been exemplified above by documenting the detrimental impact of some inconsistent boundary conditions.

While initial conditions are rather easily built, which is why we have not tackled them explicitly, constructing appropriate boundary conditions is less straightforward, as has been shown in the present study. Hopefully, the latter will help clarify the methodology to set up an age-based diagnostic approach. The first steps of it should be as follows:

1. Set out the reasons why the age, rather than other timescales (or diagnoses of another nature), is likely to be of use to help interpret the aquatic processes under consideration;
2. Select the constituent whose (mean) age is to be evaluated and explain the rationale of this choice;
3. Define the age, especially where and when the age of a particle of the constituent under study is to set or reset to zero, as well as where, when, and how this particle will cease to be taken into consideration;
4. Build the boundary conditions for the age distribution function in accordance with the outcome of the previous three steps;
5. Derive consistent boundary conditions for the concentration and age concentration using the methodology developed in this article (see also Appendix D).

Obviously, the following steps will consist in solving, analytically or numerically, the relevant partial differential problems and discuss the obtained results, which cannot be achieved in a fruitful manner if the foundations of the diagnostic approach are shaky.

Steps 1 to 3 above seem to be rather straightforward, if not trivial. However, as was seen by Delhez et al. [111], an ill-conceived diagnostic strategy may lead to the evaluation of timescales contradicting their very definition, eventually leading to dubious interpretations. Clearly, we should bear in mind the wise piece of advice of Bolin and Rodhe [55]: "To avoid misunderstandings and even erroneous conclusions, it is important to introduce precise definitions and to use them with care". It is because this word of caution has been overlooked time and again that Viero and Defina [41] referred to the field of diagnostic timescales as a modern Tower of Babel. Hopefully, the present study will be considered as a modest contribution to the deconstruction of this edifice. We strongly believe that all the developments made above are also relevant to partial ages, a recently developed generalisation of the concept of age [76,112]. This is because there is no fundamental difference between the concept of age and that of partial age: similar lines of argument should apply to both types of diagnoses.

It is impossible to address all the existing types of boundary conditions in a single paper. However, the approach advocated herein (that is, deriving the boundary conditions for the concentration and age concentration from those relevant to the age distribution function) can probably be applied to open boundary conditions other than those dealt with above, in particular, tracer-adapted versions of radiation conditions [113–116], sponge layers [117,118] and other techniques [119]. This has yet to be convincingly substantiated.

Clearly, dealing with realistic case studies is beyond the scope of the present article, for its key objective is the development of the theory to build consistent age-related boundary conditions. However, the boundary conditions used in the idealised ventilation rate assessment of Section 4 are most likely to be similar to those that would be implemented in realistic ventilation studies. This is especially true for the surface and bottom boundary conditions. Many ventilation studies [57,58] did not have to cope with lateral open boundaries. Their lateral boundaries were insulating ones, leading to trivial boundary conditions as may be seen in Section 3.1. In general, the open boundaries are believed to be the most problematic ones. To estimate water renewal of semi- enclosed domains, many authors resorted to Dirichlet boundary conditions [62,120] on open boundaries. This approach is undoubtedly the easiest one. However, we are convinced that, in the near future, more subtle boundary conditions will be worked out, involving fluxes rather than prescribed values, which will be related, in one way or another, to the boundary conditions for the momentum equations. Such boundary conditions are being developed and will be described in forthcoming articles.

Although the present study focuses on Eulerian developments, it must be realised that, in principle, all the age calculations, for idealised or realistic flow processes, can be achieved by means of Lagrangian methods, as well as Eulerian ones—as explained in van Sebille et al. [68] and some of the references therein, both approaches should converge to similar solutions. In the Lagrangian framework, it is not necessary to explicitly evaluate the concentration and the age concentration, which is why all of the above developments about boundary conditions are likely to be irrelevant to Lagrangian techniques. However, Lagrangian calculations have disadvantages—the representation of diffusive processes by means of stochastic terms is not straightforward [121,122] and the necessary computational resources are unlikely to be smaller than those required for carrying out Eulerian computations. In addition, most hydrodynamic models are Eulerian, making it somewhat easier to implement diagnoses rooted in the same framework. Finally, properties of diagnostic timescales are generally simpler to derive in the Eulerian framework than in the Lagrangian one [62,69,89,91,97,100]. All this being said, literal interpretations are much easier to produce in the Lagrangian framework, which is the reason why, in the present article, as well as in many publications of the diagnostic timescale literature, the theoretical developments and calculations are Eulerian, whereas the explanations and interpretations resort to a vocabulary rooted in the Lagrangian formalism.

The present study focused on the calculation of the age of a tracer as a means to help understand complex aquatic fluid flows. However, as underscored by Wunsch [123], there are many more aspects in tracer and timescale methods, the significance of which cannot be overestimated (e.g., time dependency, the number of space dimensions, Green's function theory, inverse methods, stochastic boundary conditions, and timescales other than those of CART or similar ones). In addition, the aforementioned article introduces a number of analytical solutions that should no longer be overlooked. In this respect, the reference to Carslaw and Jaeger [124] is a very important one. Clearly, much more work is needed in this rich field of research.

Author Contributions: Conceptualization, E.D. and A.M.; Methodology, E.D. and A.M.; Validation, I.D., J.L. and V.L.; Visualization, E.D. and I.D.; Writing—original draft, E.D. and I.D.; Writing—review & editing, E.D., I.D., J.L., V.L. and A.M. All authors have read and agreed to the published version of the manuscript.

Funding: A.M. is indebted to the European Union's Horizon 2020 research and innovation programme for the Marie Sklodowska-Curie grant agreement No 660893.

Acknowledgments: E.D. and A.M. are an honorary research associate and a postdoctoral researcher, respectively, with the Belgian Fund for Scientific Research (F.R.S.-FNRS). The authors are indebted to Valentin Vallaeys for his useful comments. Reviewers' remarks and suggestions led to significant improvements of the original version of the manuscript.

Conflicts of Interest: The authors declare no conflict of interest.

Appendix A

We set out to prove that concentration (20) and age concentration (21) satisfy boundary conditions $C(t,0) = 1$ and $\alpha(t,0) = 0$.

Concentration (20) may be rewritten as follows:

$$C(t,x) = \underbrace{\int_0^\infty \frac{x}{\sqrt{4\pi K\tau^3}} \exp\left[-\frac{(x-U\tau)^2}{4K\tau}\right] d\tau}_{=C_1(t,x)} - \underbrace{\int_t^\infty \frac{x}{\sqrt{4\pi K\tau^3}} \exp\left[-\frac{(x-U\tau)^2}{4K\tau}\right] d\tau}_{=C_2(t,x)} \tag{A1}$$

The first integral in the right-hand side of (A1) is readily seen to be independent of x, that is, $C_1(t,x) = 1$. The second one satisfies inequalities

$$0 \leq C_2(t,x) \leq \int_t^\infty \frac{x}{\sqrt{4\pi K\tau^3}} d\tau = \frac{x}{\sqrt{\pi Kt}} \tag{A2}$$

so that $C_2(t,0) = 0$, implying $C(t,0) = 1$.

Age concentration (21) leads to

$$\alpha(t,x) = \underbrace{\int_0^\infty \frac{x}{\sqrt{4\pi K\tau}} \exp\left[-\frac{(x-U\tau)^2}{4K\tau}\right] d\tau}_{=\alpha_1(t,x)} - \underbrace{\int_t^\infty \frac{x}{\sqrt{4\pi K\tau}} \exp\left[-\frac{(x-U\tau)^2}{4K\tau}\right] d\tau}_{-\alpha_2(t,x)} \tag{A3}$$

with $\alpha_1(x,t) = x/U$ and

$$0 \leq \alpha_2(t,x) \leq \int_t^\infty \frac{x}{\sqrt{4\pi K\tau}} \exp\left[-\frac{2Ux - U^2\tau}{4K}\right] d\tau = \frac{x}{U}\sqrt{\frac{4K}{\pi U^2 t}} \exp\left[\frac{2Ux - U^2 t}{4K}\right] \tag{A4}$$

As a consequence, we obtain $\alpha_1(t,0) = 0$ and $\alpha_2(t,0) = 0$, eventually yielding $\alpha(t,0) = 0$. QED.

Appendix B

We will build the dimensionless counterparts of age distribution functions (19) and (34). The previous one was established under a Dirichlet boundary condition at the departure boundary ($x = 0$), whilst the latter ensued from a Robin boundary condition.

We introduce the following dimensionless variables (identified by asterisks): $t^* = U^2t/K$, $x^* = Ux/K$, $\tau^* = Kc/U^2$ and $c^* = Kc/U^2$. Then, the dimensionless age distribution function for the Dirichlet boundary condition reads

$$c^*(t^*,x^*,\tau^*) = \frac{x^*}{\sqrt{4\pi(\tau^*)^3}} \exp\left[-\frac{(x^* - \tau^*)^2}{4\tau^*}\right] Y(t^* - \tau^*) \tag{A5}$$

and that for the Robin boundary condition is

$$c^*(t^*,x^*,\tau^*) = \left\{\frac{1}{\sqrt{\pi\tau^*}} \exp\left[-\frac{(x^* - \tau^*)^2}{4\tau^*}\right] - \frac{e^{x^*}}{2} \mathrm{erfc}\left[\sqrt{\frac{\tau^*}{4}} + \sqrt{\frac{(x^*)^2}{4\tau^*}}\right]\right\} Y(t^* - \tau^*) \tag{A6}$$

where Y is the Heaviside step function. These functions are depicted in Figure 3. Because of the particular choice of the time and space scales (K/U^2 and K/U, respectively), the above relations do not include any dimensionless parameter, implying that they are universal. In other words, they hold valid for any value of the time, space coordinate, age, velocity and diffusivity.

Appendix C

Assume, for a moment, that the vertical velocity in the simple ventilation problem of Section 4 is not zero and that W is its typical order of magnitude. Then, the vertical Peclet number, which is the ratio of vertical diffusion timescale h^2/K_z to vertical advective timescale h/W, would be $Pe_z =$

Wh/K_z. Since the Boussinesq approximation is assumed to hold valid, the velocity is divergence-free, implying that the order of magnitude of the vertical velocity satisfies $W/h = U/L$. Combining this expression with the vertical Peclet number defined above, we would obtain $Pe_z = h^2U/(LK_z)$. Thus, though it is not immediately self-evident, the Peclet number introduced in Section 4 is in line with the widely-used definition of this dimensionless number.

Appendix D

All the boundary conditions for the age distribution function set out above are special cases of the following generic form:

$$[b \cdot \nabla c + \eta c + \mu]_{x \in \Gamma} = 0 \tag{A7}$$

where $b(t, x)$ is an appropriate vector, whilst $\eta(t, x)$ and $\mu(t, x, \tau)$ are scalar functions, whose specific form depends on the nature of the boundary condition to be built. To derive the boundary conditions for the concentration and age concentration, we integrate Equation (A7) over the age, yielding

$$\left[b \cdot \nabla C + \eta C + \int_0^\infty \mu(t, x, \tau)d\tau \right]_{x \in \Gamma} = 0 \tag{A8}$$

and

$$\left[b \cdot \nabla \alpha + \eta \alpha + \int_0^\infty \tau \mu(t, x, \tau)d\tau \right]_{x \in \Gamma} = 0 \tag{A9}$$

By substituting $b(t, x) = 0$, $\eta(t, x) = 1$ and $\mu(t, x, \tau) = -C^\Gamma(t, x)\delta[\tau - a^\Gamma(t, x)]$ into relations (A8) and (A9), we readily obtain boundary conditions:

$$[C(t, x)]_{x \in \Gamma} = \int_0^\infty C^\Gamma(t, x)\delta[\tau - a^\Gamma(t, x)]d\tau = C^\Gamma(t, x) \tag{A10}$$

$$[\alpha(t, x)]_{x \in \Gamma} = \int_0^\infty \tau C^\Gamma(t, x)\delta[\tau - a^\Gamma(t, x)]d\tau = C^\Gamma(t, x)a^\Gamma(t, x) \tag{A11}$$

If the age is prescribed to be zero on the boundary ($a^\Gamma = 0$), then (A11) simplifies to $[\alpha(t, x)]_{x \in \Gamma} = 0$. Thus, we have obtained a pair of Dirichlet boundary conditions that can be applied on a departure boundary (see Section 3.1).

The simplest arrival boundary conditions, that is, Equations (39) and (40), may be derived from (A7)–(A9) by setting $b(t, x) = 0$, $\eta(t, x) = 1$ and $\mu(t, x, \tau) = 0$ (see Section 3.4).

If we take $b = n \cdot K$, $\eta = 0$ et $\mu = 0$, we obtain Neumann boundary conditions

$$[-n \cdot (K \cdot \nabla C)]_{x \in \Gamma} = 0 \tag{A12}$$

$$[-n \cdot (K \cdot \nabla \alpha)]_{x \in \Gamma} = 0 \tag{A13}$$

which may be applied on an insulating boundary.

We obtain Robin boundary conditions by setting $b = -n \cdot K$, $\eta(t, x) = -\chi(t, x)$ and

$$\mu(t, x, \tau) = \chi(t, x)C^\Gamma(t, x)\delta[\tau - a^\Gamma(t, x)], \tag{A14}$$

where $\chi(t, x)$ (≥ 0) is an appropriate velocity. The corresponding boundary conditions for the concentration and age concentration are readily seen to be

$$[n \cdot (K \cdot \nabla C) + \chi(C - C^\Gamma)]_{x \in \Gamma} = 0 \tag{A15}$$

and

$$[\boldsymbol{n} \cdot (\boldsymbol{K} \cdot \nabla \alpha) + \chi(\alpha - C^\Gamma a^\Gamma)]_{x \in \Gamma} = 0 \tag{A16}$$

As was seen in Section 3.6, conditions of this type may be implemented at water–air interface to account for gas exchanges. In this case, $\chi(t, x)$ would be the piston velocity.

Many more pairs of consistent boundary conditions for the concentration and age concentration may be derived from (A7)–(A9), which may thus be viewed as rather generic expressions.

References

1. Hall, T.M.; Plumb, R.A. Age as a diagnostic of stratospheric transport. *J. Geophys. Res. Atmos.* **1994**, *99*, 1059–1070. [CrossRef]
2. Holzer, M.; Hall, T.M. Transit-time and tracer-age distributions in geophysical flows. *J. Atmos. Sci.* **2000**, *57*, 3539–3558. [CrossRef]
3. Waugh, D.W.; Hall, T.M. Age of stratospheric air: Theory, observations, and models. *Rev. Geophys.* **2002**, *40*, 1010. [CrossRef]
4. Orbe, C.; Waugh, D.W.; Newman, P.A.; Steenrod, S. The transit-time distribution from the Northern Hemisphere midlatitude surface. *J. Atmos. Sci.* **2016**, *73*, 3785–3802. [CrossRef]
5. Dronkers, J.; Zimmerman, J.T.F. Some principles of mixing in tidal lagoons. In Proceedings of the Oceanologica Acta, N° SP, Actes Symposium International sur les lagunes côtières, SCOR/IABO/UNESCO, Bordeaux, France, 8–14 September 1982; pp. 107–117.
6. Helder, W.; Ruardij, P. A one-dimensional mixing and flushing model of the Ems-Dollard estuary: calculation of time scales at different river discharges. *Neth. J. Sea Res.* **1982**, *15*, 293–312. [CrossRef]
7. Zimmerman, J.T.F. Estuarine Residence Times. *Hydrodynamics of Estuaries*; Kjerfve, B., Ed.; CRC Press: Boca Raton, FL, USA, 1988; Volume I, pp. 75–84.
8. Delesalle, B.; Sournia, A. Residence time of water and phytoplankton biomass in coral reef lagoons. *Cont. Shelf Res.* **1992**, *12*, 939–949. [CrossRef]
9. Salomon, J.C.; Breton, M.; Guegueniat, P. A 2D long term advection—dispersion model for the Channel and southern North Sea Part B: Transit time and transfer function from Cap de La Hague. *J. Mar. Syst.* **1995**, *6*, 515–527. [CrossRef]
10. Oliveira, A.; Baptista, A.M. Diagnostic modeling of residence times in estuaries. *Water Resour. Res.* **1997**, *33*, 1935–1946. [CrossRef]
11. Jenkins, W.J. Studying subtropical thermocline ventilation and circulation using tritium and 3He. *J. Geophys. Res. Ocean.* **1998**, *103*, 15817–15831. [CrossRef]
12. Deleersnijder, E.; Wang, J.; Mooers, C.N.K. A two-compartment model for understanding the simulated three-dimensional circulation in Prince William Sound, Alaska. *Cont. Shelf Res.* **1998**, *18*, 279–287. [CrossRef]
13. Campin, J.M.; Fichefet, T.; Duplessy, J.C. Problems with using radiocarbon to infer ocean ventilation rates for past and present climates. *Earth Planet. Sci. Lett.* **1999**, *165*, 17–24. [CrossRef]
14. Andréfouët, S.; Pagès, J.; Tartinville, B. Water renewal time for classification of atoll lagoons in the Tuamotu Archipelago (French Polynesia). *Coral Reefs* **2001**, *20*, 399–408. [CrossRef]
15. Haine, T.W.N.; Hall, T.M. A generalized transport theory: Water-mass composition and age. *J. Phys. Oceanogr.* **2002**, *32*, 1932–1946. [CrossRef]
16. Monsen, N.E.; Cloern, J.E.; Lucas, L.V.; Monismith, S.G. A comment on the use of flushing time, residence time, and age as transport time scales. *Limnol. Oceanogr.* **2002**, *47*, 1545–1553. [CrossRef]
17. Waugh, D.W.; Vollmer, M.K.; Weiss, R.F.; Haine, T.W.N.; Hall, T.M. Transit time distributions in Lake Issyk-Kul. *Geophys. Res. Lett.* **2002**, *29*, doi:10.1029/2002GL016201. [CrossRef]
18. Braunschweig, F.; Martins, F.; Chambel, P.; Neves, R. A methodology to estimate renewal time scales in estuaries: the Tagus Estuary case. *Ocean. Dyn.* **2003**, *53*, 137–145. [CrossRef]
19. Waugh, D.W.; Hall, T.M.; Haine, T.W.N. Relationships among tracer ages. *J. Geophys. Res. Ocean.* **2003**, *108*. [CrossRef]
20. Andrejev, O.; Myrberg, K.; Lundberg, P.A. Age and renewal time of water masses in a semi-enclosed basin—Application to the Gulf of Finland. *Tellus Dyn. Meteorol. Oceanogr.* **2004**, *56*, 548–558.

21. Abdelrhman, M.A. Simplified modeling of flushing and residence times in 42 embayments in New England, USA, with special attention to Greenwich Bay, Rhode Island. *Estuar. Coast. Shelf Sci.* **2005**, *62*, 339–351. [CrossRef]

22. Gao, Y.; Drange, H.; Bentsen, M.; Johannessen, O.M. Tracer-derived transit time of the waters in the eastern Nordic Seas. *Tellus Chem. Phys. Meteorol.* **2005**, *57*, 332–340. [CrossRef]

23. Cornaton, F.; Perrochet, P. Groundwater age, life expectancy and transit time distributions in advective–dispersive systems: 1. Generalized reservoir theory. *Adv. Water Resour.* **2006**, *29*, 1267–1291. [CrossRef]

24. Cornaton, F.; Perrochet, P. Groundwater age, life expectancy and transit time distributions in advective–dispersive systems; 2. Reservoir theory for sub-drainage basins. *Adv. Water Resour.* **2006**, *29*, 1292–1305. [CrossRef]

25. Cucco, A.; Perilli, A.; De Falco, G.; Ghezzo, M.; Umgiesser, G. Water circulation and transport timescales in the Gulf of Oristano. *Chem. Ecol.* **2006**, *22*, S307–S331. [CrossRef]

26. Holzer, M.; Primeau, F. The diffusive ocean conveyor. *Geophys. Res. Lett.* **2006**, *33*. [CrossRef]

27. Jouon, A.; Douillet, P.; Ouillon, S.; Fraunié, P. Calculations of hydrodynamic time parameters in a semi-opened coastal zone using a 3D hydrodynamic model. *Cont. Shelf Res.* **2006**, *26*, 1395–1415. [CrossRef]

28. Kazemi, G.H.; Lehr, J.H.; Perrochet, P. *Groundwater Age*; Wiley: Hoboken, NJ, USA, 2006; p. 325.

29. Orre, S.; Gao, Y.; Drange, H.; Nilsen, J.E.Ø. A reassessment of the dispersion properties of 99Tc in the North Sea and the Norwegian Sea. *J. Mar. Syst.* **2007**, *68*, 24–38. [CrossRef]

30. Torréton, J.P.; Rochelle-Newall, E.; Jouon, A.; Faure, V.; Jacquet, S.; Douillet, P. Correspondence between the distribution of hydrodynamic time parameters and the distribution of biological and chemical variables in a semi-enclosed coral reef lagoon. *Estuar. Coast. Shelf Sci.* **2007**, *74*, 766–776. [CrossRef]

31. Cucco, A.; Umgiesser, G.; Ferrarin, C.; Perilli, A.; Canu, D.M.; Solidoro, C. Eulerian and lagrangian transport time scales of a tidal active coastal basin. *Ecol. Model.* **2009**, *220*, 913–922. [CrossRef]

32. Plus, M.; Dumas, F.; Stanisière, J.Y.; Maurer, D. Hydrodynamic characterization of the Arcachon Bay, using model-derived descriptors. *Cont. Shelf Res.* **2009**, *29*, 1008–1013. [CrossRef]

33. Lucas, L.V. Implications of estuarine transport for water quality. *Contemp. Issues Estuar. Phys.* **2010**, 273–303.

34. Maltrud, M.; Bryan, F.; Peacock, S. Boundary impulse response functions in a century-long eddying global ocean simulation. *Environ. Fluid Mech.* **2010**, *10*, 275–295. [CrossRef]

35. Cavalcante, G.H.; Kjerfve, B.; Feary, D.A. Examination of residence time and its relevance to water quality within a coastal mega-structure: The Palm Jumeirah Lagoon. *J. Hydrol.* **2012**, *468*, 111–119. [CrossRef]

36. Mouchet, A.; Deleersnijder, E.; Primeau, F. The leaky funnel model revisited. *Tellus Dyn. Meteorol. Oceanogr.* **2012**, *64*, 19131. [CrossRef]

37. Grifoll, M.; Del Campo, A.; Espino, M.; Mader, J.; González, M.; Borja, Á. Water renewal and risk assessment of water pollution in semi-enclosed domains: Application to Bilbao Harbour (Bay of Biscay). *J. Mar. Syst.* **2013**, *109*, S241–S251. [CrossRef]

38. Andutta, F.P.; Ridd, P.V.; Deleersnijder, E.; Prandle, D. Contaminant exchange rates in estuaries–New formulae accounting for advection and dispersion. *Prog. Oceanogr.* **2014**, *120*, 139–153. [CrossRef]

39. Delhez, E.J.M.; Wolk, F. Diagnosis of the transport of adsorbed material in the Scheldt estuary: A proof of concept. *J. Mar. Syst.* **2013**, *128*, 17–26. [CrossRef]

40. Rinaldo, A.; Benettin, P.; Harman, C.J.; Hrachowitz, M.; McGuire, K.J.; Van Der Velde, Y.; Bertuzzo, E.; Botter, G. Storage selection functions: A coherent framework for quantifying how catchments store and release water and solutes. *Water Resour. Res.* **2015**, *51*, 4840–4847. [CrossRef]

41. Viero, D.P.; Defina, A. Renewal Time Scales in Tidal Basins: Climbing the Tower of Babel. *Sustainable Hydraulics in the Era of Global Change*; Erpicum, S., Ed.; Taylor & Francis Group: Abingdon, UK, 2016; pp. 338–345.

42. Viero, D.P.; Defina, A. Water age, exposure time, and local flushing time in semi-enclosed, tidal basins with negligible freshwater inflow. *J. Mar. Syst.* **2016**, *156*, 16–29. [CrossRef]

43. Hiatt, M.; Castañeda-Moya, E.; Twilley, R.; Hodges, B.R.; Passalacqua, P. Channel-island connectivity affects water exposure time distributions in a coastal river delta. *Water Resour. Res.* **2018**, *54*, 2212–2232. [CrossRef]

44. Cheng, Y.; Mu, Z.; Wang, H.; Zhao, F.; Li, Y.; Lin, L. Water Residence Time in a Typical Tributary Bay of the Three Gorges Reservoir. *Water* **2019**, *11*, 1585. [CrossRef]

45. Dippner, J.W.; Bartl, I.; Chrysagi, E.; Holtermann, P.L.; Kremp, A.; Thoms, F.; Voss, M. Lagrangian Residence Time in the Bay of Gdansk, Baltic Sea. *Front. Mar. Sci.* **2019**, *6*, 725. [CrossRef]

46. Drouzy, M.; Douillet, P.; Fernandez, J.M.; Pinazo, C. Hydrodynamic time parameters response to meteorological and physical forcings: toward a stagnation risk assessment device in coastal areas. *Ocean. Dyn.* **2019**, *69*, 967–987. [CrossRef]

47. Gross, E.; Andrews, S.; Bergamaschi, B.; Downing, B.; Holleman, R.; Burdick, S.; Durand, J. The Use of Stable Isotope-Based Water Age to Evaluate a Hydrodynamic Model. *Water* **2019**, *11*, 2207. [CrossRef]

48. Huguet, J.R.; Brenon, I.; Coulombier, T. Characterisation of the Water Renewal in a Macro-Tidal Marina Using Several Transport Timescales. *Water* **2019**, *11*, 2050. [CrossRef]

49. Jiang, L.; Soetaert, K.; Gerkema, T. Decomposing the intra-annual variability of flushing characteristics in a tidal bay along the North Sea. *J. Sea Res.* **2019**, *155*, 101821. [CrossRef]

50. Mercier, C.; Delhez, E.J.M. Diagnosis of the sediment transport in the Belgian Coastal Zone. *Estuar. Coast. Shelf Sci.* **2007**, *74*, 670–683. [CrossRef]

51. Gong, W.; Shen, J. A model diagnostic study of age of river-borne sediment transport in the tidal York River Estuary. *Environ. Fluid Mech.* **2010**, *10*, 177–196. [CrossRef]

52. Ralston, D.K.; Geyer, W.R. Sediment transport time scales and trapping efficiency in a tidal river. *J. Geophys. Res. Earth Surf.* **2017**, *122*, 2042–2063. [CrossRef]

53. Delhez, E.J.M. On the concept of exposure time. *Cont. Shelf Res.* **2013**, *71*, 27–36. [CrossRef]

54. Zimmerman, J.T.F. Mixing and flushing of tidal embayments in the western Dutch Wadden Sea part I: Distribution of salinity and calculation of mixing time scales. *Neth. J. Sea Res.* **1976**, *10*, 149–191. [CrossRef]

55. Bolin, B.; Rodhe, H. A note on the concepts of age distribution and transit time in natural reservoirs. *Tellus* **1973**, *25*, 58–62. [CrossRef]

56. Takeoka, H. Fundamental concepts of exchange and transport time scales in a coastal sea. *Cont. Shelf Res.* **1984**, *3*, 311–326. [CrossRef]

57. Thiele, G.; Sarmiento, J.L. Tracer dating and ocean ventilation. *J. Geophys. Res. Ocean.* **1990**, *95*, 9377–9391. [CrossRef]

58. England, M.H. The age of water and ventilation timescales in a Global Ocean model. *J. Phys. Oceanogr.* **1995**, *25*, 2756–2777. [CrossRef]

59. Delhez, E.J.M.; Campin, J.M.; Hirst, A.C.; Deleersnijder, E. Toward a general theory of the age in ocean modelling. *Ocean. Model.* **1999**, *1*, 17–27. [CrossRef]

60. Hirst, A.C. Determination of water component age in ocean models: Application to the fate of North Atlantic Deep Water. *Ocean. Model.* **1999**, *1*, 81–94. [CrossRef]

61. Deleersnijder, E. *On the Timescales Relevant to a Linear Transport-Decay Reservoir Model*; Technical Report; Université catholique de Louvain: Ottignies-Louvain-la-Neuve, Belgium, 2019. Available online: http://hdl.handle.net/2078.1/219115 (accessed on 29 April 2020).

62. de Brye, B.; de Brauwere, A.; Gourgue, O.; Delhez, E.J.M.; Deleersnijder, E. Water renewal timescales in the Scheldt Estuary. *J. Mar. Syst.* **2012**, *94*, 74–86. [CrossRef]

63. Delhez, E.J.M.; Lacroix, G.; Deleersnijder, E. The age as a diagnostic of the dynamics of marine ecosystem models. *Ocean. Dyn.* **2004**, *54*, 221–231. [CrossRef]

64. White, L.; Deleersnijder, E. Diagnoses of vertical transport in a three-dimensional finite element model of the tidal circulation around an island. *Estuar. Coast. Shelf Sci.* **2007**, *74*, 655–669. [CrossRef]

65. Bendtsen, J.; Gustafsson, K.E.; Söderkvist, J.; Hansen, J.L. Ventilation of bottom water in the North Sea–Baltic Sea transition zone. *J. Mar. Syst.* **2009**, *75*, 138–149. [CrossRef]

66. Shah, S.H.A.M.; Primeau, F.; Deleersnijder, E.; Heemink, A. Tracing the ventilation pathways of the deep North Pacific Ocean using Lagrangian particles and Eulerian tracers. *J. Phys. Oceanogr.* **2017**, *47*, 1261–1280. [CrossRef]

67. Sun, J.; Liu, L.; Lin, J.; Lin, B.; Zhao, H. Vertical water renewal in a large estuary and implications for water quality. *Sci. Total Environ.* **2019**, 135593. [CrossRef] [PubMed]

68. Van Sebille, E.; Griffies, S.M.; Abernathey, R.; Adams, T.P.; Berloff, P.; Biastoch, A.; Blanke, B.; Chassignet, E.P.; Cheng, Y.; Cotter, C.J.; et al. Lagrangian ocean analysis: Fundamentals and practices. *Ocean. Model.* **2018**, *121*, 49–75. [CrossRef]

69. Deleersnijder, E.; Campin, J.M.; Delhez, E.J. The concept of age in marine modelling: I. Theory and preliminary model results. *J. Mar. Syst.* **2001**, *28*, 229–267. [CrossRef]

70. Deleersnijder, E.; Mouchet, A.; Delhez, E.J.M.; Beckers, J.M. Transient behaviour of water ages in the World Ocean. *Math. Comput. Model.* **2002**, *36*, 121–127. [CrossRef]

71. Delhez, E.J.M.; Deleersnijder, E. The concept of age in marine modelling: II. Concentration distribution function in the English Channel and the North Sea. *J. Mar. Syst.* **2002**, *31*, 279–297. [CrossRef]

72. Garabedian, P. *Partial Differential Equations*; Chelsea Publishing Company: New York, NY, USA, 1964; 672p.

73. Shen, J.; Haas, L. Calculating age and residence time in the tidal York River using three-dimensional model experiments. *Estuar. Coast. Shelf Sci.* **2004**, *61*, 449–461. [CrossRef]

74. Shen, J.; Lin, J. Modeling study of the influences of tide and stratification on age of water in the tidal James River. *Estuar. Coast. Shelf Sci.* **2006**, *68*, 101–112. [CrossRef]

75. Meier, H.E.M. Modeling the pathways and ages of inflowing salt-and freshwater in the Baltic Sea. *Estuar. Coast. Shelf Sci.* **2007**, *74*, 610–627. [CrossRef]

76. Liu, Z.; Wang, H.; Guo, X.; Wang, Q.; Gao, H. The age of Yellow River water in the Bohai Sea. *J. Geophys. Res. Ocean.* **2012**, *117*. [CrossRef]

77. Radtke, H.; Neumann, T.; Voss, M.; Fennel, W. Modeling pathways of riverine nitrogen and phosphorus in the Baltic Sea. *J. Geophys. Res. Ocean.* **2012**, *117*. [CrossRef]

78. Bendtsen, J.; Mortensen, J.; Rysgaard, S. Seasonal surface layer dynamics and sensitivity to runoff in a high Arctic fjord (Young Sound/Tyrolerfjord, 74 N). *J. Geophys. Res. Ocean.* **2014**, *119*, 6461–6478. [CrossRef]

79. Kärnä, T.; Baptista, A.M. Water age in the Columbia River estuary. *Estuar. Coast. Shelf Sci.* **2016**, *183*, 249–259. [CrossRef]

80. Rayson, M.D.; Gross, E.S.; Hetland, R.D.; Fringer, O.B. Time scales in Galveston Bay: An unsteady estuary. *J. Geophys. Res. Ocean.* **2016**, *121*, 2268–2285. [CrossRef]

81. Liu, R.; Zhang, X.; Liang, B.; Xin, L.; Zhao, Y. Numerical Study on the Influences of Hydrodynamic Factors on Water Age in the Liao River Estuary, China. *J. Coast. Res.* **2017**, *80*, 98–107. [CrossRef]

82. Du, J.; Park, K.; Shen, J.; Dzwonkowski, B.; Yu, X.; Yoon, B.I. Role of baroclinic processes on flushing characteristics in a highly stratified estuarine system, Mobile Bay, Alabama. *J. Geophys. Res.* **2018**, *123*, 4518–4537. [CrossRef]

83. Gao, Q.; He, G.; Fang, H.; Bai, S.; Huang, L. Numerical simulation of water age and its potential effects on the water quality in Xiangxi Bay of Three Gorges Reservoir. *J. Hydrol.* **2018**, *566*, 484–499. [CrossRef]

84. Chen, Y.; Cheng, W.; Zhang, H.; Qiao, J.; Liu, J.; Shi, Z.; Gong, W. Evaluation of the total maximum allocated load of dissolved inorganic nitrogen using a watershed—Coastal ocean coupled model. *Sci. Total Environ.* **2019**, *673*, 734–749. [CrossRef]

85. Grosse, F.; Fennel, K.; Laurent, A. Quantifying the relative importance of riverine and open-ocean nitrogen sources for hypoxia formation in the northern Gulf of Mexico. *J. Geophys. Res.* **2019**, *124*, 5451–5467. [CrossRef]

86. Li, Y.; Feng, H.; Zhang, H.; Sun, J.; Yuan, D.; Guo, L.; Nie, J.; Du, J. Hydrodynamics and water circulation in the New York/New Jersey Harbor: A study from the perspective of water age. *J. Mar. Syst.* **2019**, *199*, 103219. [CrossRef]

87. Shang, J.; Sun, J.; Tao, L.; Li, Y.; Nie, Z.; Liu, H.; Chen, R.; Yuan, D. Combined Effect of Tides and Wind on Water Exchange in a Semi-Enclosed Shallow Sea. *Water* **2019**, *11*, 1762. [CrossRef]

88. Yang, J.; Kong, J.; Tao, J. Modeling the Water-Flushing Properties of the Yangtze Estuary and Adjacent Waters. *J. Ocean. Univ. China* **2019**, *18*, 93–107. [CrossRef]

89. Beckers, J.M.; Delhez, E.; Deleersnijder, E. Some properties of generalized age-distribution equations in fluid dynamics. *SIAM J. Appl. Math.* **2001**, *61*, 1526–1544.

90. Deleersnijders, E.; Delhez, E.J.M.; Crucifix, M.; Beckers, J.M. On the symmetry of the age field of a passive tracer released into a one-dimensional fluid flow by a point-source. *Bull. de la Soc. R. des Sci. de Liège* **2001**, *70*, 5–21. Available online: https://popups.uliege.be/0037-9565/index.php?id=1392 (accessed on 29 April 2020).

91. Deleersnijder, E. *Water Renewal of a Region of Freshwater Influence (ROFI): Mathematical Properties of Some of the Relevant Diagnostic Variables*; Technical Report; Université Catholique de Louvain: Ottignies-Louvain-la-Neuve, Belgium, 2019. Available online: http://hdl.handle.net/2078.1/220841 (accessed on 29 April 2020).

92. Deleersnijder, E.; Delhez, E.J.M. Symmetry and asymmetry of water ages in a one-dimensional flow. *J. Mar. Syst.* **2004**, *48*, 61–66. [CrossRef]

93. Hall, T.M.; Haine, T.W.N. Tracer age symmetry in advective–diffusive flows. *J. Mar. Syst.* **2004**, *48*, 51–59. [CrossRef]

94. Cox, M.D. An idealized model of the world ocean. Part I: The global-scale water masses. *J. Phys. Oceanogr.* **1989**, *19*, 1730–1752. [CrossRef]

95. Goosse, H.; Campin, J.M.; Tartinville, B. The sources of Antarctic bottom water in a global ice–ocean model. *Ocean. Model.* **2001**, *3*, 51–65. [CrossRef]

96. Cornaton, F.J. Transient water age distributions in environmental flow systems: The time-marching Laplace transform solution technique. *Water Resour. Res.* **2012**, *48*. [CrossRef]

97. Delhez, E.J.M.; Deleersnijder, E.; Mouchet, A.; Beckers, J.M. A note on the age of radioactive tracers. *J. Mar. Syst.* **2003**, *38*, 277–286. [CrossRef]

98. Delhez, É.J.; Deleersnijder, É. The boundary layer of the residence time field. *Ocean. Dyn.* **2006**, *56*, 139–150. [CrossRef]

99. Deleersnijder, E. *A Conjecture about Age Inequalities*; Technical Report; Université Catholique de Louvain: Ottignies-Louvain-la-Neuve, Belgium, 2019. Available online: http://hdl.handle.net/2078.1/227647 (accessed on 29 April 2020).

100. Beckers, J.M. *YAAI: Yet Another Age Inequality*; Technical Report; Université de Liège: Liège, Belgium, 2020. Available online: http://hdl.handle.net/2268/245381 (accessed on 29 April 2020).

101. Liss, P.S. Gas Transfer: Experiments and Geochemical Implications. In *Air-Sea Exchange of Gases and Particles*; NATO ASI Series (Series C: Mathematical and Physical Sciences); Liss, P.S., Slinn, W.G.N., Eds.; Springer: Dordrecht, The Netherlands, 1983; pp. 241–298.

102. Wanninkhof, R. Relationship between wind speed and gas exchange over the ocean. *J. Geophys. Res. Ocean.* **1992**, *97*, 7373–7382. [CrossRef]

103. Jähne, B.; Haußecker, H. Air-water gas exchange. *Annu. Rev. Fluid Mech.* **1998**, *30*, 443–468. [CrossRef]

104. Haine, T.W.N. On tracer boundary conditions for geophysical reservoirs: How to find the boundary concentration from a mixed condition. *J. Geophys. Res. Ocean.* **2006**, *111*. [CrossRef]

105. Wanninkhof, R. Relationship between wind speed and gas exchange over the ocean revisited. *Limnol. Oceanogr. Methods* **2014**, *12*, 351–362. [CrossRef]

106. Deleersnijder, E. *On the Impact of the Atmosphere on the Time-Varying Age of a Passive Tracer in the Ocean*; Technical Report; Université catholique de Louvain: Ottignies-Louvain-la-Neuve, Belgium, 2017. Available online: http://hdl.handle.net/2078.1/184324 (accessed on 29 April 2020).

107. Hong, B.; Gong, W.; Peng, S.; Xie, Q.; Wang, D.; Li, H.; Xu, H. Characteristics of vertical exchange process in the Pearl River estuary. *Aquat. Ecosyst. Health Manag.* **2016**, *19*, 286–295. [CrossRef]

108. Kärnä, T.; Legat, V.; Deleersnijder, E. A baroclinic discontinuous Galerkin finite element model for coastal flows. *Ocean. Model.* **2013**, *61*, 1–20. [CrossRef]

109. Vallaeys, V.; Kärnä, T.; Delandmeter, P.; Lambrechts, J.; Baptista, A.M.; Deleersnijder, E.; Hanert, E. Discontinuous Galerkin modeling of the Columbia River's coupled estuary-plume dynamics. *Ocean. Model.* **2018**, *124*, 111–124. [CrossRef]

110. Delandmeter, P.; Lambrechts, J.; Legat, V.; Vallaeys, V.; Naithani, J.; Thiery, W.; Remacle, J.F.; Deleersnijder, E. A fully consistent and conservative vertically adaptive coordinate system for SLIM 3D v0. 4 with an application to the thermocline oscillations of Lake Tanganyika. *Geosci. Model Dev.* **2018**, *11*, 1161–1179. [CrossRef]

111. Delhez, É.J.; de Brye, B.; de Brauwere, A.; Deleersnijder, E. Residence time vs influence time. *J. Mar. Syst.* **2014**, *132*, 185–195. [CrossRef]

112. Mouchet, A.; Cornaton, F.; Deleersnijder, E.; Delhez, E.J. Partial ages: diagnosing transport processes by means of multiple clocks. *Ocean. Dyn.* **2016**, *66*, 367–386. [CrossRef]

113. Orlanski, I. A simple boundary condition for unbounded hyperbolic flows. *J. Comput. Phys.* **1976**, *21*, 251–269. [CrossRef]

114. Israeli, M.; Orszag, S.A. Approximation of radiation boundary conditions. *J. Comput. Phys.* **1981**, *41*, 115–135. [CrossRef]

115. Blumberg, A.F.; Kantha, L.H. Open boundary condition for circulation models. *J. Hydraul. Eng.* **1985**, *111*, 237–255. [CrossRef]

116. Oddo, P.; Pinardi, N. Lateral open boundary conditions for nested limited area models: A scale selective approach. *Ocean. Model.* **2008**, *20*, 134–156. [CrossRef]

117. Martinsen, E.A.; Engedahl, H. Implementation and testing of a lateral boundary scheme as an open boundary condition in a barotropic ocean model. *Coast. Eng.* **1987**, *11*, 603–627. [CrossRef]

118. Lavelle, J.; Thacker, W. A pretty good sponge: Dealing with open boundaries in limited-area ocean models. *Ocean. Model.* **2008**, *20*, 270–292. [CrossRef]

119. Blayo, E.; Debreu, L. Revisiting open boundary conditions from the point of view of characteristic variables. *Ocean. Model.* **2005**, *9*, 231–252. [CrossRef]

120. Pham Van, C.; De Brye, B.; De Brauwere, A.; Hoitink, A.; Soares-Frazao, S.; Deleersnijder, E. Numerical Simulation of Water Renewal Timescales in the Mahakam Delta, Indonesia. *Water* **2020**, *12*, 1017. [CrossRef]

121. Gräwe, U.; Wolff, J.O. Suspended particulate matter dynamics in a particle framework. *Environ. Fluid Mech.* **2010**, *10*, 21–39. [CrossRef]

122. Gräwe, U.; Deleersnijder, E.; Shah, S.H.A.M.; Heemink, A.W. Why the Euler scheme in particle tracking is not enough: the shallow-sea pycnocline test case. *Ocean. Dyn.* **2012**, *62*, 501–514. [CrossRef]

123. Wunsch, C. Oceanic age and transient tracers: Analytical and numerical solutions. *J. Geophys. Res. Ocean.* **2002**, *107*, 1. [CrossRef]

124. Carslaw, H.S.; Jaeger, J.C. *Conduction of Heat in Solids*, 2nd ed.; Clarendon Press: Oxford, UK, 1959; 510p.

Article

Numerical Simulation of Water Renewal Timescales in the Mahakam Delta, Indonesia

Chien Pham Van [1,*], Benjamin De Brye [2], Anouk De Brauwere [3], A.J.F. (Ton) Hoitink [4], Sandra Soares-Frazao [5] and Eric Deleersnijder [6]

[1]   Department of River Engineering and Disaster Management, Thuyloi University, 175 Tay Son, Dong Da, Hanoi 100000, Vietnam

[2]   Free Field Technologies (FFT), 1435 Mont-Saint-Guibert, Belgium; Benjamin.Debrye@fft.be

[3]   Karel de Grote Hogeschool, Chemie en Biomedische Laboratoriumtechnologie, Salesianenlaan 90, 2660 Antwerp, Belgium; anouk.debrauwere@gmail.com

[4]   Hydrology and Quantitative Water Management Group, Department of Environmental Sciences, Wageningen University, 6708PB Wageningen, The Netherlands; Ton.Hoitink@wur.nl

[5]   Institute of Mechanics, Materials and Civil Engineering (IMMC), Université Catholique de Louvain, B-1348 Louvain-la-Neuve, Belgium; Sandra.Soares-Frazao@uclouvain.be

[6]   Institute of Mechanics, Materials and Civil Engineering (IMMC) & Earth and Life Institute (ELI), Université Catholique de Louvain, B-1348 Louvain-la-Neuve, Belgium; eric.deleersnijder@uclouvain.be

*   Correspondence: Pchientvct_tv@tlu.edu.vn; Tel.: +84-243-563-6654

Received: 9 March 2020; Accepted: 30 March 2020; Published: 2 April 2020

Abstract: Water renewal timescales, namely age, residence time, and exposure time, which are defined in accordance with the Constituent-oriented Age and Residence time Theory (CART), are computed by means of the unstructured-mesh, finite element model Second-generation Louvain-la-Neuve Ice-ocean Model (SLIM) in the Mahakam Delta (Borneo Island, Indonesia). Two renewing water types, i.e., water from the upstream boundary of the delta and water from both the upstream and the downstream boundaries, are considered, and their age is calculated as the time elapsed since entering the delta. The residence time of the water originally in the domain (i.e., the time needed to hit an open boundary for the first time) and the exposure time (i.e., the total time spent in the domain of interest) are then computed. Simulations are performed for both low and high flow conditions, revealing that (i) age, residence time, and exposure time are clearly related to the river volumetric flow rate, and (ii) those timescales are of the order of one spring-neap tidal cycle. In the main deltaic channels, the variation of the diagnostic timescales caused by the tide is about 35% of their averaged value. The age of renewing water from the upstream boundary of the delta monotonically increases from the river mouth to the delta front, while the age of renewing water from both the upstream and the downstream boundaries monotonically increases from the river mouth and the delta front to the middle delta. Variations of the residence and the exposure times coincide with the changes of the flow velocity, and these timescales are more sensitive to the change of flow dynamics than the age. The return coefficient, which measures the propensity of water to re-enter the domain of interest after leaving it for the first time, is of about 0.3 in the middle region of the delta.

Keywords: Mahakam Delta; age; residence time; exposure time; return coefficient; CART

---

1. Introduction

Estuaries are valuable regions—their ecosystems are more diverse than those of any other aquatic region [1]—but they are under increasing pressure from multiple drivers of human-induced changes and inputs of pollutants [2]. Considerable efforts have been devoted to the understanding of the hydrodynamics of estuaries and coastal regions. Yet, due to the complex geometry of estuaries and the varying riverine and marine forcings such as river discharge, tides, waves, and wind, it is not straightforward to extract from the complicated spatio-temporally varying hydrodynamics the features that are relevant for understanding long-term transport of pollutants. Thus, their properties, which are generally summarized by the time they need to travel a certain distance [3], need to be studied in detail so as to help understand and, to a certain degree, predict the evolution of pollution in estuaries.

Different types of transport timescales can be defined. The most widely used probably are age, residence time, and exposure time [4–6]. The age, which is defined as the time that a water parcel has spent in the domain of interest since entering it through one of the boundaries [7,8], can be used for (i) quantifying the importance of hydrodynamic processes in the transport and the fate of pollutants in estuaries because most of the living biomass and masses of nutrients, contaminants, and suspended particles are carried in water flows within aquatic systems [9,10], (ii) estimating the water exchange processes [11–15], (iii) estimating the water renewal rate of semi-closed domains and inferring the horizontal circulation of shelf seas [16,17], and (iv) identifying typical trends in the water circulation and exchanges, which helps to detect the most vulnerable locations in semi-closed domains [18]. The residence time is the time required for a water parcel initially located in the domain of interest to be transported out of the domain for the first time [7,8]. This timescale can be an important indicator of water quality because it can largely affect biological processes [19–21] and can also be used for understanding phytoplankton dynamics and living biomass in semi-closed domains [4,22–24]. The exposure time is the total time spent by a water parcel in the domain of interest [4], and this timescale, combined with the residence time, allows estimating the return coefficient [5,25,26], a measure of the propensity of water parcels to re-enter the domain of interest after leaving it for the first time. These considerations suggest that age, residence time, and exposure time can be used as indicators to assess the transport rates of substances in estuaries and coastal regions [5,12,27–29] as well as (with suitable modifications) to assist in developing reliable predictive models for clustering floaters (i.e., particles of intermediate density staying close to the water–air interface because of their positive buoyancy) on the free surface of a turbulent flow and for quantifying floaters distribution dynamics [30,31].

The main objectives of the present study are to (i) compute the three abovementioned water renewal timescales in the Mahakam Delta, Borneo Island, Indonesia, and (ii) provide a preliminary assessment of temporal variation and spatial distribution of these timescales in the whole delta under different flow and tidal conditions. It is worth stressing that this is the first attempt to compute the aforementioned timescales in this delta by means of a high-resolution, unstructured-mesh, finite element model. Such simulation results are believed to be of use for improving and rationalizing spatial management of transport processes of pollutants and human activities and quantifying the evaluation of compatibility of human uses of the delta and the protection of its ecosystems.

This article is organized as follows. Section 2 introduces the Mahakam Delta channel network. Section 3 presents the unstructured-mesh, finite element model Second-generation Louvain-la-Neuve Ice-ocean Model (SLIM), water renewal timescales defined in accordance with the Constituent-oriented Age and Residence time Theory (CART, www.climate.be/cart), and the implementation of SLIM to the study area. Section 4 presents in detail the numerically-simulated age, residence and exposure times in the Mahakam Delta. Finally, in Section 5, the relevant conclusions of the present study are drawn.

2. Study Area

The domain of interest is the fluvial and tidal Mahakam Delta located at the mouth of the Mahakam River in the East Kalimantan province of Borneo, Indonesia (Figure 1). The delta includes a number of narrow and meandering channels. The width of the deltaic channels varies widely from 10 m to 3 km, while the depth ranges from 5 to 15 m. The delta discharges into the Makassar Strait, whose width varies between 200 and 300 km, with a length of about 600 km. Upstream of the delta is the Mahakam River, whose depth exhibits a large variability, with a maximum of the order of 45 m. The region of the river is characterized by a tropical rain forest climate with a dry season from May to September and a wet season from October to April. The river meanders over 900 km, and its catchment area covers about 75,000 km^2, with a mean annual river discharge of the order of 3000 m^3/s [32]. The central part of the river, located about 150 km upstream of the delta, is extremely flat. In this area, four large tributaries (Kedang Pahu, Belayan, Kedang Kepala, and Kedang Rantau) contribute greatly to the river flow, and several shallow-water lakes (e.g., Lake Jempang, Lake Melingtang, and Lake Semayang) are also connected to the river via a system of small channels. The water depth in these shallow-water lakes is of the order of 5 m.

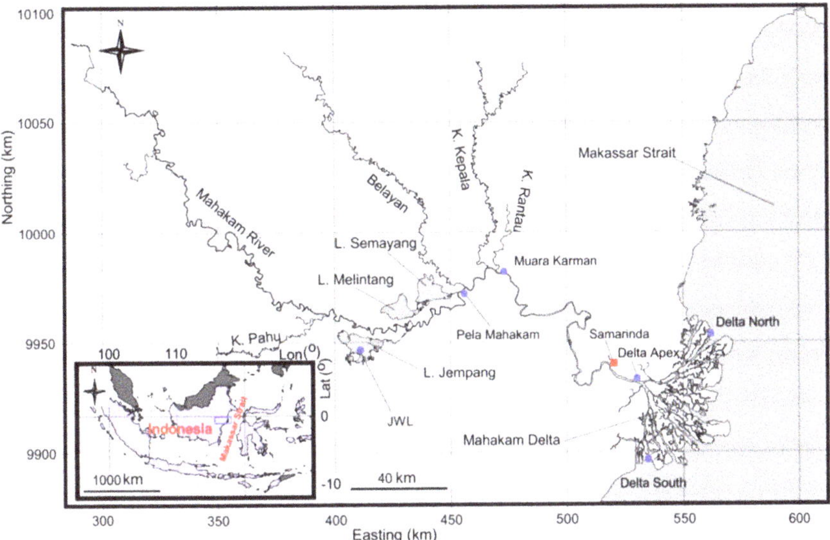

Figure 1. Map of the Mahakam river system and delta channel network, Borneo Island, Indonesia.

Different activities (e.g., shrimp farming, fishing industry, transportation, and recreation [33,34]) are performed together in the Mahakam Delta, which has thus an important role in the region. However, due to the rapid developments of aquaculture activities as well as oil and gas exploitation, the deltaic environment is highly vulnerable to degradation of water quality and loss of biotopes [33]. Recent studies (e.g., [34]) found that human activities have influenced the deltaic ecosystem for many years and led to an increase of dissolved metals therein. Thus, to design measures aimed at protecting the ecosystem and achieving good water quality in the delta, it is of interest to estimate age, residence and exposure times, thereby helping understand and quantify the transport processes that carry pollutants into and out of the delta.

3. Method

3.1. Hydrodynamics Model

The Mahakam water system clearly exemplifies a continuous riverine and marine environment that includes different interconnected regions such as river, tributaries, lakes, delta, and adjacent coastal ocean. Due to the lack of field measurements of the flow and the difficulty to obtain such measurements because of the high spatial and temporal variability of the phenomena at stake, a numerical model is resorted to. It is now becoming computationally feasible to use an integrated model of the river–sea continuum without excessive simplification of the physical processes. In this context, a combined two-dimensional depth-averaged (2D) and one-dimensional section-averaged (1D) finite-element model based on SLIM (Second-generation Louvain-la-Neuve Ice-ocean Model, www.slim-ocean.be) is implemented on the Mahakam water surface system [35,36].

The version of SLIM used herein solves 1D (section-integrated) and 2D (depth-integrated) continuity, momentum, and tracer transport equations. The governing equations are discretized in space by means of the discontinuous Galerkin finite element method (DG-FEM) with linear polynomial shape functions. The time stepping relies on the second-order diagonally implicit Runge–Kutta method [37–39]. In order to avoid the occurrence of negative water depths, the wetting and drying algorithm designed by Kärnä et al. [39] is also implemented in the hydrodynamic model. Horizontal eddy viscosity and diffusivity are parameterized in such a way that the high mesh size variability is taken into account. The bottom stress is estimated by means of a Chézy-Manning quadratic formula. Details about the equations are provided in a previous study [36].

The 2D model is implemented in the Mahakam Delta, the adjacent Makassar Strait, and three lakes (Lake Jempang, Lake Melingtang, and Lake Semayang). On the other hand, the 1D model is used in the upstream stretch of the Mahakam River, which is 300 km in length (from the city of Melak to the delta apex) and four tributaries. Figure 2 shows the grid of the computational domain, consisting of a 2D sub-domain and a 1D sub-domain. The 2D sub-domain uses an unstructured mesh that is made of 60,819 triangular elements. The longest edge of a triangle (or mesh size) varies from 5 m in the narrowest branches of the delta to around 10 km in the deepest part of the Makassar Strait. The river and its four tributaries within the 1D sub-domain have a resolution of 100 m. The grid is generated using the open-source mesh generation software GMSH (www.geuz.org/gmsh). It must be stressed that the very complex deltaic channels and coastlines are represented in detail in the computational grid, thereby capturing most of the intricacies of the domain geometry.

As for the boundary and the initial conditions, daily water discharge is prescribed at the upstream boundaries, while tidal components (elevation and velocity harmonics) are imposed at the downstream boundaries (Figure 2). The simulation of the flow is performed for different periods of time. A six months period of high flow (from November 2008 to April 2009) and a five months period of low flow (from June to October 2009) are considered so as to investigate the transport timescales under different flow conditions. A spin up period of 30 days is applied before the beginning of each of the periods of interest. The regime conditions can be reached quickly after a few days, and thus the effects of the initial conditions can be eliminated completely. To reduce the computational time, simulation of the flow in each considered period is performed firstly, and then the hydrodynamics outputs are reloaded to calculate age, residence, and exposure times.

The hydrodynamic simulation tool applied herein to the Mahakam river–sea continuum is almost similar to that described in a previous study [36]. Therefore, it is not necessary to give additional pieces of information on this. It may, however, be appropriate to stress that a thorough calibration of the Chézy-Manning coefficient was achieved, leading to the following results: (i) a constant value of 0.023 (s/m$^{1/3}$) for the Makassar Strait, (ii) a linearly increasing value in the delta region, from 0.023 (s/m$^{1/3}$) in the coastal region to 0.0275 (s/m$^{1/3}$) in the region from the delta front to the delta apex, (iii) a constant value of 0.0275 (s/m$^{1/3}$) in the Mahakam River and its four tributaries, and (iv) a larger value of 0.0305 (s/m$^{1/3}$) in the lakes.

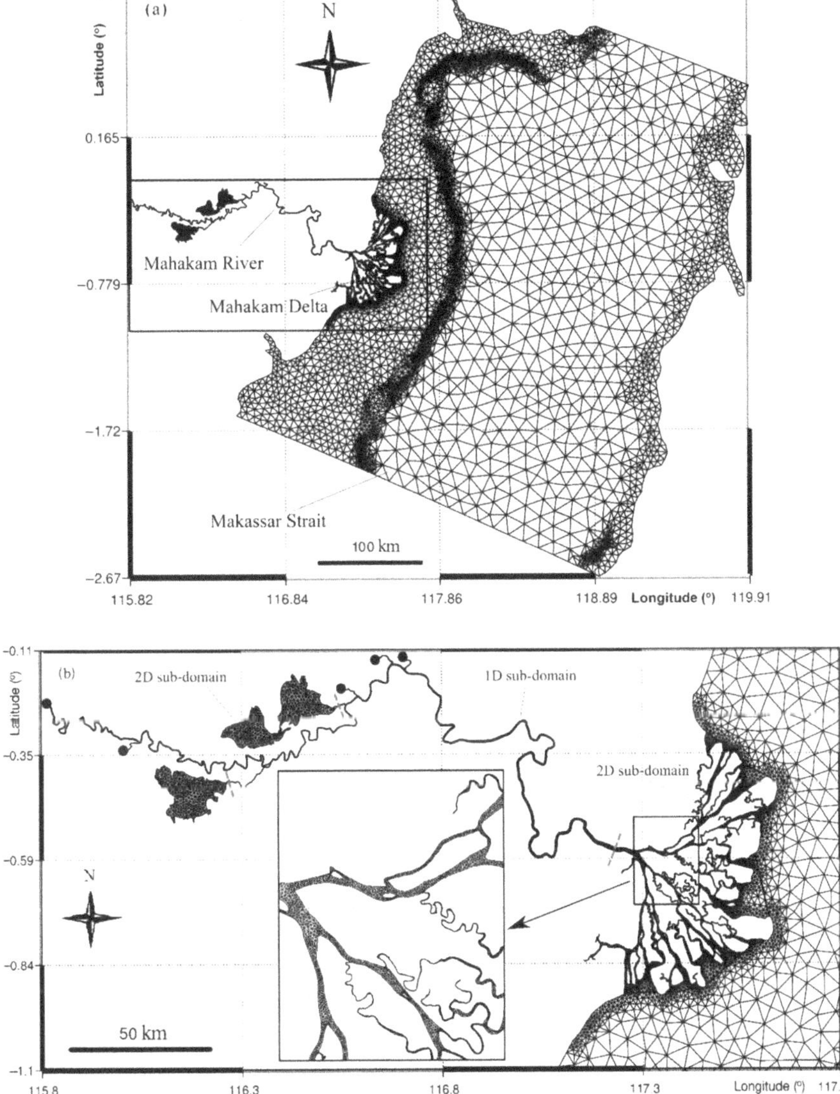

Figure 2. Grid of the computational domain: (**a**) whole computational mesh and (**b**) zoom on upstream domain and delta, showing the connection between the 1D and the 2D sub-domains (dashed-blue lines) and upstream boundary locations (black dots).

3.2. Water Renewal Timescales

3.2.1. Age of Water

The domain of interest for timescale calculations, Ω_e, is the Mahakam Delta. The water present in it consists of two components, i.e., the original water and the renewing water. The original water is the water present inside the domain at the beginning of the simulation, while the renewing water originates from the environment of the domain and progressively replaces the original water. The renewing water can enter the domain through upstream and downstream boundaries. The exact locations of the domain boundaries for timescale calculations are shown in Figure 3. The upstream boundary Γ_u is placed at the river mouth (i.e., the interface between the river and the delta), while the downstream boundary Γ_d is imposed at the delta front. According to [5], the original and the renewing waters in the delta can be treated as passive tracers and, hence, the different renewal water types can be modeled individually. If the concentrations of water originating from the upstream boundary Γ_u and the downstream boundary Γ_d of the delta are denoted C_u and C_d, respectively, the concentration of renewing water C_r in the delta, consisting of both river and sea waters, is $C_r = C_u + C_d$. Note that the concentration of a water type is defined as the water fraction originating from the boundary under consideration. Accordingly, these concentrations are dimensionless functions of time and position.

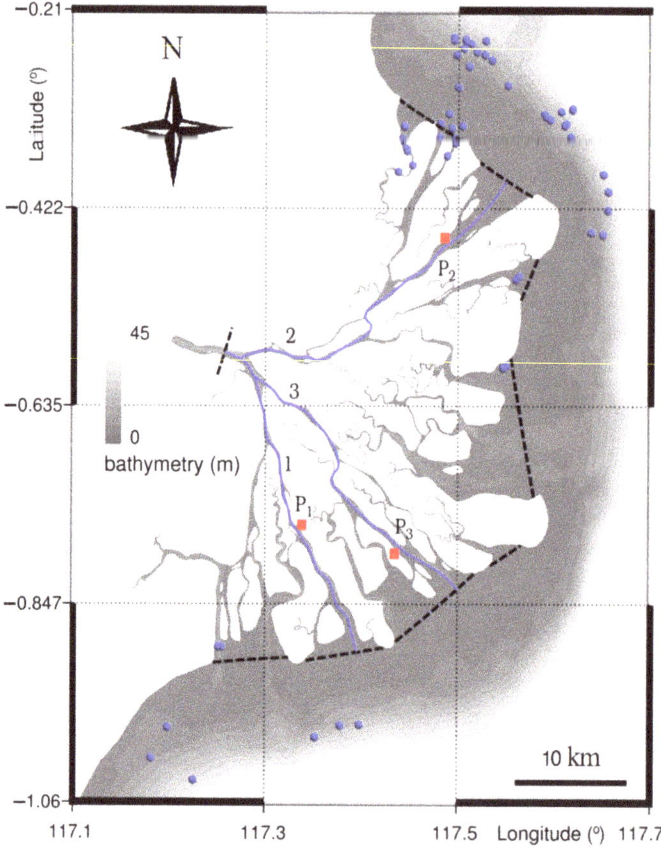

Figure 3. Bathymetry in the delta and the delta front, with field sampling sites of salinity (blue dots), upstream and downstream boundaries (black dashed lines) for computing the water renewal timescales, the three considered transects (blue lines) and three points (red squares).

We assess water renewal timescales following two approaches. In the first case, we focus on the river water (i.e., from the upstream boundary of the Mahakam Delta), assuming its concentration is dominant in most of the delta. The second approach is concerned with the whole renewing water, i.e., water originating from both the upstream and the downstream boundaries.

Three water types are considered hereinafter, which are identified by subscript i ($i = o$, u, r). The subscripts o, u, and r refer to the original water, the renewal water from the upstream boundary, and the renewal water from both the upstream and the downstream boundaries, respectively. In the framework of CART, the age of a water type can be obtained from the solution of two advection-diffusion equations [5,17,40], which govern the depth-averaged concentration C_i and the related depth-mean age concentration α_i of the water type under consideration. These equations read:

$$\frac{\partial(HC_i)}{\partial t} + \nabla \cdot (H\mathbf{u}C_i) = \nabla \cdot (H\kappa\nabla C_i) \tag{1}$$

$$\frac{\partial(H\alpha_i)}{\partial t} + \nabla \cdot (H\mathbf{u}\alpha_i) = \nabla \cdot (H\kappa\nabla\alpha_i) + HC_i \tag{2}$$

where $H = \eta + h$ is the water depth, with h being the water depth below the reference level (i.e., the mean sea level), and η is the free surface elevation; $\mathbf{u}$ is the depth-averaged horizontal velocity vector; κ is the diffusivity coefficient that is parameterized using the Okubo formulation [41], under the form:

$$\kappa = c_k\Delta^{1.15}, \tag{3}$$

with c_k being an appropriate constant, and Δ is the local characteristic length scale of the mesh. A value $c_k = 0.018$ m$^{0.85}$/s, which was calibrated from the best fit to the available salinity at 60 field sampling sites (see Figure 3) [42], is used to evaluate the diffusivity coefficient. The boundary and the initial conditions under which Equations (1) and (2) are to be solved are listed in Table 1. There are no concentration or age concentration fluxes through the impermeable boundaries. Finally, the age of water type i is calculated as follows [17]:

$$a_i = \frac{\alpha_i}{C_i}. \tag{4}$$

Table 1. Boundary and initial conditions for calculating the age in the domain of interest.

	Original Water		Renewing Water			
			River Water		Total Renewing Water	
$t = 0$	$C_o = 1$	$\alpha_o = 0$	$C_u = 0$	$\alpha_u = 0$	$C_r = 0$	$\alpha_r = 0$
$x \in \Gamma_u$	$C_o = 0$	$\alpha_o = 0$	$C_u = 1$	$\alpha_u = 0$	$C_r = 1$	$\alpha_r = 0$
$x \in \Gamma_d$	$C_o = 0$	$\alpha_o = 0$	$C_u = 0$	$\alpha_u = 0$	$C_r = 1$	$\alpha_r = 0$

At any location in the domain of interest and at any time, the concentration and the age of each water type have to satisfy the following properties [5,40]:

$$0 \leq C_i(t,\mathbf{x}) \leq 1, \; with \; i = o, \, u, \, r, \tag{5}$$

$$C_o(t,\mathbf{x}) + C_r(t,\mathbf{x}) = 1, \tag{6}$$

$$\lim_{t\to\infty} C_o(t,\mathbf{x}) = 0, \\ \lim_{t\to\infty} C_r(t,\mathbf{x}) = 1, \tag{7}$$

$$0 \leq a_i(t,\mathbf{x}) \leq t, \; with \; i = o, \, u, \, r. \tag{8}$$

Proving that inequalities and Equations (5)–(8) hold valid can be done by having recourse to demonstrations very similar to those performed in Appendices A and B of [5]. The aforementioned properties of the age suggest that the partial differential problems (from which the age of each water type and its aggregates are derived) are well posed. In other words, the behavior of the age is in accordance with physical intuition and elementary mathematical requirements.

Under the initial and the boundary conditions listed in Table 1, the concentration and the age concentration of the original water satisfy $\alpha_o(t,x) = tC_o(t,x)$, implying that the age of the original water is equal to the elapsed time, i.e., $a_o(t,x) = t$ [5,40]. This is because the original water can only leave the domain of interest as time progresses, and there is obviously no source of original water. This piece of information is of little diagnostic value. However, it is an additional element supporting the well-foundedness of the partial differential problems laid out above.

3.2.2. Residence Time

The residence time θ_r is defined as the time taken by a water parcel initially in the domain of interest to leave it for the first time. It can be obtained by solving, under the initial and the boundary conditions mentioned below, the partial differential equation:

$$\frac{\partial(H\theta_r)}{\partial t} + \nabla \cdot (H\mathbf{u}\theta_r) = \nabla \cdot [H(-\kappa)\nabla\theta_r] - H. \tag{9}$$

It is derived from the adjoint of the equation governing the concentration of the original water [21, 43]. It must be underscored that Equation (9) is not a classical depth-averaged advection–diffusion equation because of the negative sign in the diffusivity term. It must be integrated backward in time, resulting in the following equation [43]:

$$\frac{\partial(H\theta_r)}{\partial \tau} + \nabla \cdot [H(-\mathbf{u})\nabla\theta_r] = \nabla \cdot (H\kappa\nabla\theta_r) + H \tag{10}$$

with $\tau = T - t$, where T corresponds to the end of the simulation period (i.e., the time at which the reverse time integration begins).

As for the boundary condition, a no flux condition is enforced on the impermeable boundaries. According to [21], the residence time should be prescribed to be zero on the open boundaries. However, as shown in [44], a boundary layer of the residence time develops in the vicinity of the incoming open boundaries. This boundary layer cannot be resolved by the mesh employed here and hence must be treated in a suitable manner in order to prevent spurious oscillations from developing [43]. This is why it is desirable to replace the homogeneous Dirichlet boundary condition by the flux condition derived in [44] from a one-dimensional solution, which is investigated in [43]. Accordingly, the boundary conditions that are actually used in the numerical simulations are as follows:

$$\begin{aligned}
\nabla\theta_r \cdot \mathbf{n} &= \frac{e^{-u_n L^*/\kappa}}{\kappa\left(1-e^{-u_n L^*/\kappa}\right)}(\theta_r u_n - L^*) - \frac{1}{u_n}, \quad &if\ \mathbf{x} \in \Gamma_u \cup \Gamma_d\ and\ u_n > 0 \\
\theta_r &= 0, \quad &if\ \mathbf{x} \in \Gamma_u \cup \Gamma_d\ and\ u_n \leq 0
\end{aligned} \tag{11}$$

where $\mathbf{n}$ is the outward normal unit vector to the boundary, u_n (=$\mathbf{u}.\mathbf{n}$) is the normal velocity, and L^* is the width of the unresolved boundary layer. In [44], it was found to be appropriate to prescribe that the length L^* be equal to the local mesh size.

The reverse time marching integration cannot be performed without prescribing an "initial" value, i.e., $\theta_r(t = T,\mathbf{x})$, which, unfortunately, is unknown. The reverse time marching is initialized with $\theta_r(t = T,\mathbf{x}) = 0$ and, after a spin up period whose duration is equal to a couple of times the typical value of the residence time, a regime solution is attained and, hence, the "initial condition" is no longer important, which is in accordance with Delhez et al. [21].

3.2.3. Exposure Time

Due the tidal motions, a water parcel that left the domain of interest Ω_e at a given time can re-enter the domain many times before escaping it definitively. To account for this, the concept of exposure time is introduced, that is, the total time spent in the domain. The exposure time θ_e can be numerically computed by solving the following Equation [5]:

$$\frac{\partial(H\theta_e)}{\partial \tau} + \nabla \cdot [H(-\mathbf{u})\nabla\theta_e] = \nabla \cdot (H\kappa\nabla\theta_e) + H\delta_{\Omega_e} \tag{12}$$

where δ_{Ω_e} equals 1 and 0 in and out the domain of interest Ω_e, respectively.

It must be stressed that the computational domain Ω for the exposure time must be extended to explicitly represent phenomena taking place outside the domain of interest Ω_e ($\Omega_e \subset \Omega$). In the present study, the domain Ω is extended significantly both upstream and downstream of the delta, and it is taken to be identical to the computational domain used for the hydrodynamic simulations.

In the Mahakam River and four tributaries, the exposure time is obtained by solving the following one-dimensional equation:

$$\frac{\partial(A\theta_e)}{\partial \tau} + \frac{\partial[A(-u)\theta_e]}{\partial x} = \frac{\partial}{\partial x}\left(A\kappa\frac{\partial \theta_e}{\partial x}\right) \tag{13}$$

where A is the cross-sectional area, and κ is the diffusivity coefficient. The latter is parameterized under the form of Equation (3), in which the local characteristic length scale of the grid is taken to be the distance between two channel cross-sections (i.e., 100 m). Equation (13) is also solved the same way as governing equations in the 2D sub-domain (see details in [36,42]).

A water parcel at open sea boundaries, which are located at the entrance and at the outlet of the Makassar Strait, is assumed to never return into the delta, and thus a value $\theta_e = 0$ is imposed. At upstream boundaries, where the flow is assumed to be always in the downward direction, the expression $\kappa\nabla\theta_e.\mathbf{n} = 0$ is used [5]. The initial value of the exposure time is set equal to zero in the whole computational domain.

3.2.4. Finite Element Implementation

The governing equations for the age of water are solved by using a discontinuous Galerkin finite element method for the spatial operators and the second-order diagonally implicit Runge–Kutta method for the temporal operators. The computational domain is discretized into series of elements, as shown in Figure 2. The governing equations are multiplied by test functions and then integrated by parts over each element, resulting in element-wise surface and contour integral terms for the spatial operators. The surface term is solved using the DG-FEM with linear shape function, while a Roe solver is used for computing the fluxes at the interfaces between two adjacent elements in order to represent the water-wave dynamics in contour terms properly [37]. Because a second-order diagonally implicit Runge Kutta method is employed for the temporal derivative operator [35], a time step as large as 10 min can be used for our computations. The governing equations for the residence and the exposure times are solved in a manner similar to those for the age of water. It must be kept in mind, however, that the equations for the residence and the exposure times are integrated backward in time.

The necessary hydrodynamics outputs are reloaded only in the Mahakam Delta, in the forward direction for the age simulations, and in the backward in time for the residence time simulations. The abovementioned calculation procedure is very convenient and efficient in terms of computational time. The computational procedure for the exposure time is similar to that applied for the residence time. The hydrodynamic outputs are then reloaded in the whole computational domain for the backward in time integration. Then, the exposure time is computed separately and offline with the flow dynamics in the computer procedure, as for the residence time.

4. Results and Discussion

4.1. Age of Water

4.1.1. River Water

Figure 4 shows the average over the low-flow simulation period (the daily water discharge at river mouth varies from 550 to 3000 m^3/s) of the concentration and the age of the river water along the three considered transects (see Figure 3). The river water concentration is equal to unity at 26 km from the river mouth (Figure 4a). Beyond this point, the concentration significantly decreases from unity to zero, resulting in a significant increase of the age of water. As shown in Figure 5a, a similar trend of the concentration is also observed for the high-flow period (during which the daily water discharge at the river mouth varies between 3500 and 6500 m³/s). The point from which the concentration begins to decrease occurs around km 31. The difference in location of the point from which the concentration starts decreasing in the low and high flow conditions can be explained by differences of water discharges.

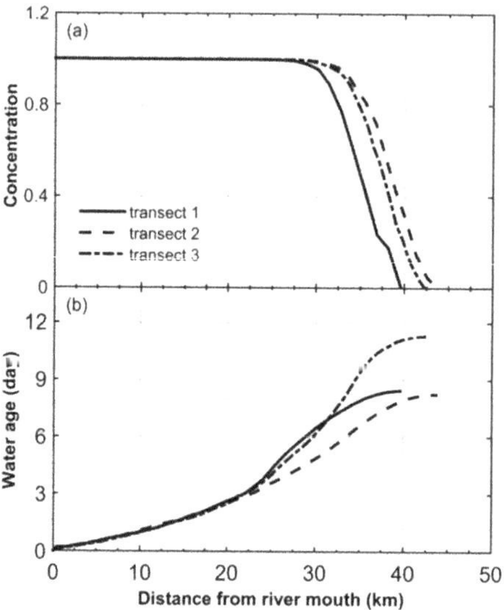

Figure 4. Calculation results of: (**a**) concentration and (**b**) water age during the low flow period, along the three considered transects (see Figure 3) for the case of the river water.

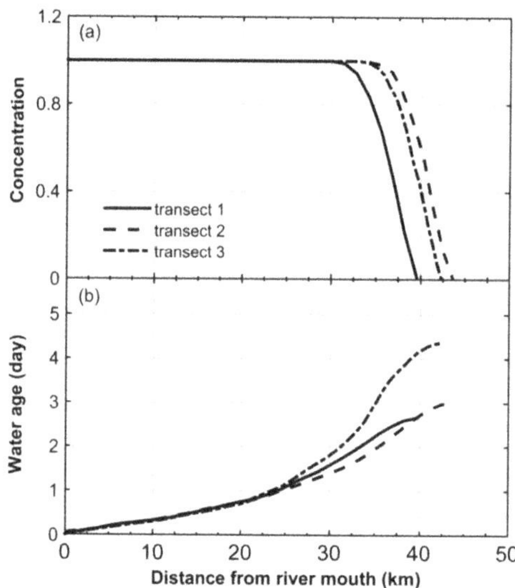

Figure 5. Calculation results of: (**a**) concentration and (**b**) water age during the high flow period, along the three considered transects (see Figure 3) for the case of the river water.

The age of the water along the three transects monotonically increases from zero at the upstream boundary (where the river water enters the domain of interest) to a maximum value at the downstream boundary (Figure 4b). A maximum value of 9 days is obtained in the main southern and northern fluvial channels (i.e., transect 1 and transect 2), while the maximum age is about 11 days in transect 3. A nonlinear increase of the age of water from the upstream to the downstream boundaries can be explained by the decrease of the flow velocity along the transects, which is due to changes in bathymetry, channel cross-section, tidal effects, and meandering and curvature of channels [36,42]. On the other hand, to investigate the effect of the tides on the water age, the standard deviation of the water age is also calculated at each location along the three considered transects, resulting in values lying between 0.5 and 4 days. This indicates that the variation of the age of water caused by tides is about 35% of the magnitude of the averaged value in the transects.

The averaged value of the water age in the high flow period also increases from zero (at the river mouth) to 4.5 days (at the downstream boundary, Figure 5b), revealing that the magnitude of the age in the high flow is three times smaller than that in the low flow. This is due to the significant increase of the flow velocity. These results reveal that the age of water is clearly linked to the river volumetric flow rate entering the delta. The standard deviation associated with the averaged value of the age of water increases from zero to 15 h along the transects, showing that the variation of the age due to the tidal variation equals 15% of the averaged value of the age, and, as expected, tidal effects on transport processes are weaker in the high flow than in the low flow regimes.

A previous study [45] of the age of river water in the Mahakam Delta showed that the mean age over a two months simulation period (May and June 2008) varied between 4 and 7 days. Indeed, the water age ranged between zero at the delta apex and 4 days at the delta front. These results are similar to those obtained for the high flow period in the present study. The slight discrepancy may be due to differences in hydrodynamics.

On downstream boundary Γ_d, the concentration of the river water (C_u) and its age concentration (α_u) are both prescribed to be zero, implying that the corresponding age ($a_u = \alpha_u/C_u$) is an indeterminate form, i.e., 0/0. In the absence of an analytical solution, explicitly evaluating this limit is impossible. However, this issue may be successfully addressed by first deriving the equation for the age. This is achieved by manipulating concentration and age concentration Equations (1) and (2) and casting the resulting relation into the appropriate form, eventually yielding:

$$\nabla a_u . \nabla C_u = \frac{C_u}{2H\kappa}\left[\frac{\partial(Ha_u)}{\partial t} - H + \nabla.(Hua_u - H\kappa\nabla a_u)\right] \tag{14}$$

On Γ_d, the concentration is prescribed to be zero. Therefore, Equation (14) simplifies to $[\nabla a_u . \nabla C_u]_{x\in\Gamma_d} = 0$. In addition, since C_u is zero on Γ_d, its gradient is parallel to **n**, the unit normal vector to the boundary. As a consequence, the age satisfies boundary condition $[\nabla a_u . \mathbf{n}]_{x\in\Gamma_d} = 0$, i.e., the derivative of the age in the direction normal to the boundary must be zero, which is why the age should have a finite value on the boundary. To the best of our knowledge, this theoretical result is novel. The simulated age profiles displayed on Figures 4 and 5 are in agreement with it.

4.1.2. Total Renewing Water

Figure 6 shows the age of the total renewing water. The total renewing water concentration along the three considered transects equals unity for both the high and the low flow periods because once the transient solution has vanished, all the original water is replaced by renewing water. As shown in Figure 6a for the low flow period, the averaged values of the total renewing water age along the transects increase from zero at boundaries to a maximum value at km 34 (as measured from the river mouth). The maximum values are 7, 6, and 8.5 days in transects 1, 2, and 3, respectively. The variation of the age of water caused by the tide along the transects ranges from 0 to 4 days, especially from km 20 to the downstream boundaries.

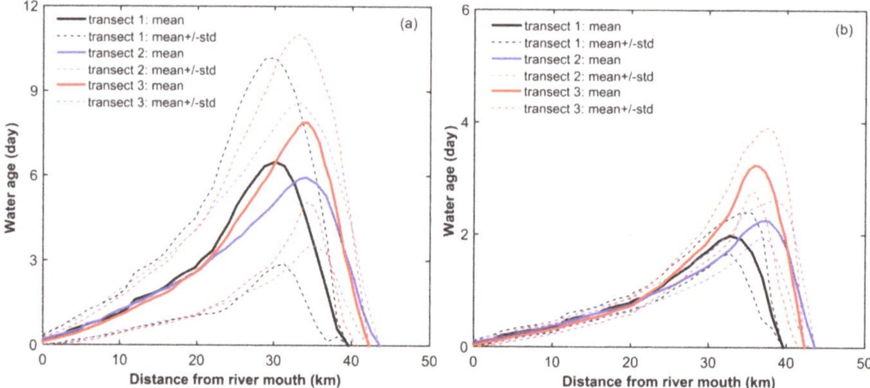

Figure 6. Calculation results of the age of the total renewing water during the: (**a**) low flow period and (**b**) high flow period, along the three considered transects (see Figure 3).

Figure 6b shows the age of water obtained in the high flow period. The age increases from zero at the boundaries to peak values (i.e., 2.0, 2.3, and 3.2 days in the transects 1, 2, and 3). The peak of the age in the high flow period is smaller by a factor of two to three than that in the low flow period, indicating again that the age of the total renewing water is also clearly linked to the water discharge entering the delta. The location of the peak value moves slightly in the seaward direction, around km 36. The standard deviation associated with mean age along the three considered transects is less than 1 day.

Figure 7a shows the spatial distribution of the averaged value of the water age for the low flow period. The mean age varies significantly in space, depending on locations and channels in the delta. The age is generally large in the middle region of the delta, where the complex hydrodynamics is caused by the combined effect of tide, river flow, and rugged geometry. A maximum value of 20 days is found in several small and narrow deltaic channels and in the creeks.

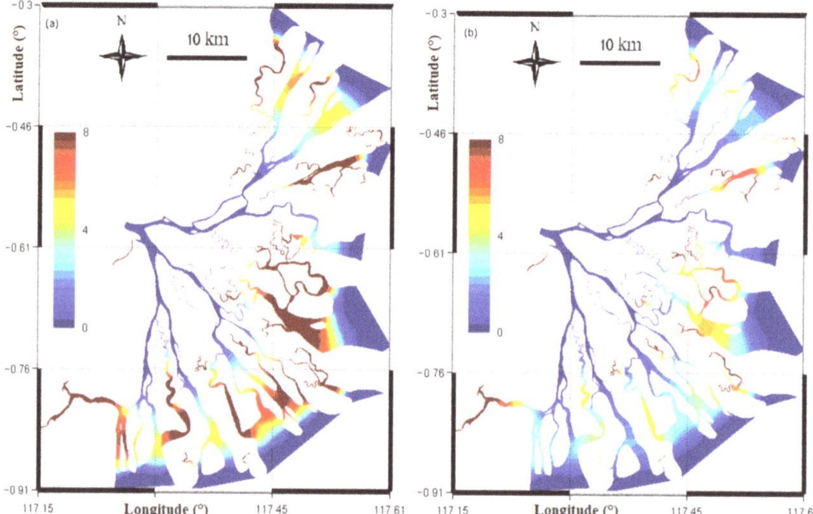

Figure 7. Distribution of the age of the total renewing water for the: (**a**) low and (**b**) high flow period. The color bar is cropped at 8 days in order to focus on the variation of the age in deltaic channels, and the unit is day.

The distribution of the averaged value of the water age for the high flow period is shown in Figure 7b. Similarly to the low flow period, the mean age also varies significantly in the deltaic region. A small value of the age (e.g., 2 days) generally occurs in the fluvial channels, e.g., transect 1, transect 2, while a larger value appears in the fluvial-tidal channels (e.g., transect 3) as well as in the creeks. The largest value of the age of water is about 15 days in the isolated tidal channels and in the creeks of the delta. This complex spatial distribution of the age mainly reflects the complex hydrodynamics, the influence of the tides, and the division of water discharge in deltaic channels. The age in the high flow period is two to three times smaller than that obtained in the low flow period.

4.2. Residence Time

The residence time must be zero on both the upstream and the downstream boundaries. However, the thin boundary layer adjacent to the incoming boundaries cannot be resolved by the present model grid, which is precisely why boundary condition (Equation (11)) is implemented. This important fact must be kept in mind when inspecting and interpreting the figures related to the residence time.

Simulated values of the residence time over the whole period of the low flow regime are shown in Figure 8 for the three considered transects, indicating that the averaged value of residence time varies in a range between 0 and 9 days in transects 1 and 2. A wider range of the residence time, from 0 to 12 days, is obtained in transect 3. The standard deviation of the averaged value of the residence time varies between zero and 3.5 days along these transects.

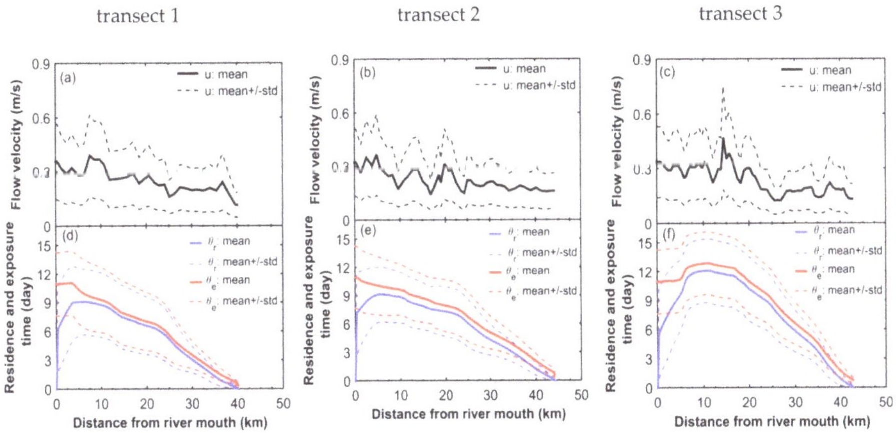

Figure 8. Simulated values of water velocity along the: (**a**) transect 1; (**b**) transect 2; (**c**) transect 3 (see Figure 3) and simulated values of residence and exposure times along; (**d**) transect 1; (**e**) transect 2; (**f**) transect 3 (see Figure 3) for the low flow period.

Figure 9 shows the averaged values of the residence time over the high flow period. The residence time generally decreases from the upstream to the downstream boundary. In the main southern and northern fluvial channels, a water parcel initially located at the delta apex takes, on average, about 4 days to leave the delta. In the meandering fluvial-tidal channel (i.e., transect 3), the water parcel spends a longer time (about 5.3 days) to leave the domain of interest, except at a distance of 5 km from the river mouth. The averaged value of the residence time in the high flow condition is two times smaller than that in the low flow conditions. The residence time standard deviation is more or less 20 h in all considered transects. As illustrated by Figure 10, part of the time variation of the residence time is caused by tides. The starting instant within a tidal cycle has a significant impact on the residence time despite the fact that the duration of a tidal cycle is significantly smaller than the order of magnitude of the residence time. A similar behavior of the residence time has been simulated in other estuaries [5,46]. Unfortunately, this intriguing property has yet to be fully explained.

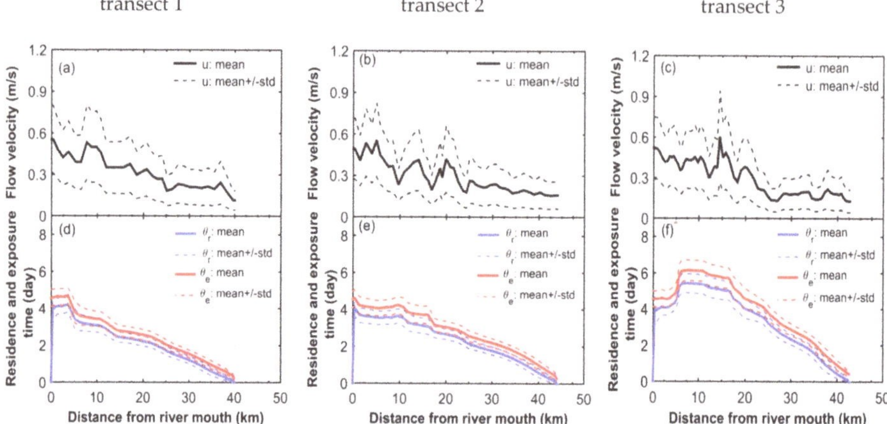

Figure 9. Simulated values of water velocity along the: (**a**) transect 1; (**b**) transect 2; (**c**) transect 3 (see Figure 3) and simulated values of residence and exposure times along; (**d**) transect 1; (**e**) transect 2; (**f**) transect 3 (see Figure 3) for the high flow period.

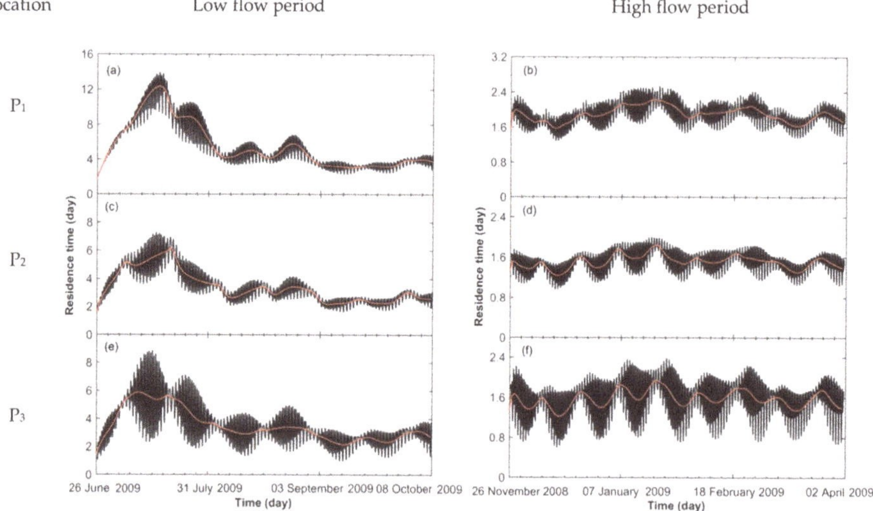

Figure 10. Simulated values of the residence time at the: (**a**) P_1; (**b**) P_2; (**c**) P_3 (see Figure 3) for the low period and simulated values of the residence time at the: (**d**) P_1; (**e**) P_2, (**f**) P_3 for the high flow period. In each panel, the back line indicates the instantaneous value, while the red curve presents the averaged value over 24 h taken to filter out tidal oscillation.

In each transect, the residence time does not monotonically decrease with the distance in the seaward direction for either the low or the high flow periods. The variation of the residence time coincides with the changes of the longitudinal velocity (see Figures 8a–c and 9a–c). This result confirms again that the residence time is a useful measure of the influence of the flow dynamics on transport processes [21].

In transect 3, the residence time increases up to a peak value of 12 days for the low flow regime (see Figure 8b) and 5.3 days for the high flow one (Figure 9b) in the first few kilometers downstream of the river mouth. The reason for this variation of the residence time can be explained by the decrease of water discharge resulting from channel bifurcations. Downstream of the delta apex bifurcation (the bifurcation right after the river mouth), only about 40% of water discharge flows into transect 2, while 60% of water discharge continuously flows in the main southern channel, i.e., transect 1 [45]. Similarly, beyond the first bifurcation in the transect 1, about 33% of water discharge in transect 1 continuously flows in the transect 3.

Figure 11 depicts the spatial distribution of the averaged value of the residence time for the low flow (Figure 11a) and for the high flow one (Figure 11b). The residence time varies significantly in space due to the changes of bathymetry and related hydrodynamics features. For instance, water parcels in tidal channels and in the creeks within the middle area of the delta take a longer time to leave the delta than water parcels in the main fluvial channels (e.g., southern and northern channels). This is because, in these channels and creeks, the water velocity is rather low. The residence time in the whole delta ranges from 0 to 15 days for both high and low flows, presenting a slightly wider range of variation in comparison with that of the water age.

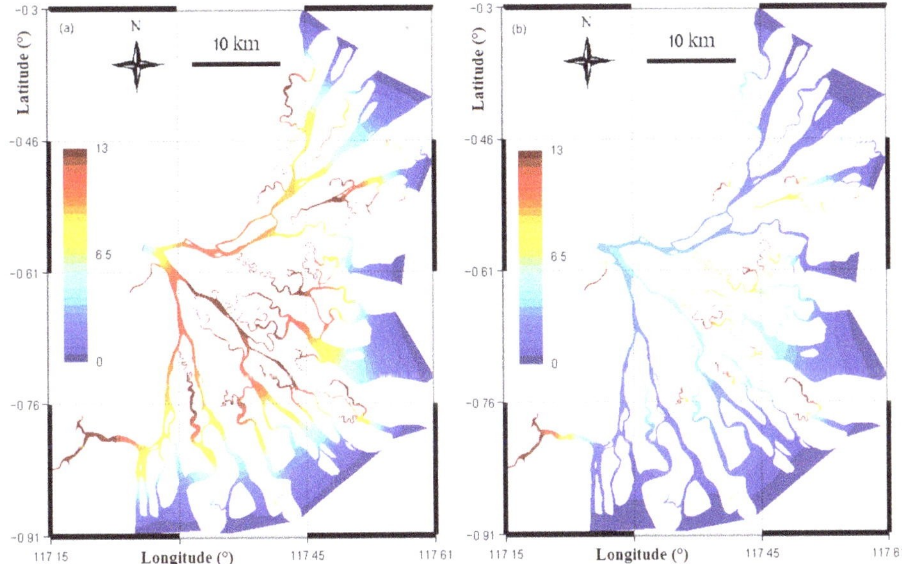

Figure 11. Distribution of the residence time in the whole delta for the: (**a**) low; (**b**) high flow period. The color bar is cropped at 13 days in order to focus on the variation of the residence time in deltaic channels, and the unit is day.

The Mahakam Delta is dominated by the mangrove ecosystem, with water containing many phytoplankton species [47,48], i.e., the basic component of the food chain that is expressed by chlorophyll concentration. Previous studies [23,49] suggest that, among different factors (e.g., nutrients, residence time, geomorphological features), the chlorophyll concentration has the highest correlation with the residence time, especially in regions where the residence time is below 50 days [23]. Indeed, phytoplankton biomass is controlled by the residence time more than by nutrient availability. As for the Mahakam Delta, study of phytoplankton and of relations between chlorophyll concentration and the residence time is relatively limited. The current simulation results of the residence time are believed to be helpful for investigating the phytoplankton in the next stages of the research.

4.3. Exposure Time

Figures 8c and 9c show the estimated exposure time along the three considered transects for the high and the low flow periods. The spatial distribution of the exposure time in the whole delta is shown in Figure 12, revealing that the exposure time is extremely different in the different periods.

In comparison with the residence time, the estimated exposure time clearly depicts two different aspects. Firstly, the exposure time at the upstream and the downstream boundaries of the delta is far from zero, indicating that the model is able to capture the reentering of water parcels caused by tidal variations. Secondly, the exposure time is always larger than the residence time in most deltaic channels, except in small creeks located in the middle delta, in which the exposure time is more or less identical to the residence time. The standard deviation of the exposure time under tidal variation in deltaic channels is similar to that of the residence time.

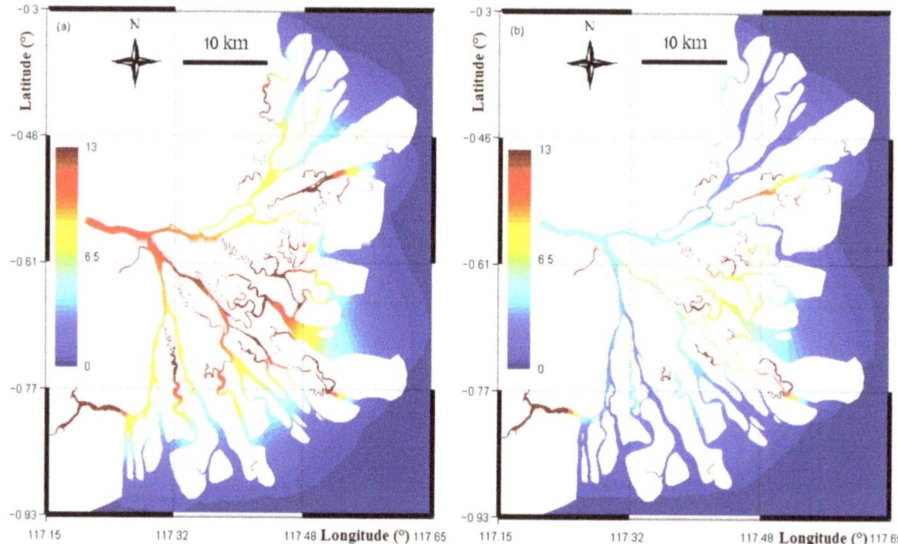

Figure 12. Distribution of the exposure time in the whole delta for the: (**a**) low; (**b**) high flow period. The color bar is cropped at 13 days in order to focus on the variation of the exposure time in deltaic channels and the unit is day.

4.4. Return Coefficient

The relative difference between the exposure and the residence time is often represented with the help of the return coefficient (*rc*) [5,26,46], which is computed as:

$$rc = \frac{\theta_e - \theta_r}{\theta_e} \tag{15}$$

where θ_e and θ_r are exposure and residence time, respectively. According to this definition, the return coefficient varies between zero and one. The lower limit of the return coefficient (*rc* = 0) occurs when the exposure time equals the residence time or when no water parcels re-enters the delta. In contrast, the upper limit of the return coefficient (*rc* = 1) is reached if the residence time is much smaller than the exposure time.

Figure 13 shows the spatial distribution of the return coefficient for the period of the low flow (Figure 13a) and for the period of the high flow (Figure 13b). The return coefficient value is averaged over each considered period. There is a slight difference of the return coefficient in different flow conditions. The return coefficient equals to unity on the upstream and the downstream boundaries, as it should be. In the middle region of the delta, the return coefficient is around 0.3, suggesting that, even far away from the open boundaries, the impact of the re-entering is non negligible.

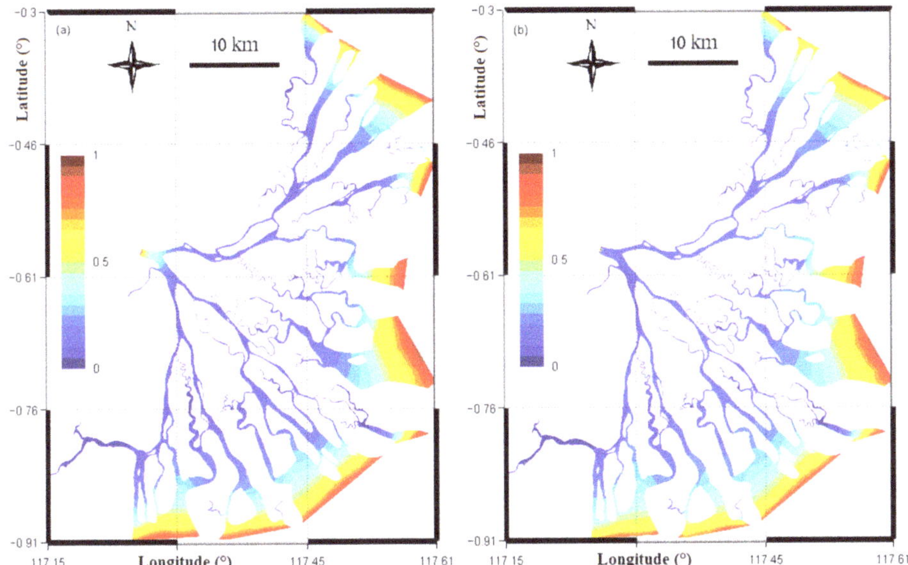

Figure 13. Distribution of the return coefficient for the: (**a**) low; (**b**) high flow period.

5. Conclusions

The Mahakam Delta presents a complex geometry and topography, with complicated flow and transport processes caused by the influence of both riverine and marine forcings. The aims of this study were to assess water renewal rates under low and high flow conditions. The results clearly showed that, firstly, there was a clear link between water renewal timescales and the river volumetric flow rate. In the main deltaic channels, the age of the renewing water from the upstream delta and the age of renewing water from both the upstream and the downstream of the delta were respectively 5 and 3.5 days for the high flow, while these values were generally three times greater in the low flow. The residence and the exposure times were 6 days in the high flow and about 12 days in the low flows. The magnitude of the three water renewal timescales under different flow conditions is of the order of one spring-neap tidal cycle in the whole delta.

Secondly, the age of renewing water from the upstream delta monotonically increased from the river mouth to the delta front, while the age of renewing water from both the upstream and the downstream of the delta monotonically increased from the river mouth and the delta front to the middle delta. The residence and the exposure times did not decrease monotonically with the distance in the seaward direction, and variations of these timescales coincided with the changes of the flow velocity, revealing that these timescales are more sensitive to the change of flow dynamic than the age.

Thirdly, along deltaic channels, the variation of the water renewal timescales caused by the tide was about 35% of the order of magnitude of the averaged value. The results also clearly showed that the model was able to capture the reentering of water parcels caused by tidal variations, e.g., return coefficient was close to unity at the delta apex and delta front, while its value was about 0.3 in the middle region of the delta.

Author Contributions: Conceptualization: C.P.V., B.d.B., A.d.B., A.J.F.(T.)H., S.S.-F., E.D.; methodology: E.D.; software: C.P.V., B.d.B.; formal analysis: C.P.V., B.d.B., A.d.B., E.D.; writing—original draft preparation: C.P.V., E.D.; writing—review and editing: C.P.V., B.d.B., A.d.B., A.J.F.(T.)H., S.S.-F., E.D.; visualization: C.P.V., B.d.B.; Funding acquisition: S.S.-F., E.D. All authors have read and agreed to the published version of the manuscript.

Funding: This research received no external funding.

Acknowledgments: Eric Deleersnijder and Sandra Soares-Frazão are honorary research associates with the Belgium Fund for Scientific Research (F.R.S.—FNRS). C.P.V. and E.D. are indebted to Insaf Draoui for her useful comments on an early version of the manuscript.

Conflicts of Interest: The authors declare no conflict of interest.

References

1. Jay, D.A.; Greyer, W.R.; Montgomery, D.R. An ecological perpective on estuarine classification. In *A Synthetic Approach to Research and Practice*; Island Press: Washington, DC, USA, 2000; pp. 149–175.
2. Halpern, B.S.; Walbridge, S.; Selkoe, K.A.; Kappel, C.V.; Micheli, F.; D'Agrosa, C.; Bruno, J.F.; Casey, K.S.; Ebert, C.; Fox, H.E.; et al. A Global Map of Human Impact on Marine Ecosystems. *Science* **2008**, *319*, 948–952. [CrossRef] [PubMed]
3. Shi, H.; Yu, X. Application of transport timescales to coastal environmental assessment: A case study. *J. Environ. Manag.* **2013**, *130*, 176–184. [CrossRef] [PubMed]
4. Monsen, N.; Cloern, J.; Lucas, L.; Monismith, S. A comment on the use of flushing time, residence time, and age as transport time scales. *Limnol. Oceanogr.* **2002**, *47*, 1545–1553. [CrossRef]
5. De Brye, B.; De Brauwere, A.; Gourgue, O.; Delhez, E.J.M.; Deleersnijder, E. Water renewal timescales in the Scheldt Estuary. *J. Mar. Syst.* **2012**, *94*, 74–86. [CrossRef]
6. Viero, D.P.; Defina, A. Water age, exposure time, and local flushing time in semienclosed, tidal basins with negligible freshwater inflow. *J. Mar. Syst.* **2016**, *156*, 16–29. [CrossRef]
7. Bolin, B.; Rodhe, H. A note on the concepts of age distribution and transit time in natural reservoirs. *Tellus* **1973**, *25*, 58–62. [CrossRef]
8. Zimmerman, J.T.F. Mixing and flushing of tidal embayments in the western Dutch wadden sea part I: Distribution of salinity and calculation of mixing time scales. *Neth. J. Sea Res.* **1976**, *10*, 149–191. [CrossRef]
9. Huang, W.; Liu, X.; Chen, X.; Flannery, M.S. Estimating river flow effects on water ages by hydrodynamic modeling in Little Manatee River estuary, Florida, USA. *Environ. Fluid Mech.* **2010**, *10*, 197–211. [CrossRef]
10. Gross, E.; Andrews, S.; Bergamaschi, B.; Downing, B.; Holleman, R.; Burdick, S.; Durand, J. The use of stable isotope-based water age to evaluate a hydrodynamic model. *Water* **2019**, *11*, 2207. [CrossRef]
11. Takeoka, H. Fundamental concepts of exchange and transport time scales in a coastal sea. *Cont. Shelf Res.* **1984**, *3*, 311–326. [CrossRef]
12. Liu, W.-C.; Chen, W.-B.; Kuo, J.-T.; Wu, C. Numerical determination of residence time and age in a partially mixed estuary using three-dimensional hydrodynamic model. *Cont. Shelf Res.* **2008**, *28*, 1068–1088. [CrossRef]
13. Ren, Y.; Lin, B.; Sun, J.; Pan, S. Predicting water age distribution in the Pearl River Estuary using a three-dimensional model. *J. Mar. Syst.* **2014**, *139*, 276–287. [CrossRef]
14. Kärnä, T.; Baptista, A.M. Water age in the Columbia River estuary. *Estuar. Coast. Shelf Sci.* **2016**, *183*, 249–259. [CrossRef]
15. Yang, J.; Kong, J.; Tao, J. Modeling the water-flushing properties of the Yangtze estuary and adjacent waters. *J. Ocean Univ. China* **2019**, *18*, 93–107. [CrossRef]
16. Shang, J.; Sun, J.; Tao, L.; Li, Y.; Nie, Z.; Liu, H.; Chen, R.; Yuan, D. Combined effect of tides and wind on water exchange in a semi-enclosed shallow sea. *Water* **2019**, *11*, 1762. [CrossRef]
17. Deleersnijder, E.; Campin, J.-M.; Delhez, E.J.M. The concept of age in marine modelling I. Theory and preliminary model results. *J. Mar. Syst.* **2001**, *28*, 229–267. [CrossRef]
18. De Serio, F.; Armenio, E.; Ben Meftah, M.; Capasso, G.; Corbelli, V.; De Padova, D.; De Pascalis, F.; Di Bernardino, A.; Leuzzi, G.; Monti, P.; et al. Detecting sensitive areas in confined shallow basins. *Environ. Model. Softw.* **2020**, *126*, 104659. [CrossRef]
19. Braunschweig, F.; Martins, F.; Chambel, P.; Neves, R. A methodology to estimate renewal time scales in estuaries: The Tagus Estuary case. *Ocean Dyn.* **2003**, *53*, 137–145. [CrossRef]
20. Shen, J.; Haas, L. Calculating age and residence time in the tidal York River using three-dimensional model experiments. *Estuar. Coast. Shelf Sci.* **2004**, *61*, 449–461. [CrossRef]
21. Delhez, E.J.M.; Heemink, A.W.; Deleersnijder, E. Residence time in a semi-enclosed domain from the solution of an adjoint problem. *Estuar. Coast. Shelf Sci.* **2004**, *61*, 691–702. [CrossRef]
22. Cloern, J.E.; Cole, B.E.; Wong, R.L.J.; Alpine, A.E. Temporal dynamics of estuarine phytoplankton: A case study of San Francisco Bay. *Hydrobiologia* **1985**, *129*, 153–176. [CrossRef]

23. Delesalle, B.; Sournia, A. Residence time of water and phytoplankton biomass in coral reef lagoons. *Cont. Shelf Res.* **1992**, *12*, 939–949. [CrossRef]

24. Lucas, L.; Thompson, J.; Brown, L. Why are diverse relationships observed between phytoplankton biomass and transport time? *Limnol. Oceanogr.* **2009**, *54*, 381–390. [CrossRef]

25. Arega, F.; Armstrong, S.; Badr, A.W. Modeling of residence time in the East Scott Creek Estuary, South Carolina, USA. *J. Hydro-Environ. Res.* **2008**, *2*, 99–108. [CrossRef]

26. De Brauwere, A.; De Brye, B.; Blaise, S.; Deleersnijder, E. Residence time, exposure time and connectivity in the Scheldt Estuary. *J. Mar. Syst.* **2011**, *84*, 85–95. [CrossRef]

27. Andutta, F.P.; Ridd, P.V.; Deleersnijder, E.; Prandle, D. Contaminant exchange rates in estuaries—New formulae accounting for advection and dispersion. *Prog. Oceanogr.* **2014**, *120*, 139–153. [CrossRef]

28. Huguet, J.R.; Brenon, I.; Coulombier, T. Characterisation of the water renewal in a macro-tidal Marina using several transport timescales. *Water* **2019**, *11*, 2050. [CrossRef]

29. Cheng, Y.; Mu, Z.; Wang, H.; Zhao, F.; Li, Y.; Lin, L. Water residence time in a typical tributary bay of Three Gorges reservoir. *Water* **2019**, *11*, 1585. [CrossRef]

30. Lovecchio, S.; Marchioli, C.; Soldati, A. Time persistency of floating particle clusters in free-surface turbulence. *Phys. Rev. E* **2013**, *88*, 033003. [CrossRef]

31. Gutiérrez, P.; Aumaitre, S. Clustering of floaters on the free surface of a turbulent flow: An experimental study. *Eur. J. Mech. B/Fluids* **2016**, *60*, 24–32. [CrossRef]

32. Allen, G.P.; Chambers, J.L.C. *Sedimentation in the Modern and Miocene Mahakam Delta*; Indonesian Petroleum Association: Jakarta, Indonesia, 1998; p. 236. Available online: https://books.google.be/books?id=2j9OAQAAIAAJ (accessed on 1 January 1998).

33. Chaîneau, C.-H.; Mine, J.; Suripno. The integration of biodiversity conservation with oil and gas exploration in sensitive tropical environments. *Biodivers. Conserv.* **2010**, *19*, 587–600. [CrossRef]

34. Budiyanto, F. Study of metal contaminant level in the Mahakam Delta: Sediment and dissolved metal perpectives. *J. Coast. Dev.* **2013**, *16*, 147–157.

35. De Brye, B. Multiscale Finite-Element Modelling of River-Sea Continua. Ph.D. Thesis, Université Catholique de Louvain, Louvain-la-Neuve, Belgium, 2011.

36. Pham Van, C.; De Brye, B.; Spinewine, B.; Deleersnijder, E.; Hoitink, A.J.F.; Sassi, M.G.; Hidayat, H.; Soares-Frazão, S. Simulations of flow in the tropical river-lake-delta system of the Mahakam land-sea continuum, Indonesia. *Environ. Fluid Mech.* **2016**, *16*, 603–633. [CrossRef]

37. Comblen, R.; Lambrechts, J.; Remacle, J.-F.; Legat, V. Practical evaluation of five partly discontinuous finite element pairs for the non-conservative shallow water equations. *Int. J. Numer. Methods Fluids* **2010**, *63*, 701–724. [CrossRef]

38. De Brye, B.; De Brauwere, A.; Gourgue, O.; Kärnä, T.; Lambrechts, J.; Comblen, R.; Deleersnijder, E. A finite-element, multi-scale model of the Scheldt tributaries, river, estuary and ROFI. *Coast. Eng.* **2010**, *57*, 850–863. [CrossRef]

39. Kärnä, T.; De Brye, B.; Gourgue, O.; Lambrechts, J.; Comblen, R.; Legat, V.; Deleersnijder, E. A fully implicit wetting-drying method for DG-FEM shallow water models, with an application to the Scheldt Estuary. *Comput. Methods Appl. Mech. Eng.* **2011**, *200*, 509–524. [CrossRef]

40. Gourgue, O.; Deleersnijder, E.; White, L. Toward a generic method for studying water renewal, with application to the epilimnion of Lake Tanganyika. *Estuar. Coast. Shelf Sci.* **2007**, *74*, 628–640. [CrossRef]

41. Okubo, A. Oceanic diffusion diagrams. *Deep Sea Res.* **1971**, *18*, 789–802. [CrossRef]

42. Pham Van, C.; Gourgue, O.; Sassi, M.; Deleersnijder, E.; Hoitink, A.J.F.; Soares-Frazao, S. Modelling fine-grained sediment transport in the Mahakam land-sea continuum, Indonesia. *J. Hydro-Environ. Res.* **2016**, *13*, 103–120. [CrossRef]

43. Blaise, S.; De Brye, B.; De Brauwere, A.; Deleersnijder, E.; Delhez, E.J.M.; Comblen, R. Capturing the residence time boundary layer—Application to the Scheldt Estuary. *Ocean Dyn.* **2010**, *60*, 535–554. [CrossRef]

44. Delhez, E.J.M.; Deleersnijder, E. The boundary layer of the residence time field. *Ocean Dyn.* **2006**, *56*, 139–150. [CrossRef]

45. De Brye, B.; Schellen, S.; Sassi, M.; Vermeulen, B.; Karna, T.; Deleersijder, E.; Hoitink, T. Preliminary results of a finite-element, multi-scale model of the Mahakam Delta (Indonesia). *Ocean Dyn.* **2011**, *61*, 1107–1120. [CrossRef]

46. Andutta, F.P.; Helfer, F.; De Miranda, L.B.; Deleersnijder, E.; Thomas, C.; Lemckert, C. An assessment of transport timescales and return coefficient in adjacent tropical estuaries. *Cont. Shelf Res.* **2016**, *124*, 49–62. [CrossRef]

47. Budhiman, S.; Salama, S.M.; Vekerdy, Z.; Verhoef, W. Deriving optical properties of Mahakam Delta coastal waters, Indonesia using *in situ* measurements and ocean color model inversion. *ISPRS J. Photogramm. Remote Sens.* **2012**, *68*, 157–169. [CrossRef]

48. Suroso, B.; Hutabarat, J.; Afiati, N. The Potential of Tiger Prawn Fry from Delta Mahakam, East Kalimantan Indonesia. *Int. J. Sci. Eng.* **2013**, *6*, 43–46. [CrossRef]

49. Furnas, M.J.; Mitchell, A.W.; Gilmartin, M.; Revelante, N. Phytoplantonk biomass and primary production in semi-enclosed reef lagoons of the central Great Barrier Reef, Australia. *Coral Reefs* **1990**, *9*, 1–10. [CrossRef]

Article

The Use of Stable Isotope-Based Water Age to Evaluate a Hydrodynamic Model

Edward Gross [1,2,*], Stephen Andrews [2], Brian Bergamaschi [3], Bryan Downing [3], Rusty Holleman [1], Scott Burdick [2] and John Durand [1]

[1]   Center for Watershed Sciences, University of California, Davis, One Shields Avenue, Davis, CA 95616, USA; cdholleman@ucdavis.edu (R.H.); jrdurand@ucdavis.edu (J.D.)

[2]   Resource Management Associates, 1756 Picasso Avenue, Suite G, Davis, CA 95618, USA; steve@rmanet.com (S.A.); scott@rmanet.com (S.B.)

[3]   U.S. Geological Survey, Sacramento, CA 95819, USA; bbergama@usgs.gov (B.B.); bdowning@usgs.gov (B.D.)

*   Correspondence: edgross@ucdavis.edu; Tel.: +1-510-847-4061

Received: 4 September 2019; Accepted: 18 October 2019; Published: 23 October 2019

Abstract: Transport time scales are common metrics of the strength of transport processes. Water age is the time elapsed since water from a specific source has entered a study area. An observational method to estimate water age relies on the progressive concentration of the heavier isotopes of hydrogen and oxygen in water that occurs during evaporation. The isotopic composition is used to derive the fraction of water evaporated, and then translated into a transport time scale by applying assumptions of representative water depth and evaporation rate. Water age can also be estimated by a hydrodynamic model using tracer transport equations. Water age calculated by each approach is compared in the Cache Slough Complex, located in the northern San Francisco Estuary, during summer conditions in which this region receives minimal direct freshwater inflow. The model's representation of tidal dispersion of Sacramento River water into this backwater region is evaluated. In order to compare directly to isotopic estimates of the fraction of water evaporated ("fractional evaporation") in addition to age, a hydrodynamic model-based property tracking approach analogous to the water age estimation approach is proposed. The age and fractional evaporation model results are analyzed to evaluate assumptions applied in the field-based age estimates. The generally good correspondence between the water age results from both approaches provides confidence in applying the modeling approach to predict age through broader spatial and temporal scales than are practical to assess using the field method, and discrepancies between the two methods suggest aspects of both approaches that may be improved. Model skill in predicting water age is compared to skill in predicting salinity. Compared to water age, salinity observations are shown to be a less useful diagnostic of transport in this low salinity region in which salt inputs are poorly constrained.

Keywords: San Francisco Estuary; Sacramento–San Joaquin Delta; water age; transport time scales; hydrodynamic model; tidal hydrodynamics; stable isotopes

1. Introduction

Time scales are metrics of the time associated with processes such as physical transport or biogeochemical reactions [1]. Comparison of transport time scales with biogeochemical time scales provide insight to the relative importance of transport. When the time scale of a biogeochemical process is much shorter than a transport time scale, the biogeochemical process is typically more important than transport processes in determining constituent concentrations. Intermediate cases in which transport time scales and biogeochemical time scales are similar can have desired ecological outcomes. For example, primary and secondary productivity can be maximized when these time scales are similar [2].

Water age is a transport time scale quantifying the time elapsed since a water parcel entered a study area [3]. In contrast, residence time quantifies the time required for a water parcel starting at a specific time and location to leave the study area [3]. Water age can be directly useful in estimating biogeochemical rates [4]. Water age can be estimated based on the fractional evaporation of water inferred from variation in the stable isotopes of hydrogen (^{2}H) and oxygen in water (^{18}O) along transects through the Cache Slough Complex (CSC) in the northern San Francisco Estuary (SFE) [4]. These estimates showed a broad range of water ages across the region. However, the analysis was based on several approximations, including a constant and uniform water depth and an evaporation rate averaged over the two-months prior to data collection.

A tracer based modeling approach [5,6] has been widely applied to estimate water age. A specific application of this flexible approach provides the algebraic mean age of all water parcels from a specific source present at a given time and location. This approach is generally used to represent the age of water volume or conservative substances. The mean water age will generally be larger than the "radio age" of decaying tracers such as a radioactive tracer [7] (Delhez et al., 2003). Since the observational data uses stable isotopes (not decaying isotopes) whose ratio is influenced by evaporation, the mean water age and isotopic age are believed to be conceptually consistent. We specifically estimate the mean age ("age") of Sacramento River water advected down the Sacramento River to the seaward (downstream) end of the CSC and mixed through the CSC by tidal mixing processes during the summer and early fall of 2014. We compare this "predicted" age from the hydrodynamic model with the "isotopic age" estimated from the water isotope data. In addition, we estimate fractional evaporation using an extension of the published age approach [6] and compare these to estimates from stable isotope measurements [4].

The main objective of this study is to investigate the utility of water isotope-based age estimates in evaluating the representation of transport processes by a hydrodynamic model. The isotopic-age is a potential alternative to salinity observations to calibrate or validate a hydrodynamic model. In dominantly freshwater systems with agricultural return flows, conductivity (and thus practical measurements of salinity) can be substantially influenced by local sources of salts, nitrate, and phosphate [8]. These sources are often poorly quantified, limiting the utility of salinity observations for model calibration. In this study, we compare predicted water age and isotopic water age to the CSC. An additional objective is the evaluation of depth and evaporation rate assumptions applied to estimated isotopic water age [4]. The desired outcome of this study is that both the modeling and field approaches can be used with increased confidence to study relationships between water age and biogeochemical concentrations and transformation rates.

2. Materials and Methods

2.1. Site Description

Our study area includes the Cache Slough Complex (CSC) and neighboring regions at the northwest extreme of the San Francisco Estuary and within the Sacramento–San Joaquin Delta. While the contemporary Delta has little of the historic marsh and other wetland habitat present during historical conditions [9], the northwest portion of the Delta is unique due to the presence of extensive freshwater tidal wetlands and adjacent floodplain habitat. The CSC (Figure 1) comprises several largely natural channels including Cache Slough, Lindsey Slough, and Prospect Slough; straight manmade conveyance channels including Shag Slough, Liberty Cut, the Toe Drain and the Sacramento Deep Water Ship Channel; and former agricultural land restored to tidal action, including Liberty Island and Little Holland Tract (Figure 1). Within it are varying habitats including subtidal channels, intertidal mudflat areas, and vegetated marsh. Both the geometry and bathymetry of the CSC continue to evolve due to deliberate tidal restoration and slow deterioration of unmaintained levees.

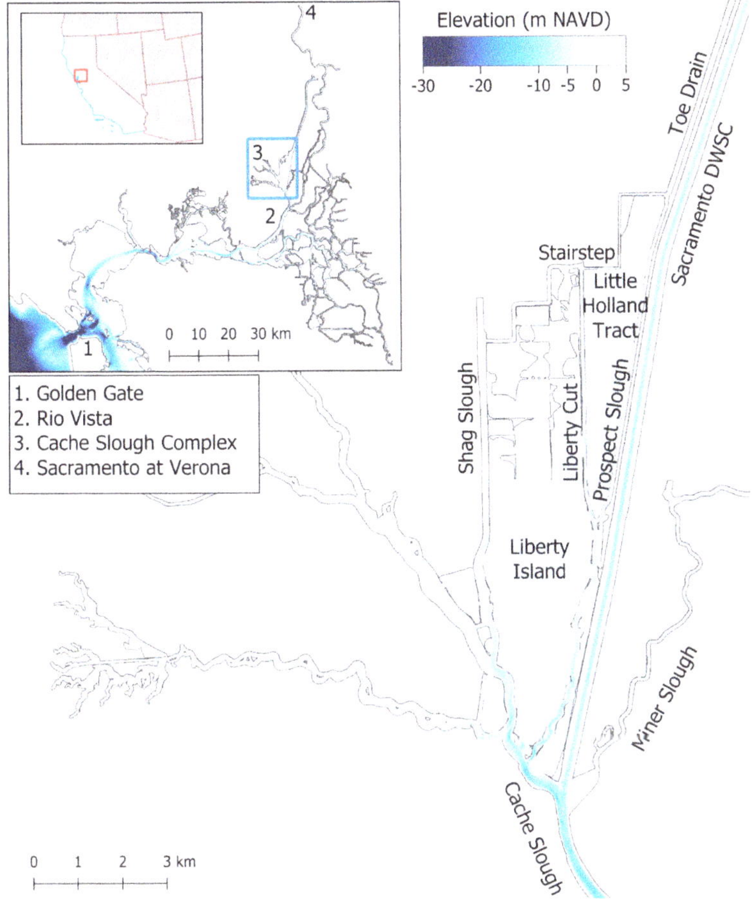

Figure 1. The bathymetry and names of channels in the Cache Slough Complex. The inset image in the upper left shows the northern portion of the San Francisco Estuary with relevant geographical labels. The location of the estuary in the state of California is shown in the small inset image in the upper left corner.

The CSC has mixed diurnal and semidiurnal tides with a typical greater diurnal range of 1.2 m. The climate is Mediterranean, classified as dry-summer subtropical (Csb) in the Köppen Climate classification. The wet season is typically November through April. During the dry season, the CSC receives low direct freshwater input, while during the wet season, episodic flood events deliver large inflows. Consumptive use by agriculture in neighboring regions results in net landward flow in portions of the CSC. Due to low net flows in dry periods, landward sediment fluxes are noted in this region, dominantly consisting of Sacramento River-associated sediment arriving via Miner Slough [10].

2.2. Isotopic Water Age

The isotopic composition is defined by the ratio of stable isotopes in a water sample. Specifically, δ^2H quantifies the normalized deviation in the ratio of ^{2}H (Deuterium) to ^{1}H (hydrogen) relative to a standard ratio and, similarly, δ^{18}O quantifies the normalized deviation in the ratio of ^{18}O to ^{16}O from a standard ratio. Lighter isotopes are preferentially evaporated, leading to an evaporative signal in δ^2H and $\delta^{18}O$ used to infer the amount of evaporation undergone by a water sample. In the dataset used in this study [4], δ^2H varied by approximately 20‰, while $\delta^{18}O$ varied by approximately 4‰.

The fractional evaporation represents the fraction of water that has evaporated relative to water with a known original isotopic composition. Water age (τ) is estimated from a given fractional evaporation (χ) as

$$\tau = \chi \frac{H}{E} \tag{1}$$

where H is a representative water depth and E is a representative evaporation rate. In [4], H was assumed to be 3 m and E was calculated as 0.0054 m d^{-1} from a two-month average evaporation rate measured at the Hastings Tract station of the California Irrigation Management Information System [11].

2.3. Hydrodynamic Model

The three-dimensional UnTRIM model engine [12,13] was applied to simulate flow and transport in the CSC. UnTRIM solves the discretized Reynolds-averaged shallow water equations on an unstructured grid. It resolves the relevant physical processes resulting in transport of dissolved constituents such as salt, allows for wetting and drying of computation cells [14], and sub-grid scale representation of bathymetry [13]. Vertical turbulent mixing in the model is parameterized using a k-ε closure, which solves one equation for turbulent kinetic energy (k) and another for turbulent dissipation (ε) using published parameter values [15]. Bed friction is parameterized using a quadratic stress formula and bed roughness height, z_0.

The computational mesh and bathymetry for the UnTRIM San Francisco Estuary model are shown in Figure 2. Cell side lengths range from less than 5 m to more than 1000 m, and 1 m layer spacing was used in the vertical. Bathymetry was specified using a digital elevation model of the San Francisco Estuary developed using a large number of bathymetric surveys [16–19]. The model grid was refined substantially in the CSC relative to the grids applied in previous applications [19,20]. The higher resolution in the channels of the CSC allows better representation of tidal and residual velocities.

The simulation period was chosen as 12 May 2014 through 2 October 2014. This allows over four months for Sacramento River water to mix into the CSC prior to comparison to the continuous underway measurement observations collected on 1 October 2014. The model spin-up period is more than twice as long as isotopic water age estimates for the region [4]. Water year 2014 was classified as dry in the Sacramento Valley [21]. The net flows were quite small prior to the data collection with substantial contributions from agricultural withdrawals and return flows (Table 1). River and diversion boundary conditions were prescribed at locations shown in Figure 2 using data obtained from United States Geological Survey [22] and California Department of Water Resources [21] monitoring sites. Local agricultural diversions, return flows, and groundwater seepage (collectively referred to as net channel depletions) were prescribed at 257 locations throughout the Delta, shown in Figure 2, using estimates from [23]. Observed water levels at Point Reyes were used as the offshore boundary condition, and specified offshore salinity was 33.5 psu (practical salinity unit; 1 PSU equals 1‰). Spatially variable evaporation, precipitation and wind speed were applied based on observations at several meteorological stations (Figure 2).

Table 1. Average observed flows during September 2014, immediately prior to isotope data collection.

Flow Category	Flow (m^{3}s^{-1})
Total Delta inflow	251.29
Total Delta exports	−100.91
Total Delta agricultural withdrawals	32.84
Total Delta agricultural returns	23.46

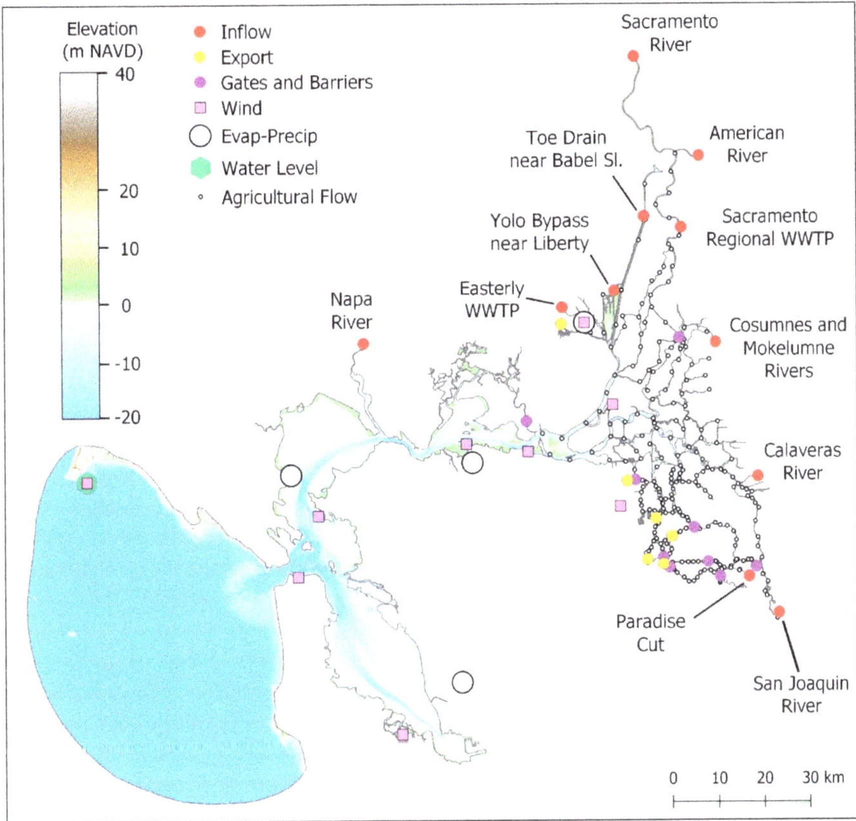

Figure 2. San Francisco Estuary model extent and bathymetry. Model boundary condition locations are shown for river inflows, diversions, agricultural flows, and hydraulic structures. Station locations used in setting regional wind and evaporation-precipitation model inputs are also shown.

Model output was compared against observed flow, stage, and salinity data collected at continuous monitoring stations throughout the estuary [12] (Figure 3). A model skill metric [24] was computed at each calibration location, as in previous San Francisco Estuary calibration efforts [25]. The skill metric is calculated as

$$Skill = 1 - \frac{\sum_{i=1}^{n}|P_i - O_i|^2}{\sum_{i=1}^{n}\left(|P_i - \overline{O}| + |O_i - \overline{O}|\right)^2} \tag{2}$$

where P are the predicted values, O are the observations, n is the total number of observations and $\overline{O}$ is the average of the observations.

We also summarize model performance using target diagrams [26], also used in San Francisco Estuary applications [27]. The predicted longitudinal and vertical salinity structure was compared to monthly observations collected by the USGS on a longitudinal transect from the Golden Gate to Rio Vista [28].

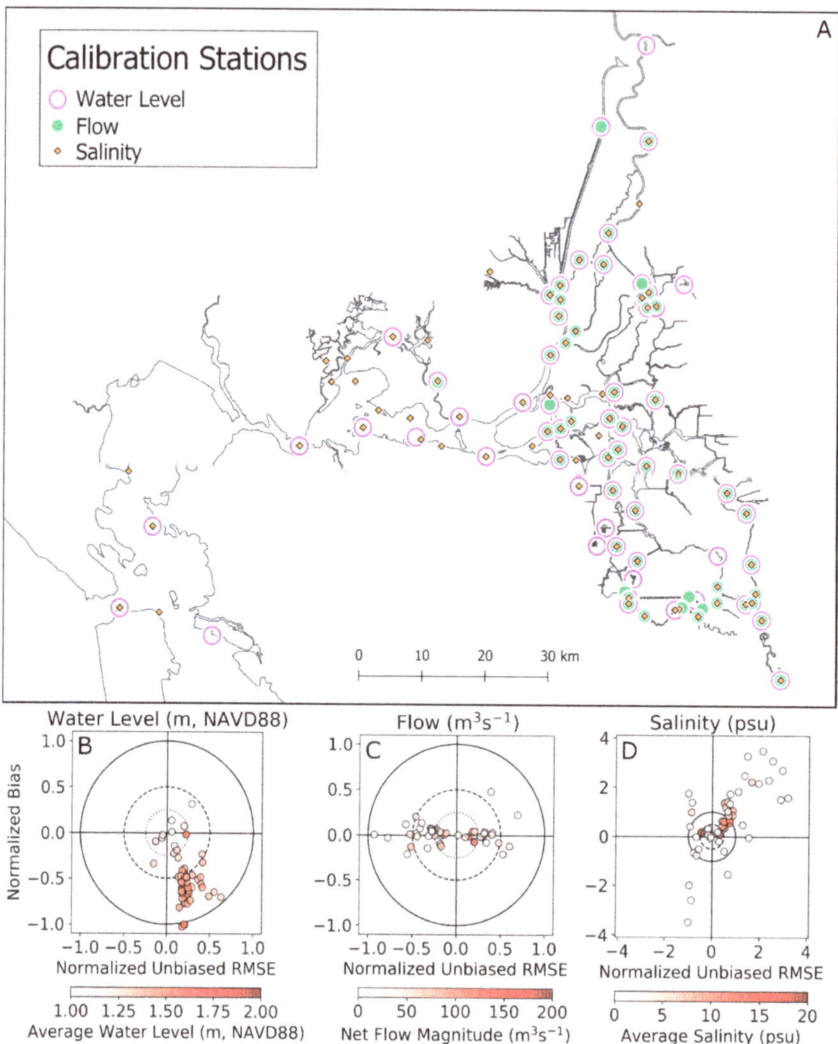

Figure 3. (**A**) Model calibration stations and (**B–D**) model calibration target diagrams for (**A**) water level, (**B**) flow, and (**C**) salinity in 2014. The normalized bias metric for eight of the 57 salinity stations lies outside of the axis bounds.

2.4. Predicted Water Age Calculation

In order to estimate mean age of a source of water, the transport of two conservative tracers was simulated in the UnTRIM model [6]. The first tracer is used to tag water entering at the Sacramento River flow boundary condition and is calculated using a three-dimensional advection diffusion equation

$$\frac{\partial C}{\partial t} + \nabla \cdot (\boldsymbol{u}C) = \frac{\partial}{\partial z}\left(K_T \frac{\partial C}{\partial z}\right)$$

(3)

where C is the tracer concentration with dimensions mass per volume, $\boldsymbol{u}$ is a three-dimensional velocity vector with dimensions length per time, and K_T is the vertical eddy diffusivity with dimensions length squared per time. Horizontal eddy diffusion is neglected.

This equation is discretized with a conservative finite volume approach [29]. The discretized equation can be represented as

$$C^{n+1} = \mathcal{A}(C^n)$$ (4)

where $\mathcal{A}$ represents a discrete advection-diffusion operator [29].

A second equation is used to represent "age-concentration" [6]. The governing equation of age-concentration can be written in a manner similar to Equation (3)

$$\frac{\partial \alpha}{\partial t} + \nabla \cdot (\boldsymbol{u}\alpha) = \frac{\partial}{\partial z}\left(K_T \frac{\partial \alpha}{\partial z}\right) + C$$ (5)

where α is the age-concentration with dimensions time-mass per volume. Its discretized form can be written as

$$\alpha^{n+1} = \mathcal{A}(\alpha^n) + \Delta t C^n$$ (6)

where Δt is the computational time step [30]. Then the age can be estimated as the ratio of the age-concentration and concentration.

$$a = \frac{\alpha}{C}$$ (7)

where a has dimensions of time. In our application, the concentration C in each cell and time step represents the portion of water at that location that entered the domain as Sacramento River inflow and a is the mean age of that water. This is referred to as the "mean age" because the age calculated at any cell and any time step represents the algebraic mean of the water parcels of the source water present at that time and location [6]. The initial conditions of the scalar transport equations are zero concentration and age-concentration throughout the domain. The boundary conditions of C are zero at all boundaries except for the upstream Sacramento River boundary where concentration is 1. The boundary conditions of age-concentration are zero at all boundaries. After concentration C and age-concentration α are predicted throughout the domain and simulation period, the mean age a is then calculated in a post-processing analysis.

2.5. Fractional Evaporation Calculation

In order to estimate the mean experience of a water quality property (e.g., depth, light level, or temperature) experienced by the tracer C, a property tracking equation is used in addition to the equations in the age calculation. The property-age-concentration is governed by the equation

$$\frac{\partial \beta}{\partial t} + \nabla \cdot (\boldsymbol{u}\beta) = \frac{\partial}{\partial z}\left(K_T \frac{\partial \beta}{\partial z}\right) + \psi C$$ (8)

where β is the property-age-concentration and ψ is the instantaneous value of the property. Its discretized form can be written as

$$\beta^{n+1} = \mathcal{A}(\beta^n) + \Delta t \psi^n C^n$$ (9)

where Δt is the computational time step. Then the mean property experienced by the tracer can be estimated as the ratio of the property-age-concentration and age-concentration.

$$b = \frac{\beta}{\alpha}$$ (10)

where b is the mean property experienced by the tracer. The initial condition of the property-age-concentration transport equation is zero throughout the domain. The boundary conditions of β are zero at all boundaries.

We will use this approach to estimate what has been referred to as the evaporation to inflow (E/I) ratio [4] and we refer to as fractional evaporation, represented by χ in Equation (1). This can also be understood as fractional evaporation from a water source with a specific isotopic composition. In our specific application ψ in Equation (9) represents the fraction of the water column that evaporates in one day. For example, an evaporation rate of 0.01 m d^{-1} and a water column depth of 1 m, would result in ψ of 0.01. The value of ψ is updated at each time step and water column in the simulation. In our application, β in Equation (9) is the daily fractional evaporation-age-concentration with dimensions time-mass per volume. β divided by the age-concentration in Equation (10) is an estimate of mean daily fractional evaporation (b) that the tracer at a given location has experienced since entering the domain. In order to convert this to a total fractional evaporation we multiply by the age (a) from Equation (7). This estimate can be directly compared to the fractional evaporation estimated from isotope data in [4].

The novel approach described by Equations (8)–(10) is analogous to the age calculation approach [6]. Note by inspection of Equations (6) and (9) that if the instantaneous property ψ is constant in time and space then Equation (10) reduces to $b = \psi$. In other words, if the instantaneous property is a constant and uniform value of ψ, the mean experience of the tracer (C from Equation (3)) of that property is equivalent to ψ. If there is no transport and $\psi(x, y, z, t)$ is variable but C is uniformly 1 everywhere (for example, if the tracer was used to tag water in a lake), then the transport terms in Equations (5) and (8) are zero and Equation (10) is simply the time integrated value of ψ at a given location. Hence, in these simple cases the governing equations (Equations (4)–(10)) produce behavior consistent with an intuitive understanding of the mean experience of a water property by a tracer.

3. Results

3.1. Hydrodynamic Model Calibration

The hydrodynamic model predicted stage, flow and salinity accurately at the majority of stations in the model domain (Table 2). The target diagrams in Figure 3 show less normalized bias and similar normalized unbiased RMSE (unRMSE) in flow and similar normalized bias in water level relative to [31]. However, our results had larger normalized bias in water level with our results tending to underpredict mean water level in the Delta. The tidally-averaged water level is sensitive to the bottom friction parameters chosen. Larger bottom friction decreases bias in water level but increases errors in flow predictions. The salinity target diagram shows larger errors than other applications of the model [25], particularly at the low salinity stations, because the simulation period used here was a dry period of a drought year in which the uncertainty in tributary inflows and agricultural withdrawal and return flows makes salinity prediction challenging. The uncertainty in these flows and in the salinity associated with agricultural return flows makes salinity an imperfect water quality constituent to evaluate model performance in freshwater regions. In contrast to the typically poor salinity predictions in the interior Delta, comparisons to USGS transects in Figure 4 show good prediction of salinity from the Golden Gate to Rio Vista, located in the western Delta. This is consistent with good calibration results achieved with the in other applications, most recently for the two simulation years documented in [32].

Water year 2014 was classified as critically dry in the Sacramento Valley [33]. For this reason, salinity conditions were fairly static during the simulation period with a slow increasing trend (Figure 5). While the spatial pattern of predicted salinity is similar to the observed salinity, largely due to the salinity being set from observed salinity on 13 May, the model predicts more salt intrusion to the Delta than is observed. While this is a small spatial shift as seen in Figure 4, it results in a substantial relative error in salinity in the brackish portion of the estuary. This bias of overprediction of salinity in the northern Delta is more evident by comparison to the USGS continuous underway measurements in the Cache Slough Complex [4] in Figure 6. The mean observed salinity was 0.188 psu while the mean predicted salinity was 0.132 psu, corresponding to a bias of 0.55 psu. However, as seen at Rio Vista in the salinity error in the model that was present outside of the Cache Slough Complex, so the salinity data is of little use to evaluate the representation of transport processes inside the Cache Slough Complex. The correlation coefficient was −0.06, indicating that the model also does not capture spatial gradients in salinity which is in part because the estimated flow and salinity associated with agricultural returns [23] is uncertain. Furthermore, the calculation of salinity from conductivity measurements is uncertain because the mixture of salt ions is different in agricultural return flows and seawater. Partially for these reasons, we turn to the water isotope data [4] to provide a better test of the representation of local dynamics in the Cache Slough Complex.

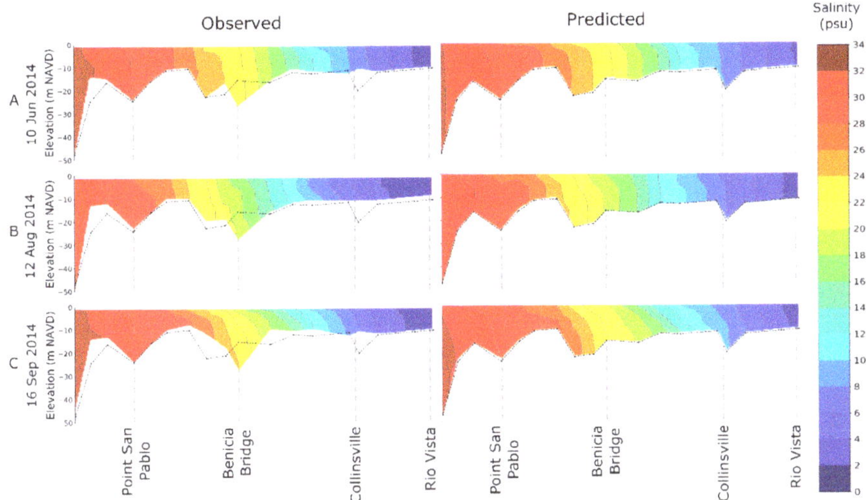

Figure 4. Observed and predicted salinity transects from the Golden Gate to Rio Vista at all dates available during the simulation period: (**A**) 10 June 2014, (**B**) 12 August 2014, and (**C**) 16 September 2014. The black line indicates model bathymetry along transect.

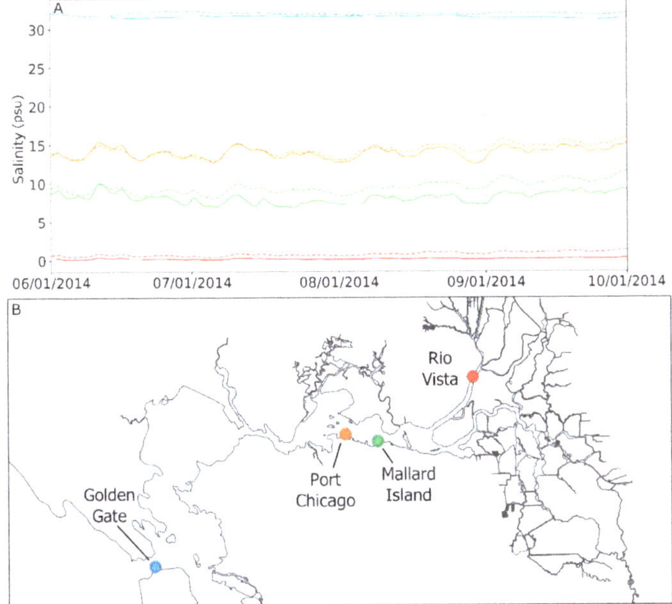

Figure 5. Observed (solid lines) and predicted (dashed lines) tidally-averaged salinity at four stations along the axis of the northern San Francisco Estuary. Line colors of panel (**A**) correspond to marker colors of panel (**B**).

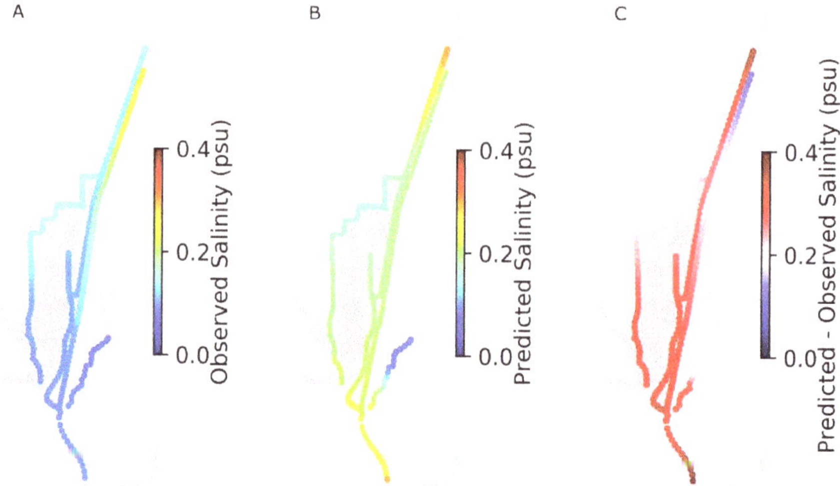

Figure 6. (**A**) Observed salinity; (**B**) predicted salinity; (**C**) predicted-observed salinity in the Cache Slough Complex during the U.S. Geological Survey (USGS) field study on 1 October 2014.

Table 2. Model performance metrics averaged across all continuous monitoring stations.

Parameter	R^2	Skill
Water level	0.990	0.969
Flow	0.969	0.985
Salinity	0.690	0.724

3.2. Sacramento River Water Age

The depth-averaged predicted Sacramento River water age has been calculated at the time and location of each isotopic age estimate. The predicted Sacramento River water age is zero where the tracer enters the model boundary at Verona. In contrast, the isotopic water age is relative to the specific isotopic composition at an origin point in Miner Slough, where isotopic water age is correspondingly defined as zero. The predicted water age in Miner Slough is 4.2 days at the origin of isotopic age. In order to allow direct comparison of predicted age and isotopic age, this offset is subtracted from the predicted age (Figure 7). For clarity, when observations were available from both the initial transect and return trip in a channel, only the initial transect data is shown. The distribution of predicted age is similar to the isotopic age. The most notable differences are lower predicted age in the Stairstep channel and higher predicted age in Liberty Cut. There is a bias toward overprediction of age (Figure 8) of 0.7 days and a standard error of 6.9 days.

The predicted water age is the mean age of Sacramento River water tracer that entered the domain during the simulation period. The simulation does not provide information regarding the age of water in the domain at the beginning or before the simulation period. The water in the model domain at any point in time is a mixture of water that was present at the beginning of the simulation which has unknown age and provenance and water that entered during the simulation via the Sacramento River that has an estimated age. Other sources of inflow during this period are small. Figure 8b shows that the Sacramento River water tracer has spread through the study area to become the dominant source of water everywhere except in the Toe Drain.

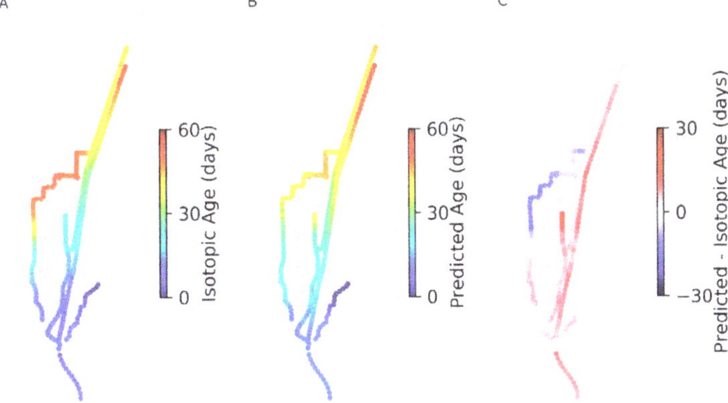

Figure 7. (**A**) Water age estimated from stable isotope observations; (**B**) hydrodynamic model predicted water age; (**C**) difference between predicted and isotopic water age.

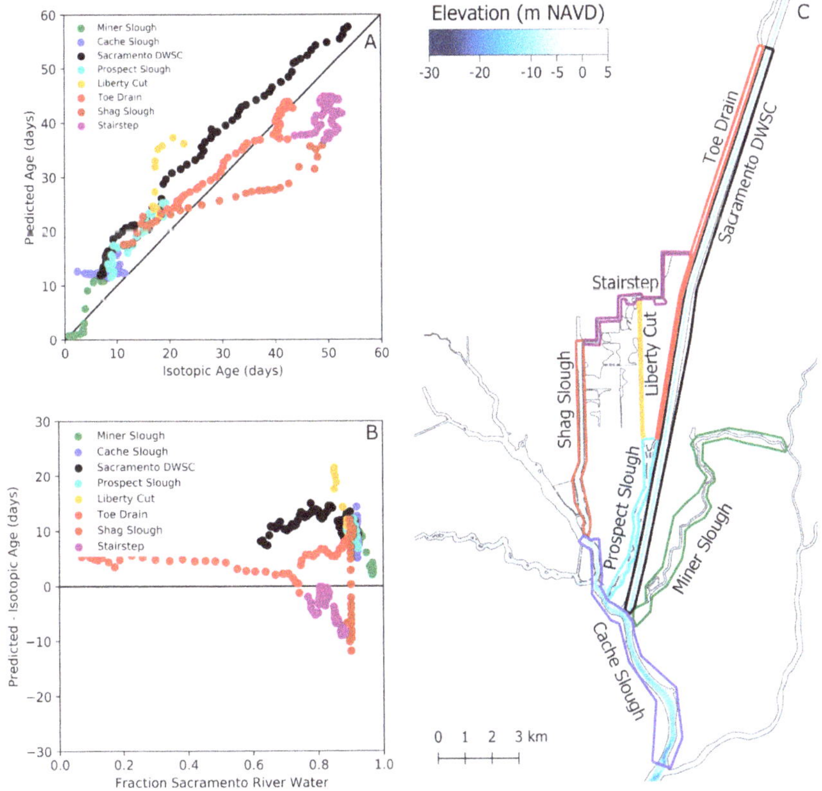

Figure 8. (**A**) Relationship between predicted and isotopic age at the time and location of continuous underway measurements. (**B**) The difference between predicted and isotopic water age plotted against fraction of Sacramento River water at each sampling location. (**C**) The polygons defining regions referenced in panels (**A**,**B**).

3.3. Sacramento River Water Fractional Evaporation

The depth-averaged predicted Sacramento River fractional evaporation has been calculated at each time and locations of isotope data ("isotopic fractional evaporation"). The minimum reported isotopic fractional evaporation is 0.034 in Miner Slough. The model fractional evaporation values are offset to be identical at that one point, which requires subtracting 0.029 from each predicted fractional evaporation value in order to compare directly to the observations. The distribution of predicted fractional evaporation is similar to the isotopic fractional evaporation. The overall bias in fractional evaporation is an overprediction of 0.005 and the standard error is 0.020. Additional performance metrics for comparisons to underway measurement data for salinity and isotopic water age and fractional evaporation estimates are given in Table 3.

Table 3. Model performance metrics in comparisons to underway measurements.

Parameter	R^2	Skill
Salinity	0.00397	−3.330
Water Age	0.867	0.841
Fractional Evaporation	0.884	0.559

The most notable difference in Figures 9 and 10 is higher predicted fractional evaporation in the Toe Drain. Figure 8 indicates that the Sacramento River water tracer has not fully spread through the Toe Drain during the simulation period indicating that both the age and fractional evaporation of the total water mass in that channel cannot be predicted accurately by the model. For example, in parts of the Toe Drain, only 10% of the water in the model derives from Sacramento River water that entered the domain during the simulation. While we can estimate the age and fractional evaporation of the 10% of the water at that location that entered the domain as Sacramento River water, we cannot estimate the age or fractional evaporation of the other 90% of the water volume at that location. Therefore, isotopic and predicted fractional evaporation are not expected to match closely in much of the Toe Drain. Through the Toe Drain and nearby regions, the flow direction and magnitude depend strongly on small but uncertain agricultural diversion and return flows during the study period.

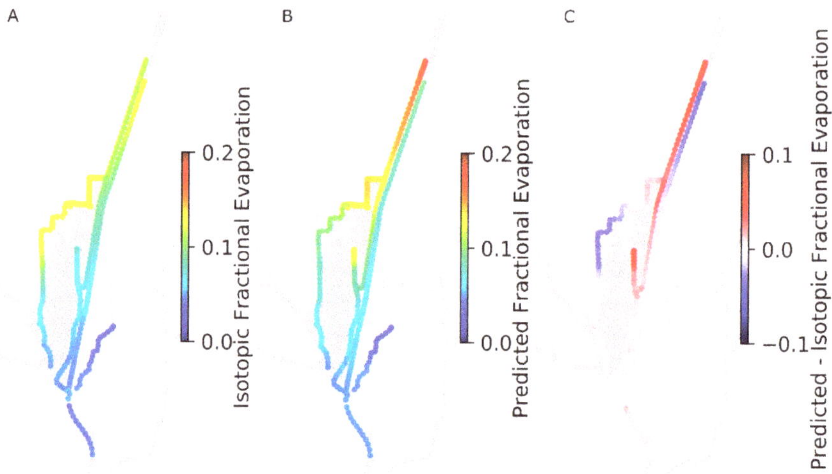

Figure 9. (**A**) Fractional evaporation estimated from stable isotope observations; (**B**) predicted fractional evaporation of Sacramento River water; (**C**) predicted minus isotopic fractional evaporation.

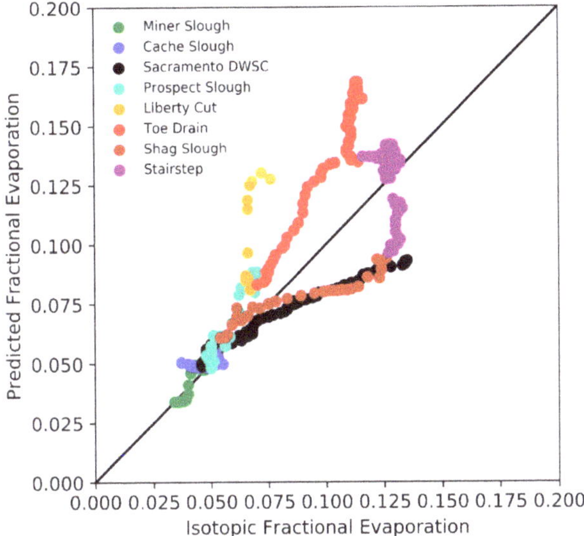

Figure 10. Relationship between predicted and isotopic fractional evaporation at the time and location of continuous underway measurements.

3.4. Evaluation of Assumptions in Age Estimates

The water isotope-based age estimate approach [4] requires specification of a representative depth and evaporation rate. A fixed depth of 3 m was assumed and a two-month average of an observed evaporation rates (0.0054 m d^{-1}) was applied. These assumptions can be examined. The predicted daily evaporation rate is calculated as a moving average over the period of the predicted age preceding the field study. For example, if a point in Cache Slough has a predicted age of 10 days at the time of the field study on 1 October 2014, the observed evaporation rate would be the average from 22 September 2014 through 1 October 2014 to estimate the average evaporation rate experienced by that water parcel since entering the study area at Miner Slough. The averaged evaporation over the period associated with the age at each point is shown in Figure 11A. The bias in assumed evaporation rate was −0.00059 m d^{-1} and the standard error was 0.00048 m d^{-1}. The evaporation rate over the period corresponding to age was smaller than the assumed evaporation rate at most locations.

The representative depth (Figure 11B) is estimated from Equation (1), reorganized to solve for H, the water depth that the water parcel at that location has experienced over the period corresponding to the water age, which we have referred to as "representative depth". This representative depth uses the tracer-based predictions of age in Figure 7B, fractional evaporation shown in Figure 8B and the evaporation rate shown in Figure 11A. The average predicted representative depth is 2.99 m in close agreement to the assumption of 3 m in [4]. However, as expected, the predicted representative depth is larger than 3 m in the deep portions of the model domain including the Sacramento Deep Water Ship Channel and shallower than 3 m in the Stairstep region and the Toe Drain because the local depths are different than 3 m and the water in those regions has relatively high age, thus has experienced the local depths for a substantial amount of time.

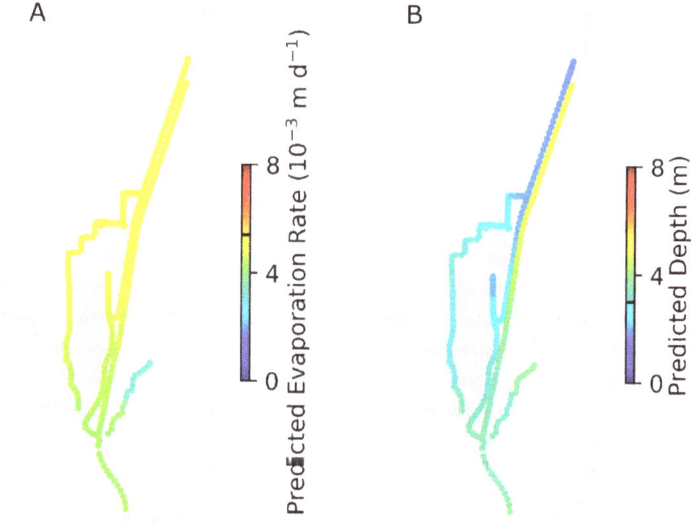

Figure 11. (**A**) Evaporation rate estimated using predicted age, with assumed evaporation rate of 0.00054 m d^{-1} shown with a black line in the colorbar. (**B**) Predicted representative depth estimated from predicted fractional evaporation, evaporation rate and predicted age, with the assumed depth of 3 m shown with a black line in the colorbar.

4. Discussion

The Cache Slough Complex is an ecologically important region that receives little direct freshwater inflow during summer and fall. Most of the water in the CSC in that period arrives from the Sacramento River and has mixed landward from the confluence of Miner Slough, Cache Slough and the Sacramento River. Therefore, water age is almost entirely dependent on tidal mixing processes in this region. Furthermore, model calibration using salinity data is difficult in the CSC because it is influenced by agricultural withdrawals and return flows. Uncertainty in these agricultural flows and the salinity associated with agricultural return flows is large [22]. For this reason, any calibration errors in salinity in that region likely derive from errors in boundary conditions in the hydrodynamic model. Furthermore, errors in oceanic salt intrusion can accumulate far from the region of interest. For example, if estuarine circulation processes are not well represented at any point between the ocean and the CSC, salinity predictions in the CSC may be inaccurate even if local mixing processes are represented accurately by the model. Therefore, the isotopic water age estimates [4] provide a unique opportunity to evaluate the ability of a hydrodynamic model to represent tidal mixing processes in the CSC. The results presented here suggest that the hydrodynamic model represents these mixing processes well. The significant error in the Stairstep Channel and Liberty Cut may be occurring due to inaccurate prediction of tidal residual flows or uncertainty in agricultural withdrawal and return flows. Ongoing observational and modeling studies in the CSC may provide further insight.

Figure 6 indicates that observed salinity from the continuous underway measurements [4] is typically lower than hydrodynamic model predicted salinity. However, the substantial overestimate of salinity at the seaward end of the domain suggests that the errors in predicted salinity may be largely due to errors outside the study area, leading to too much salt intrusion into the Sacramento River and Cache Slough. Estimating salt intrusion in these dry summer conditions in the San Francisco Estuary is generally difficult because uncertain agricultural diversions are roughly as large as net inflows to the Delta leading to small and uncertain net Delta outflow as well as uncertainty in the distribution of flows throughout the Delta [34]. In contrast the age analysis indicates the extent to which landward transport of water entering the CSC from Miner Slough are represented by the model, thus is less sensitive to errors in flow in other parts of the model domain.

While the isotope-based age estimates require assumptions of a representative water depth, evaporation rate, the fractional evaporation (E/I) can be estimated directly from the isotope data with no additional assumptions. The reduced number of assumptions in the fractional evaporation data allows a more confident evaluation of model performance. In order to predict fractional evaporation from tracer results a novel approach was suggested as an extension of the tracer based mean water age approach of [6]. The predicted fractional evaporation generally matched the observations well but was less accurate in the portions of the CSC furthest landward. The circulation in these regions and isotopic composition may be substantially influenced by small agricultural diversion and return flows.

It should be noted that the field-based isotopic method also has significant potential sources of uncertainty. The isotopic ratio of agricultural return flows is uncertain and may contribute to differences between isotopic and predicted fractional evaporation.

The work here emphasizes that the term "water residence time" as used in [4] is consistent with the definition of "mean water age" as defined in [6]. Furthermore, the model results confirm that the dominant water source in the CSC is the Sacramento River. Therefore, a single end-member of isotopic composition for the water source, as assumed in [4] is appropriate. However, this assumption may be less accurate in some of the landward reaches of the CSC due to agricultural return flows of different isotopic composition.

The evaporation rate over the period associated with predicted water age is typically smaller than the two-month average evaporation rate used in [4]. This is because predicted age was lower than 2 months in most of the domain and observed evaporation at Hastings Tract was lower prior to the field data collection on October 1, 2014 than it was in August and September.

With the predicted evaporation shown in Figure 11, the model predictions include all variables in Equation (1), except for the depth. Therefore, the depth that is consistent with the predicted factional evaporation and age estimates can be calculated from Equation (1). This depth estimate is typically larger than the assumed depth of 3 m [4] in deep regions such as the Sacramento Deep Water Ship Channel and lower in shallow regions like the Toe Drain. Note that the equivalent depth of evaporation increases with distance up the Sacramento Deep Water Ship Channel not because of local depth changes but because the fraction of evaporation that has occurred within the Sacramento River Deep Water Ship Channel increases with landward distance.

5. Conclusions

Our comparison demonstrates that the isotope-based water age approach in [4] is useful for validating model-based transport time scale estimates. The water isotope-based approach is useful for estimating transport time scales in regions with a clearly defined source of water and substantial fractional evaporation over relevant time scales of transport. The information from isotopic composition of the water is more useful than salinity data for evaluation of representation of transport processes in these far landward reaches of the estuary because it is less sensitive to representation of mixing processes seaward of the study area. Similarly, in freshwater regions with negligible salinity and salinity gradients, age remains a practical diagnostic of transport. Finally, this validation of the model-predicted age of source water suggests the approach can be used to relate age of source water and associated biogeochemical constituents with local observed concentrations of those constituents to estimate biogeochemical rates to extend the estimates in [4] over broader spatial and temporal scales.

Author Contributions: Conceptualization, E.G., R.H.; Methodology, E.G., R.H.; Software, E.G., S.A., S.B.; Validation, R.H. S.A.; Investigation, E.G., S.A., S.B.; Resources, B.B., B.D.; Data Curation, S.B., B.D.; Writing—original draft preparation, E.G.; Writing—Review and editing, E.G. R.H., S.A., B.B., J.D.; Visualization, E.G., S.B.; Supervision, J.D.; Project administration, J.D.; Funding acquisition, J.D.

Funding: This research was funded by California Department of Fish and Wildlife under Proposition 1—Delta Water Quality and Ecosystem Restoration Grant Program.

Acknowledgments: The authors thank Richard Rachiele for guidance and advice on model calibration. The authors thank Ludmilla Rechiman for contributions to model development, Thomas Handley for collecting and providing bathymetry data and Eli Ateljevich and Jon Burau for helpful discussions on model development and calibration.

Conflicts of Interest: The authors declare no conflict of interest.

References

1. Lucas, L.V. Implications of estuarine transport for water quality. In *Contemporary Issues in Estuarine Physics*; Valle-Levinson, A., Ed.; Cambridge University Press: Cambridge, UK, 2010; pp. 273–307, ISBN 978-0-511-67656-7.

2. Cloern, J.E. Habitat Connectivity and Ecosystem Productivity: Implications from a Simple Model. *Am. Nat.* **2007**, *169*, E21–E33. [CrossRef] [PubMed]

3. Monsen, N.E.; Cloern, J.E.; Lucas, L.V.; Monismith, S.G. A comment on the use of flushing time, residence time, and age as transport time scales. *Limnol. Oceanogr.* **2002**, *47*, 1545–1553. [CrossRef]

4. Downing, B.D.; Bergamaschi, B.A.; Kendall, C.; Kraus, T.E.C.; Dennis, K.J.; Carter, J.A.; Von Dessonneck, T.S. Using Continuous Underway Isotope Measurements to Map Water Residence Time in Hydrodynamically Complex Tidal Environments. *Environ. Sci. Technol.* **2016**, *50*, 13387–13396. [CrossRef] [PubMed]

5. Delhez, E.J.M.; Campin, J.-M.; Hirst, A.C.; Deleersnijder, E. Toward a general theory of the age in ocean modelling. *Ocean Model.* **1999**, *1*, 17–27. [CrossRef]

6. Deleersnijder, E.; Campin, J.-M.; Delhez, E.J.M. The concept of age in marine modelling I. Theory and preliminary model results. *J. Mar. Syst.* **2001**, *28*, 229–267. [CrossRef]

7. Delhez, É.J.M.; Deleersnijder, É.; Mouchet, A.; Beckers, J.-M. A note on the age of radioactive tracers. *J. Mar. Syst.* **2003**, *38*, 277–286. [CrossRef]

8. Monitoring and Assessing Water Quality—Volunteer Monitoring|Monitoring & Assessment|US EPA. Available online: https://archive.epa.gov/water/archive/web/html/index-18.html (accessed on 14 December 2018).

9. Atwater, B.F. *San Francisco Bay: The urbanized estuary: Investigations into the Natural History of San Francisco Bay and Delta with Reference to the Influence of Man*; Conard, S.G., Dowden, J.N., Hedel, C.W., MacDonald, R.L., Savage, Eds.; The Division: San Francisco, CA, USA, 1979; ISBN 978-0-934394-01-7.

10. Morgan-King, T.L.; Schoellhamer, D.H. Suspended-Sediment Flux and Retention in a Backwater Tidal Slough Complex near the Landward Boundary of an Estuary. *Estuaries Coasts* **2013**, *36*, 300–318. [CrossRef]

11. [CIMIS] California Irrigation Management Information System. Available online: https://cimis.water.ca.gov/ (accessed on 5 July 2016).

12. Casulli, V.; Walters, R.A. An unstructured grid, three-dimensional model based on the shallow water equations. *Int. J. Numer. Methods Fluids* **2000**, *32*, 331–348. [CrossRef]

13. Casulli, V.; Stelling, G.S. Semi-implicit subgrid modelling of three-dimensional free-surface flows. *Int. J. Numer. Meth. Fluids* **2011**, *67*, 441–449. [CrossRef]

14. Casulli, V. A high-resolution wetting and drying algorithm for free-surface hydrodynamics. *Int. J. Numer. Meth. Fluids* **2009**, *60*, 391–408. [CrossRef]

15. Warner, J.C.; Sherwood, C.R.; Arango, H.G.; Signell, R.P. Performance of four turbulence closure models implemented using a generic length scale method. *Ocean Model.* **2005**, *8*, 81–113. [CrossRef]

16. Wang, R.; Ateljevich, E. San Francisco Bay and Sacramento-San Joaquin Delta DEM. Available online: http://baydeltaoffice.water.ca.gov/modeling/deltamodeling/modelingdata/DEM.cfm (accessed on 18 June 2019).

17. Foxgrover, A.C.; Finlayson, D.P.; Jaffe, B.E.; Fregoso, T.A. *Bathymetry and Digital Elevation Models of Coyote Creek and Alviso Slough, South San Francisco Bay, California*; US Department of the Interior, US Geological Survey: Washington, DC, USA, 2018.

18. Fregoso, T.A.; Wang, R.-F.; Alteljevich, E.; Jaffe, B.E. San Francisco Bay Delta Bathymetric/Topographic digital elevation model (DEM). In *2016 SF Bay Delta DEM 10-m*. Available online: https://www.sciencebase.gov/catalog/item/58599681e4b01224f329b484 (accessed on 14 December 2018).

19. Delta Bathymetry. Available online: https://gis.water.ca.gov/app/bathymetry/ (accessed on 14 December 2018).

20. Kimmerer, W.J.; Gross, E.S.; Slaughter, A.M.; Durand, J.R. Spatial Subsidies and Mortality of an Estuarine Copepod Revealed Using a Box Model. *Estuaries Coasts* **2019**, *42*, 218–236. [CrossRef]

21. [CDWR] California Department of Water Resources. Available online: http://cdec.water.ca.gov/reportapp/javareports?name=wsihist (accessed on 5 July 2016).

22. [USGS] Water Data for USA. Available online: https://waterdata.usgs.gov/nwis (accessed on 14 December 2018).
23. [CDWR] DCD: Delta Channel Depletion. Available online: http://water.ca.gov/Library/Modeling-and-Analysis/Bay-Delta-Region-models-and-tools/DCD (accessed on 14 December 2018).
24. Dayflow. Available online: http://water.ca.gov/Programs/Environmental-Services/Compliance-Monitoring-And-Assessment/Dayflow-Data (accessed on 14 December 2018).
25. Willmott, C.J. On the Validation of Models. *Phys. Geogr.* **1981**, *2*, 184–194. [CrossRef]
26. Andrews, S.W.; Gross, E.S.; Hutton, P.H. Modeling salt intrusion in the San Francisco Estuary prior to anthropogenic influence. *Cont. Shelf Res.* **2017**, *146*, 58–81. [CrossRef]
27. Jolliff, J.K.; Kindle, J.C.; Shulman, I.; Penta, B.; Friedrichs, M.A.M.; Helber, R.; Arnone, R.A. Summary diagrams for coupled hydrodynamic-ecosystem model skill assessment. *J. Mar. Syst.* **2009**, *76*, 64–82. [CrossRef]
28. MacWilliams, M.; Bever, A.; Gross, E.; Ketefian, G.; Kimmerer, W. Three-Dimensional Modeling of Hydrodynamics and Salinity in the San Francisco Estuary: An Evaluation of Model Accuracy, X2, and the Low-Salinity Zone. *San Fr. Estuary Watershed Sci.* **2015**, *13*. [CrossRef]
29. [USGS] U.S. Geological Survey. Available online: https://sfbay.wr.usgs.gov/access/wqdata/index.html (accessed on 14 December 2018).
30. Casulli, V.; Zanolli, P. High resolution methods for multidimensional advection–diffusion problems in free-surface hydrodynamics. *Ocean Model.* **2005**, *10*, 137–151. [CrossRef]
31. Rayson, M.D.; Gross, E.S.; Hetland, R.D.; Fringer, O.B. Time scales in Galveston Bay: An unsteady estuary. *J. Geophys. Res.* **2016**, *121*, 2268–2285. [CrossRef]
32. Martyr-Koller, R.C.; Kernkamp, H.W.J.; van Dam, A.; van der Wegen, M.; Lucas, L.V.; Knowles, N.; Jaffe, B.; Fregoso, T.A. Application of an unstructured 3D finite volume numerical model to flows and salinity dynamics in the San Francisco Bay-Delta. *Estuar. Coast. Shelf Sci.* **2017**, *192*, 86–107. [CrossRef]
33. Kimmerer, W.; Wilkerson, F.; Downing, B.; Dugdale, R.; Gross, E.; Kayfetz, K.; Khanna, S.; Parker, A.; Thompson, J. Effects of Drought and the Emergency Drought Barrier on the Ecosystem of the California Delta. *San Fr. Estuary Watershed Sci.* **2019**, *17*. [CrossRef]
34. CDEC Report Products. Available online: http://cdec.water.ca.gov/reportapp/javareports?name=WSIHIST (accessed on 14 December 2018).

Article

Age of Water Particles as a Diagnosis of Steady-State Flows in Shallow Rectangular Reservoirs

Benjamin Dewals [1,*], Pierre Archambeau [1], Martin Bruwier [1], Sebastien Erpicum [1], Michel Pirotton [1], Tom Adam [2], Eric Delhez [3] and Eric Deleersnijder [4]

[1] Research Unit Urban & Environmental Engineering (UEE), Hydraulics in Environmental and Civil Engineering (HECE), University of Liège, 4000 Liège, Belgium; pierre.archambeau@uliege.be (P.A.); martin.bruwier@engie.com (M.B.); s.erpicum@uliege.be (S.E.); michel.pirotton@uliege.be (M.P.)
[2] SGI Ingénieurs, 5032 Isnes, Belgium; tom.adam@swecobelgium.be
[3] Research Unit Aerospace & Mechanical Engineering (A&M), Mathematical Modelling & Methods, University of Liège, 4000 Liège, Belgium; e.delhez@uliege.be
[4] Institute of Mechanics, Materials and Civil Engineering (IMMC) & Earth and Life Institute (ELI), Université Catholique de Louvain, 1348 Louvain-la-Neuve, Belgium; eric.deleersnijder@uclouvain.be
* Correspondence: b.dewals@uliege.be

Received: 20 July 2020; Accepted: 5 October 2020; Published: 11 October 2020

Abstract: The age of a water particle in a shallow man-made reservoir is defined as the time elapsed since it entered it. Analyzing this diagnostic timescale provides valuable information for optimally sizing and operating such structures. Here, the constituent-oriented age and residence time theory (CART) is used to obtain not only the mean age, but also the water age distribution function at each location. The method is applied to 10 different shallow reservoirs of simple geometry (rectangular), in a steady-state framework. The results show that complex, multimodal water age distributions are found, implying that focusing solely on simple statistics (e.g., mean or median age) fails to reflect the complexity of the actual distribution of water age. The latter relates to the fast or slow pathways that water particles may take for traveling from the inlet to the outlet of the reservoirs.

Keywords: age distribution function; shallow reservoir; water age; numerical modeling

1. Introduction

Shallow reservoirs are common hydraulic structures. They are used as storm water retention or treatment ponds [1–7], sedimentation tanks [8–10] or service reservoirs [11]. They are also of use in aquaculture [12]. Many of these shallow reservoirs are flat-bottomed, and rectangular or quasi-rectangular in shape [9–11]. Despite this simple geometry, the flow fields are particularly complex. As shown in several laboratory studies [7,13–17], the flow patterns may involve a straight jet, a jet reattached to one of the side-walls, or even a meandering jet, despite steady inflow conditions. The observed flow pattern depends mainly on the expansion ratio (width of the rectangular reservoir compared to the inlet channel width), the reservoir length-to-width ratio and the hydraulic boundary conditions at the inlet and outlet. The complex nature of flow fields in rectangular shallow reservoirs was also investigated in a range of numerical studies [15,18–22].

The operational performance of shallow reservoirs is strongly affected by the type of flow pattern, because the flow field has a direct influence on the reservoir sediment trapping efficiency [4], on the spatial distribution of sediment deposits [8,23], on pollutant removal efficiency [1], and on the disinfection efficiency in service reservoirs [11]. Accurately predicting the flow field in shallow reservoirs is therefore of high engineering relevance for guiding their optimal design, sizing and management.

Given the complexity of turbulent flow fields in shallow reservoirs, gaining a complete understanding of the flow processes may be intricate if only primitive hydrodynamic variables (i.e., depth, velocity and eddy viscosity fields) are examined. Therefore, it is appropriate to also consider complementary diagnostic strategies, such as those relying on timescales. The foundations of the theory of timescales for reservoirs were laid by Bolin and Rodhe [24]. Accordingly, the age, the residence time and the transit time are usually defined as the time elapsed since entering the domain of interest, the time needed to leave it, and the total time spent in it, respectively [25,26].

A limited number of studies have used such timescales to assess the performance of shallow reservoirs, mainly based on the concept of residence time (Table 1). These include the following:

- Persson [27] applied a 2D depth-averaged model to estimate the residence time distribution in ponds of various layouts, based on the advection–diffusion equation for a tracer. The results revealed that a subsurface berm or an island placed in front of the inlet reduces short-circuiting, and improves the effective volume and degree of mixing;
- Sonnenwald et al. [6] used a 3D computational model to obtain residence time distributions for vegetated storm water treatment ponds, showing that the presence of vegetation results in residence times close to those of plug flow conditions;
- By means of tracer studies with laboratory-scale models, Guzman et al. [3] evaluated the residence time distribution for 54 topographies of storm water detention ponds and treatment wetlands, to compare the hydraulic performances of the various designs.

The three abovementioned studies focused solely on timescales evaluated at the outlet of the reservoir. None of them provided the spatial distribution of diagnostic timescales in the reservoir, though this is of relevance for the design of reservoirs since it enables the pinpointing of stagnation regions. Zhang et al. [11] did apply a commercial CFD model to compute the spatial distribution of water age in three service reservoirs of rectangular, square and circular shape. However, to reduce computational time, Zhang et al. [11] simulated only one half of each reservoir, thereby assuming a priori that the flow field is symmetric. This simplifying hypothesis conflicts with experimental observations, which reveal the existence of large asymmetric recirculations in shallow reservoirs [13–15]. Therefore, the approach of Zhang et al. [11] fails to capture the influence of such flow patterns on the spatial distribution of water age.

Here, the focus is on evaluating the water age, which reflects the rate at which water is renewed in the reservoirs under consideration. Indeed, the main performance criteria of shallow reservoirs, such as the rate of sediment settling, pollutant removal or disinfection, are closely related to the water renewal rate. Our computations provide spatially distributed information on water age across the whole reservoir. Moreover, in the Eulerian framework adopted herein and in earlier studies, the water particles present in a given computational cell do not all have the same age because, due to diffusion, they have not all followed the same path from the reservoir inlet to the considered cell. Therefore, here, we do not only compute the mean water age, as highlighted in Table 1.

Table 1. Contribution of the present study compared to the existing literature.

	Persson [27]	Sonnenwald et al. [6]	Guzman et al. [3]	Zhang et al. [11]	Present Study
Modeling approach	2D	3D	Lab	3D	2D
Considered layouts	13	4 × 3	45+	3	10
Full domain considered (i.e., no a priori assumption of symmetric flow)	✓	✓	✓		✓
Residence time distribution (at the outlet)	✓	✓	✓		✓
Mean water age throughout the reservoir				✓	✓
Age distribution throughout the reservoir					✓

Unlike previous studies (Table 1), we also evaluate the distribution function of water age in each computational cell, i.e., the histogram of the age of the water particles present in the cell. To do so, the constituent-oriented age and residence time theory (CART, www.climate.be/cart) is used. It is a conceptual toolbox that enables formulating partial differential problems for the evaluation of diagnostic timescales [28]. As a case study, 10 rectangular shallow reservoirs of various dimensions were analyzed. The flow and diffusivity fields were simulated over the whole domain, thereby ensuring that the symmetry or asymmetry of the flow emerges as a result of the computation [18], without prior assumption in this respect.

Studying the spatial variability in the statistical distribution of water age in man-made reservoirs is novel and of academic and engineering interest. Our new computational results show that using such a position-dependent water age distribution function turns out to be a valuable means of diagnosing flow characteristics in shallow reservoirs. Indeed, they highlight remarkable features, such as the maximum values of mean age inside the domain, which are considerably larger than at the reservoir outlet, as well as multimodal distributions of water age, which can be related to specific patterns in the flow fields (e.g., flow recirculations, stagnation regions).

The data as well as the mathematical and numerical models are described in Section 2, while in Section 3 the computational results are presented and discussed. Conclusions are drawn in Section 4.

2. Data and Method

The considered reservoir configurations are first introduced (Section 2.1) and the flow model used for computing the corresponding flow fields is briefly presented (Section 2.2). Next, the mathematical framework used for evaluating the water age distribution function is outlined (Section 2.3) as well as some key properties of the mean water age (Section 2.4). Finally, the computational procedure set up for applying the diagnostic strategy to the shallow reservoirs of interest is described (Section 2.5).

2.1. Characteristics of the Considered Reservoirs

In total, 10 cases of steady-state water flows in laboratory-scale rectangular shallow reservoirs of various sizes are considered (Figure 1 and Table 2). The geometric characteristics of the reservoirs and the hydraulic conditions are exactly the same as those analyzed by Camnasio et al. [18]. The reservoir bottom is smooth and horizontal, while the width of the inlet and outlet channels is kept constant and equal to $b = 0.25$ m (Figure 1). In the considered configurations, the reservoir width B varies between 0.6 m and 4 m, and the reservoir length L between 3 m and 6 m (Table 2). In all cases, the inlet and outlet channels are located on the reservoir centerline.

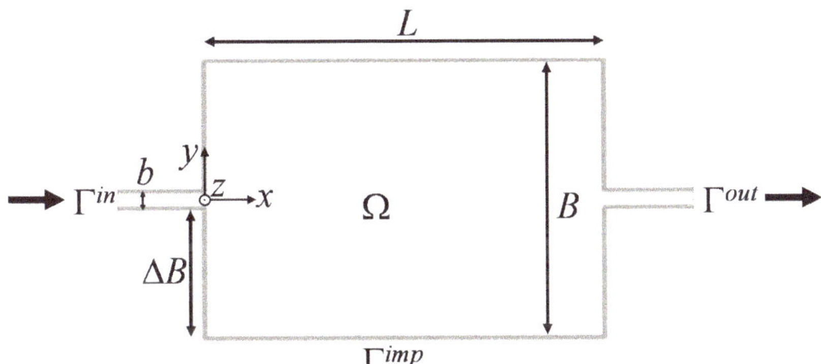

Figure 1. Geometry of the rectangular shallow reservoirs and main notations. Ω is the domain of interest, while Γ^{in}, Γ^{out} and Γ^{imp} refer to the inlet, outlet and impermeable boundaries, respectively.

Table 2. Tested configurations.

Test ID	L (m)	B (m)	SF (–)	Type of Flow Pattern	Test ID in Camnasio et al. [18]
1	3	4	3.6	Detached jet (D)	5
2	4	3	5.8	Detached jet (D)	6
3	5	4	6.0	Near-transition (NT)	4
4	5.8	4	6.9	Near-transition (NT)	2
5	6	4	7.2	Reattached jet (R)	1
6	4	2	7.6	Reattached jet (R)	7
7	4	1	12.5	Reattached jet (R)	8
8	6	1	18.8	Reattached nearly-plug-flow (R-NPF)	9
9	6	0.75	24.0	Reattached nearly-plug-flow (R-NPF)	10
10	6	0.6	29.7	Reattached nearly-plug-flow (R-NPF)	11

As shown by Dufresne et al. [14] and Goltsman and Saushin [29], the type of flow pattern developing in rectangular shallow reservoirs depends on the value of a so-called shape factor, SF, expressing the relative length of the reservoir compared to its expansion ($\Delta B = (B - b)/2$) and to the width of the inlet channel: SF $= L/(\Delta B^{0.6}\, b^{0.4})$. Critical values of SF were determined by Dufresne et al. [14] based on laboratory experiments. A detached jet flowing straight from the inlet to the outlet is observed for relatively short reservoirs (SF < 6.2). In contrast, for longer reservoirs (SF > 6.8), the jet reattaches on either of the side-walls. In the case of particularly long reservoirs, a flow pattern close to plug flow is observed in the downstream portion of the reservoir. For $6.2 \leq$ SF ≤ 6.8, a transition zone was reported, in which both detached and reattached flow patterns may occur.

The set of 10 configurations considered here (Table 2) covers the different types of steady flow patterns that can be observed in rectangular shallow reservoirs. In contrast, since the focus is set here on steady flow configurations, the case of meandering jets is not considered [21,30].

Considering that the reservoir bottom and walls are smooth is certainly a simplifying assumption. It is well known that bottom roughness may alter the flow pattern observed in such reservoirs, as shown by Choufi et al. [17]. Indeed, varying the roughness changes the relative importance of friction forces and other governing forces, such as horizontal shear. This effect is comparable to the influence of varying the flow shallowness, which leads to the definition of frictional and non-frictional regimes [30,31]. Nonetheless, the flow model used here was already shown to provide reasonable predictions for a broad range of flow types, covering both the non-frictional and frictional regimes [21].

Similarly, the bottoms of real-world reservoirs are generally not characterized by a uniform roughness and are not entirely flat, e.g., due to spatially varied sediment depositions. Again, the flow model used here is able to reproduce the effect of spatial variability in the reservoir bottom's characteristics [19]. However, we stick here to simple flat and smooth reservoirs to highlight the main features of water age distribution as a function of the reservoir geometry.

2.2. Flow Model and Computed Flow Fields

The current study relies directly on the flow fields computed earlier by Camnasio et al. [18]. In their model, the ratio of the flow depth to the horizontal size of the reservoir (aspect ratio) was assumed to be sufficiently small that the vertical variations of all the variables, except pressure, may be ignored. Therefore, the shallow-water equations were used to compute the depth-averaged horizontal velocity, $\mathbf{u}$, and the flow depth, H. These variables depend on the horizontal position vector $\mathbf{x} = (x, y)$, where x and y are Cartesian coordinates. The horizontal eddy viscosity and diffusivity are evaluated by means of a two-length-scale depth-averaged k–ε turbulence model, as described by Camnasio et al. [18].

The complete system of partial differential equations solved, coupling the shallow-water equations and the equations for k and ε, are written out in full by Camnasio et al. [18]. The numerical computations were performed on a Cartesian grid (spacing of 0.025 m), with a second-order accurate finite volume scheme implemented in the academic model WOLF 2D.

Although the aspect ratio of the reservoirs under study is small, so that it is generally appropriate to deal with depth-averaged variables, the occurrence of secondary currents (e.g., in curved portions of the jet or in recirculation cells) cannot be excluded completely. Such flow processes should be investigated based on a non-hydrostatic 3D flow model, which is out of the scope of the present research. Nonetheless, the validity of the model considered here for predicting horizontal flow fields in shallow reservoirs was repeatedly proven [18,19,21,32]. Particularly, the model assessment included a grid sensitivity analysis [32] based on the grid convergence index introduced by Roache et al. [33], as well as a demonstration that the model predicts accurately the velocity profiles in similar shallow reservoirs, as considered here [18,19]. Therefore, the validation of the flow model is not discussed further.

Examples of the computed flow fields are displayed in Figure 2, while the flow and eddy viscosity fields for the 10 considered configurations are provided in the Supplemental Material (Figures S1 and S2). The flow patterns are in agreement with the empirical criteria of Dufresne et al. [14]. In Tests 1 and 2, the jet is detached (Figure 2a), whereas a reattached jet is found in Tests 5, 6 and 7 (Figure 2c). Tests 8, 9 and 10 also involve a reattached jet; but in these cases, the reservoirs are so long that a nearly-plug-flow is observed in the downstream part of the reservoir (Figure 2d). Tests 3 and 4 are very close to the empirical transition zone; but for both of them, the computed flow fields show a detached jet (Figure 2b) [18].

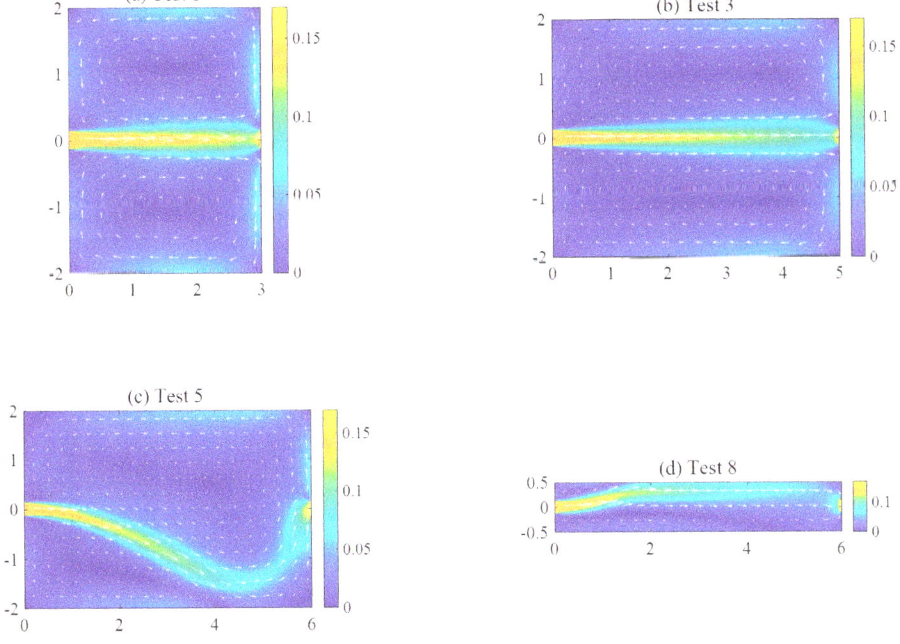

Figure 2. Examples of computed flow fields: (**a**) detached jet, (**b**) detached jet, near transition, (**c**) reattached jet, and (**d**) reattached jet with plug flow in the downstream part of the reservoir [18]. The color scale represents the velocity magnitude (m/s).

2.3. Water Age Distribution Function

The age of a water particle is defined as the time elapsed since it entered the domain of interest by crossing its inlet. However, the evolution of a single particle is rather irrelevant. A sufficiently large number of them must be considered. Accordingly, the water particles present at a given instant in an elemental control volume are unlikely to have followed the same paths to reach it. In other words, they are likely to have different ages, hence the need to introduce the histogram of their ages.

Hereinafter, the domain of interest (the reservoir under study) is denoted by Ω. Its boundary, Γ, is made up of three parts (Figure 1), the inlet, outlet and impermeable boundaries, which are referred to as Γ^{in}, Γ^{out} and Γ^{imp}, respectively.

The so-called age distribution function, the core variable of the age theory resorted to herein [28,34], will help us assess the variability of the water particles' travel times. To introduce this function, the first step consists in defining an elemental control domain ($\delta\Omega$) well suited to the depth-averaged approach. The corresponding space coordinate intervals are $[x, x + \delta x]$ and $[y, y + \delta y]$, where the space increments are such that $\delta x, \delta y \to 0$. The volume of $\delta\Omega$ is $\delta V (\mathbf{x}) = \delta x\, \delta y\, H (\mathbf{x})$. The age distribution function is denoted as $c (\mathbf{x}, \tau)$, where the independent variable τ is the age ($0 \leq \tau \leq \infty$). In accordance with Delhez et al. [28], the following definition is adopted: in the abovementioned elemental control domain, the mass of the water whose age lies in the interval $[\tau, \tau + \delta\tau]$ tends to $\rho\, c (\mathbf{x}, \tau)\, \delta V\, \delta\tau$ in the limit $\delta V, \delta\tau \to 0$ (with ρ as the water density). The physical dimension of the age distribution function $c (\mathbf{x}, \tau)$ is time^{-1}.

From mass budget considerations, Delhez et al. [28] derived the general equation governing the age distribution function of any constituent of a fluid mixture, taking into account advection, diffusion and reactive processes. Specifically, the mass flux evaluated in the age direction is due to ageing, i.e., the fact that the age of every water particle increases at the same pace as time progresses. Here, a simplified version thereof was used because the focus is set on steady-state and depth-averaged variables. In addition, water is regarded as a passive tracer (i.e., no reactions are to be considered). Accordingly, the water age distribution function obeys:

$$\frac{\partial(Hc)}{\partial\tau} = -\nabla\cdot(Hc\mathbf{u} - H\mathbf{K}\cdot\nabla c) \tag{1}$$

where $\mathbf{K}(\mathbf{x})$ denotes the diffusivity tensor, which must be symmetric and positive definite. Equation (1) is of a parabolic nature. Formally, Equation (1) is similar to an "evolution" equation, in which the independent variable τ plays a role equivalent to that of time in a classical advection–diffusion equation. A detailed derivation of this equation, as well as some of the mathematical properties of its solution, may be found in Deleersnijder and Dewals [35].

As pointed out above, the age is defined as the time elapsed since entering the shallow reservoir under study. Therefore, in the domain interior Ω, no water particle has zero age, yielding the "initial condition" (i.e., for $\tau = 0$):

$$c(\mathbf{x}, 0) = 0 \tag{2}$$

The age of a water particle is set to zero at the moment it enters the domain through inlet Γ^{in}. The incoming-flux condition reads [36]:

$$\left[(Hc\mathbf{u} - H\mathbf{K}\cdot\nabla c)\cdot\mathbf{n}\right]_{\mathbf{x}\in\Gamma^{in}} = \left[H(\mathbf{u}\cdot\mathbf{n})\delta(\tau - 0)\right]_{\mathbf{x}\in\Gamma^{in}} \tag{3}$$

where δ is the Dirac delta function. On the inlet, not all the water particles have zero age due to diffusion. A tailgate is placed at the downstream end of the outlet channel, so that the fluxes there are assumed to be of an advective nature. This leads to the boundary condition:

$$\left[(H\mathbf{K}\cdot\nabla c)\cdot\mathbf{n}\right]_{\mathbf{x}\in\Gamma^{out}} = 0 \tag{4}$$

On the impermeable boundary, a no-flux condition is prescribed,

$$\left[(HC\mathbf{u} - H\mathbf{K}\cdot\nabla c)\cdot\mathbf{n}\right]_{\mathbf{x}\in\Gamma^{imp}} = 0 \tag{5}$$

which, given the definition of the impermeable boundary, reduces to this simpler formulation:

$$\left[(-H\mathbf{K}\cdot\nabla c)\cdot\mathbf{n}\right]_{\mathbf{x}\in\Gamma^{imp}} = 0 \tag{6}$$

2.4. Mathematical Properties of the Mean Water Age

The partial differential problem constituted, of Equation (1) and auxiliary conditions (2)–(4) and (6), ensures a number of key properties of the age distribution function $c\,(\mathbf{x},\tau)$, namely $c\,(\mathbf{x},\tau)$ is non negative, $\tau\,c\,(\mathbf{x},\tau)$ tends to zero in the limit $\tau\to\infty$, and $c\,(\mathbf{x},\tau)$ satisfies the following integral constraint expressing that the concentration of water $C(\mathbf{x})$ is equal to unity at any location in the domain [28]:

$$C(\mathbf{x}) = \int_0^\infty c(\mathbf{x},\tau)\,d\tau = 1 \tag{7}$$

In accordance with Delhez [28], the mean age of the water at location $\mathbf{x}$ is defined as the first-order moment of the water age distribution function:

$$a(\mathbf{x}) = \int_0^\infty \tau\,c(\mathbf{x},\tau)\,d\tau \tag{8}$$

This expresses the so-called age-averaging hypothesis [34].

A steady-state equation governing the mean age can be derived by taking the first-order moment of Equations (1)–(4) and (6). However, in this research, this equation was not solved directly, but the mean age was derived from the computation of the age distribution function $c\,(\mathbf{x},\tau)$. The obtained values of $a(\mathbf{x})$ exhibit the following properties [35]:

- The solution is unique;
- $a(\mathbf{x})$ is non-negative;
- The value of the mean age on the incoming boundary is not zero, unless diffusion is zero. Indeed, on Γ^{in}, there is a mixture of water particles that are entering the domain and particles that have been moving for some time in the domain and were brought back to the incoming boundary by diffusion;
- The maximum of the mean age is not necessarily on the outgoing boundary Γ^{out}, but it may be located inside the domain of interest. This contrasts with a one-dimensional setting, in which the maximum of mean age occurs at the outgoing boundary because there is only one path from the reservoir inlet to the outlet;
- On the outgoing boundary Γ^{out}, the average value of the mean age satisfies:

$$\overline{\xi} = \tfrac{1}{S^{out}}\int_{\Gamma^{out}} H(\mathbf{x})\,a(\mathbf{x})\,d\Gamma^{out} = \tfrac{V}{Q} - \int_{\Gamma^{out}} H(\mathbf{x})\left[a(\mathbf{x}) - \overline{\xi}\right]\mathbf{u}\cdot\mathbf{n}\,d\Gamma^{out},$$

$$\text{with}\quad S^{out} = \int_{\Gamma^{out}} H(\mathbf{x})\,d\Gamma^{out}; \quad V = \int_\Omega H(\mathbf{x})\,d\Omega; \quad Q = \int_{\Gamma^{out}} H(\mathbf{x})\,\mathbf{u}\cdot\mathbf{n}\,d\Gamma^{out}. \tag{9}$$

where V and Q denote the volume of the water contained in the reservoir and the volumetric flow rate entering (or leaving) it, respectively. This underscores the importance of the ratio V/Q, which may be regarded as the order of magnitude of the mean age on the outgoing boundary (also called transit time). In a one-dimensional setting in a steady state, the mean age at the outgoing boundary is equal to V/Q irrespective of the diffusivity and cross-sectional area profiles.

2.5. Computational Procedure

From a computational point of view, obtaining the age distribution function $c(\mathbf{x},\tau)$ from the partial differential problem (5)–(9) is unlikely to be the optimal method due to the Dirac delta impulse prescribed as incoming boundary condition. Therefore, as proposed earlier [37], an alternate approach was set up here. In line with the theory of linear system dynamics, stating that the impulse response can be evaluated as the derivative of the step response, i.e., by prescribing a step function as inflowing boundary condition instead of the Dirac delta impulse. The step response is of course much easier to compute numerically than the impulse response.

As such, two steps were followed to evaluate the age distribution function. First, the step response, $b(\mathbf{x}, \tau)$, was obtained by solving the following partial differential problem, which is akin to a standard advection–diffusion problem (in which τ plays a role equivalent to that of the time):

$$\frac{\partial(Hb)}{\partial \tau} = -\nabla\cdot(Hb\mathbf{u} - H\,\mathbf{K}\cdot\nabla b) \tag{10}$$

$$b(\mathbf{x}, 0) = 0 \tag{11}$$

with

$$\left[(Hb\mathbf{u} - H\,\mathbf{K}\cdot\nabla b)\cdot\mathbf{n}\right]_{\mathbf{x}\in\Gamma^{in}} = \left[H\,\mathbf{u}\cdot\mathbf{n}\right]_{\mathbf{x}\in\Gamma^{in}} \tag{12}$$

$$\left[(H\,\mathbf{K}\cdot\nabla b)\cdot\mathbf{n}\right]_{\mathbf{x}\in\Gamma^{out}\cup\Gamma^{imp}} = 0 \tag{13}$$

Here, the incoming flux is independent of τ instead of a Dirac delta function in problem (5)–(9). In the computations, an isotropic eddy diffusivity ($\mathbf{K} = \kappa\,\mathbf{I}$) was used, with the value of κ taken as equal to the eddy viscosity.

In a second step, the age distribution function $c\,(\mathbf{x}, \tau)$ was computed by evaluating the derivative of $b\,(\mathbf{x}, \tau)$:

$$c(\mathbf{x}, \tau) = \frac{\partial}{\partial \tau}b(\mathbf{x}, \tau) \tag{14}$$

In line with Equation (7), the cumulative distribution function $b\,(\mathbf{x}, \tau)$ tends to unity for $\tau \to \infty$. Finally, once the age distribution function is computed, the mean water age $a(\mathbf{x})$ can be evaluated thanks to Equation (8). Note that an alternate approach consists in simulating the problem in Laplace space [38].

3. Results

3.1. Model Verification: Prismatic Channel

For the purpose of model verification, the age profile in a one-dimensional setting, for which an analytical solution is available, was first computed. A prismatic rectangular channel of length $L = 1$ m and unit width is considered, in which the section-averaged flow velocity is $U = 1$ m/s. In a steady state, the value of U is uniform along the reservoir. A uniform along-flow diffusivity K is assumed. The age is

$$a(x) = \frac{L}{U}\left(\frac{x}{L} + \frac{1 - e^{-Pe(1-x/L)}}{Pe}\right) = \frac{V}{Q}\left(\frac{x}{L} + \frac{1 - e^{-Pe(1-x/L)}}{Pe}\right) \tag{15}$$

where $Pe = U\,L/K$ is the Peclet number.

In Figure 3a, this analytical solution is compared to the water age computed from Equations (10)–(14) and Equation (8). By varying the value of the diffusivity K, this comparison was performed for Peclet numbers ranging over nearly three orders of magnitude (from 0.3 up to 100). The computations and the analytical solution match very well. The results for smaller values of K (i.e., Pe lower than 100) cannot be distinguished visually from those obtained with $Pe = 100$. Hence they are not represented.

Note that the mean water age is indeed different from zero at the inflow boundary, as emphasized in Section 2.4. Moreover, the smaller the relative importance of diffusion is (i.e., the larger the Peclet number), the smaller the mean water age on the inflowing boundary will be.

Another remarkable feature of the results is that they all converge towards the same value, $a(L) = L/U = V/Q$, at the exit side, irrespective of the value of the Peclet number. However, a benefit of applying our model is that it gives access not only to the mean water age $a(x)$, but also to the age distribution function $c(x, \tau)$, for which an analytical solution written in closed form is not available. Figure 3b reveals that, although the mean age $a(L)$ is the same for various values of Peclet number, the age distribution functions differ substantially—the stronger the diffusion, the more the distribution is skewed towards large values of age, but nonetheless smaller values of age are more frequent.

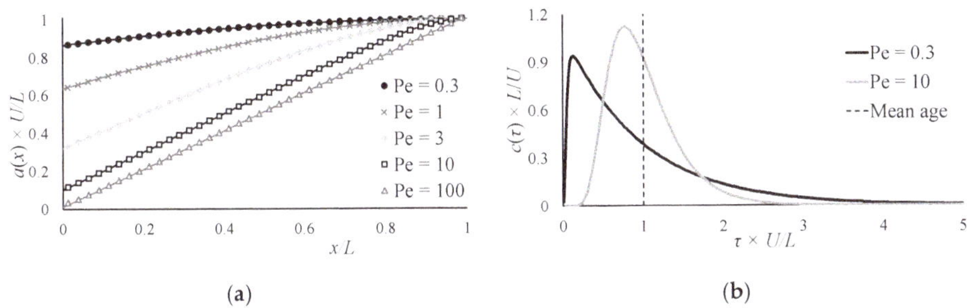

Figure 3. (a) Computed (symbols) and analytical (solid lines) profiles of water mean age $a(x)$ in a one-dimensional prismatic channel, and (b) corresponding age distribution functions.

3.2. Distribution of Water Age at Reservoir Outlet

The two-step procedure described in Section 2.5 was applied to all the configurations listed in Table 2. Examples of step responses computed for Test 5 and Test 6 are shown in Figures S3 and S4 in the Supplemental Material. Here, the water age distribution on the centerline of the reservoir outlet section is first examined. For all configurations, the water age distribution and cumulative distribution at the reservoir outlet are shown in logarithmic scale in Figure 4. The following observations can be made:

- In the nearly-plug-flow configurations (Tests 8, 9 and 10), the water age distribution at the outlet is unimodal, but strongly skewed towards the higher values of τ;
- Looking sequentially at the results obtained for Tests 7, 6 and 5, all corresponding to configurations with a reattached jet, it appears that the water age distribution at the outlet shifts gradually from a unimodal to a bimodal distribution;
- In all cases with a detached jet, the water age distribution at the outlet is bimodal (Tests 4, 3 and 2), or even multimodal in the case of Test 1.

The occurrence of bi- and multimodal distributions is attributed to the large flow recirculations, which develop on both sides of the main jet in the case of a reattached jet (Tests 6 and 5) and, to an even greater extent, in the case of detached jets. The water reaching the reservoir outlet contains particles which have followed distinct pathways: a portion of them followed the main jet (fast route), whereas others remained trapped for some time in the large flow recirculations (slow route). Since the computed age distributions are finely discretized as a function of τ, we discard the hypothesis that the multimodal nature of the computed distributions is a bias from numerical computation. Overall, these results show that a reduced number of indicators (e.g., mean age and standard deviation) would not be sufficient to reflect the complexity and variety of the actual water age distributions.

Figure 4 also displays the mode, the median and the mean of water age. These quantities are compared in scatter plots given in Figures S5 and S6 in the Supplemental Material. In all configurations, the mean age exceeds the median age, which in turn exceeds the mode of the age distribution. The median age is generally 5% to 35% larger than the mode of the distribution. The mode and the median age are the closest to each other in the configurations with a detached jet (Tests 1 and 2), with differences not exceeding 5% to 10%.

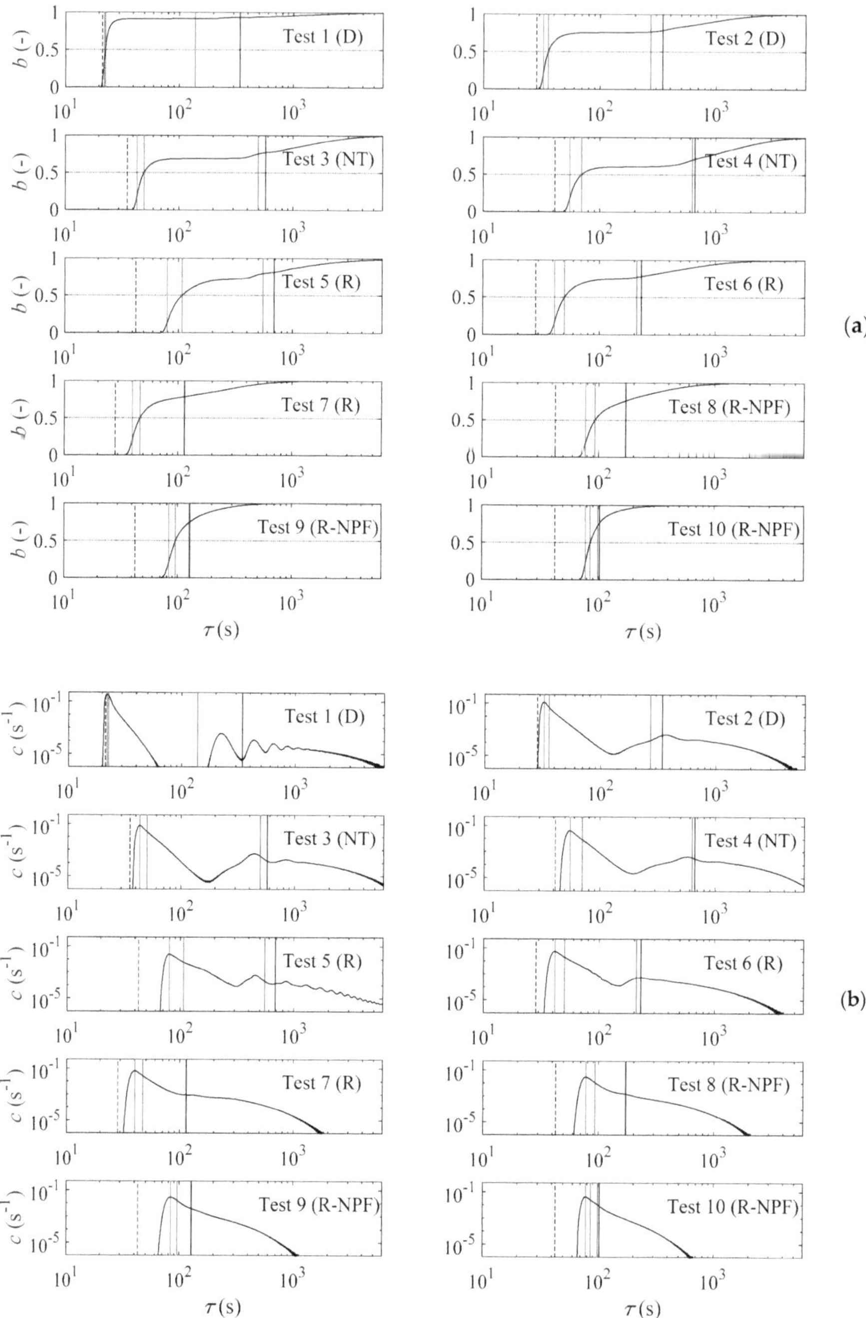

Figure 4. Water age distribution at the reservoir outlet in all configurations: (**a**) cumulated distributions b and (**b**) distributions c. Abbreviations "D", "NT", "R" and "R-NPF" refer respectively to detached jet, near-transition, reattached and reattached with nearly-plug-flow. The blue (–), green (–) and red (–) vertical lines (–) show the mode, median and mean of the age distribution, respectively. The plain (–) and dashed (–) black vertical lines represent the characteristic times V/Q and $L/(Q/S^{in})$, respectively.

The mean age differs considerably from the mode and the median age. Indeed, the mean ages are found to be up to ten times higher than the corresponding median values. The tests with particularly large flow recirculations lead to the largest differences between mean and median age. This is attributed to the particularly high age of the water particles reaching the outlet after staying trapped for some time in those recirculations. This is also perfectly consistent with the considerable positive skewness of the corresponding water age distributions. In contrast, in the nearly-plug-flow configurations (Tests 8, 9 and 10), the mean age exceeds the median age by no more than a factor of two.

Finally, Figure 4 also displays two characteristic times which are straightforward to derive from the reservoir geometry and the boundary conditions:

- V/Q, which would correspond to the transit time through the reservoir in the case of a perfect plug flow all over the reservoir;
- $L/(Q/S^{in})$, would be equal to the transit time through the reservoir if the jet at the inlet was not diffused at all in the crosswise direction.

The truth is of course in-between these two extremes. Indeed, Figure 4 reveals that the characteristic times V/Q and $L/(Q/S^{in})$ tend to provide an envelope for the mode, median and mean ages. In the case of nearly-plug-flow, the characteristic time V/Q is found to approximate remarkably well the mean water age, with overestimations of no more than 5%. Furthermore, in one case of a reattached jet in a particularly elongated reservoir (Test 7), V/Q matches the mean age with a difference of less than 1%. In all other configurations, V/Q remains a reasonable approximation for the mean age, with overestimations of no more than 22%. Only in the case corresponding to a detached jet in a relatively wide basin (Tests 1), V/Q overestimates the mean age by a factor 2.5.

The second characteristic time, $L/(Q/S^{in})$, is generally between 1.5 and 2.5 times smaller than the median age and the mode of the age distribution. Only in the configurations with a detached jet (Tests 1 and 2), $L/(Q/S^{in})$ approximates well the mode and the median age, with an underestimation of the order of 10% to 20% only. Note that only for Test 1, a small portion of the water particles show a transit time even smaller than the characteristic value $L/(Q/S^{in})$. This results from the velocity profile over the jet cross-section. Indeed, the central part of the jet at the inlet has a velocity higher than Q/S^{in}. In all other cases, the characteristic time $L/(Q/S^{in})$ substantially underestimates the actual mode, median and mean values of the age distribution.

Consequently, $L/(Q/S^{in})$ appears as a useful time scale for the configurations with a detached jet flowing directly from the inlet to the reservoir outlet (Tests 1 and 2), whereas the characteristic time scale V/Q approximates well the mean age at the outlet in the case of elongated reservoirs with nearly-plug-flow conditions (Tests 8, 9 and 10) or close to (Tests 6 and 7).

3.3. Distribution of Water Age within the Reservoirs

Here, the water age distributions at specific locations within the reservoir are analyzed. A configuration with a reattached jet, namely Test 5, is considered. It shows a particularly complex flow pattern. The corresponding water age distributions are displayed in Figure 5. Similar results for Test 6 are displayed in Figure S7 in the Supplemental Material.

In both configurations (Tests 5 and 6), Point B is located on the centerline of the main jet, at a distance from the reservoir inlet equal to 3.5 times the inlet width. The water age distribution at this location is fairly simple, unimodal and with a slight skewness towards the greater water ages. Unsurprisingly, the mean, median and mode for age are all very close, as the water particles at Point B have all followed a similar route. Indeed, these three quantities differ from each other by no more than 1% to 3% in Test 5, and 2% to 5% in Test 6. This contrasts strongly with the results obtained at the reservoir outlet (Section 3.2) and at other locations within the reservoir, which all show much "wider" age distributions.

Points E and F are located in the smaller recirculation zone on the right side of the main jet. These distributions are multimodal, reflecting the coexistence in these recirculations of water particles that have followed distinct paths, such as either coming "directly" from the reservoir inlet or having been temporarily "trapped" in the recirculation zone.

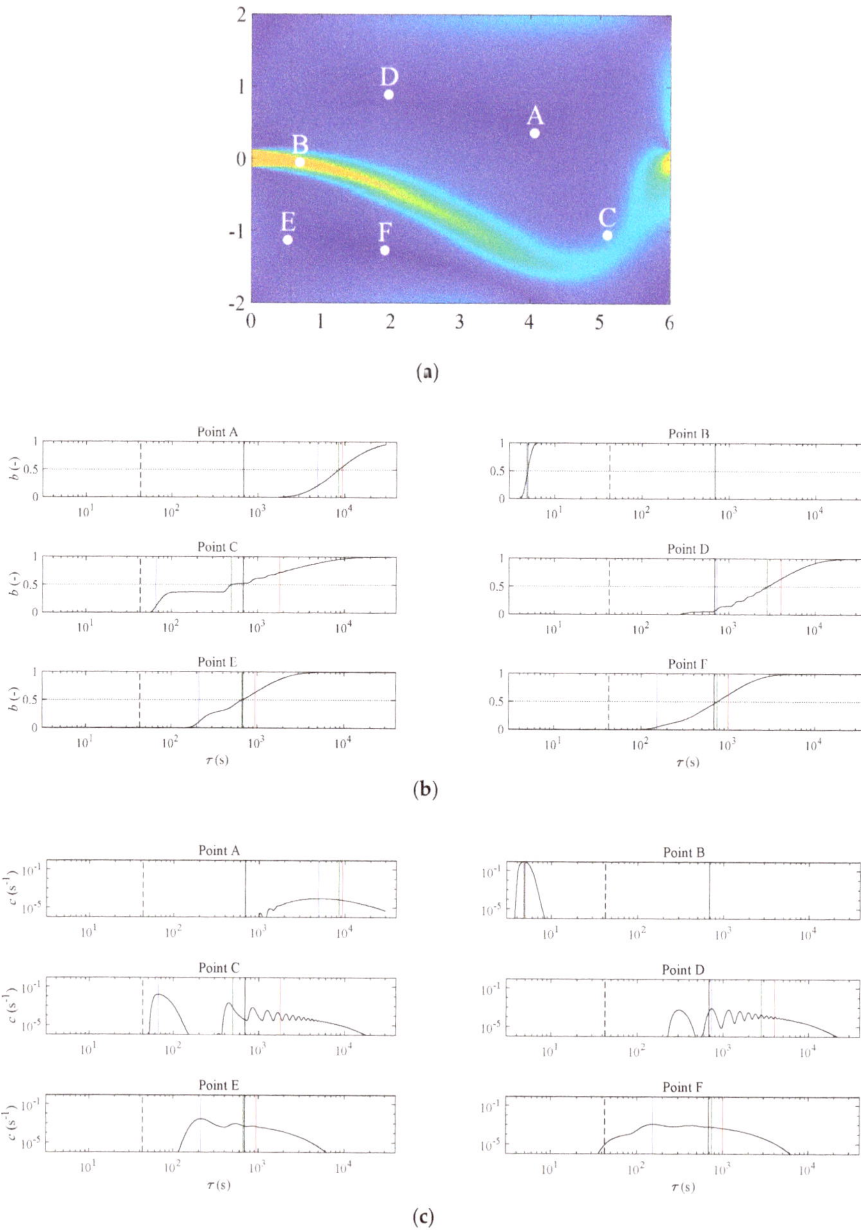

Figure 5. Water age distribution at six locations in the reservoir for Test 5: (**a**) considered locations A, B, C, D, E and F, (**b**) cumulated distributions b and (**c**) water age distributions c. The blue (–), green (–) and red (–) vertical lines show the mode, median and mean of the age distribution, respectively. The plain (–) and dashed (–) black vertical lines represent the characteristic times V/Q and $L/(Q/S^{in})$, respectively.

The results are to a great extent similar at Point D, which is located in the larger recirculation on the left side of the jet. In the case of Test 5, the water age distribution at Point D shows multiple peaks.

Points A are situated close to the center of the larger recirculation on the left side of the jet. The corresponding distributions are also multimodal, with a particularly large "spreading", revealing the coexistence of water particles characterized by ages ranging over at least two orders of magnitude. Note that, despite the long computation time dedicated to the simulation, the water age distribution captured by the computational model at Point A in the case of Test 5 is not complete. However, this does not undermine the conclusions, since the key features of the distribution are well captured.

Finally, Point C is located along the jet, at a distance of about seven outlet widths from the reservoir outlet. The water age distribution at this location approaches the water age distribution at the reservoir outlet.

3.4. Mean Age in the Reservoirs

Figure 6 represents the spatial distribution of the mean water age a, scaled by the characteristic time V/Q introduced in Section 3.2. This non-dimensional mean water age takes maximum values in the large flow recirculations, irrespective of whether the main jet is reattached (e.g., Test 5) or not (e.g., Test 1).

The maximum values of the non-dimensional water age tend not to exceed a value close to four, except in Test 5. In this case, the non-dimensional mean water age reaches about 20 in the core of the large flow recirculation.

For each test, Figure 7 shows a histogram of the computed non-dimensional mean water ages $a/(V/Q)$ across the 10 considered reservoirs. The displayed range of $a/(V/Q)$ does not extend beyond 10 because, although existing for Test 5, the number of values above 10 is so small that they are not visible in the graphs.

In the most elongated reservoir (Test 10), the values $a/(V/Q)$ are mostly below unity, in accordance with Deleersnijder and Dewals [35]. In the other tests corresponding to nearly-plug-flow (Tests 8 and 9), the maximum values of $a/(V/Q)$ remain between 1 and 3.

In the cases with a reattached jet (Tests 5, 6 and 7), the coexistence of fast and slow routes is clearly reflected in the histograms of $a/(V/Q)$. Indeed, two groups of values of $a/(V/Q)$ can be distinguished in the histograms: one corresponding to $a/(V/Q) > 2$ (in the larger flow recirculations) and another one to $a/(V/Q) < 2$ (main jet and smaller recirculation), consistently with Figure 6e–g.

The other tests also show a distinctive bimodal pattern in the histograms of $a/(V/Q)$. The most populated mode shows values for $a/(V/Q)$ of the order of 3 to 5, and corresponds to the larger flow recirculations on either sides of the main jet. Besides this, a smaller peak can be seen for very small values of $a/(V/Q)$, corresponding to the main jet whose characteristic time is closer to $L/(Q/S^{in})$ than V/Q.

4. Conclusions

In this research, the complexity of the distribution function of water age within shallow reservoirs of simple rectangular geometry is highlighted. These water age distribution functions are often multimodal as a result of the coexistence of fast and slow routes within the reservoir. The existence of distinct routes results from various large-scale turbulent structures developing in such reservoirs. As a result, a limited number of statistical values (e.g., mean and standard deviation of age) would fail to reflect the complexity of the actual water age distributions in such reservoirs.

To evaluate the water age distribution, the "step response" was first computed and, next, the result was derived to obtain the age distribution function. This procedure was applied to 10 configurations of rectangular shallow reservoirs, for which steady flow fields, previously computed with a validated model, were readily available. They involve a variety of flow patterns, including detached or reattached jets, large flow recirculations and nearly-plug-flow.

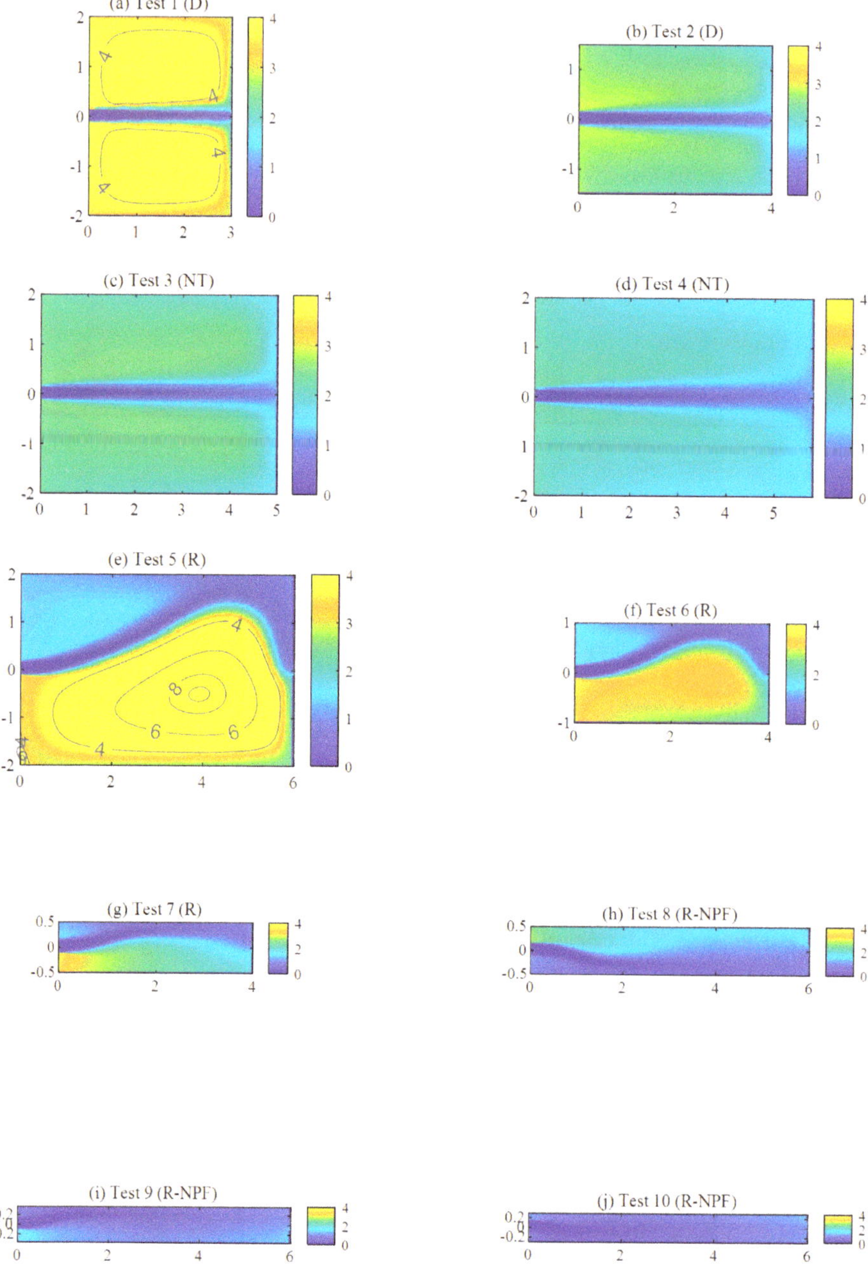

Figure 6. Computed mean age a, scaled by the characteristic time V/Q. Contours are represented only for values of $a/(V/Q) \geq 4$, i.e., beyond the range of the color scale. Subfigures (**a**–**j**) correspond to the 10 reservoir configurations.

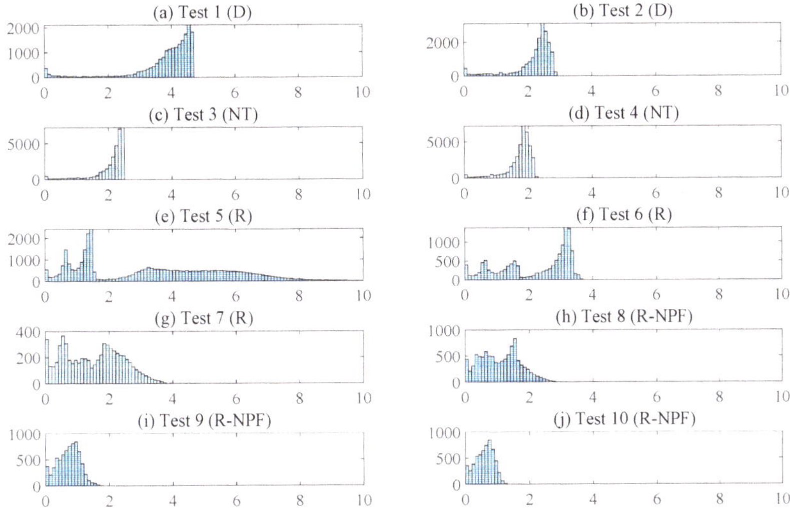

Figure 7. Histograms of the non-dimensional mean water age $a/(V/Q)$ across the 10 considered reservoirs. Subfigures (**a–j**) correspond to the 10 reservoir configurations.

Two characteristic time scales are worth considering V/Q (with V the reservoir volume and Q the inflow discharge) and $L/(Q/S^{in})$ (with L the reservoir length and S^{in} the inlet cross-section), and their relative merits are discussed as a function of the reservoir dimensions. The former time scale provides an estimate of practical relevance for the mean water age at the outlets of elongated reservoirs, whereas the latter is a useful proxy for the mode and the median age at the outlets of relatively short reservoirs. The spatial distribution of the mean age within the reservoir is shown to vary strongly with the reservoir's geometry, and these variations are attributed to characteristics of the flow fields.

The insights provided by this numerical research may prove important for guiding the optimal geometric and hydraulic design of reservoirs used for water storage, for trapping sediments or as service reservoirs. Indeed, the prediction of water age and its distribution enables the pinpointing of issues such as preferential sedimentation or uneven disinfection time in the case of service reservoirs. Specifically, knowing the water age and its distribution enables a comparison of the corresponding time scales to other time scales of relevance, such as the characteristic time of sedimentation (as a function of the settling velocity of the sediment particles of interest), or the time during which a disinfectant remains active in a service reservoir [11]. Comparing these time scales for a range of design options enables assessing which are the most favorable geometric and hydraulic design choices. Maintenance costs may also be cut down thanks to a deeper understanding of water age distribution in such reservoirs.

Future research should focus on computing not only the water age distribution, but also the age of sediments transported either as bed load or as suspended load [39,40]. Comparing the results of our Eulerian modeling framework with those of a Lagrangian approach [41,42], including stochastic terms to represent the impact of turbulent diffusion [43–45], would lead to valuable insights into the relative performance of each technique. Moreover, our diagnostic tools should be extended to accommodate 3D flow fields in shallow reservoirs [8,20,46], since three-dimensional turbulent structures and secondary currents are likely to alter the distribution of water age. Finally, examining the performances of alternate turbulence closures would be another valuable piece of further research.

Supplementary Materials: The following are available online at http://www.mdpi.com/2073-4441/12/10/2819/s1, Figure S1: Flow fields computed by Camnasio et al. (2014). The colour scale represents the flow velocity magnitude (m/s), Figure S2: Fields of eddy viscosity (m^2s^{-1}) computed by Camnasio et al. (2014), Figure S3: Step response

$b(\mathbf{x},\tau)$ computed for various values of τ in the reservoir corresponding to Test 5, Figure S4: Step response $b(\mathbf{x}, \tau)$ computed for various values of τ in the reservoir corresponding to Test 6, Figure S5: Comparison of age statistics and characteristic times (linear scale), Figure S6: Comparison of age statistics and characteristic times (log scale), Figure S7: Water age distribution at six locations in the reservoir for Test 7: (**a**) considered locations A, B, C, D, E and F, (**b**) cumulated distributions b and (**c**) water age distributions c. The blue (–), green ( ) and red (–) vertical lines show the mode, median and mean of the age distribution, respectively. The plain (–) and dashed (–) black vertical lines represent the characteristic times V/Q and $L/(Q/S^{in})$, respectively.

Author Contributions: Conceptualization, B.D., P.A. and E.D. (Eric Deleersnijder); Formal analysis, B.D., P.A., S.E., E.D. (Eric Delhez) and E.D. (Eric Deleersnijder); Investigation, M.B. and T.A.; Methodology, B.D., P.A. and E.D. (Eric Deleersnijder); Software, P.A. and M.B.; Validation, B.D., P.A., M.B. and E.D. (Eric Deleersnijder); Visualization, B.D.; Writing—original draft, B.D. and E.D. (Eric Deleersnijder); Writing—review and editing, B.D., P.A., S.E., M.P., E.D. (Eric Delhez) and E.D. (Eric Deleersnijder). All authors have read and agreed to the published version of the manuscript.

Funding: This research received no external funding.

Acknowledgments: The last author is indebted to Jean Van Schaftingen for providing a useful piece of information about the theory of partial differential equations. The authors also acknowledge the support of Mohamad Rammal for the numerical resolution of section-averaged models. E.J.M.D. and E.D. are honorary research associates with the Belgian Fund for Scientific Research (F.R.S.-FNRS). Part of the present research work has been done when E.D. was a part-time professor with the Delft Institute of Applied Mathematics, Delft University of Technology, The Netherlands.

Conflicts of Interest: The authors declare no conflict of interest.

References

1. Brink, I.C.; Kamish, W. Associations between stormwater retention pond parameters and pollutant (Suspended solids and metals) removal efficiencies. *Water SA* **2018**, *44*, 45–53. [CrossRef]

2. Dominic, J.A.; Aris, A.Z.; Sulaiman, W.N.A.; Tahir, W.Z.W.M. Discriminant analysis for the prediction of sand mass distribution in an urban stormwater holding pond using simulated depth average flow velocity data. *Environ. Monit. Assess.* **2016**, *188*, 1–15. [CrossRef]

3. Guzman, C.B.; Cohen, S.; Xavier, M.; Swingle, T.; Qiu, W.; Nepf, H. Island topographies to reduce short-circuiting in stormwater detention ponds and treatment wetlands. *Ecol. Eng.* **2018**, *117*, 182–193. [CrossRef]

4. MoayeriKashani, M.; Hin, L.S.; Ibrahim, S. Experimental investigation of fine sediment deposition using particle image velocimetry. *Environ. Earth Sci.* **2017**, *76*, 655. [CrossRef]

5. Sebastian, C.; Becouze-Lareure, C.; Lipeme Kouyi, G.; Barraud, S. Event-based quantification of emerging pollutant removal for an open stormwater retention basin—Loads, efficiency and importance of uncertainties. *Water Res.* **2014**, *72*, 239–250. [CrossRef] [PubMed]

6. Sonnenwald, F.; Guymer, I.; Stovin, V. Computational fluid dynamics modelling of residence times in vegetated stormwater ponds. *Proc. Inst. Civ. Eng. Water Manag.* **2018**, *171*, 76–86. [CrossRef]

7. Stovin, V.R.; Saul, A.J. Computational fluid dynamics and the design of sewage storage chambers. *Water Environ. J.* **2000**, *14*, 103–110. [CrossRef]

8. Isenmann, G.; Dufresne, M.; Vazquez, J.; Mose, R. Bed turbulent kinetic energy boundary conditions for trapping efficiency and spatial distribution of sediments in basins. *Water Sci. Technol.* **2017**, *76*, 2032–2043. [CrossRef]

9. Liu, X.; Xue, H.; Hua, Z.; Yao, Q.; Hu, J. Inverse calculation model for optimal design of rectangular sedimentation tanks. *J. Environ. Eng.* **2013**, *139*, 455–459. [CrossRef]

10. Tarpagkou, R.; Pantokratoras, A. CFD methodology for sedimentation tanks: The effect of secondary phase on fluid phase using DPM coupled calculations. *Appl. Math. Model.* **2013**, *37*, 3478–3494. [CrossRef]

11. Zhang, J.-M.; Lee, H.P.; Khoo, B.C.; Peng, K.Q.; Zhong, L.; Kang, C.-W.; Ba, T. Shape effect on mixing and age distributions in service reservoirs. *J. Am. Water Works Assoc.* **2014**, *106*, E481–E491. [CrossRef]

12. Oca, J.; Masalo, I. Design criteria for rotating flow cells in rectangular aquaculture tanks. *Aquac. Eng.* **2007**, *36*, 36–44. [CrossRef]

13. Camnasio, E.; Orsi, E.; Schleiss, A. Experimental study of velocity fields in rectangular shallow reservoirs. *J. Hydraul. Res.* **2011**, *49*, 352–358. [CrossRef]

14. Dufresne, M.; Dewals, B.J.; Erpicum, S.; Archambeau, P.; Pirotton, M. Classification of flow patterns in rectangular shallow reservoirs. *J. Hydraul. Res.* **2010**, *48*, 197–204. [CrossRef]

15. Kantoush, S.A.; De Cesare, G.; Boillat, J.L.; Schleiss, A.J. Flow field investigation in a rectangular shallow reservoir using UVP, LSPIV and numerical modelling. *Flow Meas. Instrum.* **2008**, *19*, 139–144. [CrossRef]

16. Peltier, Y.; Erpicum, S.; Archambeau, P.; Pirotton, M.; Dewals, B. Experimental investigation of meandering jets in shallow reservoirs. *Environ. Fluid Mech.* **2014**, *14*, 699–710. [CrossRef]

17. Choufi, L.; Kettab, A.; Schleiss, A.J. Bed roughness effect on flow field in rectangular shallow reservoir. [Effet de la rugosité du fond d'un réservoir rectangulaire à faible profondeur sur le champ d'écoulement]. *Houille Blanche* **2014**, *5*, 83–92.

18. Camnasio, E.; Erpicum, S.; Archambeau, P.; Pirotton, M.; Dewals, B. Prediction of mean and turbulent kinetic energy in rectangular shallow reservoirs. *Eng. Appl. Comput. Fluid Mech.* **2014**, *8*, 586–597. [CrossRef]

19. Camnasio, E.; Erpicum, S.; Orsi, E.; Pirotton, M.; Schleiss, A.J.; Dewals, B. Coupling between flow and sediment deposition in rectangular shallow reservoirs. *J. Hydraul. Res.* **2013**, *51*, 535–547. [CrossRef]

20. Esmaeili, T.; Sumi, T.; Kantoush, S.A.; Haun, S.; Rüther, N. Three-dimensional numerical modelling of flow field in shallow reservoirs. *Proc. Inst. Civ. Eng. Water Manag.* **2016**, *169*, 229–244. [CrossRef]

21. Peltier, Y.; Erpicum, S.; Archambeau, P.; Pirotton, M.; Dewals, B. Can meandering flows in shallow rectangular reservoirs be modeled with the 2D shallow water equations? *J. Hydraul. Eng.* **2015**, *141*, 04015008. [CrossRef]

22. Peng, Y.; Zhou, J.G.; Burrows, R. Modeling free-surface flow in rectangular shallow basins by using lattice boltzmann method. *J. Hydraul. Eng.* **2012**, *137*, 1680–1685. [CrossRef]

23. Schleiss, A.J.; Franca, M.J.; Juez, C.; De Cesare, G. Reservoir sedimentation. *J. Hydraul. Res.* **2016**, *54*, 595–614. [CrossRef]

24. Bolin, B.; Rodhe, H. A note on the concepts of age distribution and transit time in natural reservoirs. *Tellus* **1973**, *25*, 58–62. [CrossRef]

25. Monsen, N.E.; Cloern, J.E.; Lucas, L.V.; Monismith, S.G. A comment on the use of flushing time, residence time, and age as transport time scales. *Limnol. Oceanogr.* **2002**, *47*, 1545–1553. [CrossRef]

26. Takeoka, H. Fundamental concepts of exchange and transport time scales in a coastal sea. *Cont. Shelf Res.* **1984**, *3*, 311–326. [CrossRef]

27. Persson, J. The hydraulic performance of ponds of various layouts. *Urban Water* **2000**, *2*, 243–250. [CrossRef]

28. Delhez, E.J.M.; Campin, J.-M.; Hirst, A.C.; Deleersnijder, E. Toward a general theory of the age in ocean modelling. *Ocean Model.* **1999**, *1*, 17–27. [CrossRef]

29. Goltsman, A.; Saushin, I. Flow pattern of double-cavity flow at high Reynolds number. *Phys. Fluids* **2019**, *31*, 065101. [CrossRef]

30. Peltier, Y.; Erpicum, S.; Archambeau, P.; Pirotton, M.; Dewals, B. Meandering jets in shallow rectangular reservoirs: POD analysis and identification of coherent structures. *Exp. Fluids* **2014**, *55*, 1740. [CrossRef]

31. Chu, V.H.; Liu, F.; Altai, W. Friction and confinement effects on a shallow recirculating flow. *J. Environ. Eng. Sci.* **2004**, *3*, 463–475. [CrossRef]

32. Dufresne, M.; Dewals, B.J.; Erpicum, S.; Archambeau, P.; Pirotton, M. Numerical investigation of flow patterns in rectangular shallow reservoirs. *Eng. Appl. Comput. Fluid Mech.* **2011**, *5*, 247–258. [CrossRef]

33. Roache, P.J. Perspective: A method for uniform reporting of grid refinement studies. *J. Fluids Eng. Trans. Asme* **1994**, *116*, 405–413. [CrossRef]

34. Deleersnijder, E.; Campin, J.-M.; Delhez, E.J.M. The concept of age in marine modelling I. *Theory and preliminary model results. J. Mar. Syst.* **2001**, *28*, 229–267. [CrossRef]

35. Deleersnijder, E.; Dewals, B. *Mathematical Properties of the Position-Dependent, Steady-State Water Age in a Shallow Reservoir*; Working Note; Université Catholique de Louvain: Louvain-la-Neuve, Belgium, 2020; p. 18. Available online: http://hdl.handle.net/2078.1/230041 (accessed on 4 October 2020).

36. Deleersnijder, E.; Draoui, I.; Lambrechts, J.; Legat, V.; Mouchet, A. Consistent boundary conditions for age calculations. *Water* **2020**, *12*, 1274. [CrossRef]

37. Varni, M.; Carrera, J. Simulation of groundwater age distributions. *Water Resour. Res.* **1998**, *34*, 3271–3281. [CrossRef]

38. Cornaton, F.; Perrochet, P. Groundwater age, life expectancy and transit time distributions in advective-dispersive systems; 2. Reservoir theory for sub-drainage basins. *Adv. Water Resour.* **2006**, *29*, 1292–1305. [CrossRef]

39. Ralston, D.K.; Geyer, W.R. Sediment transport time scales and trapping efficiency in a Tidal river. *J. Geophys. Res. Earth Surf.* **2017**, *122*, 2042–2063. [CrossRef]

40. Gong, W.; Shen, J. A model diagnostic study of age of river-borne sediment transport in the tidal York River Estuary. *Environ. Fluid Mech.* **2010**, *10*, 177–196. [CrossRef]

41. Liu, W.-C.; Chen, W.-B.; Hsu, M.-H. Using a three-dimensional particle-tracking model to estimate the residence time and age of water in a tidal estuary. *Comput. Geosci.* **2011**, *37*, 1148–1161. [CrossRef]

42. Arega, F.; Badr, A.W. Numerical age and residence-time mapping for a small tidal creek: Case study. *J. Waterw. Port Coast. Ocean Eng.* **2010**, *136*, 226–237. [CrossRef]

43. Spivakovskaya, D.; Heemink, A.W.; Milstein, G.N.; Schoenmakers, J.G.M. Simulation of the transport of particles in coastal waters using forward and reverse time diffusion. *Adv. Water Resour.* **2005**, *28*, 927–938. [CrossRef]

44. Scott, C.F. Particle tracking simulation of pollutant discharges. *J. Environ. Eng.* **1997**, *123*, 919–927. [CrossRef]

45. Heemink, A.W. Stochastic modelling of dispersion in shallow water. *Stoch. Hydrol. Hydraul.* **1990**, *4*, 161–174. [CrossRef]

46. Zahabi, H.; Torabi, M.; Alamatian, E.; Bahiraei, M.; Goodarzi, M. Effects of geometry and hydraulic characteristics of shallow reservoirs on sediment entrapment. *Water* **2018**, *10*, 1725. [CrossRef]

 *water*

Article

Water Residence Time in a Typical Tributary Bay of the Three Gorges Reservoir

Yao Cheng [1], Zheng Mu [1], Haiyan Wang [2], Fengxia Zhao [1], Yu Li [1] and Lei Lin [3,*]

[1] School of Water Conservancy and Hydroelectric Power, Hebei University of Engineering, Handan 056002, China

[2] National Marine Environmental Forecasting Center, Beijing 100081, China

[3] State Key Laboratory of Estuarine and Coastal Research, East China Normal University, Shanghai 200062, China

* Correspondence: llin@sklec.ecnu.edu.cn

Received: 13 June 2019; Accepted: 29 July 2019; Published: 31 July 2019

Abstract: Tributary bays of the Three Gorges Reservoir (TGR) are suffering from environmental problems, e.g., eutrophication and algae bloom, which could be related to the limited water exchange capacity of the tributary bays. To understand and quantify the water exchange capacity of a tributary bay, this study investigated the water residence time (RT) in a typical tributary bay of TGR, i.e., the Zhuyi Bay (ZB), using numerical simulation and the adjoint method to obtain the RT. The results show that RT of ZB with an annual mean of 16.7 days increases from the bay mouth to the bay top where the maximum can reach 50 days. There is a significant seasonal variation in RT, with higher RT (average 20 days) in spring and autumn and lower RT (average < 5 days) in the summer. The sensitivity experiments show that the TGR water level regulation has a strong influence on RT. The increase in the water level could increase RT of ZB to some extent. Density currents induced by the temperature difference between the mainstream and tributaries play an important role in the water exchange of ZB, while the impacts of the river discharges and winds on RT are insignificant.

Keywords: residence time; Three Gorges Reservoir; tributary bay; density current; water level regulation

---

1. Introduction

The fate of chemical and biological species in aquatic systems is determined by the combination of (passive) transport and species-specific transformations [1]. Processes controlling planktonic biomass and contaminant distributions in semi-enclosed bodies are linked to the water transport timescales. Therefore, water transport timescales are important indicators for analyzing and estimating pollution threats to the aquatic ecosystem [2–4]. The water transport timescales can be also adapted to quantify the water exchange capacity of semi-enclosed water systems and quantify the transport rate of contaminants that are taking place in both the dissolved and adsorbed phases [5,6].

There are several defined transport timescales, namely (1) the flushing time, (2) the age, (3) the residence time (RT), and (4) the exposure time. The flushing time is a bulk or integrative parameter that describes the general exchange characteristics of a waterbody without identifying the underlying physical processes. The age is defined to be the time elapsed since the particle under consideration left the region in which its age is prescribed to be zero. The RT of a water parcel in a control region ω is the time taken by that water parcel to leave ω for the first time. The exposure time which is actually a concept extension of RT is defined as the cumulative time spent by the particle in the control domain, irrespective of its possible excursions out of ω [7–11]. Among these, RT is a very helpful and popular concept widely used in hydro-ecological research. For instance, the export rate of nutrients was proved to be strongly negatively related to the RT [12,13]. Shifts in bacterioplankton community composition along the salinity gradient of the Parker River estuary and Plum Island Sound, in northeastern Massachusetts, were related to RT [14]. A linear relationship is established between the RT and phytoplankton biomass expressed as chlorophyll-a concentrations in coral reef lagoons [15]. The available nutrient supply for algae growth and bloom is determined not only by the nutrient loads, but also by the retention of nutrients, which is related to the RT of a system [12]. Moreover, the RT is a key parameter for the occurrence of algal blooms, as they require that phytoplankton cells remain in favorable conditions for long enough [16–18].

The construction of the Three Gorges Dam (TGD) is one of the most extensive anthropogenic impacts on surface water in China. As the third longest river in the world and the longest river in Asia, the Yangtze River, stretching from the Tibetan Plateau to eastern China, spans for a total length of 6300 km and drains an area of 1,800,000 km^2 [19]. Its annual flow is 951.3 km^3. The TGD, located at the end of the upper Yangtze River, is 185 m high. Construction began in 1998 and was completed in 2003. The TGR is currently one of the largest reservoirs in the world, with a capacity of 39.3 billion m^3 over a length of 663 km and an average width of 1.1 km [20,21].

As the largest hydropower project in the world, the Three Gorges Project has brought remarkable benefits, including flood control, electricity generation, and shipping capacity improvement. Meanwhile, the impacts of the Three Gorges Reservoir (TGR) on the ecosystem and environment have been widely discussed [22–26]. After the impoundment in 2003, the TGR was formed along the Yangtze River, starting from Chongqing to the dam site at Yichang. Approximately 40 tributaries were transformed into tributary bays and became a part of the TGR. The total area of these bays accounts for 1/3 of the whole surface area of the TGR. It has dramatically changed the aquatic ecosystem from a continuous lotic ecosystem to a huge reservoir, which exhibits complex hydrodynamic processes [27–30]. As a consequence, increasingly serious eutrophication and algal blooms usually occurred in spring and autumn in tributary bays in the TGR which were induced by multiple factors, including the ecological and hydrodynamic environment change [31–34]. The decline of ecological health and ecosystem services function for river basins has become a worrying environmental problem [35,36]. Given that the water RT is a good indicator quantifying the water exchange capacity, the magnitude of water RT and its variation could be helpful to understand the cause of the seasonal eutrophication and algal blooms in the tributary bays from the hydrodynamic point of view. However, the water RT and its spatiotemporal variations in tributary bays of the TGR were not well documented.

In this study, the water RT in one of the TGR tributary bays (Zhuyi Bay) was characterized for the first time using the adjoint method. The impact of dynamic factors on the variation of RT was examined, including the water elevation regulation, tributary discharge, mainstream discharge, baroclinic forcing (i.e., the stress induced by the water density difference), and wind.

2. Model Setup

2.1. Study Area

The Zhuyi River (ZR) is a primary tributary of the north bank of the Yangtze River (Figure 1), which is located in the middle of the TGR and is 156 km away from the TGD. ZR has a watershed area of 153.6 km^2, a length of 31.4 km, and an annual average discharge of 2.54 m^3/s which is much lower than that of the Yangtze River's ~11,000 m^3/s. After the impoundment of the TGR, a 6-km-long bay was formed, which was influenced by TGR regulation, which can modify the water level from 145 m to 175 m. Hereafter, this area is called Zhuyi Bay (ZB) in this paper. The ZB's average depth is 36.23 m and its maximum water depth is 80 m when the TGR is at the highest level in the winter (up to 175 m). The wind over ZB is weak with an annual mean wind speed of 1.5 m/s.

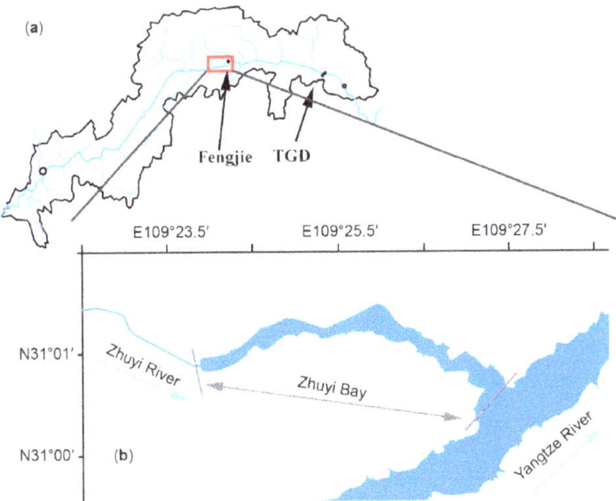

Figure 1. Illustration of the domain of interest, with (**a**) the Yangtze River basin. (**b**) the location of the Zhuyi Bay (ZB).

2.2. Hydrodynamic Model

A three-dimensional (3D) hydrodynamic model, i.e., the Marine Environment Research and Forecasting Model (MERF) [37], was used in this study. The model used a finite-difference scheme on a staggered C-grid in the horizontal direction. In the vertical direction, the terrain-following σ-coordinate system was adopted to accurately represent both the free surface and the bottom topography. The model included baroclinic processes with complete thermodynamics and the turbulence parameterization scheme proposed by Munk and Anderson [38]. In spite of the use of the relatively simple turbulence closure scheme, the model results show the hydrodynamic model basically represented the hydrodynamic characteristics of the study area (Section 3.1) and the closure scheme was of high computational efficiency. Therefore, the turbulence parameterization scheme is deemed to be sufficient for this study. Advection terms in the model were discretized using a total variation diminishing (TVD) scheme with second-order accuracy [39]. The second-order accurate and semi-implicit scheme was employed to solve for the surface elevation. Diffusion terms are discretized using the space centered finite-difference method and solved implicitly and explicitly in the vertical and horizontal directions, respectively. The governing equations and more detailed information on the model are presented in Liu et al. [37].

As shown in Figure 2, the model domain included the entire ZB and partial YR. The domain is discretized by 147 × 83 grids with a uniform grid spacing of 50 m. Uniform 10 σ-layers were used in the vertical direction. The time step is set to 5 s. The daily river discharge, water elevation and temperature of YR are prescribed at the boundaries OB1 and OB2 (Figure 2). The daily river discharge and water temperature of Zhuyi River in 2014 are prescribed at OB3 (Figure 2). Daily winds stress and heat flux are prescribed at the water surface. The wind stress and heat flux are calculated by bulk formulas [40]. The daily water level data were provided by the China Three Gorges Corporation. The daily meteorological data are obtained from the Fengjie Meteorological Bureau. The daily river water temperature data are interpolated from the monthly mean data. In addition, to validate the hydrodynamic model, we also conducted in situ observations at site OBS in ZB for 12 months. We measured the velocity and water temperature for one day for each month using the Nortek Velocimeter (Vector-300 m) and YSI (Yellow Springs Instruments Inc., Yellow springs, OH, USA) EXO2 multisensor sonde, respectively.

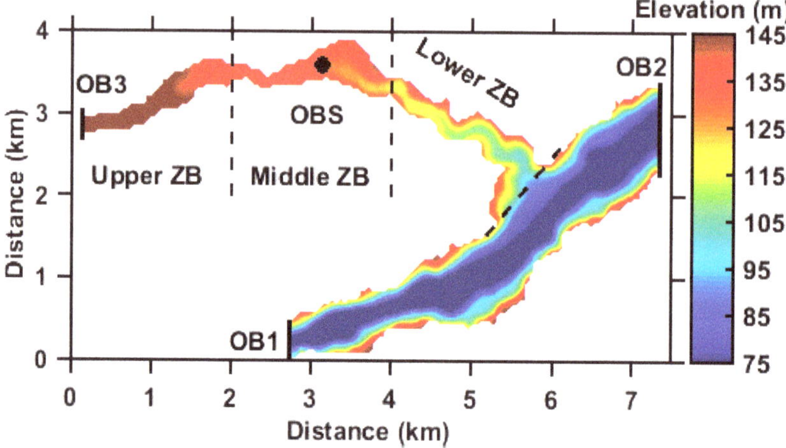

Figure 2. Computational domain and the river elevation (m).

Assuming that the interannual variability of the hydrodynamics can be ignored compared to the significant seasonal cycle and the annual TGR regulation is basically similar, we regard the forcing data of 2014 as a climatological forcing to drive the model and attain the climatological hydrodynamics of ZB. After 3 years spin-up, the modeled velocity and diffusion coefficient results in one year are outputted to drive the RT model introduced in the next section.

2.3. Diagnosing RT by the Adjoint Method

In this study, the adjoint method for deriving RT developed by Delhez et al. [41] is used. The governing equation of RT derived by Delhez et al. [41] is presented below:

$$\frac{\partial \overline{\theta}}{\partial t} + \delta_\omega(x) + \mathbf{v} \cdot \nabla \overline{\theta} + \nabla\left[\mathbf{K} \cdot \nabla \overline{\theta}\right] = 0 \tag{1}$$

where $\overline{\theta}$ denotes RT, $\mathbf{v}$ is the velocity field, $\mathbf{K}$ is the diffusion tensor, and $\delta_\omega(x)$ is the characteristic function of the domain of interest ω, and $\delta_\omega(x) = \begin{cases} 1 & \forall x \in \omega \\ 0 & \forall x \notin \omega \end{cases}$. Equation (1) must be integrated backward in time with the reversed flow to solving $\overline{\theta}$ [41].

Based on Equation (1), the RT model was established. The adjoint model is dealt with using the same finite-difference method and the same grids and layers as the hydrodynamic model. A total variation diminishing scheme with a Superbee and HSIMT (High-order Spatial Interpolation at the Middle Temporal level) alternating flux limiter (TVDal) developed by Lin and Liu is used in the advection term discretization [39]. The initial RT field is set to zero [42]. The ZB mouth connecting ZB and YR is regarded as the open boundary of the control region. The homogenous Dirichlet boundary conditions $\overline{\theta} = 0$ are prescribed on this boundary [42,43]. The impermeable boundary, i.e., $\mathbf{n} \cdot \nabla \overline{\theta} = 0$, is used at the closed boundaries (i.e., the water-air interface, the bottom and the lateral boundaries) [41,44]. We ran the RT model backward in time using previously saved velocity and diffusion coefficients fields. The RT value in ZB reaches a stable value after 2 years spin-up, and the RT in the third year is used for analysis.

3. Results

3.1. Validation of the Hydrodynamic Model

Due to the TGR regulation, the water level decreases from about 173 m elevation in January to 145 m elevation in June, remains at this level for about two months, and then increases from 145 m elevation in August to 175 m elevation in November and December. Due to the prescription on the open boundary and the fast gravity wave speed, the simulated water level change is highly consistent with the measured data, with the maximum error of less than 2 cm (Figure 3). The hydrodynamic model for ZB is further validated using observed velocity and water temperature at site OBS in ZB (Figure 2) in 2014. The simulated velocity shows a pattern of density current where the velocity directions in the upper layer and lower layer are opposite for most of the months. The magnitude of the current is basically less than 4 cm/s. The simulated flow field agrees well with the pattern of observed current (Figure 4). The observed water temperature profiles show that the water temperature in ZB increases from January to July and decreases from August to December and the annual mean difference between the surface layer and the bottom layer is ~0.2 °C (Figure 5). The monthly variation and the vertical distribution of the modeled water temperature are also consistent with the observations (Figure 5). Overall, there is a good agreement between the model results and the observations, indicating that the model used in this study can basically reproduce the main features of the hydrodynamics of ZB.

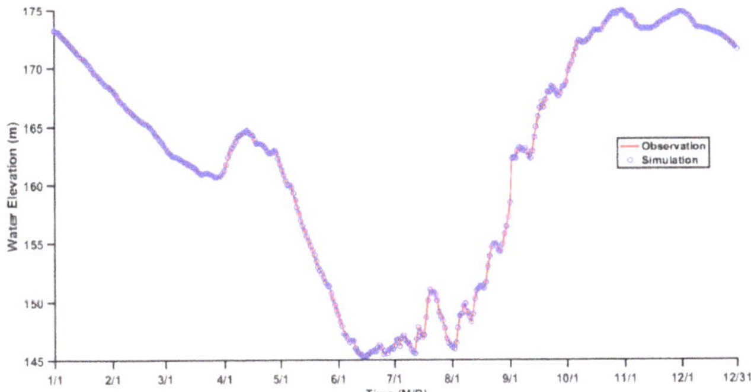

Figure 3. The observed and simulated daily water level at site OBS in 2014.

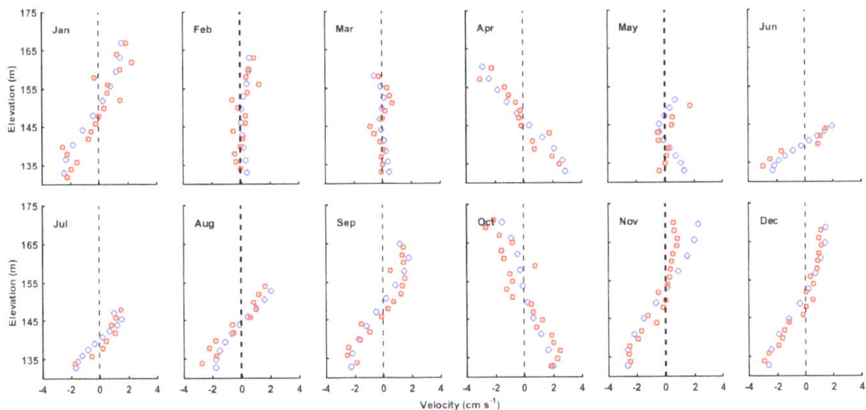

Figure 4. Comparison of the observed and simulated velocity profiles at site OBS for the 12 months in 2014. Blue circles denote simulation, and red circles denote observation.

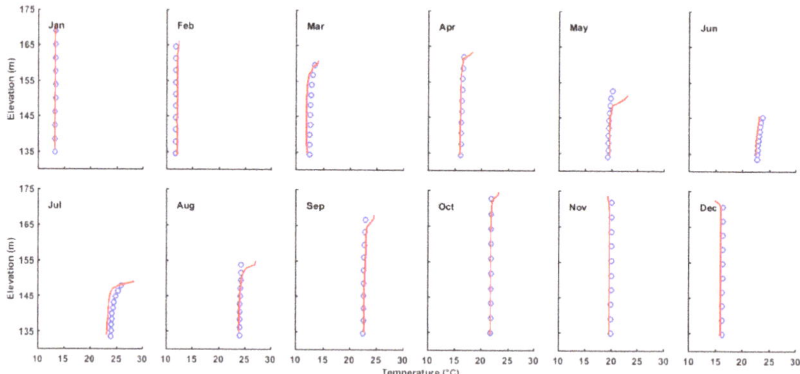

Figure 5. Comparison of observed and simulated water temperature profiles at site OBS for the 12 months in 2014. Blue circles denote simulation, and red lines denote observation.

3.2. Annual Mean RT of ZB

The annual mean RT of ZB is presented in Figure 6. The volume-averaged RT is 16.7 days. There is a clear longitudinal variation of RT which increases from the bay mouth to the bay top. The vertical mean RT ranges from 0 to 21 days in the lower ZB, 21 to 41 days in the middle ZB, and 41 to 53 days in the upper ZB (Figure 6a). The averaged RT of the upper, middle, and lower ZB are 37.9, 24.5, and 6.0 days, respectively. A relatively rapid spatial variation in RT occurs in the middle ZB with a relatively narrow channel. The RT distributions in the surface and bottom layers are basically consistent with the vertical mean (Figure 6b,c). The vertical difference between the surface and bottom RT was less six days in the entire ZB (Figure 6d). The mean surface RT of ZB was about 16.6 days which is slightly lower than the mean bottom RT (18.7 days). As shown in Figure 7, RT is vertically uniform along the deep channel section basically, while the surface RT was slightly larger than the bottom RT in the upper ZB, while it was reversed in the middle and lower ZB. The magnitude of current is the order of 1 cm/s. For the horizontal length of 6 km, the timescale of horizontal transport is about 69 days. The mixing coefficient is about the order of 0.003 m²/s. For the average water depth of 20 m, the timescale of vertical mixing is about 1.5 days, which is much smaller than that of the horizontal transport. Therefore, the annual mean RT exhibits little vertical variability.

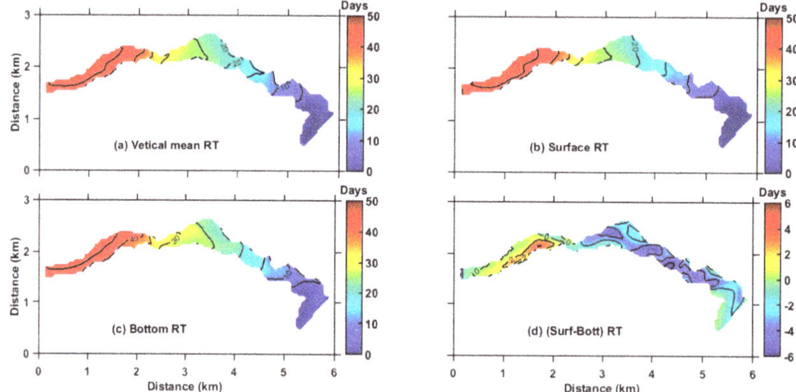

Figure 6. Vertical mean (**a**), surface (**b**), and bottom (**c**) RT (days) averaged over 2014. The contour interval is five days. (**d**) difference between the surface and bottom RT. Positive value denotes larger RT in surface layers. The contour interval is two days.

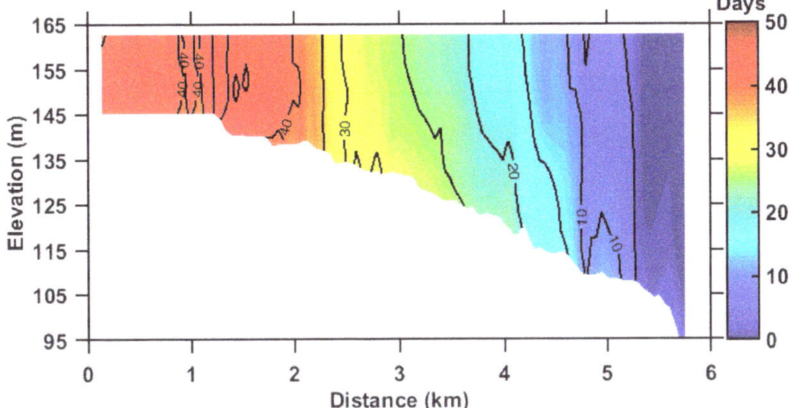

Figure 7. Vertical profile of the annual mean RT (days) along the deep channel section. The contour interval is five days.

3.3. Seasonal Variation of RT

The RT of ZB exhibits a remarkable monthly variation (Figure 8). The smallest average RT value is less than 10 days which occurs in the late spring and early summer (May-July), while in other seasons, the average RT is more than 15 days. The average RT in February-April and August-October are longest which are even more than 20 days. This seasonal variation in RT implies that soluble nutrients in ZB could have a relatively short retention time in summer. In contrast, nutrient released in the spring and autumn has a longer retention time in the ZB, which could provide favorable conditions for the algae bloom in spring.

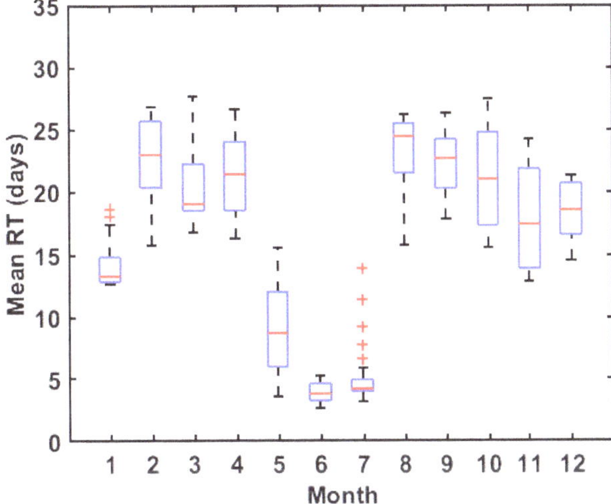

Figure 8. Monthly variation of RT averaged over the entire Bay. Red lines denote medians of RT in the month, blue rectangles denote the first and third quartiles, dashed lines denote the upper and lower whiskers, and red crosses denote the outliers.

The RT variations along the bay for different months are shown in Figure 9. The RT decreases from the bay top to the bay mouth in all months. More significant monthly variation occurs at the upper ZB. In the upper ZB, the largest RT can reach more than 60 days in winter and RT in June is only 10 days.

At the bay mouth, RT is basically less than 10 days for all months. In addition, there are minor vertical differences in the entire ZB, and the vertical difference is less than 10 days (results are not shown).

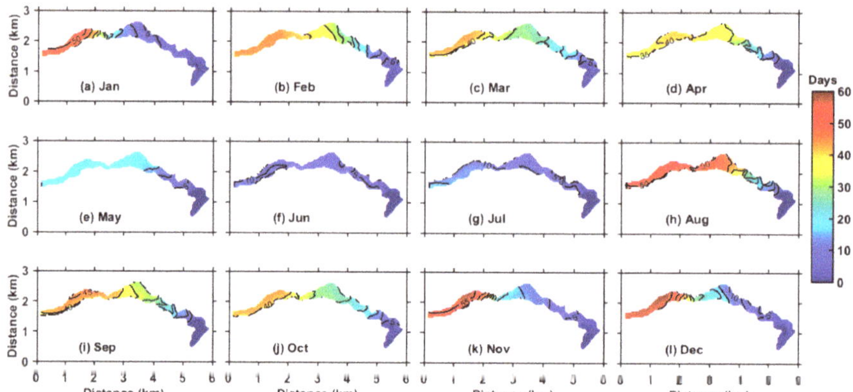

Figure 9. Vertical mean RT (days) averaged of each month, the contour interval is five days.

4. Discussion

4.1. Relationship between RT and TGR Regulation

The TGR regulation changes the water level of tributary bays, which in turn influences the water depth and velocity in tributary bays. The daily variation in RT of ZB shows a good correlation to the water level (Figure 10). Therefore, it is of interest to examine the impact of TGR regulation on the RT.

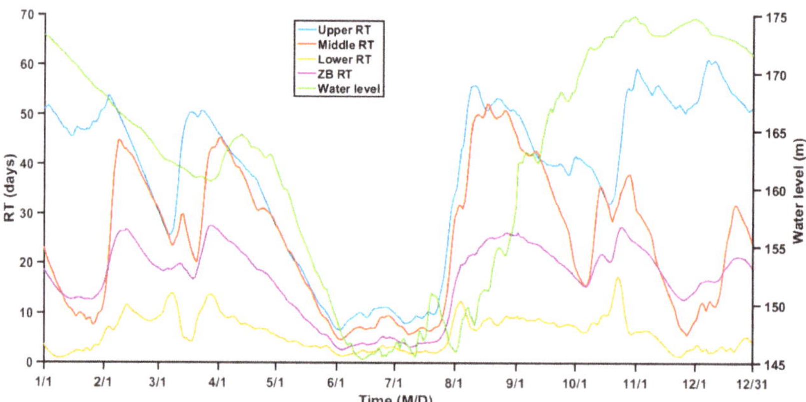

Figure 10. Time series of daily mean RT in different ZB regions and the daily water level.

As shown in Figure 10, the water level declines from January to June, remains low for two months (i.e., June and July), and elevates form August to December. A similar variation happens to the RT in ZB which decreases rapidly before June and increase after July (Figure 10). Average RT of ZB in June and July are lowest in a year, corresponding to the lowest water level in these two months. The Pearson correlation between water level and the RT of the upper, middle, lower, and ZB average is 0.735, 0.153, 0.141 and 0.552, respectively, and the correlations are all significant at the 0.01 level, indicating a significant influence of TGR regulation on the tributary bay RT, though the correlation between the water level and RT is different for different regions of ZB.

To further examine the impact of TGR regulation on the RT, a sensitivity experiment in which the water level is maintained at an annual mean level of 162.30 m is conducted. The results of the experiment show that the RT of ZB is significantly modified where the water level was maintained constant (Figure 11). The average RT in June and July increases to two times comparing to the original RT, which suggests that the low water level are favorable for the water exchange of the tributary bay while the high water level maintained by TGR could increase the water RT in the tributary bay.

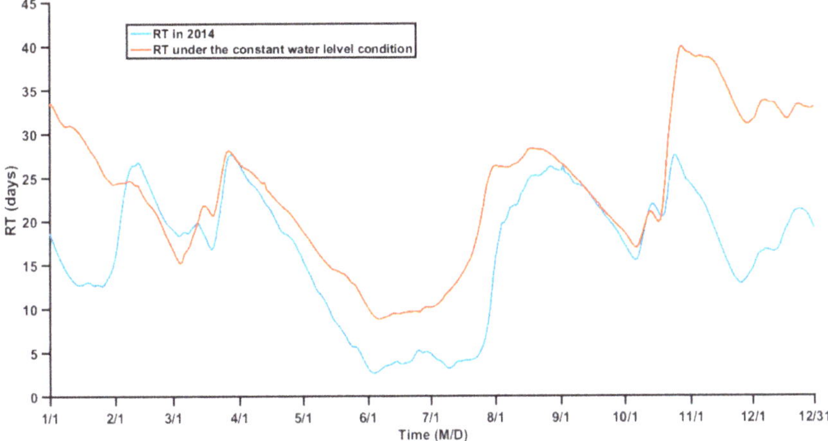

Figure 11. Comparison of ZB RT in 2014 and RT under the constant water level.

In the spring and autumn, the water level of TGR is relatively high which could induce a higher RT of ZB than that of the other seasons, especially in the upper and middle reaches (Figure 9), which would result in water environment problems. Previous research has shown that the algae bloom often occurred in spring and autumn at the upper and middle reach of the tributary bay [45–48]. The high RT in this region of tributary suggested that the nutrient could have the longer retention time and could not be diluted immediately, which could be beneficial to phytoplankton growth, and algae bloom occurred as the result.

4.2. Influences of Dynamic Processes on the RT

The above analyses suggest that the water level of TGR plays an important role in the seasonal variation in the RT of ZB. To further understand the impacts of other dynamic factors on the tributary bay RT, we conducted a set of sensitivity experiments. In the experiments, we artificially modified the driving forces and calculate the RT again. Four cases (Table 1) were designed to examine the influences of driving forces, including local winds, baroclinic forcing, and river discharge on the RT. When running each of these cases, only one driving force from the control case (hereafter referred to as Case 0, whose results were presented in Section 4) is changed. In the experiment of Case 4 with no baroclinic forcing, the water temperature in the entire domain was set to be uniform and constant. The uniform and constant temperature formed a field of uniform water density, thus the baroclinic forcing is zero in Case 4. Both the hydrodynamic model and the RT model were run again to obtain a new ZB RT for each case (Table 1). The new RTs for different cases were compared to the RT of the control case to quantify the impact of different dynamic processes on the RT of ZB.

The variation of RT is generally believed to be highly related to the river discharge [49,50]. The effects of the tributary discharge and Yangtze River discharge on the RT are examined. Both the change of the tributary discharge and Yangtze River discharge account for a minor percentage of the variation of the RT (Figure 12). When the Zhuyi River discharge was decreased by 50% (Case 1), the RT in different regions increased only several days. When the Yangtze River discharge was decreased by 50% (Case 2), there was a similar pattern to Case 1. The little change of RT in Cases 1 and 2 indicate that the tributary RT is not dominated by the river discharge, which may be due to the weak discharge of the Zhuyi River and the weak impact of the Yangtze River flow on the water exchange of ZB.

Table 1. Configurations of sensitivity experiments for Case 0 to Case 4.

Case	Tributary Discharge	Yangtze River Discharge	Local Winds	Baroclinic Forcing
0	Yes	Yes	Yes	Yes
1	0.5 times	Yes	Yes	Yes
2	Yes	0.5 times	Yes	Yes
3	Yes	Yes	No	Yes
4	Yes	Yes	Yes	No

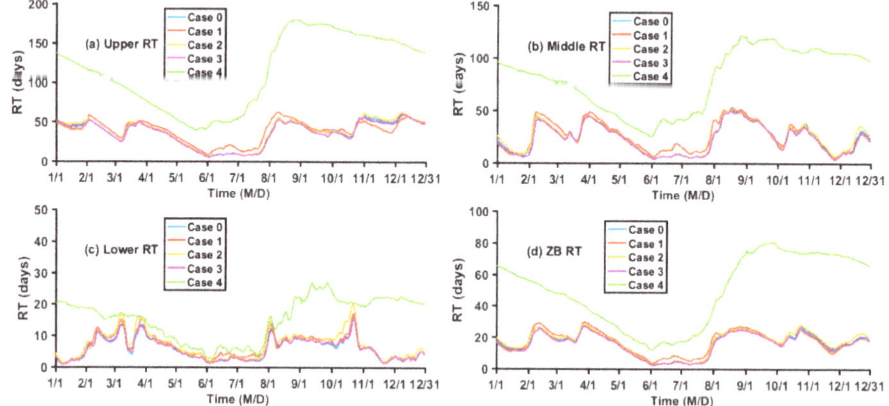

Figure 12. Time series of vertical mean RT in different cases.

The influence of wind on estuarine and semi-enclosed sea circulation has long been recognized [51–53]. However, the comparison between Case 3 and Case 0 suggests that the impact of winds on the tributary RT is also insignificant (Figure 12). The average speed of wind over ZB is only 1.45 m/s, which is much weaker than the wind over estuaries and semi-enclosed seas, and thus the influence of wind on tributary RT is very limited as suggested by the sensitivity experiment.

The exclusion of baroclinic forcing (Case 4) significantly increases the RT in ZB (Figure 12). Comparing to the control case, RT in the upper, middle and lower ZB increase 77.3 days, 56.2 days and 9.7 days, respectively. The average RT of ZB increases from 44% to 503%. The experiment suggests that the baroclinic current induced by the density difference is critical for the RT in ZB and thus plays a dominant role in the water exchange of ZB.

There is an apparent seasonal variation in the temperature difference between ZB and Yangtze River (Figure 13). The water temperature difference can induce the density current (see Figure 4) and enhance the water exchange between the tributary and Yangtze River. The density-induced circulation accelerated the water exchange was also found in the James River [54], the Bohai Sea [55] and the Seto Inland Sea [56]. The larger difference of water temperature in June and July could enhance the density current. The intense density current along with a smaller volume induces the lowest RT in June and July. In addition, the seasonal variation in the difference in water temperature could account for the seasonal variation in RT in ZB for the experiment with the constant water level. Therefore, the density-induced circulation is an important dynamic process for the water exchange of ZB.

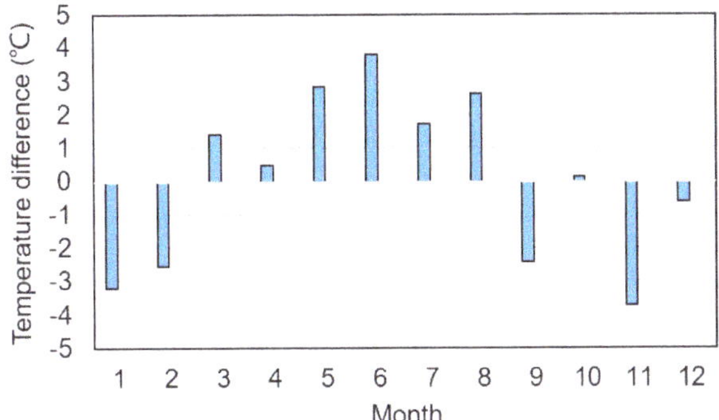

Figure 13. The observed water temperature difference between ZB and the Yangtze River. Positive value indicates that ZB temperature is higher than the Yangtze River. Negative value indicates that ZR temperature is lower than the Yangtze River.

4.3. The Potential Relationship between RT and Algal Blooms in the Tributary

Previous studies have shown that the algal bloom showed a longitudinal characteristic in tributary bays of TGR. Algal blooms often occurred in the upper and middle reaches of tributary bays [48,57]. The relationship between phytoplankton biomass and transport time [2] may explain this phenomenon. There is a clear longitudinal variation in RT which increases from the lower ZB to the upper ZB. The longer RT in the upper and middle bay indicates that the algal growth could be faster than the loss (i.e., transported out of the bay) in this region. By contrast, the shorter RT in the lower bay suggests that the loss may be faster than the growth and the phytoplankton biomass decreases in this region. Therefore, the possibility of algae bloom in the lower bay is very small.

In addition, previous studies also showed that the algae bloom in tributary bays of TGR often occurred in the spring and autumn [32,58] when the RT is longer and the longitudinal gradient of RT is also larger compared to summer (Figure 8). The algal growth may be faster than the loss in these months. Therefore, possibility of algae bloom in these months is relatively large. The seasonal variation in RT can provide a hydrodynamic explanation for the seasonal algal bloom. However, the mechanism of algal blooms still needs to be further investigated by means of biogeochemical analyses and phytoplankton ecosystem numerical modeling.

5. Conclusions

To understand the water exchange capacity of the tributary bay of TGR, this study investigated the spatiotemporal variation in RT in one of typical tributary bay of TGR, i.e., ZB, using the numerical simulation and the adjoint method for obtaining RT. The impacts of the water level, river discharge, wind, and baroclinic forcing on the variation of RT were further discussed based on sensitivity experiments. The main conclusions are summarized as follows. (1) The annual mean RT of ZB is 16.7 days. There were minor vertical differences (<6 days) in the entire ZB due to the water being well mixed in ZB. (2) The RT of ZB increases from the bay mouth (<5 days) to the bay top (>50 days). (3) There is a significant seasonal variation of RT, with high RT in the spring and autumn and lower RT in the summer. (4) The TGR water level regulation has a strong influence on tributary RT. The increase in the water level could increase the RT in the tributary bay. (5) The impact of the tributary discharge, the Yangtze River discharge, and wind on RT are minimal, while the baroclinic forcing induced by the temperature difference between the mainstream and tributary exerts a significant impact on RT.

Author Contributions: Conceptualization, Y.C. and L.L.; methodology, H.W. and L.L.; investigation, Z.M., F.Z. and Y.L.; writing—original draft preparation, Y.C. and L.L.

Funding: The study was financially supported by the National Key R&D Program of China (2017YFC0505305), the National Natural Science Foundation of China (51509066), the China Postdoctoral Science Foundation Funded Project (2017M621409), and the University Science and Technology Research Project of Hebei, China (ZD2019005).

Acknowledgments: The authors appreciated the editor and two anonymous reviewers for their constructive comments and suggestions on the revision of this paper.

Conflicts of Interest: The authors declare no conflict of interest.

Abbreviations

Acronym	Full Name
TGR	Three Gorges Reservoir
TGD	Three Gorges Dam
ZB	Zhuyi Bay
RT	Residence time

References

1. De Brauwere, A.; de Brye, B.; Blaise, S.; Deleersnijder, E. Residence time, exposure time and connectivity in the Scheldt Estuary. *J. Mar. Syst.* **2011**, *84*, 85–95. [CrossRef]
2. Lucas, L.V.; Thompson, J.K.; Brown, L.R. Why are diverse relationships observed between phytoplankton biomass and transport time? *Limnol. Oceanogr.* **2009**, *54*, 381–390. [CrossRef]
3. McLusky, D.S.; Elliott, M.; Elliott, M. *The Estuarine Ecosystem: Ecology, Threats and Management*; Oxford University Press: Oxford, UK, 2004.
4. Wolanski, E.; Elliott, M. *Estuarine Ecohydrology: An Introduction*; Elsevier: Amsterdam, The Netherlands, 2015.
5. Delhez, É.J.M.; Wolk, F. Diagnosis of the transport of adsorbed material in the Scheldt Estuary: A proof of concept. *J. Mar. Syst.* **2013**, *128*, 17–26. [CrossRef]
6. Andutta, F.P.; Ridd, P.V.; Deleersnijder, E.; Prandle, D. Contaminant exchange rates in estuaries—New formulae accounting for advection and dispersion. *Prog. Oceanogr.* **2014**, *120*, 139–153. [CrossRef]
7. Bolin, B.; Rodhe, H. A note on the concepts of age distribution and transit time in natural reservoirs. *Tellus* **1973**, *25*, 58–62. [CrossRef]
8. Zimmerman, J.T.F. Mixing and flushing of tidal embayments in the western dutch wadden sea part I: Distribution of salinity and calculation of mixing time scales. *Neth. J. Sea Res.* **1976**, *10*, 149–191. [CrossRef]
9. Takeoka, H. Fundamental concepts of exchange and transport time scales in a coastal sea. *Cont. Shelf Res.* **1984**, *3*, 311–326. [CrossRef]
10. Delhez, E.J.M.; Campin, J.-M.; Hirst, A.C.; Deleersnijder, E. Toward a general theory of the age in ocean modelling. *Ocean Model.* **1999**, *1*, 17–27. [CrossRef]
11. Delhez, É.J.M. On the concept of exposure time. *Cont. Shelf Res.* **2013**, *71*, 27–36. [CrossRef]
12. Nixon, S.W.; Ammerman, J.W.; Atkinson, L.P.; Berounsky, V.M.; Billen, G.; Boicourt, W.C.; Boynton, W.R.; Church, T.M.; Ditoro, D.M.; Elmgren, R.; et al. The fate of nitrogen and phosphorus at the land-sea margin of the north atlantic ocean. *Biogeochemistry* **1996**, *35*, 141–180. [CrossRef]
13. Dettmann, E.H. Effect of water residence time on annual export and denitrification of nitrogen in estuaries: A model analysis. *Estuaries* **2001**, *24*, 481–490. [CrossRef]
14. Crump, B.C.; Hopkinson, C.S.; Sogin, M.L.; Hobbie, J.E. Microbial biogeography along an estuarine salinity gradient: Combined influences of bacterial growth and residence time. *Appl. Environ. Microbiol.* **2004**, *70*, 1494–1505. [CrossRef]
15. Delesalle, B.; Sournia, A. Residence time of water and phytoplankton biomass in coral reef lagoons. *Cont. Shelf Res.* **1992**, *12*, 939–949. [CrossRef]
16. Lucas, L.V.; Koseff, J.R.; Cloern, J.E.; Monismith, S.G.; Thompson, J.K. Processes governing phytoplankton blooms in estuaries. I: The local production-loss balance. *Mar. Ecol. Prog. Ser.* **1999**, *187*, 1–15. [CrossRef]
17. Lucas, L.V.; Koseff, J.R.; Monismith, S.G.; Cloern, J.E.; Thompson, J.K. Processes governing phytoplankton blooms in estuaries. II: The role of horizontal transport. *Mar. Ecol. Prog. Ser.* **1999**, *187*, 17–30. [CrossRef]
18. Valiela, I.; McClelland, J.; Hauxwell, J.; Behr, P.J.; Hersh, D.; Foreman, K. Macroalgal blooms in shallow estuaries: Controls and ecophysiological and ecosystem consequences. *Limnol. Oceanogr.* **1997**, *42*, 1105–1118. [CrossRef]

19. Chen, C.; Li, J.; Shen, H.; Wang, Z. Yangtze river of China: Historical analysis of discharge variability and sediment flux. *Geomorphology* **2001**, *41*, 77–91. [CrossRef]

20. Nilsson, C.; Reidy, C.A.; Dynesius, M.; Revenga, C. Fragmentation and flow regulation of the world's large river systems. *Science* **2005**, *308*, 405. [CrossRef]

21. Yang, S.L.; Zhang, J.; Dai, S.B.; Li, M.; Xu, X.J. Effect of deposition and erosion within the main river channel and large lakes on sediment delivery to the estuary of the yangtze river. *J. Geophys. Res. Earth Surf.* **2007**, *112*, 111–119. [CrossRef]

22. Wu, J.; Huang, J.; Han, X.; Xie, Z.; Gao, X. Three-gorges dam—Experiment in habitat fragmentation? *Science* **2003**, *300*, 1239–1240. [CrossRef] [PubMed]

23. Shen, G.; Xie, Z. Three Gorges Project: Chance and challenge. *Science* **2004**, *304*, 681. [CrossRef]

24. Stone, R. Three Gorges Dam: Into the unknown. *Science* **2008**, *321*, 628–632. [CrossRef] [PubMed]

25. Fu, B.-J.; Wu, B.-F.; Lü, Y.-H.; Xu, Z.-H.; Cao, J.-H.; Niu, D.; Yang, G.-S.; Zhou, Y.-M. Three Gorges Project: Efforts and challenges for the environment. *Prog. Phys. Geogr.* **2010**, *34*, 741–754. [CrossRef]

26. Xu, X.; Tan, Y.; Yang, G. Environmental impact assessments of the Three Gorges Project in China: Issues and interventions. *Earth Sci. Rev.* **2013**, *124*, 115–125. [CrossRef]

27. Holbach, A.; Norra, S.; Wang, L.; Yijun, Y.; Hu, W.; Zheng, B.; Bi, Y. Three Gorges Reservoir: Density pump amplification of pollutant transport into tributaries. *Environ. Sci. Technol.* **2014**, *48*, 7798–7806. [CrossRef] [PubMed]

28. Zhao, Y.; Zheng, B.; Wang, L.; Qin, Y.; Li, H.; Cao, W. Characterization of mixing processes in the confluence zone between the Three Gorges Reservoir mainstream and the daning river using stable isotope analysis. *Environ. Sci. Technol.* **2016**, *50*, 9907–9914. [CrossRef]

29. Cheng, Y.; Wang, Y.; Zhou, H.; Dang, C. The influence of the Three Gorges Reservoir regulation on a typical tributary heat budget. *Environ. Earth Sci.* **2018**, *77*, 764. [CrossRef]

30. Cheng, Y.; Wang, Y.; Zhou, H.; Hu, M.; Jiang, R.; Bao, Y.; Dang, C. Heat budget contribute rate in the Three Gorges Reservoir tributary bay between mainstream and tributary using stable isotope analysis. *Water Supply* **2019**, *19*, 553–564. [CrossRef]

31. Wang, L.; Cai, Q.; Tan, L.; Kong, L. Phytoplankton development and ecological status during a cyanobacterial bloom in a tributary bay of the Three Gorges Reservoir, China. *Sci. Total Environ.* **2011**, *409*, 3820–3828. [CrossRef]

32. Liu, L.; Liu, D.; Johnson, D.M.; Yi, Z.; Huang, Y. Effects of vertical mixing on phytoplankton blooms in xiangxi bay of Three Gorges Reservoir: Implications for management. *Water Res.* **2012**, *46*, 2121–2130. [CrossRef]

33. Yang, Z.; Cheng, B.; Xu, Y.; Liu, D.; Ma, J.; Ji, D. Stable isotopes in water indicate sources of nutrients that drive algal blooms in the tributary bay of a subtropical reservoir. *Sci. Total Environ.* **2018**, *634*, 205–213. [CrossRef] [PubMed]

34. Cheng, Y.; Li, Y.; Ji, F.; Wang, Y. Global sensitivity analysis of a water quality model in the Three Gorges Reservoir. *Water* **2018**, *10*, 153. [CrossRef]

35. Zhou, Z.; Li, X.; Chen, L.; Li, B.; Liu, T.; Ai, B.; Yang, L.; Liu, B.; Chen, Q. Macrobenthic assemblage characteristics under stressed waters and ecological health assessment using ambi and m-ambi: A case study at the xin'an river estuary, yantai, China. *Acta Oceanol. Sin.* **2018**, *37*, 77–86. [CrossRef]

36. Li, H.; Li, Z.; Li, Z.; Yu, J.; Liu, B. Evaluation of ecosystem services: A case study in the middle reach of the heihe river basin, northwest China. *Phys. Chem. Earth Parts A/B/C* **2015**, *89–90*, 40–45. [CrossRef]

37. Liu, Z.; Lin, L.; Xie, L.; Gao, H. Partially implicit finite difference scheme for calculating dynamic pressure in a terrain-following coordinate non-hydrostatic ocean model. *Ocean Model.* **2016**, *106*, 44–57. [CrossRef]

38. Munk, W.H. Note on the theory of the thermocline. *J. Mar. Res.* **1948**, *7*, 276–295.

39. Lin, L.; Liu, Z. Tvdal: Total variation diminishing scheme with alternating limiters to balance numerical compression and diffusion. *Ocean Model.* **2019**, *134*, 42–50. [CrossRef]

40. Fairall, C.W.; Bradley, E.F.; Rogers, D.P.; Edson, J.B.; Young, G.S. Bulk parameterization of air-sea fluxes for tropical ocean-global atmosphere coupled-ocean atmosphere response experiment. *J. Geophys. Res. Oceans* **1996**, *101*, 3747–3764. [CrossRef]

41. Delhez, É.J.M.; Heemink, A.W.; Deleersnijder, É. Residence time in a semi-enclosed domain from the solution of an adjoint problem. *Estuar. Coast. Shelf Sci.* **2004**, *61*, 691–702. [CrossRef]

42. Delhez, E.J.M. Transient residence and exposure times. *Ocean Sci.* **2006**, *2*, 1–9. [CrossRef]

43. Delhez, É.J.M.; Deleersnijder, É. The boundary layer of the residence time field. *Ocean Dyn.* **2006**, *56*, 139–150. [CrossRef]

44. De Brye, B.; de Brauwere, A.; Gourgue, O.; Delhez, E.J.M.; Deleersnijder, E. Water renewal timescales in the Scheldt Estuary. *J. Mar. Syst.* **2012**, *94*, 74–86. [CrossRef]

45. Zeng, H.; Song, L.; Yu, Z.; Chen, H. Distribution of phytoplankton in the Three-Gorge Reservoir during rainy and dry seasons. *Sci. Total Environ.* **2006**, *367*, 999–1009. [CrossRef]

46. Zhou, G.; Bi, Y.; Zhao, X.; Chen, L.; Hu, Z. Algal growth potential and nutrient limitation in spring in Three-Gorges Reservoir, China. *Fresenius Environ. Bull.* **2009**, *18*, 1642–1647.

47. He, Q.; Kang, L.; Sun, X.; Jia, R.; Zhang, Y.; Ma, J.; Li, H.; Ai, H. Spatiotemporal distribution and potential risk assessment of microcystins in the Yulin River, a tributary of the Three Gorges Reservoir, China. *J. Hazard. Mater.* **2018**, *347*, 184–195. [CrossRef]

48. Dai, H.; Mao, J.; Jiang, D.; Wang, L. Longitudinal hydrodynamic characteristics in reservoir tributary embayments and effects on algal blooms. *PLoS ONE* **2013**, *8*, e68186. [CrossRef]

49. Hagy, J.D.; Boynton, W.R.; Sanford, L.P. Estimation of net physical transport and hydraulic residence times for a coastal plain estuary using box models. *Estuaries* **2000**, *23*, 328–340. [CrossRef]

50. Shen, J.; Haas, L. Calculating age and residence time in the tidal york river using three-dimensional model experiments. *Estuar. Coast. Shelf Sci.* **2004**, *61*, 449–461. [CrossRef]

51. Scully, M.E. Wind modulation of dissolved oxygen in Chesapeake Bay. *Estuaries Coasts* **2010**, *33*, 1164–1175. [CrossRef]

52. Li, Y.; Li, M. Wind-driven lateral circulation in a stratified estuary and its effects on the along-channel flow. *J. Geophys. Res. Oceans* **2012**, *117*. [CrossRef]

53. Du, J.; Shen, J. Water residence time in Chesapeake Bay for 1980–2012. *J. Mar. Syst.* **2016**, *164*, 101–111. [CrossRef]

54. Shen, J.; Lin, J. Modeling study of the influences of tide and stratification on age of water in the tidal James River. *Estuar. Coast. Shelf Sci.* **2006**, *68*, 101–112. [CrossRef]

55. Liu, Z.; Wang, H.; Guo, X.; Wang, Q.; Gao, H. The age of Yellow River water in the Bohai Sea. *J. Geophys. Res. Oceans* **2012**, *117*. [CrossRef]

56. Wang, H.; Guo, X.; Liu, Z. The age of Yodo River water in the Seto Inland Sea. *J. Mar. Syst.* **2019**, *191*, 24–37. [CrossRef]

57. Li, J.; Yang, W.; Li, W.; Mu, L.; Jin, Z. Coupled hydrodynamic and water quality simulation of algal bloom in the Three Gorges Reservoir, China. *Ecol. Eng.* **2018**, *119*, 97–108. [CrossRef]

58. Ye, L.; Han, X.; Xu, Y.; Cai, Q. Spatial analysis for spring bloom and nutrient limitation in xiangxi bay of Three Gorges Reservoir. *Environ. Monit. Assess.* **2007**, *127*, 135–145. [CrossRef] [PubMed]

 *water*

Article

Study on the Transport of Terrestrial Dissolved Substances in the Pearl River Estuary Using Passive Tracers

Bo Hong [1,2,*], **Guangyu Wang** [1], **Hongzhou Xu** [3,*] **and Dongxiao Wang** [2,4]

[1] School of Civil and Transportation Engineering, South China University of Technology, Wushan Rd., Tianhe District, Guangzhou 510641, China; mswgy@mail.scut.edu.cn

[2] Southern Laboratory of Ocean Science and Engineering (Guangdong, Zhuhai), Zhuhai, 519000, China; dxwang@scsio.ac.cn

[3] Institute of Deep-sea Science and Engineering, Chinese Academy of Sciences, Luhuitou Rd., Sanya 572000, China

[4] State Key Laboratory of Tropical Oceanography, South China Sea Institute of Oceanology, Chinese Academy of Science, Guangzhou 510301, China

* Correspondence: bohong@scut.edu.cn (B.H.); hzxu@sidsse.ac.cn (H.X.)

Received: 12 November 2019; Accepted: 23 April 2020; Published: 26 April 2020

Abstract: Highly populated river deltas are experiencing marine environment degradation resulting from the tremendous input of terrestrial dissolved substances (TeDS). The Pearl River Delta is one of the deltas with degradation of the water quality and ecological condition. The Pearl River Estuary (PRE) was investigated to reveal the fate and transport timescales of TeDS in order to provide guidance on water resource management and pollutant transport prediction. By using passive tracers in a calibrated 3D numerical model, the TeDS transports from five different outlet groups were investigated systematically. The TeDS transport time was computed by using the concept of water age, which is a measure of the time that has elapsed since the tracer was transported from the upstream boundary to the downstream concerned area. The tracer impacted area was defined by the area with tracer concentrations > 0.2 (arbitrary unit). The domains that were impacted by the tracer coming from each outlet group were identified separately. In the wet season, the impacted area was larger than in other seasons. The most prominent variations appeared in the Jiaomen–Hengmen–Hongqili (JHH) and Modaomen (MD) outlets. The hydrodynamic conditions controlled the offshore spreading of the TeDS. Assuming the TeDS were conservative, it took approximately 10–20 days for the TeDS to be transported from the head water to the entrance of the outlet. For the TeDS coming from the head water of the Humen outlet, it took approximately 40 (80) days for the TeDS to be transported out of the mouth of the Lingding Bay during the wet (dry) season. For the case of the TeDS coming from the head water of the JHH outlets, it took approximately 20 (40) days for the TeDS to be transported out of the Lingding Bay during the wet (dry) season. For the MD, Jiti and Yamen–Hutiao outlets, it usually took approximately 10 days for the TeDS to be transported from the head water to the inner shelf. The correlation coefficient between the river flow and tracer concentrations was 0.78, and between the river flow and transport time it was −0.70 at a station in the lower Lingding Bay. At the estuary mouth, the impacts of other forcing fields got stronger.

Keywords: transport process; passive tracers; water age; terrestrial dissolved substances; Pearl River Estuary

---

1. Introduction

In recent decades, a number of studies have shown evidence of coastal marine environment degradation resulting from tremendous inputs of terrestrial dissolved substances (TeDS), which include pollutants, nutrients, organic matter, etc. [1–4]. In developing countries, 80% of domestic sewage is discharged into rivers, lakes and oceans without any treatment [5]. A growing number of anthropogenic activities create more environmental pressures on river deltas compared to other regions [6,7]. Understanding the fate and transport dynamics of TeDS in river deltas is especially urgent in dealing with coastal marine pollution and eutrophication and making efficient policies to alleviate the impact of human activities.

The transport of water and dissolved substances in an estuary is influenced by numerous factors, such as river discharge, tides, winds, bottom topography, the Coriolis effect, etc. It is usually difficult to track TeDS coming from different source regions, since hydrodynamic processes usually result in a mixture of them. Meanwhile, it is also challenging to estimate the transport timescales of TeDS, because the calculation of spatially and temporally varied TeDS transport times is not straightforward when using traditional data like velocity, salinity or nutrient concentration. However, accurately mapping the fate and transport timescales of these biogeochemical substances is essential to both oceanographic research and environmental management. Using artificial tracers in numerical modeling is one of the feasible methods that can be used to adequately quantify the transport of TeDS in complicated coastal marine environments.

The advantages of using artificial tracers in marine environments have been demonstrated in several previous studies. For example, Du and Shen [8] used passive tracers (dye) to understand the influence of estuarine circulation on the redistribution of riverine materials from different sources. Hirst [9] examined the penetration and fate of North Atlantic deep water in a global ocean model. Delhez and Deleersnijder [10] simulated the advection–dispersion of tracers discharged at the nuclear fuel reprocessing plant of Cap de La Hague and demonstrated the advantages of the method in the description of temporal variations of the mean age. Both particle trajectories and passive tracers were used in a three-dimensional circulation model by Döös and Engqvist [11] to estimate the potential fate and distribution of radio-nuclides released in the coastal region of the Baltic Sea. Hong et al. [12] used passive tracers to investigate the water exchange between Baltimore Harbor and Chesapeake Bay. Additionally, transport times are frequently used to diagnose circulation and mixing in estuaries and the ventilation rates of lakes, estuaries and ocean basins. Shen and Haas [13] used a three-dimensional numerical model to calculate the age distribution for the substances released from the head waters of the York River Estuary under different hydrodynamic conditions. Through age tracer modeling, Gustafsson and Bendtsen [14] quantified the timescales of downward (upward) mixing of surface (bottom) water in a shallow fjord. Hong and Shen [15] also used passive tracer transport timescales to estimate the sea-level rise effect on the gravitational circulation of Chesapeake Bay. The vertical transport timescale was used as an index to estimate bottom water dissolved oxygen replenishment in Chesapeake Bay by Hong and Shen [16]. Karna and Baptista [17] applied the concept of water age to investigate water renewal timescales in the Columbia River Estuary. A recent study by Du et al. [18] used passive particles to study the fate and retention of pollutants in Galveston Bay. They used a new transport timescale, called the local exposure time, to quantify the spatially varying susceptibility to released pollutants.

In recent decades, concerns about the fate and transport timescales of TeDS have risen in highly populated river delta areas. Being an estuary with complex river networks and profound anthropogenic impacts, the Pearl River Estuary (PRE) is a good candidate to investigate TeDS transport processes. The PRE is a large and productive estuary located in the Guangdong–Hong Kong–Macao Greater Bay Area of southern China (Figure 1). The PRE has eight outlets (Humen, Jiaomen, Hongqili, Hengmen, Modaomen, Jitimen, Hutiaomen and Yamen) and several bays and sub-estuaries (Lingding Bay, Modaomen estuary, Jitimen estuary, Huangmao Bay, and Shenzhen Bay). The Lingding Bay is the largest bay in the PRE. The water body of the PRE receives a high load of anthropogenic organic pollutants from increased industry, agricultural activities, fish dike farming and sewage effluents [19–21]. The degradation of the water quality and ecological condition of the PRE has been frequently reported [21–24]. Data released by the 2016 Guangdong Marine Environment Bulletin indicated that the total mass of pollutants released into the Pearl River in 2016 was 2.03×10^6 tons, which includes chemical oxygen at 1.52×10^6 tons, ammonia at 2.82×10^4 tons, nitrate at 4.12×10^5 tons, nitrite nitrogen at 2.74×10^4 tons and total phosphorus at 2.40×10^4 tons. Because of the large number of rivers and streams and frequent precipitations events in the Pearl River Delta (PRD), surface runoff is a major source of contaminants and nutrient loads in the adjacent coastal water. Large amounts of wastes and nutrients are continuously generated throughout the PRD, resulting in a substantial increase in the number of contamination sources and a deterioration of water quality and ecological conditions in the PRE.

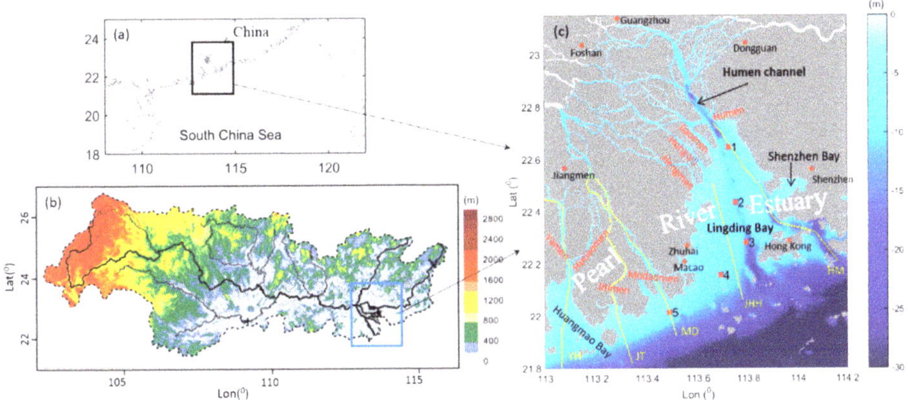

Figure 1. (**a**) The Pearl River Estuary (PRE) on the southern coast of China; (**b**) the Pearl River basin. The color scale represents the elevation above mean sea level in meters; (**c**) the bathymetry of the PRE. The PRE includes eight outlets (Humen, Jiaomen, Hongqili, Hengmen, Modaomen, Jitimen, Hutiaomen and Yamen) that bring fresh water to the estuary. The Lingding Bay is the largest bay in the PRE. Stations (red squares, named 1–5 from upstream to downstream) and transects (yellow lines, named HM, JHH, MD, JT and YH from east to west) are used for later analyses. Major cities (Guangzhou, Foshan, Dongguan, Shenzhen, Jiangmen, Zhuhai, Hong Kong and Macao) around the river delta are also marked (red circles). Among these cities, Guangzhou and Shenzhen are megalopolises.

To date, the fate and transport timescales of TeDS in the PRE have not been fully investigated. With a drainage area of approximately 4.5×10^5 km^2, the Pearl River discharges both freshwater and TeDS into the continental shelf of the northern South China Sea (SCS) through eight outlets (Figure 1c). It is usually difficult to isolate pollutants coming from the different outlets of the PRE. Once the TeDS have entered an outlet, how to locate their impacted area? This type of information, however, would help to identify the dominant sources of pollutants in the inner shelf. Additionally, once the pollution event has happened in the upstream, how long will it take for pollutants to be transported into the concerned area and what are their major pathways? To answer these questions, a systematic assessment of TeDS transport processes in response to various hydrodynamic conditions is required. In this study, a three-dimensional hydrodynamic model with a passive tracer transport module was applied to investigate the transport processes of TeDS under realistic hydrodynamic conditions in the PRE. By using passive tracers, both the temporal and spatial variations of TeDS and their corresponding transport timescales could be calculated within assumptions. The transport of TeDS can be modulated by both physical and biochemical processes that vary over time scales ranging from minutes to decades. The physical and biochemical processes are tightly coupled. However, several previous studies have indicated that the contributions of physical processes are especially important in controlling the transport of nutrients and pollutants, and that a complete understanding of the physical processes is essential to quantify the impact of biochemical processes on estuarine systems [16,25,26]. To isolate the biochemical processes from the physical processes in the PRE, the biochemical processes were excluded from the three-dimensional numerical model and the TeDS were simulated by passive tracers only. This was an efficient way to evaluate the modulation of physical processes on the fate and transport timescales of the TeDS. Detailed information on the TeDS simulation can be found in Section 2. The findings of this study will provide important information and function as a scientific reference for the management of water resources and marine environment protection in the Guangdong–Hong Kong–Macao Greater Bay Area.

The paper is organized as follows. Section 2 describes the model configuration and strategy of modeling the tracer transport processes to understand the transport of TeDS. Section 3 presents the results. Discussion and conclusions are presented in Sections 4 and 5, respectively.

2. Materials and Methods

2.1. Numerical Model Description

The three-dimensional Hydrodynamic-Eutrophication Model (HEM-3D) developed by the Virginia Institute of Marine Science was used for this study. The hydrodynamic portion of the HEM-3D model is the Environmental Fluid Dynamics Code (EFDC) [27]. It uses boundary-fitted curvilinear grids in the horizontal and sigma grids in the vertical. This model has been successfully applied in Chesapeake Bay [8,15], estuaries in Virginia, USA [13,28,29], and the PRE [30]. In its application in the PRE, the model was fully calibrated for tides, salinity and current velocity against all available observations [30]. The same model configurations and external forcing fields were used in this study. As shown in Figure 2, the model domain covered the entire PRE and the adjacent upstream river network. The horizontal resolution of the model grids ranged from approximately 40 m inside the PRE to approximately 1000 m in the shelf area. The open boundary was extended far from the PRE mouth to exclude potential numerical noise from the open boundary conditions. The radiation open boundary condition was used at the open ocean boundary. There were 20 vertical sigma layers with high resolutions near the surface and bottom layers, respectively. The model was initialized by interpolating the cruise-observed winter-time salinity and temperature profiles to the model grids. The red dots in Figure 2 mark the locations where the upstream river flow entered the model domain. These open boundaries were located in an area with crisscrossed water channels. Because no long-term data were available to directly specify the open boundary conditions in these crisscrossed water channels, both the daily river discharge data observed at stations in the major tributaries of the Pearl River (blue dots in Figure 2, data obtained from the Pearl River Information Center of Water Resources of PR China) and the ratios of freshwater distribution among the eight outlets [31,32] were used to determine the distribution of river discharge in each water channel. More detailed information about the model configuration can be found in Hong et al. [30].

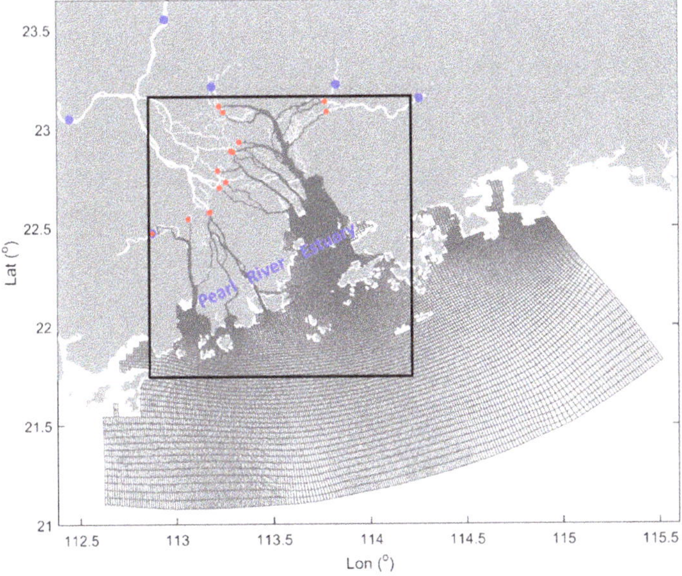

Figure 2. Model grids. The red dots mark the locations where the upstream river flow entered the model domain and the blue dots represent the gauge stations where in situ daily river flow measurements were obtained in the major tributaries of the Pearl River.

2.2. Strategy for Modeling the Transport of Terrestrial Dissolved Substances

The Pearl River is the second largest river in China. The total river discharge reaches a maximum in summer ($\sim$2.1 $\times$ 10^4 m^3 s^{-1}) and a minimum in winter ($\sim$3.4 $\times$ 10^3 m^3 s^{-1}) [33]. From the hydrological perspective, pollutants and nutrients from the Pearl River basin are transported to the PRE and finally enter the open shelf of the northern SCS. TeDS from the Pearl River basin are transported to the estuary through eight outlets (marked in Figure 1c). According to their geographic locations, these outlets can be divided into five groups, i.e., the Humen (HM) outlet, the Jiaomen–Hengmen–Hongqili (JHH) outlets, the Modaomen (MD) outlet, the Jiti (JT) outlet, and the Yamen–Hutiao (YH) outlets. The transport of TeDS from upstream to the open sea was simulated by the numerical model, with passive tracers entering the domain through upstream open boundaries. The locations where the upstream river flow entered the model domain (red dots, Figure 2) were also where the passive tracers were released. For a specific outlet group, the TeDS were simulated by the model run, with passive tracers entering the domain from the corresponding river inflow spots that entered a given outlet group. In each case, the tracers came from different outlet groups, but the hydrodynamic conditions were identical. The model was run from 2006 to 2007, and only the results from 2007 were used for analyses in order to exclude the impact of the model spin-up. There are two aspects that must be addressed in the monitoring of TeDS transport. One is the amount of TeDS, which was represented in a simplified way by the tracer concentration. The other is the TeDS transport timescale, which was calculated using the concept of water age. The age of a particle of water constituent was defined as the time elapsed since the particle under consideration left the region in which its age was prescribed to be zero [34,35]. Besides the water age, the flushing time and residence time are the two other fundamental concepts of transport time [36–39]. The flushing time is usually regarded as a bulk or integrative property that describes the overall exchange or renewal capability of a water body [40,41]. The flushing time can be used to estimate the overall flushing capability of a water body and it establishes the time scale for the physical transport of river-borne material [42]. The residence time of a water parcel is defined as the time needed for the water parcel to reach the outlet [43]. The residence time is usually used to measure the time that a water parcel remains in a waterbody [36]. The residence time is not suitable for measuring the timescales of a water parcel that was transported from the upstream head point to the downstream concerned area. In our case, the water ages were computed as a measure of the time that had elapsed since the TeDS were transported from the upstream open boundary to the downstream concerned area. Using the water age allowed us to show the spatial and temporal variations of TeDS transport time in the PRE.

The movement of the TeDS was traced by artificial passive tracers governed by the following equation [10,34]:

$$\partial c/\partial t + V \cdot \nabla c - \nabla \cdot (K \cdot \nabla c) = 0 \tag{1}$$

Here, c is the tracer concentration, V is the velocity vector, K is the diffusivity tensor, $\nabla = i\partial/\partial x + j\partial/\partial y + k\partial/\partial k$ and t is time. There was no sinking of the tracer within the study area. According to Deleersnijder et al. [34] and Delhez et al. [35], the evolution of age concentration ($\alpha(t, x, y, z)$) is described as

$$\partial \alpha/\partial t + V \cdot \nabla \alpha - \nabla \cdot (K \cdot \nabla \alpha) = c \tag{2}$$

The transport time (water age) can be calculated as $\tau = \alpha/c$. At the upstream boundary where the tracer entered the computation domain, the tracer concentration was prescribed as 1 (arbitrary unit) and the water age was prescribed as 0 (days) in each model layer. When the tracer concentration in the downstream estuary was 0.5, it meant that at this location the tracer concentration was 50% of the concentration measured at the upstream boundary. The river discharge was converted to velocity and used as the normal velocity at the upstream boundary.

Unlike other estuaries with a single river discharge spot from the head, the PRE has a very complicated river network and several river discharge spots (see red dots in Figure 2). When simulating the TeDS transport time (water age), the tracers were released from the upstream boundary of each outlet group. Different outlet groups were calculated separately. Although there was more than one release spot per outlet group, these spots were regarded as one unit that brought TeDS to this outlet group. It should be noted that the tracer concentrations and transport times from the model results represented the spatial and temporal distributions of a single nutrient or contaminant, not the mixture of different nutrients or different contaminants. However, for different nutrients or contaminants, the concentration distribution in the estuary depended on the total loading at the release spots of a given outlet group (regarded as a unity, tracer concentration = 1). A more precise method would be to investigate each release spot separately. However, for the crisscrossed river network of the PRE, dividing these spots into five outlet groups was a more efficient method to conduct a systematic assessment. Our results indicated that the water age difference within the head water of a given outlet group was minor. Our strategy was therefore feasible and efficient for investigating the transport timescales of TeDS coming from different outlets.

The artificial passive tracers defined here represented a generic constituent with boundary concentrations specified to be 1 and kept constant with time. The boundary tracer concentration could be taken to represent the normalized TeDS concentration, which was normalized by the incoming TeDS concentration at the river boundary. The simulated tracer concentrations could be used to quantify the redistribution of TeDS in the PRE. The five outlet groups divided the sources into five groups by location. Many real TeDS undergo transformations and are accompanied by other source/sink processes along their journey from the river head to the open sea. As biogeochemical processes were not included in this model, the high tracer concentrations meant only potentially high TeDS concentrations in the estuary. The transport time (water age) defined here represented the time elapsed since the dissolved substances entered the domain through the upstream boundary. Because the modeling all of the relevant TeDS transport processes was not feasible, we chose to model the tracer behavior to provide diagnostic information relevant to the transport of TeDS under the impact of physical forcing. Reliance on generic tracers and their associated diagnostics is now standard practice in similar studies (e.g., Shen and Hass [13]; Kärnä and Baptista [17]; deBrye et al. [44]; Li et al. [45]).

The spatial and temporal variations of the tracer concentrations and transport times were used in the following to quantify the transport of TeDS in the PRE under realistic hydrodynamic conditions. The advantage of this method was that the results could be used to index any kind of dissolved substances coming from upstream, including pollutants or nutrients. Once they entered the aquatic system, the tracers tracked their pathway and revealed their responses to physical processes in the calibrated 3D numerical model. The spatial–temporal variations of the tracer concentrations provided a clear map of the TeDS concentration variations controlled by the physical forcing fields. Although biochemical processes were excluded from our model, the results are still very instructive for water resource management and pollutant transport prediction.

3. Results

3.1. Distribution of Terrestrial Dissolved Substances Coming from Different Outlets

Analyses of the transport of TeDS coming from different outlets were conducted first to identify areas that could be influenced by the TeDS from each outlet. Because the monthly mean river discharge in 2007 was slightly lower than the long-term mean data (Figure 3), the hydrodynamic conditions in 2007 can be regarded as a normal year versus a typical wet or dry year. The monthly mean results in January and June were selected to represent the typical dry and wet season, respectively. The TeDS were simulated by tracers released from different outlet groups under identical hydrodynamic conditions. The tracer concentrations during the typical wet and dry seasons are shown in Figure 4. Obvious seasonal variations can be clearly identified. High concentrations of tracers appeared during the wet season and occupied a much larger area than during the dry season. The high tracer concentrations indicated potentially high TeDS concentrations in the PRE.

It can be discerned from the tracer distribution (Figure 4) that TeDS coming from the head water of the HM outlet mainly impact the Humen channel and the upper eastern portion of the PRE. Even during the wet season, the tracer mainly lingered along the eastern coast of the Lingding Bay. TeDS coming from the head water of the JHH outlets potentially have the largest impact area among all of the outlet groups. These tracers occupied a large portion of the Lingding Bay. More importantly, TeDS from the JHH outlets have a potentially strong impact on the inner shelf. These TeDS can be transported southeastward toward the open sea during the wet season and westward along the coastline during the dry season. At Shenzhen Bay and in the area around Hong Kong Island, both the HM and JHH outlets are potentially important sources of TeDS. The MD outlet provides an important potential source of TeDS to the coastal area, especially during the wet season, as at that time large amounts of tracer were injected toward the shelf and spread southeastward to the open sea. TeDS from the JT and YH outlets are expected to have limited impacts and mainly remain in the area close to the mouth of the outlets.

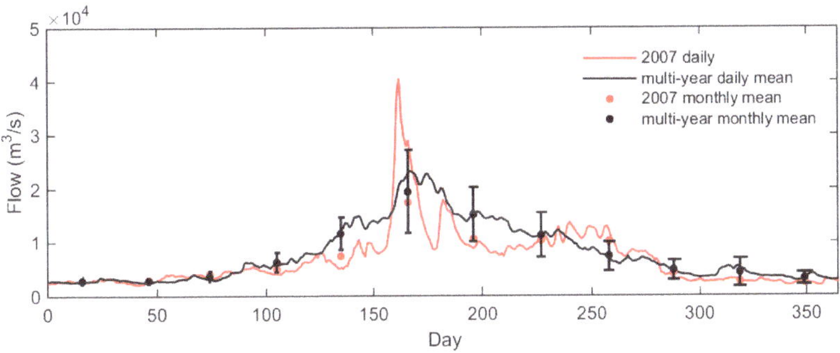

Figure 3. Daily and monthly flow of the Pearl River (summation of river flow measured at blue gauge stations in Figure 2) in 2007. The multi-year (2000–2014) mean daily and monthly river discharge and its standard deviation are also plotted for comparison.

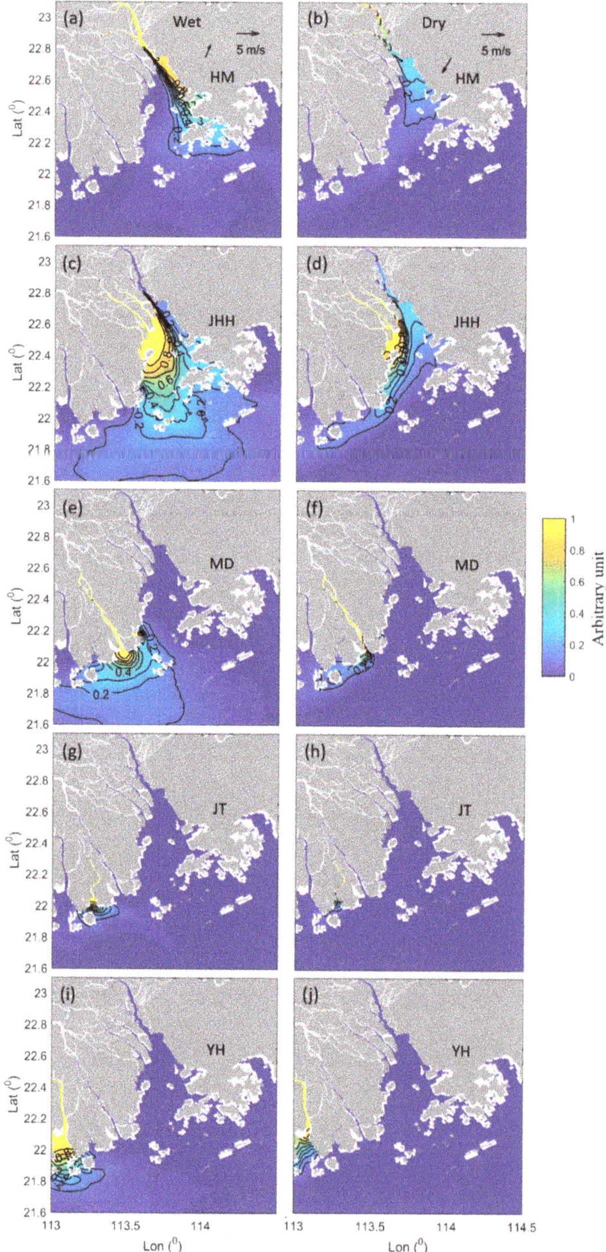

Figure 4. Tracer concentrations (arbitrary unit) in the case of tracers entering the domain from the HM (**a,b**), JHH (**c,d**), MD (**e,f**), JT (**g,h**) and YM (**i,j**) outlets, respectively. High tracer concentrations indicate potentially high TeDS concentrations. Left panels: typical wet season (monthly mean result in June). Right panels: typical dry season (monthly mean result in January). The corresponding monthly mean wind velocity (m/s) is superimposed on the top panel.

To clarify the domains that can be influenced by each outlet, we calculated the total percentage of time that the tracer concentrations were greater than 0.2 (arbitrary unit) in a year. It could be expected that the direct influence of each outlet group disappeared in areas when the percentage was zero, while the major influenced spots could be identified as those with a high percentage. The results (Figure 5) revealed that the influence of the HM outlet only dominated in the upper eastern part of the Lingding Bay, as the lower bay experienced tracer concentrations greater than 0.2 (arbitrary unit) only 20%–30% of the time during one year. For tracers coming from the JHH outlets, the major influenced area (with a percentage greater than 60%) covered a significant portion of the Lingding Bay. This indicates that high concentrations of TeDS could be observed very frequently in this area. In the coastal area adjacent to the PRE mouth, the percentage was around 40%. The westward and southeastward transport showed relatively higher percentages than other directions did. The MD outlet resulted in a high percentage along the coastline. This indicates that the impacts of TeDS decrease toward the offshore regions. The percentages of westward and southeastward tracer transport were relatively higher than that in other directions. The JT and YH outlet groups had a limited impact, with significant seaward spreading occurring less than 10% of the time during one year.

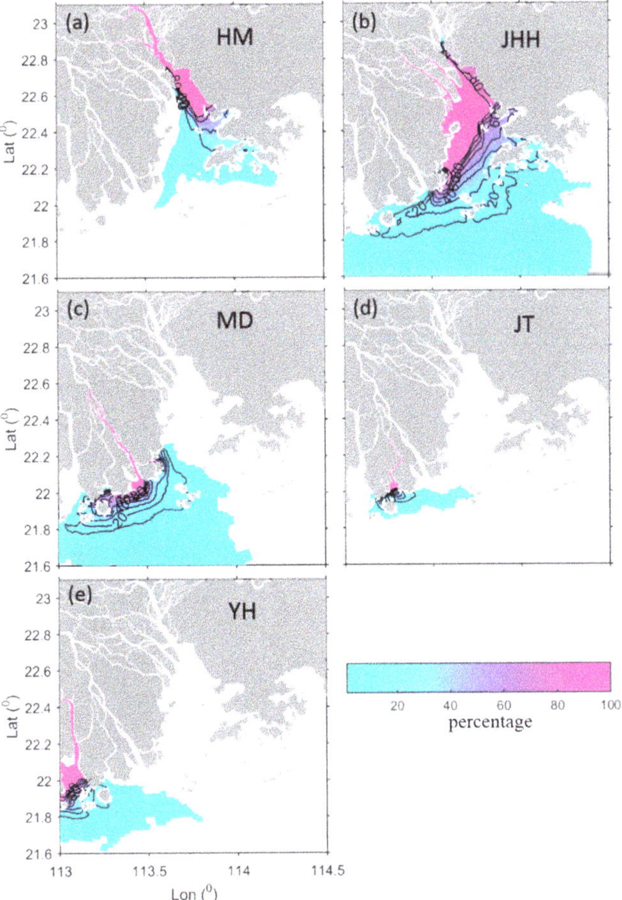

Figure 5. Contours showing the total percentage of time that the tracer concentration was greater than 0.2 (arbitrary unit) for the year 2007. The shaded area shows the percentage > 0. Each panel shows a case of tracers coming from the HM (**a**), JHH (**b**), MD (**c**), JT (**d**) and YM (**e**) outlets.

3.2. Transport Time of Terrestrial Dissolved Substances from Each Outlet

The transport time (water age) of the tracers coming from the head waters of each outlet group could be calculated (Figure 6). The spots for tracers entering the domain in each case are marked by red dots, which are part of the river inflow spots shown in Figure 2. Significant spatial variations can be observed. For the case of tracers coming from the head waters of the HM outlet during the wet season (Figure 6a), it took approximately 20 days for the tracers to be transported from the head waters to the entrance of the outlet, and approximately 40 days from the head waters to the mouth of Lingding Bay. During the dry season (Figure 6b), such transport will take approximately 60 days and 80 days, respectively. During the dry season, transport requires a long time in the HM channel, which means that the TeDS will move slowly from the head waters to the entrance of the HM outlet. This indicates that the TeDS could stay in the channel for a long time. Previous studies indicated that low dissolved oxygen levels (reaching the threshold of hypoxia) can be observed all year in the upper reaches of the PRE, extending from the Guangzhou Channel to downstream of the HM outlet [19,46], which is consistent with our results showing a long transport time of tracer in this area. Our numerical model experiments indicated that the downstream water age differences between the individual spot release locations in the model run were minor. Therefore, although there are six tracer releasing spots for the HM outlet in our calculations, they can be regarded as one unit.

In the case of the tracers released from the head waters of the JHH outlets, it took less than 10 days to reach the outlet entrance and about 20 days to reach the Lingding Bay mouth during the wet season, and approximately 20 and 40 days to reach the outlet entrance and Lingding Bay mouth, respectively, during the dry season. This indicates that the seaward transport of TeDS from the JHH outlets is faster than their transport from the HM outlet. It is interesting that the dry season hydrodynamic conditions resulted in a large difference in water age between the western and eastern parts of Lingding Bay. Transport along the eastern part took an extra 20 days to reach the open shelf. The MD, JT and YH outlets discharged the tracers directly at the inner shelf; during the wet season it took approximately 10 days for the tracers to be transported from the head water of these outlets to the inner shelf, and approximately 20–30 days for the TeDS to be spread over the shelf. During the dry season, the seaward spreading of the tracers from these outlets was limited to the inner shelf.

Transects of the transport time are plotted in Figure 7 to show the various transport time scales of TeDS coming from different outlets. The locations of these transects are indicated in Figure 1c, which follow the major pathways of tracers according to the results in Section 3.1. The results are averaged for June or January are used to represent the typical results in wet or dry season conditions. The vertical structure of the transport time (water age) can be clearly discerned. The HM (Figure 7a,b) and JHH transects (Figure 7c,d) show the transport time of TeDS coming from the HM and JHH outlets, respectively. The transport process during the wet season is faster than during the dry season. The dry season transport takes 50 days longer than the wet season to progress from the head water to the inner shelf. Along the MD outlet transect, less than 5 days are required for the TeDS to be transported from the head water to the entrance of the MD outlet during the wet season. The water age difference, and therefore the TeDS transport time, between the MD mouth and the open shelf is approximately 10 days. The offshore transport adjacent to the MD mouth is relatively slow and the tracer concentrations are relatively high (Figure 4e). Hypoxia can occasionally be observed in this area in summer [47], which may be due to the slow transport process of nutrients. During the dry season, transport from the head water to the MD mouth takes approximately 10 days. Transport along the JT and YH outlets shows similar behavior to the MD outlet. The dry season transport time along the YH outlet is very slow compared with the wet season. It can be expected that the dry season longitudinal circulation in this outlet is the weakest of all the studied outlets.

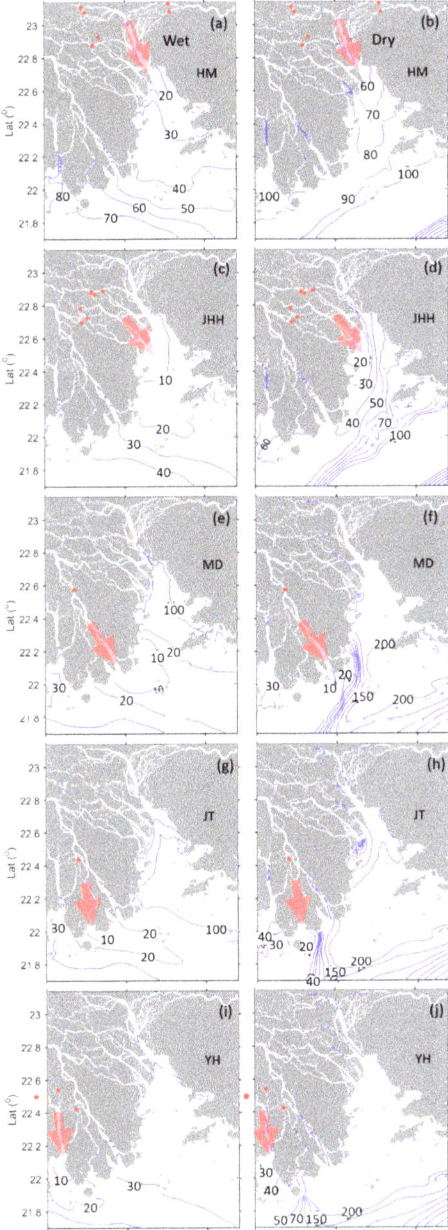

Figure 6. Transport time (days) of tracers coming from the HM (**a**,**b**), JHH (**c**,**d**), MD (**e**,**f**), JT (**g**,**h**) and YH (**i**,**j**) outlets. Left panel: typical wet season (monthly mean result in June). Right panel: typical dry season (monthly mean result in January). The spots for tracers entering the domain are marked by the red dots in each case, and are also part of the river inflow locations shown in Figure 2.

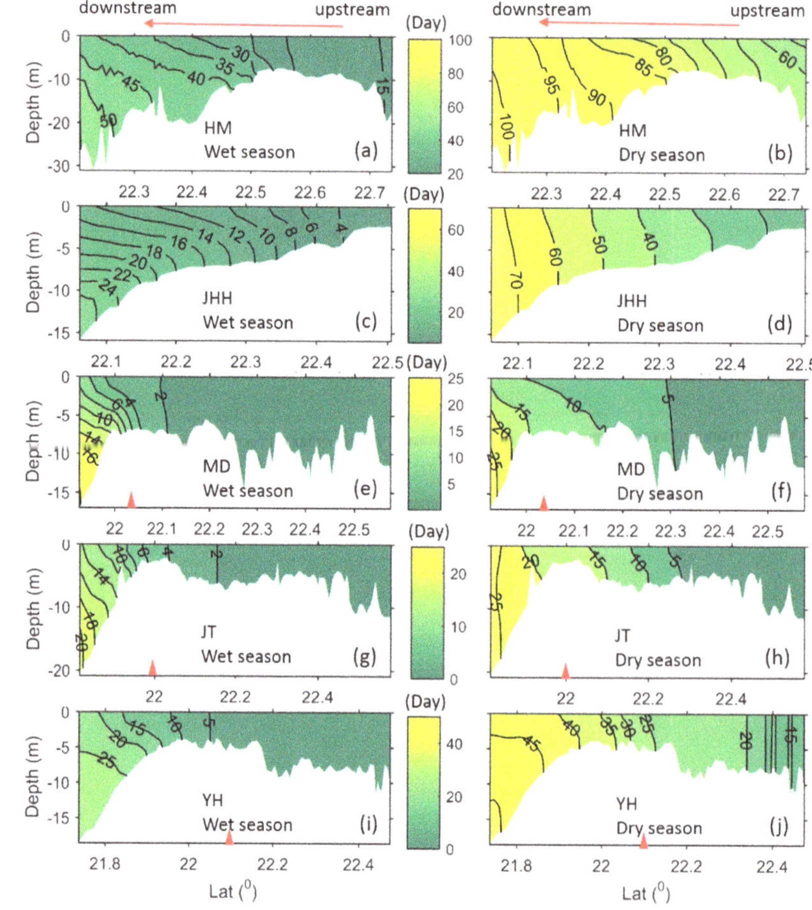

Figure 7. Transport time (days) along the five transects: (**a,b**) HM; (**c,d**) JHH; (**e,f**) MD; (**g,h**) JT; (**i,j**) YH. For each transect, the wet season (left column, averaged in June) and dry season (right column, averaged in January) conditions are plotted. The locations of these transects are indicated in Figure 1c, which follows the major pathways of the tracers according to the results in Section 3.1. The upstream and downstream directions are indicated. The red triangles (in (**e–j**)) indicate the entrance of each outlet.

4. Discussion

4.1. Variations in Impacted Area

The impacted area is defined as an area with tracer concentrations > 0.2. If the initial tracer concentration at the upstream releasing point is defined as 1, the tracer concentration will have decayed to 0.2 (which is 20% of the initial concentration, implying that the impact of the TeDS has been greatly decreased) at the edge of the impacted area. For comparison, the area with a tracer concentration > 0.1 was also calculated. Figure 8a–e shows the variations in the impacted area for each outlet group and Figure 8f shows the total impacted area in the PRE. The total impacted area was calculated by adding the impacted areas of each outlet group, with the overlapped area excluded. The increase in the area can be clearly observed during the wet season. The most prominent variations appear in the JHH and MD outlets. The increase in the impacted area (black lines in Figure 8) during the wet season can be several thousand km^2 for the JHH (Figure 8b) and MD (Figure 8c) outlets. For the other outlet groups, the increase can be several hundred km^2 (Figure 8a,d,e). Compared with the river flow (shown in Figure 8f), the flow peak leads the area peak by approximately 14 days. This time scale is consistent with the mean wet season TeDS transport time shown in Figure 6.

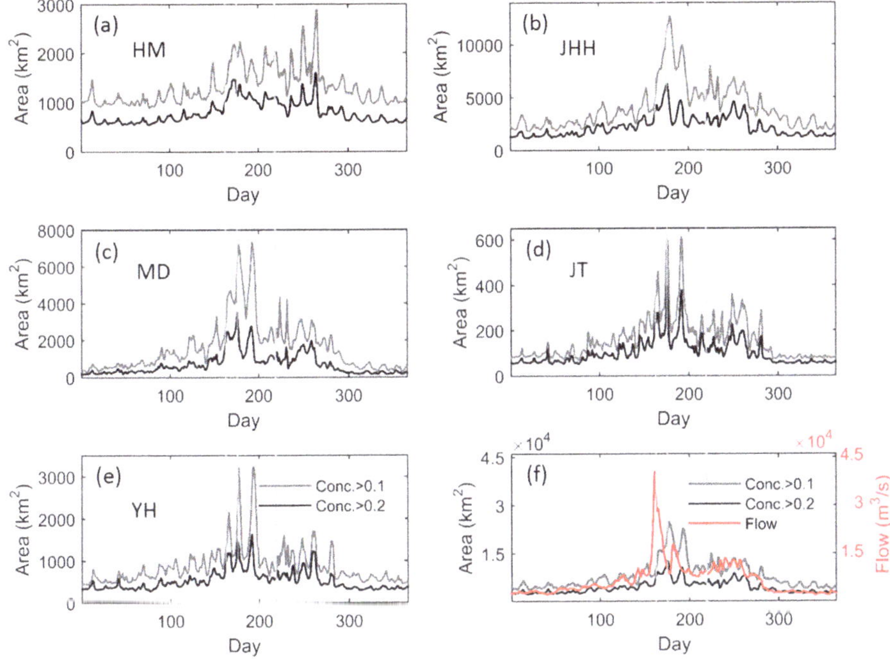

Figure 8. Tracer impacted area (km^2), calculated by the area with a tracer concentration greater than 0.2 (black lines). For comparison, the area with a tracer concentration > 0.1 (gray lines) was also calculated. (**a**) HM; (**b**) JHH; (**c**) MD; (**d**) JT; (**e**) YH. The areas were calculated from the model run of each case. (**f**) The total impacted area was calculated by adding the impacted area of each outlet group, with the overlapped area excluded. The river flow is also superimposed.

4.2. Effect of Hydrodynamic Condition

From the results of the impact areas shown in Figure 8, we selected days 163, 176 and 219 as representative days to show the variation in the impacted areas (marked by vertical gray lines in Figure 8). On day 163, the river flow reaches its peak. On day 176, the total impacted area reaches its peak (Figure 8f). Day 219 was selected to show the condition without river flow pulses. The distributions of the tracer at the selected days are demonstrated in Figure 9. The wind conditions on the corresponding day are also given in each panel. The river flow starts to increase from day 156 and reaches its peak on day 163. High tracer concentrations can be observed, especially at the JHH, MD and YH outlets. At day 176, large amounts of the tracers are transported offshore to the open sea. The strong southwesterly wind generates offshore Ekman transport, which plays an important role. On day 219, the river flow is relatively stable and the effect of the river pulse has disappeared. The tracers are transported out of the estuary along the west coast of the PRE. The southeasterly wind forcing also prevents the offshore transport of the tracers. These results indicate that TeDS coming from different outlet groups follow different pathways and that their impacted areas vary under different hydrodynamic conditions.

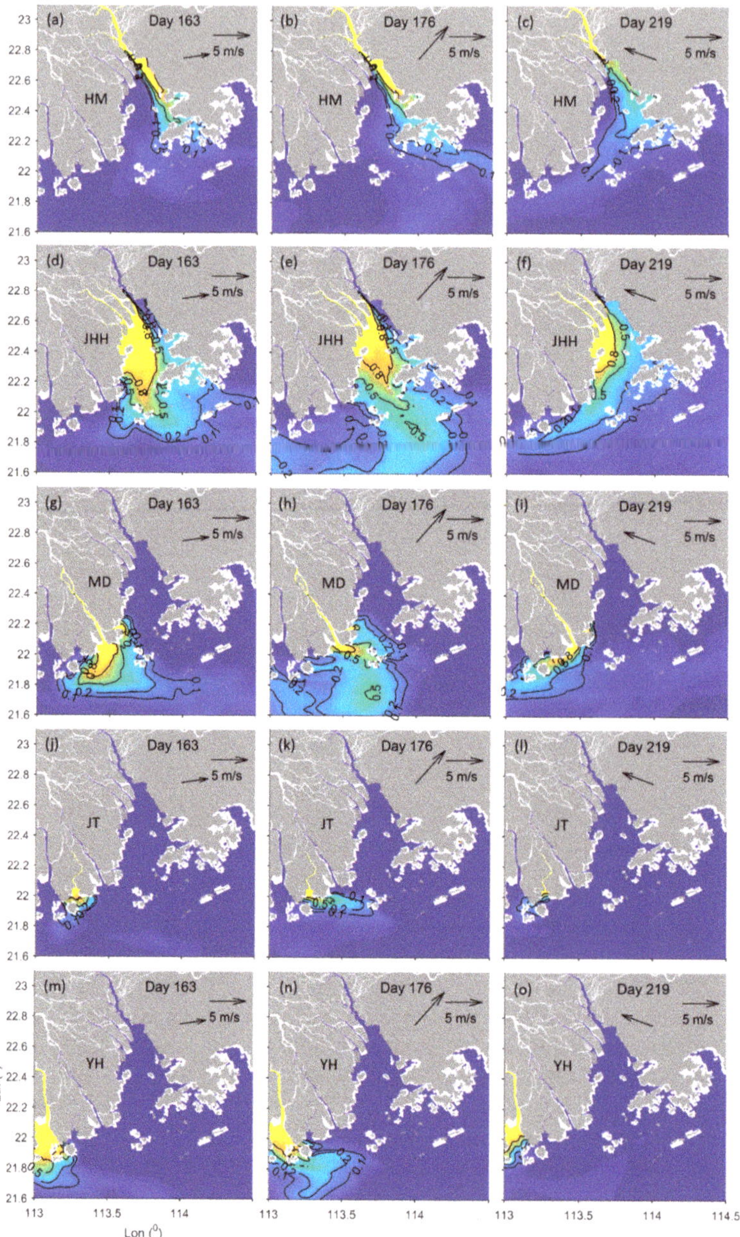

Figure 9. Transport of tracers on day 163 (left column), 176 (mid column) and 219 (right column), indexed by the tracer concentration (arbitrary unit) coming from the HM (**a–c**), JHH (**d–f**), MD (**g–i**), JT (**j–l**) and YH (**m–o**) outlet groups. The labeled isolines are 0.8, 0.5, 0.2 and 0.1. The wind vector on the corresponding day is also presented.

We noticed that the pathways, tracer concentrations and the impacted area showed large variations under different hydrodynamic conditions. The circulation and salinity conditions on day 163, 176 and 219 are presented in Figure 10. On day 176 (Figure 10c), the current directions of the lower estuary and the inner shelf are mostly eastward, with a strong southeastward flow appearing in the area adjacent to the PRE mouth. Such eastward flow disappears on days 163 and 219. However, the southwestward flow is strong along the west coast of the PRE on days 163 and 219 (Figure 10a,e), while the circulation on day 219 is much weaker. The distribution of salinity (Figure 10b,d,f) is used to show the freshwater distribution. It seems that the southwesterly wind facilitates offshore transport of the river plume, while the southeasterly wind tends to push the river plume flow westward along the west coast of the PRE. More detailed analyses on the wind effects will be performed in a separate paper. In this study, we focused on identifying the relative contributions of each outlet group and their impacted domains.

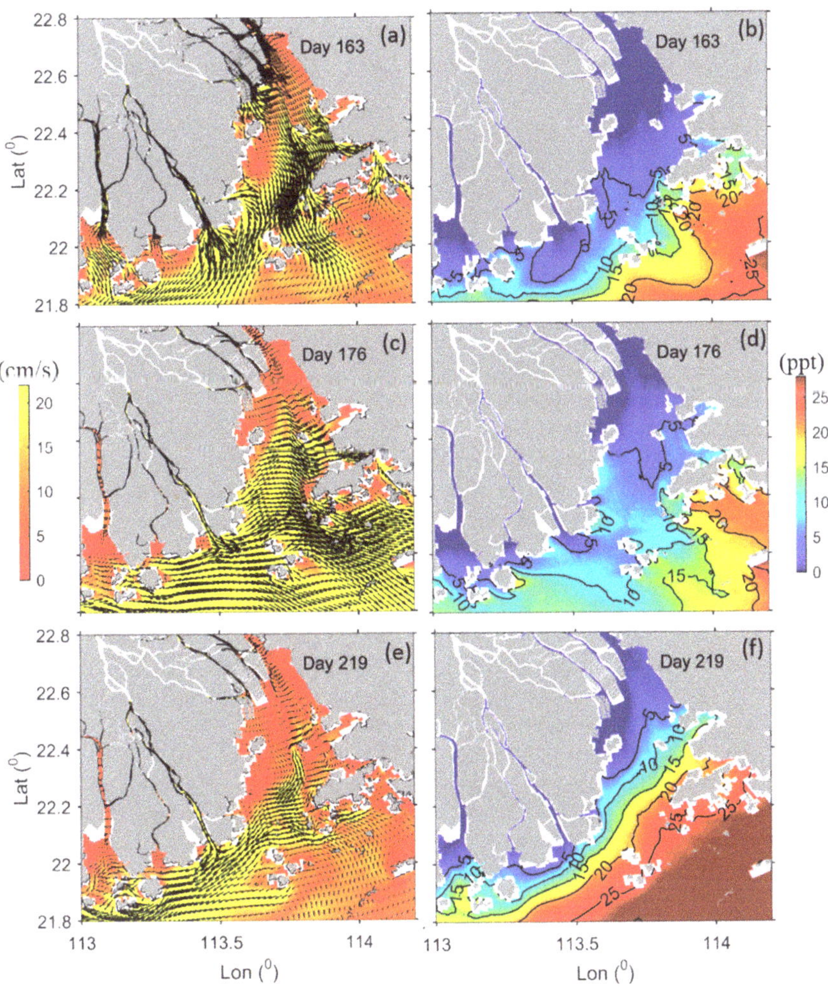

Figure 10. Daily averaged current (**a,c,e**; cm/s) and salinity (**b,d,f**; ppt) from the model results on day 163 (top row), 176 (center row) and 219 (bottom row). The shaded plots in (**a,c,e**) are the current speed, showing the strength of circulation.

Table 1 presents statistics on the tracer occupation days at stations 1–5 (marked in Figure 1c). The days with daily mean tracer concentrations > 0.2 are counted. Stations 1, 2, 3 and 4 were selected to represent the conditions from the upper estuary to lower estuary. Stn. 5 is located in the inner shelf adjacent to the MD outlet. It can be seen that the tracer coming from the JHH outlets has the largest number of occupation days. From the upper estuary to the inner shelf, the tracer occupation days range from 365 days to 154 days. The impact of the HM outlet mainly lies in the upper reaches of the estuary. In the lower estuary (Stns. 3, 4), its impact ranges from 20 days to only a few days. The impact of the MD outlet on Stn. 5 reaches up to 293 days. At Stn. 4, the impact of the MD outlet can occasionally be observed, with only 10 days' occupation in 2007. The impact of the JT and YH outlets cannot be observed at Stns. 1–5, which implies that the TeDS coming from these outlets are seldom transported into Lingding Bay and the inner shelf adjacent to the MD outlet.

Table 1. Days with tracer concentrations greater than 0.2 (arbitrary unit) in 2007 at stations 1, 2, 3, 4 and 5. Tracers coming from the Humen (HM), Jiaomen–Hengmen–Hongqili (JHH), Modaomen (MD), Jiti (JT) and Yamen–Hutiao (YH) outlets are calculated separately.

Station	HM	JHH	MD	JT	YH
Stn. 1	365	365	0	0	0
Stn. 2	71	365	0	0	0
Stn. 3	21	197	0	0	0
Stn. 4	2	170	10	0	0
Stn. 5	0	154	293	0	0

4.3. Correlation Analysis

Owing to the high levels of nutrients at the HM and JHH outlets [19,22,24], and the fact that the TeDS coming from these outlets are firstly discharged into Lingding Bay then the open sea, we conducted a model run with the tracer coming from both HM and JHH simultaneously. This was reasonable because the impacts of the MD, JT and YH outlets are very minor inside Lingding Bay. The model was run from 2005–2009. Figure 11a,b presents the correlation analyses of the river flow, tracer concentration and transport time at Stn. 3, a station in lower Lingding Bay. The correlation coefficient of the river flow and tracer concentration is 0.78 (at the 95% confidence level). According to Section 4.1, a 14-day time lag exists between the river flow and the impacted area. We conducted the time-lag correlation between the river flow and tracer concentration at Stn. 3. The highest correlation coefficient (=0.79) can be obtained with a tracer concentration time series lag of 2 days. A negative correlation between the river flow and transport time was found, with a correlation coefficient of −0.70. The river pulses could reduce the transport time (water age) in a very short time. The curvilinear fits of the river flow, tracer concentration (Figure 11c–e) and transport time (Figure 11f–h) at Stns. 2, 3 and 4 shows that from the upper to the lower estuary, the correlation of the river flow with the TeDS transport process weakens. At Stn. 4, the diagram is more scattered than at Stns. 2 and 3, which means that the impacts of other forcing fields (like wind, open sea circulation, etc.) are getting stronger. Being a bell-shaped estuary, the hydrodynamic circulation in the lower PRE is different from the traditional estuarine circulation, and the interaction with the open shelf is very active and complicated. It is worthwhile to thoroughly investigate the external forcing effect on the estuary–ocean exchange of the PRE, which will be conducted in a separate paper.

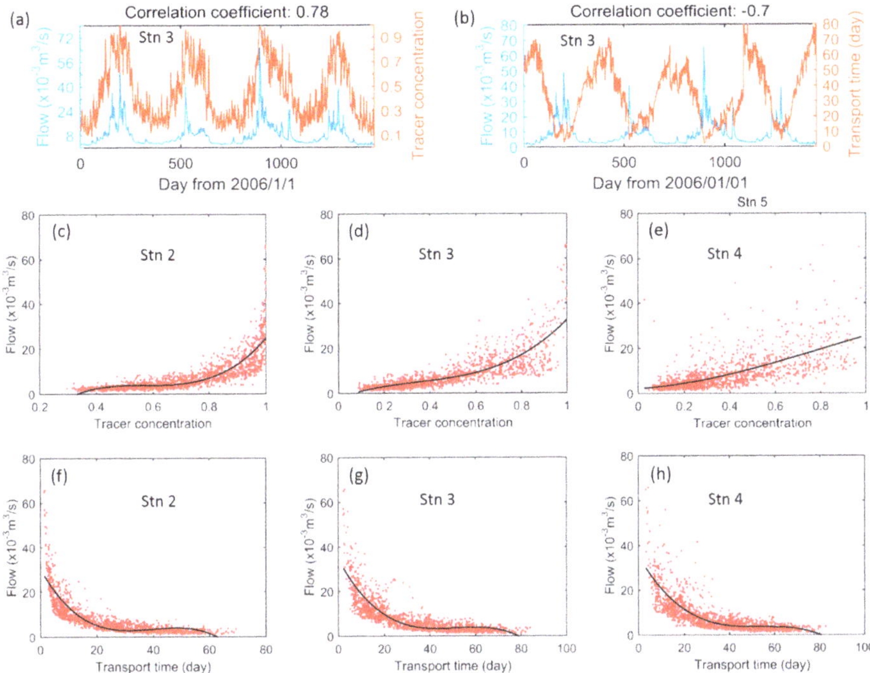

Figure 11. (**a**) Time series of the flow and tracer concentrations at station 3. The correlation coefficient is 0.78; (**b**) Time series of the flows and transport times at station 3. The correlation coefficient is 0.70. (**c–e**) Scatter diagram of the flows and tracer concentrations at stations 2, 3 and 4; (**f–h**) scatter diagram of the flow and transport times at stations 2, 3 and 4. The polynomial fitting is superimposed for (**c–h**).

5. Conclusions

Understanding the fate and transport processes of TeDS in the PRE is essential for dealing with the coastal marine pollution and alleviating the impact of human activities on marine environments. In this study, the outlets of the PRD are divided into five groups according to their geographic locations. By adding passive tracers in a calibrated three-dimensional numerical model, the potential TeDS transport from each outlet group was investigated systematically under realistic forcing conditions. The major conclusions include: (1) During the wet season, the potential impacted area is much larger than during other seasons of the year, and the offshore transport of TeDS to the open sea can be frequently discerned. During the dry season, the tracers were transported mainly along the west coast of the PRE. (2) The JHH outlet has the largest impact among all of the outlet groups. The TeDS coming from the HM outlet are likely to mainly impact the Humen channel and the upper eastern portion of the PRE. Inside Lingding Bay, the impact of the tracers from the MD outlet can occasionally be observed, while the impact of the JT and YH outlets can seldom be observed. (3) For tracers coming from the head water of the HM outlet, it takes approximately 20 days for the tracer to be transported from the head water to the entrance of the outlet and approximately 40 days for the tracer to be transported to the mouth of Lingding Bay during the wet season. During the dry season, these times increase to 60 and 80 days, respectively. For the case of tracers released from the head water of the JHH outlets, the potential seaward transport of TeDS is fast. It takes less than 10 days for tracers to reach the outlet entrance and approximately 20 days to reach Lingding Bay mouth during the wet season, while in the dry season it takes 20 and 40 days, respectively. For the MD, JT and YH outlets, it usually takes approximately 10 days for the tracer to be transported from the head water to the inner shelf, and it takes 20–30 days for the tracer to be spread over the shelf during the wet season. Transport from the

MD outlet is fast, but at the inner shelf adjacent to the MD outlet mouth, the offshore transport is slow. (4) The correlation coefficients between the river flow and tracer concentration (transport time) is 0.78 (−0.70) inside the PRE. In the areas adjacent to the estuary mouth, the impacts of other forcing fields are stronger. Although biochemical processes are excluded from our model, the results are still very instructive for water resource management and pollution transport prediction in the Pearl River Delta.

Author Contributions: Conceptualization, B.H.; methodology, B.H.; validation, B.H., G.W. and H.X.; formal analysis, B.H., G.W.; investigation, B.H., H.X.; resources, H.X., D.W.; data curation, B.H., H.X.; writing—original draft preparation, B.H., H.X., G.W.; writing—review and editing, B.H., H.X.; visualization, B.H., G.W.; funding acquisition, B.H., D.W. All authors have read and agreed to the published version of the manuscript.

Funding: This research was funded by the Key Research Program of Frontier Sciences, CAS (No. QYZDJ-SSW-DQC022) and National Natural Science Foundation of China, Grant number 41976014, 51761135021, 41666001).

Acknowledgments: We would like to thank the anonymous reviewers and the Editor for their efforts to improve the manuscript.

Conflicts of Interest: The authors declare no conflict of interest.

References

1. Kemp, W.; Boynton, W.; Adolf, J.; Boesch, D.; Boicourt, W.; Brush, G.; Cornwell, J.; Fisher, T.; Glibert, P.M.; Hagy, J.; et al. Eutrophication of Chesapeake Bay: Historical trends and ecological interactions. *Mar. Ecol. Prog. Ser.* **2005**, *303*, 1–29. [CrossRef]

2. Murphy, R.R.; Kemp, W.M.; Ball, W.P. Long-Term Trends in Chesapeake Bay Seasonal Hypoxia, Stratification, and Nutrient Loading. *Chesap. Sci.* **2011**, *34*, 1293–1309. [CrossRef]

3. Nixon, S.W. Coastal marine eutrophication: A definition, social causes, and future concerns. *Ophelia* **1995**, *41*, 199–219. [CrossRef]

4. Paerl, H.W.; Valdes, L.M.; Peierls, B.L.; Adolf, J.E.; Harding, L.W. Anthropogenic and climatic influences on the eutrophication of large estuarine ecosystems. *Limnol. Oceanogr.* **2006**, *51*, 448–462. [CrossRef]

5. WWAP (World Water Assessment Programme). *United Nations World Water Development Report 2015: Water for a Sustainable World*; United Nations Educational, Scientific and Cultural Organization: Paris, France, 2015.

6. Liu, B.; Peng, S.; Liao, Y.; Long, W. The causes and impacts of water resources crises in the Pearl River Delta. *J. Clean. Prod.* **2018**, *177*, 413–425. [CrossRef]

7. Rahman, M.; Penny, G.; Mondal, M.; Zaman, M.; Kryston, A.; Salehin, M.; Nahar, Q.; Islam, M.; Bolster, D.; Tank, J.; et al. Salinization in large river deltas: Drivers, impacts and socio-hydrological feedbacks. *Water Secur.* **2019**, *6*, 100024. [CrossRef]

8. Du, J.; Shen, J. Transport of Riverine Material From Multiple Rivers in the Chesapeake Bay: Important Control of Estuarine Circulation on the Material Distribution. *J. Geophys. Res. Biogeosci.* **2017**, *122*, 2998–3013. [CrossRef]

9. Hirst, A.C. Determination of water component age in ocean models: Application to the fate of North Atlantic Deep Water. *Ocean Model.* **1999**, *1*, 81–94. [CrossRef]

10. Delhez, E.J.M.; Deleersnijder, E. The concept of age in marine modeling II. Concentration distribution function in the English Channel and the North Sea. *J. Mar. Syst.* **2002**, *31*, 279–297. [CrossRef]

11. Döös, K.; Engqvist, A. Assessment of water exchange between a discharge region and the open sea—A comparison of different methodological concepts. *Estuar. Coast. Shelf Sci.* **2007**, *74*, 709–721. [CrossRef]

12. Hong, B.; Panday, N.; Shen, J.; Wang, H.V.; Gong, W.; Soehl, A. Modeling water exchange between Baltimore Harbor and Chesapeake Bay using artificial tracers: Seasonal variations. *Mar. Environ. Res.* **2010**, *70*, 102–119. [CrossRef] [PubMed]

13. Shen, J.; Haas, L. Calculating age and residence time in the tidal York River using three-dimensional model experiments. *Estuar. Coast. Shelf Sci.* **2004**, *61*, 449–461. [CrossRef]

14. Gustafsson, K.E.; Bendtsen, J. Elucidating the dynamics and mixing agents of a shallow fjord through age tracer modelling. *Estuar. Coast. Shelf Sci.* **2007**, *74*, 641–654. [CrossRef]

15. Hong, B.; Shen, J. Responses of estuarine salinity and transport processes to potential future sea-level rise in the Chesapeake Bay. *Estuar. Coast. Shelf Sci.* **2012**, *104*, 33–45. [CrossRef]

16. Hong, B.; Shen, J. Linking dynamics of transport timescale and variations of hypoxia in the Chesapeake Bay. *J. Geophys. Res. Oceans* **2013**, *118*, 6017–6029. [CrossRef]

17. Kärnä, T.; Baptista, A.M. Water age in the Columbia River estuary. *Estuar. Coast. Shelf Sci.* **2016**, *183*, 249–259. [CrossRef]

18. Du, J.; Park, K.; Yu, X.; Zhang, Y.J.; Ye, F. Massive pollutants released to Galveston Bay during Hurricane Harvey: Understanding their retention and pathway using Lagrangian numerical simulations. *Sci. Total. Environ.* **2020**, *704*, 135364. [CrossRef]

19. He, B.; Dai, M.; Zhai, W.-D.; Guo, X.; Wang, L. Hypoxia in the upper reaches of the Pearl River Estuary and its maintenance mechanisms: A synthesis based on multiple year observations during 2000–2008. *Mar. Chem.* **2014**, *167*, 13–24. [CrossRef]

20. Dai, M.; Guo, X.; Zhai, W.-D.; Yuan, L.; Wang, B.; Wang, L.; Cai, P.; Tang, T.; Cai, W. Oxygen depletion in the upper reach of the Pearl River estuary during a winter drought. *Mar. Chem.* **2006**, *102*, 159–169. [CrossRef]

21. Yan, M.; Nie, H.; Xu, K.; He, Y.; Hu, Y.; Huang, Y.; Wang, J. Microplastic abundance, distribution and composition in the Pearl River along Guangzhou city and Pearl River estuary, China. *Chemosphere* **2019**, *217*, 879–886. [CrossRef]

22. Dai, M.; Gan, J.; Han, A.; Kung, H.S.; Yin, Z. Physical dynamics and biogeochemistry of the Pearl River plume. In *Biogeochemical Dynamics at Major River-Coastal Interfaces*; Bianchi, T.S., Allison, M.A., Cai, W., Eds.; Cambridge University Press (CUP): Cambridge, UK, 2013; pp. 321–352.

23. Hu, J.; Li, S.; Geng, B. Modeling the mass flux budgets of water and suspended sediments for the river network and estuary in the Pearl River Delta, China. *J. Mar. Syst.* **2011**, *88*, 252–266. [CrossRef]

24. Qiu, D.; Huang, L.; Zhang, J.; Lin, S. Phytoplankton dynamics in and near the highly eutrophic Pearl River Estuary, South China Sea. *Cont. Shelf Res.* **2010**, *30*, 177–186. [CrossRef]

25. Wilson, R.E.; Swanson, R.L.; Crowley, H.A. Perspectives on long-term variations in hypoxic conditions in western Long Island Sound. *J. Geophys. Res. Space Phys.* **2008**, *113*. [CrossRef]

26. Scully, M.E. Wind Modulation of Dissolved Oxygen in Chesapeake Bay. *Chesap. Sci.* **2010**, *33*, 1164–1175. [CrossRef]

27. Hamrick, J.M.; Wu, T.S. Computational design and optimization of the EFDC/HEM3D surface water hydrodynamic and eutrophication models. In *Next Generation Environmental Models and Computational Methods*; Delich, G., Wheeler, M., Eds.; Society for Industrial and Applied Mathematics: Philadelphia, PA, USA, 1997; pp. 143–161.

28. Gong, W.; Shen, J.; Hong, B. The influence of wind on the water age in the tidal Rappahannock River. *Mar. Environ. Res.* **2009**, *68*, 203–216. [CrossRef] [PubMed]

29. Hong, B.; Shen, J.; Xu, H. Upriver transport of dissolved substances in an estuary and sub-estuary system of the lower James River, Chesapeake Bay. *Front. Earth Sci.* **2018**, *12*, 583–599. [CrossRef]

30. Hong, B.; Liu, Z.; Shen, J.; Wu, H.; Gong, W.; Xu, H.; Wang, D. Potential physical impacts of sea-level rise on the Pearl River Estuary, China. *J. Mar. Syst.* **2020**, *201*, 103245. [CrossRef]

31. PRWRC (Pearl River Water Resource Conservancy). *The Pearl River Records 1 (Zhujiang Zhi)*; Guangdong Science & Technology Press: Guangzhou, China, 1991. (In Chinese)

32. Cheng, Z.L. Decadal variation of hydrological status in stream network area and the eight outlets of Pearl River Delta. *Acta Sci. Nat. Univ. Sunyatseni* **2001**, *40* (Suppl. 2), 29–31.

33. Zhao, H. *The Evolution of the Pearl River Estuary*; China Ocean Press: Beijing, China, 1990; p. 357.

34. Deleersnijder, E.; Campin, J.; Delhez, E. The concept of age in marine modeling I. Theory and preliminary model results. *J. Marine Syst.* **2001**, *28*, 229–267. [CrossRef]

35. Delhez, E.J.; Campin, J.-M.; Hirst, A.C.; Deleersnijder, E. Toward a general theory of the age in ocean modelling. *Ocean Model.* **1999**, *1*, 17–27. [CrossRef]

36. Du, J.; Shen, J. Water residence time in Chesapeake Bay for 1980–2012. *J. Mar. Syst.* **2016**, *164*, 101–111. [CrossRef]

37. Huang, W.; Liu, X.; Chen, X.; Flannery, M.S. Estimating river flow effects on water ages by hydrodynamic modeling in Little Manatee River estuary, Florida, USA. *Environ. Fluid Mech.* **2009**, *10*, 197–211. [CrossRef]

38. Alber, M.; Sheldon, J. Use of a Date-specific Method to Examine Variability in the Flushing Times of Georgia Estuaries. *Estuar. Coast. Shelf Sci.* **1999**, *49*, 469–482. [CrossRef]

39. Hagy, J.D.; Boynton, W.; Sanford, L.P. Estimation of Net Physical Transport and Hydraulic Residence Times for a Coastal Plain Estuary Using Box Models. *Estuaries* **2000**, *23*, 328. [CrossRef]

40. Geyer, W.R.; Morris, J.T.; Pahl, F.G.; Jay, D.A. Interaction between physical processes and ecosystem structure: A comparative approach. In *Estuarine Science: A Synthetic Approach to Research and Practice*; Hobbie, J.E., Ed.; Island Press: Washington, DC, USA, 2000; pp. 177–210.

41. Oliveira, A.; Baptista, A. Diagnostic modeling of residence times in estuaries. *Water Resour. Res.* **1997**, *33*, 1935–1946. [CrossRef]

42. Rayson, M.; Gross, E.S.; Hetland, R.D.; Fringer, O.B. Time scales in Galveston Bay: An unsteady estuary. *J. Geophys. Res. Oceans* **2016**, *121*, 2268–2285. [CrossRef]

43. Zimmerman, J. Mixing and flushing of tidal embayments in the western Dutch Wadden Sea part I: Distribution of salinity and calculation of mixing time scales. *Neth. J. Sea Res.* **1976**, *10*, 149–191. [CrossRef]

44. De Brye, B.; De Brauwere, A.; Gourgue, O.; Delhez, E.J.; Deleersnijder, E. Water renewal timescales in the Scheldt Estuary. *J. Mar. Syst.* **2012**, *94*, 74–86. [CrossRef]

45. Li, Y.; Feng, H.; Zhang, H.; Sun, J.; Yuan, D.; Guo, L.; Nie, J.; Du, J. Hydrodynamics and water circulation in the New York/New Jersey Harbor: A study from the perspective of water age. *J. Mar. Syst.* **2019**, *199*, 103219. [CrossRef]

46. Harrison, P.J.; Yin, K.; Lee, J.; Gan, J.; Liu, H. Physical–biological coupling in the Pearl River Estuary. *Cont. Shelf Res.* **2008**, *28*, 1405–1415. [CrossRef]

47. Cui, Y.; Wu, J.; Ren, J.; Xu, J. Physical dynamics structures and oxygen budget of summer hypoxia in the Pearl River Estuary. *Limnol. Oceanogr.* **2018**, *64*, 131–148. [CrossRef]

MDPI

St. Alban-Anlage 66

4052 Basel

Switzerland

Tel. +41 61 683 77 34

Fax +41 61 302 89 18

www.mdpi.com

Water Editorial Office

E-mail: water@mdpi.com

www.mdpi.com/journal/water